IRON-BASED SUPERCONDUCTORS

edited by
Nan-Lin Wang
Hideo Hosono
Pengcheng Dai

MATERIALS, PROPERTIES, AND MECHANISMS

IRON-BASED SUPERCONDUCTORS

PAN STANFORD PUBLISHING

Published by

Pan Stanford Publishing Pte. Ltd.
Penthouse Level, Suntec Tower 3
8 Temasek Boulevard
Singapore 038988

Email: editorial@panstanford.com
Web: www.panstanford.com

British Library Cataloguing-in-Publication Data
A catalogue record for this book is available from the British Library.

Iron-Based Superconductors: Materials, Properties, and Mechanisms

ISBN 978-981-4303-22-4 (Hardcover)
ISBN 978-981-4303-23-1 (eBook)

Printed in the USA

Contents

Preface

Superconductivity has just celebrated its 100th birthday, and yet despite its advanced age, it has continued to be a field of active research. Much of this is due to continued discovery of new materials and new phenomena. Heike Kamerlingh Onnes won the Nobel Prize for his discovery of superconductivity in mercury, and coined the word "superconductivity." However, the microscopic origin of superconductivity had not been understood until the discovery of Bardeen-Cooper-Schrieffer (BCS) theory in 1957. The BCS theory is based on Landau's Fermi liquid for metals, and stipulates that superconductivity occurs because electrons in the superconducting state form Cooper pairs through the electron-phonon interactions to overcome the Coulomb repulsive force. The superconductors that can be described within the BCS theory are called conventional or BCS superconductors. In 1986, a new family of superconductors with high critical temperature (high-T_c) based on copper-oxides was discovered by Johannes Bednorz and Karl Müller, opening a new era of superconductivity. Instead of being metallic, the undoped parent compounds of copper oxide superconductors are Mott insulators, and it is generally believed that superconductivity in these materials is different from the BCS mechanism. For this reason, these materials are called unconventional superconductors. However, in spite of more than 25 years of intensive work, there is still no consensus on the microscopic origin of superconductivity in the copper oxides. In early 2008, another family of unconventional superconductors — iron-based superconductors — was discovered by Hosono's group. This discovery immediately caused great excitement in the community because this is only the second class of high-T_c superconductors in addition to the copper oxides. Since the original discovery paper four years ago, more than 4000 papers have been published on the subject. Although it is generally agreed that superconductivity in both iron- and copper-based superconductors originates from electron-forming Cooper pairs in the superconducting state, much is under debate concerning the microscopic origins for the pairing interaction. In this book, we summarize the recent progress in the field and highlight important accomplishments over the past four years.

The first two chapters focus on the materials and their characterizations. The story of how the superconductivity in Fe-pnictides was discovered is described by Hosono in the first chapter. The crystal structure and properties of parent compounds and growth of the single crystals and epitaxial thin films are also presented. The second chapter is devoted to the Fe-chalcogenide system $K_{1-x}Fe_{2-y}Se_2$, which appears to exhibit very different properties from that of Fe-pnictide compounds. Those overviews are followed by the physical properties investigations and theoretical understandings. The experimental explorations of electronic structures, Fermi surface topologies, the charge excitations at various energy scales, and superconducting gap symmetry in terms of angle-resolved photoemission, quantum oscillation, and optical spectroscopy techniques are overviewed in Chapters 3, 4, and 5, respectively. The magnetic structure and spin excitations probed by neutron scattering and nuclear magnetic resonance techniques are summarized in Chapters 6 and 7. Theoretical understanding of the charge and spin excitations, paring symmetry, and pairing mechanism based on different approaches, i.e., weak coupling versus strong coupling, are overviewed in the last three chapters. We hope that the book could serve as the primary reference for graduate students and researchers working in the field.

Iron-based superconductors are currently still under intensive research. During the period of our editing this review volume, new developments have been made in various aspects in this field. A number of review articles, special issues in several journals and a monograph on the physics of iron-based superconductors also appeared in literature. To benefit our readers, we list several of them:

1. Y. Izyumov and E. Kurmaev, *High-T_c Superconductors Based on FeAs Compounds*, Springer Series in Materials Science, vol. **143**,

Springer-Verlag Berlin Heidelberg (2010). (This is the first monograph on the physics of iron-based superconductors.)

2. D.C. Johnson, The puzzle of high temperature superconductivity in layered iron pnictides and chalcogenides, *Adv. Phys.*, **59**, 803 (2010). (It is a comprehensive review covering papers up to early 2010.)
3. M. D. Lumsden and A. D. Christianson, Magnetism in Fe-based superconductors, *J. Phys.: Condens. Matter*, **22**, 203203 (2010).
4. G. R. Stewart, Superconductivity in iron compounds, *Rev. Mod. Phys.*, **83**, 1589 (2011).
5. R. Pottgen and D. Johrendt, 100 Years of Superconductivity, *Zeitschrift fur Kristallographie*, **226**, 309 (2011). (Superconducting materials were reviewed from crystal chemical view.)
6. G. Crabtree, L. Greene, and P. Johnson (edited), Celebrating 100 years of superconductivity: special issue on the iron-based superconductors, *Rep. Prog. Phys.*, **74**, 12 (2011). (This special issue consists of 13 review articles. Experimental and theoretical approaches to pairing mechanism are highlighted.)
7. H. Hosono, H. Fukuyama, and H. Akai (edited), Special issue on iron-based superconductors, *Solid State Commun.*, **152**, 631–746 (2012). (This special issue consists of 18 articles. Experimental and theoretical approaches to the physical properties are highlighted.)
8. H. Hiramatsu, T. Katase, T. Kamiya, and H. Hosono, Thin film growth and device fabrication of iron-based superconductors, *J. Phys. Soc. Jpn.*, **81**, 011011 (2012).

Chapter 1

Iron-Based Superconductors: Discovery and Progress in Materials

Hideo Hosono
Frontier Research Center & Materials and Structures Laboratory, Tokyo Institute of Technology, 4259 Nagatsuta, Midori-ku, Yokohama, Kanagawa 226-8503, Japan
hosono@msl.titech.ac.jp

$LaFeO_{1-x}F_x$ with $T_c = 26$ K was reported in early 2008. Since then, more than 4,000 papers have been published and the maximum T_c reached 56 K, which is next to high-T_c cuprates. Iron-based superconductors have several unique properties mostly arising from their multi-orbital nature. The presence of a vast number of parent materials is a characteristic of this system. This chapter describes the background research to the discovery, the crystal structure and properties of parent materials, and growth of the single crystals and epitaxial thin films.

1.1 Introduction

Superconductivity is the most dramatic and clear-cut phenomenon which occurs in condensed matters and its potential impact on our society is huge toward realization of low carbon emission society.

Iron-Based Superconductors: Materials, Properties, and Mechanisms
Edited by Nan-Lin Wang, Hideo Hosono, and Pengcheng Dai

ISBN 978-981-4303-22-4 (Hardcover), 978-981-4303-23-1 (eBook)
www.panstanford.com

Although the theoretical framework describing superconductivity was established by the BCS theory, there is no solid theoretical guideline which is truly convincing for experimentalists in this field to approach high-T_c materials even now. These are the reasons why superconductivity has been attracted much attention of many scientists for a century since the discovery of superconductivity in Hg by Kamerlingh Onnes.

Prior to the discovery of cuprate superconductors in 1986, BCS theory seemed to fully explain superconductivity. However, Müller and Bednorz's discovery[1] led to a critical temperature (T_c) above the liquid nitrogen temperature, (e.g., $T_c = 90$ K of $YBa_2Cu_3O_x$),[2] and inevitably required a significant modification to the theory. A breakthrough discovery occurred in traditional metal-based superconductors in 2001, i.e., the finding of MgB_2 with $T_c =$ 40 K by Akimitsu's group,[3] which was basically explained by BCS theory.

The new iron-based superconductor ($LaFeAsO_{1-x}F_x$: $T_c =$ 26 K) found by Hosono's group[4] in January 2008 rekindled intensive research in this area and opened a new frontier for superconductivity. In the early stage, this new superconductor family was called "Pnictide Superconductors". However, researchers now call them "Iron-based Superconductors" (FeSCs) because detailed measurements and evaluations have clarified that they all have similar electronic structure where the electrons derived from Fe ion play an essential role in superconductivity.

An electron has a charge and a spin. Magnetism emerges from the long-range ordering of electron spins, while superconductivity occurs on the formation of Cooper pairs of electrons, which are mostly diamagnetic due to spin singlet state (a spin triplet superconductor is rare such as $SrRuO_3$[5]). Therefore, the relationship between magnetism and superconductivity had long been believed to be ambivalent, and the use of the magnetic element with a large spin moment, typically Fe, Co and Ni, was avoided in the field of superconductivity. Hence, the discovery of a high-T_c superconductor based on iron significantly impacted research in this field. Material scientists and physicists encountered a new frontier of potential elements for superconducting material exploration.

1.2 Small History on Discovery and Progress in Parent Materials

We have been concentrated on exploration of new transparent oxide conductors and semiconductors since 1993.[6] Our study on transparent p-type semiconducting materials with high hole mobilities began with $CuAlO_2$,[7] and led to the discovery of LaCuChO (Ch: chalcogen) with high performance.[8] As a succession of this finding, we focused on exploration of novel magnetic semiconductors by replacing Cu^+ with a closed 3d shell structure with magnetic elements with an open 3d shell such as Fe^{2+} or Mn^{2+}. When we began this subject, we did not target the superconductivity, but fortunately we routinely measured the resistivity of our newly prepared materials down to 2 K because another our concern was to convert a cement component $12CaO{\cdot}7Al_2O_3$(C12A7) from insulating state to metallic state.[9,10] During these experiments, simultaneous substitution of Fe^{2+} or Ni^{2+} for Cu^+ and P^{3-} or As^{3-} for Ch^{2-}, led to discover of low-T_c superconducting materials: LaFePO (T_c = 5 K),[11] LaNiPO (T_c = 3 K)[12] and LaNiAsO (T_c = 2.4 K)[13] as summarized in Fig. 1.1. The T_c values of these materials did not markedly increase by impurity doping. Although non-doped LaFeAsO did not exhibit superconductivity, it displayed anti-ferromagnetism at a Neel temperature (T_N) of ~150 K, and changed into a superconductor with T_c = 26 K upon substitution of $O^{2-}$by F^-, i.e., electron doping as shown in Fig. 1.2. This discovery led to the opening of the FeSCs field, which is currently a significant part of superconductivity research.

Our group of Tokyo Tech and Takahashi group of Nihon Univ had been continuing a cooperative research on LaFePO since 2007. When we discovered the relatively high T_c in $LaFeAsO_{1-x}F_x$ in October, 2007, the target was switched from LaFePO to $LaFeAsO_{1-x}F_x$ and we found a raise of T_c to 43 K under a pressure of ~4GPa. This temperature exceeded the T_c (40 K) of MgB_2 and is next to the high-T_c cuprates. We submitted a letter paper describing this result to *Nature* on February 26[14] and started to synthesize the $LuFeAsO_{1-x}F_x$ at the same time. It was evident that high pressure effectively works to raise the T_c in this system. We thought if La ion with the largest radius among rare earth ions could be replaced

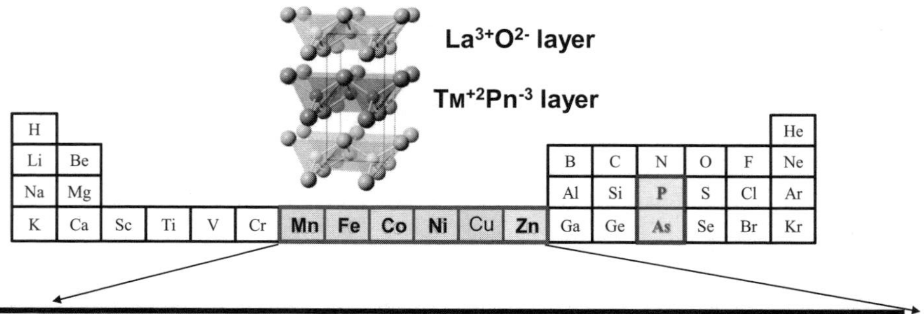

TM^{2+}*(electron configuration)*	$Mn(3d^5)$		$Fe(3d^6)$		$Co(3d^7)$		$Ni(3d^8)$		(Cu)	$Zn(3d^{10})$	
Pn	P	As	P	As	P	As	P	As		P	As
Elect. Prop.	Mott Insulator		Superconductor		Metal		Superconductor		–	Semiconductor	
Magnetism	AF				FM				–	Non-magnetic	
E_g	~1 eV		–		–		–		–	~1.5 eV	
T_C (SC) $T_{N,C}$(Mag.)	> 400 K		Undoped: 4 K	F-doped: 26 K	43 K	66 K	Undoped: 3 K	Undoped: 2.4 K		–	
Ref.	Yanagi *et al.* JAP (2009)		Kamihara *et al.* JACS (2006), Kamihara *et al.* JACS (2008)		Yanagi *et al.* PRB (2008)		Watanabe *et al.* IC (2007), Watanabe *et al.* JSSC (2007)		–	Kayanuma *et al.* PRB (2007), Kayanuma *et al.* TSF (2008)	

Figure 1.1 Electromagnetic properties of LaTMPnO (where TM:3d transition metal, Pn = P, As). See also Color Insert.

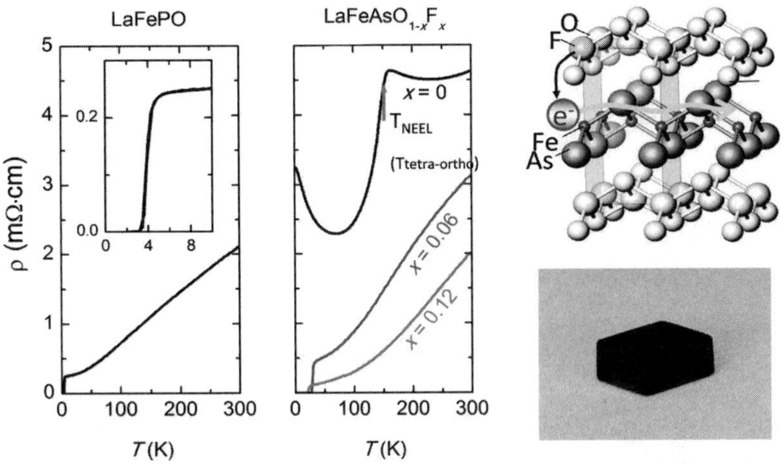

Figure 1.2 Resistivity-temperature curves for LaFePO and $LaFeAsO_{1-x}F_x$ along with crystal structure and the sample photo.

by Lu^{3+}, a nonmagnetic rare earth ion with the smallest radius, the highest T_c would have been realized. However, this trial was totally unsuccessful. The compound LuFeAsO itself was not obtained (even to date). During this struggle, higher T_c's were posted on a preprint server (http://arxiv.org/archive/cond-mat) by research groups in China led by X. Cheng[15] or Z.-X. Zhao[16] in early March. They obtained higher T_c by replacing La ion with Ce, or Sm ion with an open 4f-shell. These reports were shocking for us because we had been intentionally avoided the use of magnetic rare earth ion based on an experimental result (presented at 15th Conference on Intercalation at Korea in May, 2007[17]) that superconductivity disappears when La ion in LaFePO (T_c = 4 K) was replaced by Ce.[18] That is, we misunderstood without careful checking whether or not the 4f orbitals of rare earth ions are close to the Fermi level which is primarily composed of Fe 3d orbitals. According to a band calculation, the valence band width of LaFePO is ~20% larger than that of LaFeAsO.[19] Thus, Ce 4f electron strongly interacts with the carriers in LaFePO and resulted in the conversion to heavy fermion,[20] while in CeFeAsO such a conversion did not occur owing to that Ce 4f level is not so close to the E_F. At the end of May, another shocking paper[21] reporting the superconductivity in $Ba_{1-x}K_xFe_2As_2$

was posted on the cond.mat. by Dirk Johrendt's group in Germay because we had been continuing the electron-doping to $BaFe_2As_2$ after discovering a T_c (4 K) in $LaNi_2As_2$ in March.[22]

Figure 1.3 summarizes the progress of Fe (Ni)-based superconducting materials at the initial stage covering the discovery of major parent materials. The abscissa denotes the date of a paper received in a journal (before 2008) or posted on the preprint-server (after 2008).

1.3 Crystal Structure of Parent Materials

Although numerous FeSCs have been reported, their parent materials may be classified into five types in terms of crystal structures (Fig. 1.4). The structural unit common to these materials is the square net of Fe^{2+} (formal charge) coordinated tetrahedrally by four pnictgen (Pn) or chalcogen (Ch) atoms. Unlike cuprate superconductors, where the parent materials are Mott insulators, this layer exhibits a metallic conductivity without doping. An insulating blocking layer composed by M, MO or MF etc., where M indicates a metallic element such as an alkali metal, alkaline earth, or rare earth element that lies between FePn layers. The local structure of the FePn layer is affected directly by the atomic (or ionic) size of M because M elements in the blocking layer bond to Pn elements. The features of each parent material for FeSCs are described below.

1.3.1 *1111-Type Materials (LnFePnO, Ln: Lanthanide)*

Compounds with the same structure as LaFeAsO, the earliest versions of FeSCs is called the "1111-type" after atomic composition ratio. Figure 1.4a shows their crystal structure, which is a ZrCuSiAs-type structure with a tetragonal P4/nmm space group. Over 300 materials belong to this group, including the p-type wide-gap semiconducting LaCuChO (Ch = S, Se, Te) mentioned in Section 2. Although LaFePO and LaFeAsO along with their crystal structures were reported more than 10 years ago,[23] superconductivity was discovered in 2006[12] for the former and 2008[4] for the latter. Their structural and electronic two-dimensionality is highest among the

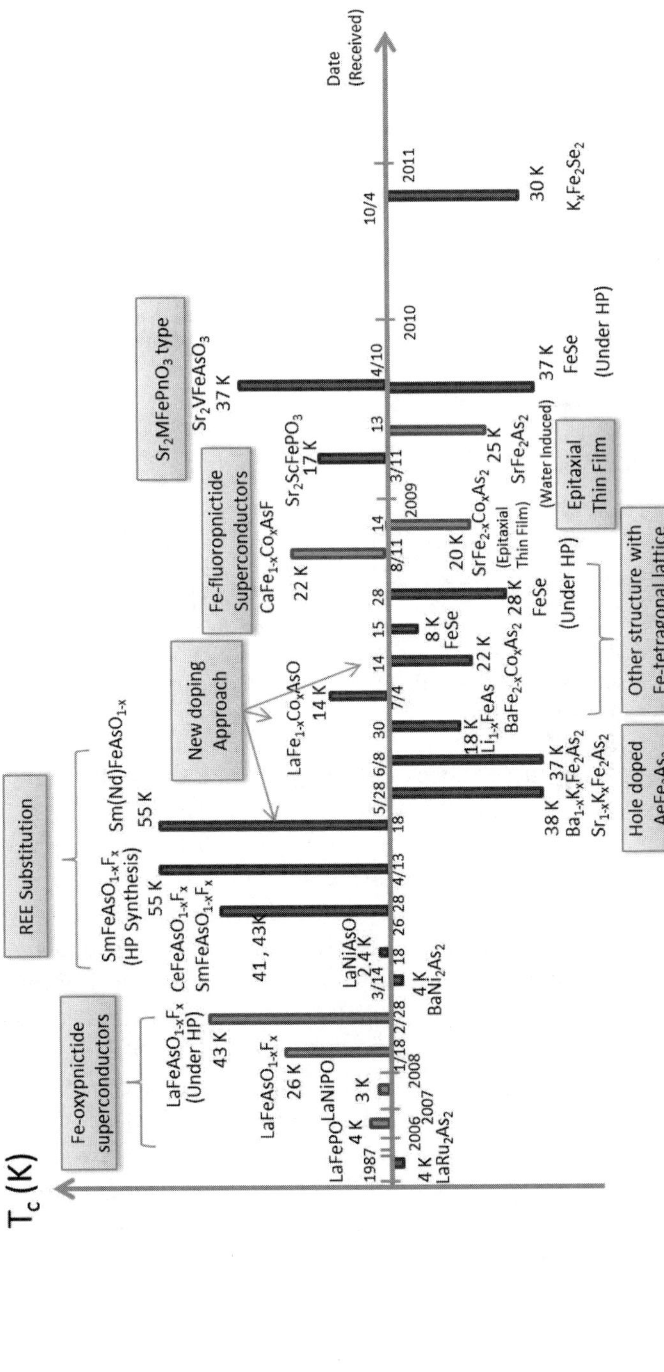

Figure 1.3 Time series of papers reporting superconductivity in each parent compound of iron/nickel pnictide (chalcogenide). Date after March 2008 denotes day posted on *Condensed Matter* Preprint server.

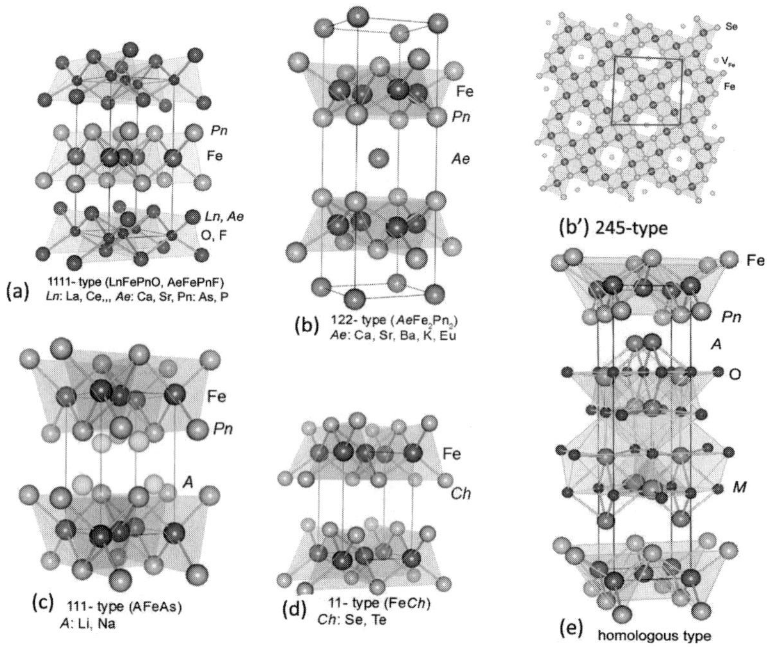

Figure 1.4 Crystal structure of parent compounds. (a) 1111-type, (b) 122-type, (b′) 245-type, (c) 111-type, (d) 11-type, and (e) the homologous-type. See also Color Insert.

five types of parent materials except a type with a thick block, and only this group has T_c values above 50 K.

The unit cell of 1111-type compounds ($a \sim 0.4$ nm, $c \sim 0.8$–0.9 nm) contains two molecular formulae, and is composed of an alternating stack of positively charged LnO layers and negatively charged FeAs layers along the c-axis. The FeAs layer is constructed with an array of edge-shared $FeAs_4$ tetrahedrons constituting the planar square net of Fe. As mentioned above, the local structure of the FeAs layer is the same in each of FeSCs. The distance between the adjacent FeAs layers corresponds to the length of the c-axis ($\sim$0.8–0.9 nm). The formal valence state of each atom is Ln^{3+}, Fe^{2+}, As^{3-}, or O^{2-}. A Fe^{2+} contains 6 electrons in five 3d orbitals and each of them participates to the Fermi level. The lanthanide elements from La to Gd can occupy the Ln site for a 1111-type of material with Pn = P[24], whereas Ho and Y can also occupy the Ln site for Pn = As.[25]

AeFeAsF (Ae = Ca and Sr) with an AeF blocking layer belongs to this type.[26] Structural chemistry of layered d-metal pnictides oxides were reviewed in references.[27,28]

1.3.2 *122-Type Materials ($AeFe_2Pn_2$, Ae: Alkaline Earth or Eu)*

122-type materials have a "$ThCr_2Si_2$" crystal structure with a tetragonal I4/mmm space group.[29] This structure can accommodate a variety of elements[30] like perovskites in oxide system. $CeCu_2Si_2$, a prototype heavy fermion superconductor, belongs to this group.

In the case of $AeFe_2P_2$, lanthanides (La-Pr, Eu) as well as alkaline earth elements can occupy the *Ae* site, while in $AeFe_2As_2$ the *Ae* site can be occupied by alkaline earth, alkali metal, and Eu^{2+}. Figure 1.4b shows the crystal structure for the 122-type. The layer composed of *Ae* ions, which is thinner than the Ln-O layer of the 1111-type, is sandwiched by the FeAs conducting layers. The separation between the adjacent FeAs layers of the 122-type (0.5–0.6 nm) is shorter than that of the 1111-type (0.8–0.9 nm). Because the nearest FeAs layers face each other with a mirror plane, the lattice parameter c is twice the FeAs-FeAs distance. The lattice parameter a($\sim$0.4 nm) is almost the same as that of the 1111-type. Consequently, both 1111 and 122-type materials have similar Fe-Fe distance in the FeAs layer. Since large-sized single crystals of several millimeters can be obtained using Sn or FeAs as a flux, the physical properties of 122-type are well evaluated relative to other types of FeSCs. Johrendt's group was the first to report superconductivity for 122-type materials.[21]

Recently, new 122- type compounds showing relatively high T_c value, e.g., 30 K for $K_xFe_2Se_2$[31] and 27 K for $Ce_{0.8}(FeSe_{0.98})_2$,[32] have been reported. Although these materials were initially classified as the 11-type intercalated by alkali metal, they have the 122-type structure with a tetragonal I4/mmm space group. These materials contain large Fe vacancy, $\sim$1/4 of Fe ion sites, and the composition of the parent material may be described as $A_2Fe_4Se_5$ (A = K, Rb, Cs and Tl) with high Neel temperature ($\sim$500 K) and a band gap (0.37 eV and 0.03 eV).[33] Thus, this family of materials may be regarded as a new parent compound called "245-type" (Fig. 1.4b').

1.3.3 *111-Type Materials (AFePn, A: Alkali Metal)*

While *Ae* ions (formal charge of +2) exists in every two FePn molecules in the 122-type, 111-type compounds contain two *A* ions (*A*: Li^+, Na^+) per two FePn molecules. The crystal structure of this type is known as the "CeFeSi" type, with a tetragonal P4/nmm space group (Fig. 1.4c). This type is compatible with the structure of the 1111-type where all the oxygen atoms are removed and the Ln site is occupied by Li^+ or Na^+. Wang *et al.*[34] (T_c = 18 K: LiFeAs) and Tapp *et al.*[35] (T_c = 18 K: NaFeAs) first reported superconductivity for 111-type materials.

1.3.4 *11-Type Materials* ($Fe_{1+x}Se$)

The 11-type has the simplest crystal structure composed of FeCh layer, and first reported by M.K.Wu group in 2008.[36] The crystal structure is known as "α-PbO" type with a tetragonal P4/nmm space group (Fig. 1.4d). A typical 11-type superconductor is β-FeSe (T_c = 8 K).[36] According to the phase diagram of Fe-Se, the β-type exists in a narrow composition region (x = 0.010–0.025 in $Fe_{1+x}Se$), and the stable form near room temperature is α-FeAs with a hexagonal lattice and the NiAs type structure. To obtain β-FeSe, the material is kept at 400°C for prolonged duration. Medvedov *et al.*[37] suggested the 11-type may exhibit a high T_c (= 37 K) under 8.9 GPa.

1.3.5 *Homologous-Type Materials:* (Fe_2As_2)($Ae_{n+1}M_mO_y$)

The distance between the adjacent FePn/Ch layers shortens in the order of the 1111, 122, 111 and 11-types. The homologous types of iron pnictide oxide have a thick blocking layer composed of a quasi-perovskite structure assembled by MO_5 pyramids and *Ae* (see Fig. 1.4e for the homologous-type (Fe_2As_2)($Sr_4Sc_2O_6$). The FeAs-FeAs distance is 1.55 nm and 2.45 nm for (Fe_2As_2)($Sr_4Sc_2O_6$) and (Fe_2As_2)($Ca_6(Sc,Ti)_5O_y$), respectively. Shimoyama's group systematically has been studied this type of materials.[38–40] It is natural to expect from an analogy with cuprate that T_c would be increase with the thickness of superconducting layers. Figure 1.5 shows the variation in T_c with the inter Fe-layer separation. The highest

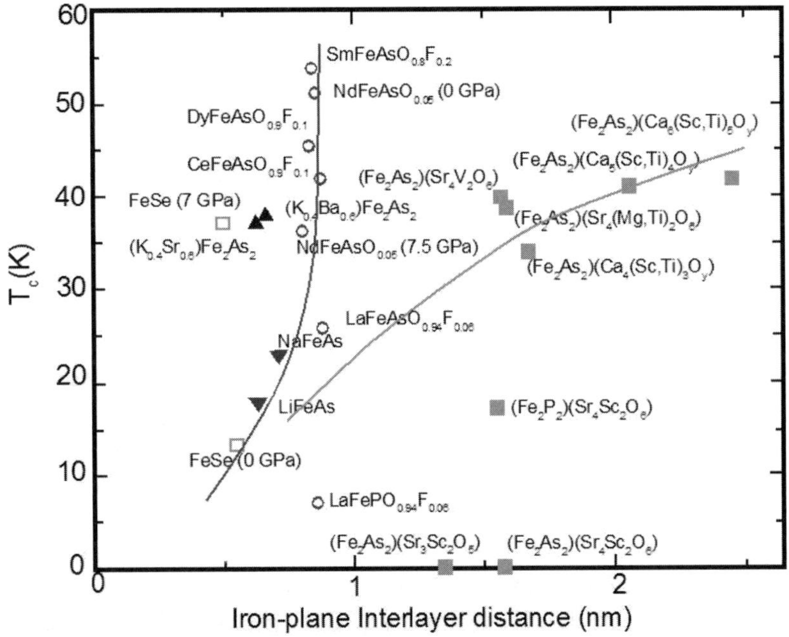

Figure 1.5 Correlation between T_c and iron-plane interlayer distance for various layered FeSCs.[40]

T_c attained is 43 K for $(Fe_2As_2)(Ca_6(Sc,Ti)_5O_y$ but T_c is saturated around ~1.5 nm spacing.

Recently, Nohara *et al.*[41] reported $Ca_{10}(Pt_4As_8)(Fe_{2-x}Pt_xAs_2)_5$ with 38 K along with crystal structural analysis. This material may be regarded as an extension but the spacing layers look metallic conducting unlike the previous materials. This finding was immediately followed by Lohnert *et al.*[42]

1.4 Parent Material and Superconductivity

1.4.1 *Doping Effect*

Intrinsic parent materials do not exhibit superconductivity, but become superconductors with relatively high T_c only by doping. In this section, the doping effect is described with a focus on the 1111 and 122-types.

1.4.1.1 The 1111-type

The parent materials for the 1111-type have tetragonal crystal structure at room temperature, but transform into orthorhombic structure at lower temperatures. Pauli para-magnetism is shown at room temperature, but changes into anti-ferromagnetism at a slightly lower temperatures than that of the structural transitions.[43,44] The magnetic structure of CaFeAsF (antiferromagnetic metal phase) measured by neutron diffraction at low temperature is shown in Fig. 1.6.[45,46] Stripe-type antiferromagnetic ordering (parallel along a-axis but anti-parallel along b-axis) of the magnetic moment in the FeAs plane and anti-parallel ordering between the adjacent FeAs planes are observed.

Electrons are doped into the bulk when an element with more valence electrons is substituted, while holes are doped by substituting an element with fewer valence electrons. In both 1111 and 122-types, it is possible to substitute Fe or As in the conducting layer and La, Ba, O or F in the blocking layer. The former and latter are called "direct doping" and "indirect doping," respectively. With electrons doping by substituting O with F in LnFeAsO, the tetragonal–orthorhombic transition temperature and Neel temperature decrease, and superconductivity is induced in succession. The critical temperature (T_c) increases, reaches a maximum, and then decreases as the dopant level increases. The superconductivity occurs in the tetragonal phase and not in the orthorhombic phase due presumably to antiferromagnetic ordering in the orthorhombic phase in the 1111-type as shown in Fig. 1.7a.[47–52] In contrast to previous reports, Martinelli *et al.*[53] recently reported serious retention of the tetragonal-orthorhombic structural transition by doping of F into $SmFeAsO_{1-x}F_x$ and proposed a new phase diagram they observed the presence of the orthorhombic phase below 130 K in the composition region showing superconductivity ($x \sim 0.2$, $T_c >$ 50 K) using high-resolution synchrotron X-ray diffraction analysis. The orthorhombic distortion was too small to be detected by a conventional X-ray diffraction instrument. The 1111-type contains

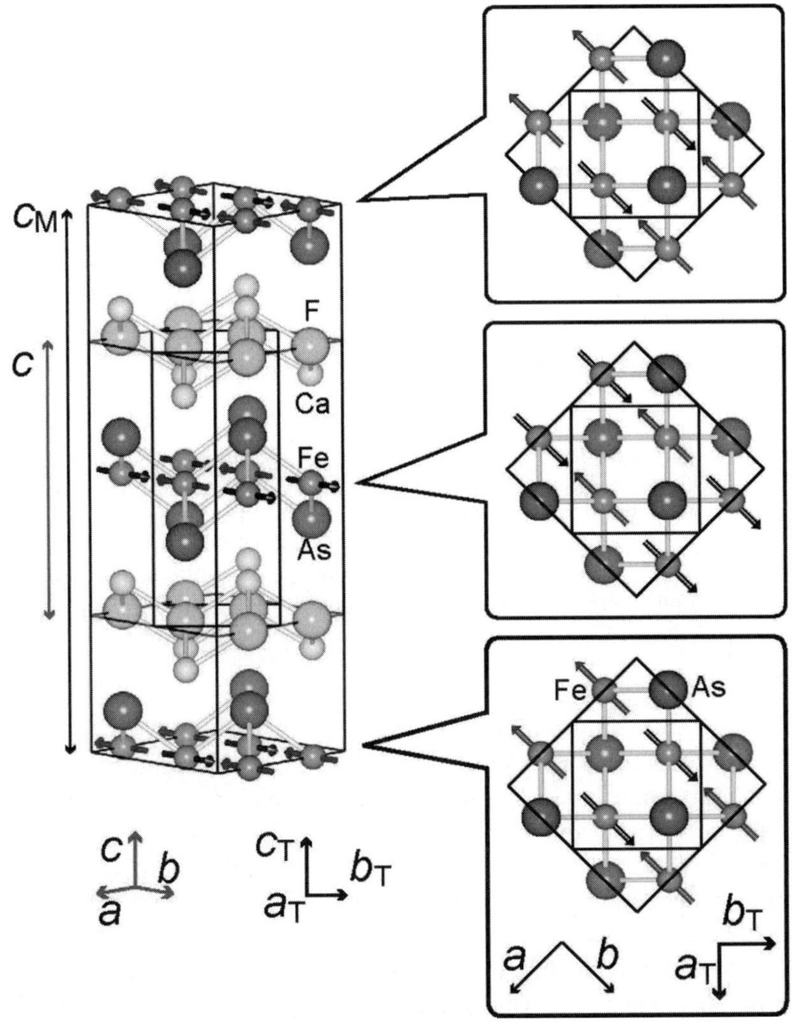

Figure 1.6 Crystal and magnetic structure of CaFeAsF. Crystal structure is tetragonal at room temperature (lattice parameters $a_T = b_T \sim 0.4$ nm, $c_T \sim 0.9$ nm), but transforms into an orthorhombic (lattice parameters $a > b \sim \sqrt{2}a_T$, $c \sim c_T$) at 120 K. In the antiferromagnetic phase (<118 K), the electron spins of Fe order parallel to the b-axis and opposite to the a-axis direction, aligning the total moment in the a-axis direction. Because spins order with inversion between neighboring FaAs planes, the c-axis length of the magnetic sub-lattice is twice of that of the crystal lattice.

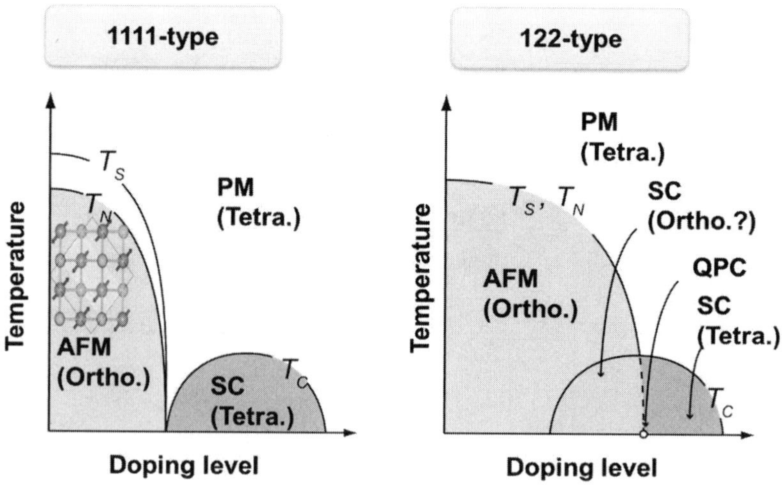

Figure 1.7 Temperature versus composition phase diagram for (a) 1111-type and (b) 122-type. (Ts: structural transition temperature, T_N: magnetic transition (Neel) temperature, AFM: antiferromagnetic phase, PMM: paramagnetic metal phase, SC: superconducting phase).

at least four components (more than five if dopants are included), and precisely controlling its composition is very difficult. Even if starting materials are mixed at the correct composition ratio, the prepared substance contains an impurity phase whose composition differs from the nominal one for many polycrystalline materials. For example, the impurity phases of SmOF, SmAs and/or Sm_2O_3 are formed, and their concentrations increase as the x increases in $SmFeAsO_{1-x}F_x$, which shows the highest T_c.[16,50–52,54,55] This situation is similar to that of other 1111-type materials containing rare-earth elements. The impurity phase by reactions, LnOF, is very stable, and consequently, difficult to erase. Kohler *et al.*[56] demonstrated the conspicuous difference in composition between nominal and resulting $SmFeAsO_{1-x}F_x$ from quantitative analysis using an electron probe micro-analysis (EPMA); they noted that the measured F-content in a specimen with the nominal composition of $SmFeAsO_{0.75}F_{0.25}$ is ~0.12. Karpinski *et al.*[57] reported that without exception, single crystals of $LnFeAsO_{1-x}F_x$(Ln = La, Pr, Nd, Sm, Gd) contain considerable oxygen deficiencies.

The first FeSC reported was formed by electron doping as $LaFeAsO_{1-x}F_x$ where F occupied the O site as (F_O+e').[58] Adding to the substitution of F (T_c = 55 K for $SmFeAsO_{1-x}F_x$), various procedures for the electron doping have been reported, including the formation of an oxygen vacancy (V_O+e': T_c = 55 K for $SmFeAsO_{1-x}$) using a high-pressure synthesis,[59,60] substitution of Th for Ln (Th_{Ln}+e': T_c = 56 K for $Gd_{1-x}Th_xFeAsO$),[61,62] and substitution of Co, Ni or Ir for Fe (Co_{Fe}+e': T_c = 18 K for $PrFe_{1-x}Co_xAsO$, T_c = 23 K for $CaFe_{1-x}Co_xAsF$, Ni_{Fe}+2e': T_c = 11 K for $LaFe_{1-x}Ni_xAsO$, Ir_{Fe}+e': T_c = 18 K for $SmFe_{1-x}Ir_xAsO$).[26,63–65] The optimum doping level is 0.1–0.2/Fe atom for each doping mode, and indirect doping appears to be more effective than direct doping in achieving a high T_c. Figures 1.8(a) and (b) show the ρ–T plots of undoped LnFeAsO (parent) and oxygen vacancy-doped LnFeAsO (superconductor), respectively.[51] The change in slope at 100–150 K shown in the parent material indicates structural and magnetic phase transitions. Doping suppressed these changes and superconductivity emerges. Figure 1.9 shows the variation of maximal T_c attained for Ln = Pr, Nd, Sm, Gd and Dy in Ln-1111. The low T_c of Eu-1111 is tentatively attributed to the change in the formal valence of Eu (+3 to +2). Recently, a new electron-doping method was proposed, substitution of H^- for an O^{2-} site.[49] An advantage of this method is higher solubility limit of H^- at the O^{2-} site than F^- in the 1111 compounds. Figure 1.10 shows the phase diagram obtained for Sm1111 system. Electron-doping was possible up to 45%, which is ~4 times larger than that for F-substitution. As a consequence, the dome of superconductivity including over-doped regime was clarified, and it turned out that T_c remains at a high temperature in the very high electron-doped state.

Emergence of superconductivity by hole-doping was reported using substitution of Ln with Sr ($Nd_{1-x}Sr_xFeAsO$, T_c = 13.5 K at x = 0.2), but its shielding volume fraction by superconductivity is small.[67,68]

1.4.1.2 The 122-type

Compared to the 1111-type, doping methods to generate 122-type superconductors are more diverse. $Ba_{1-x}K_xFe_2As_2$ (T_c = 38 K at

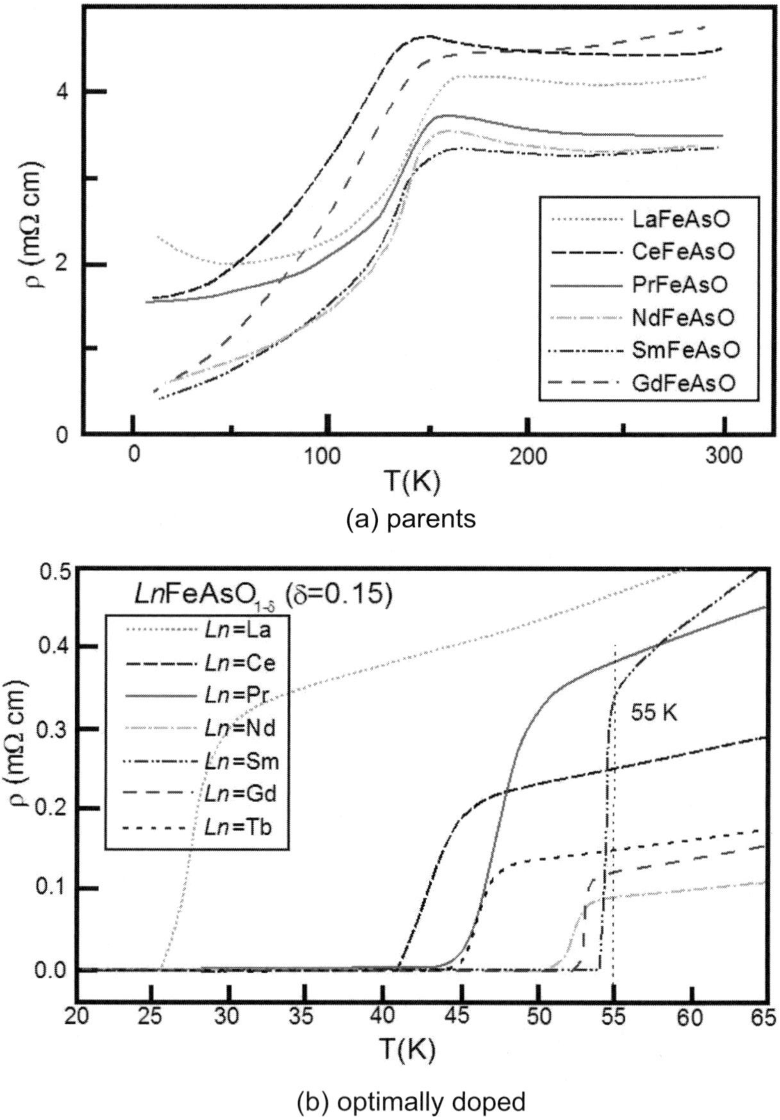

Figure 1.8 Temperature dependence of resistivity for select typical 1111-type (a) parent materials and (b) oxygen-deficient superconductors.[51]

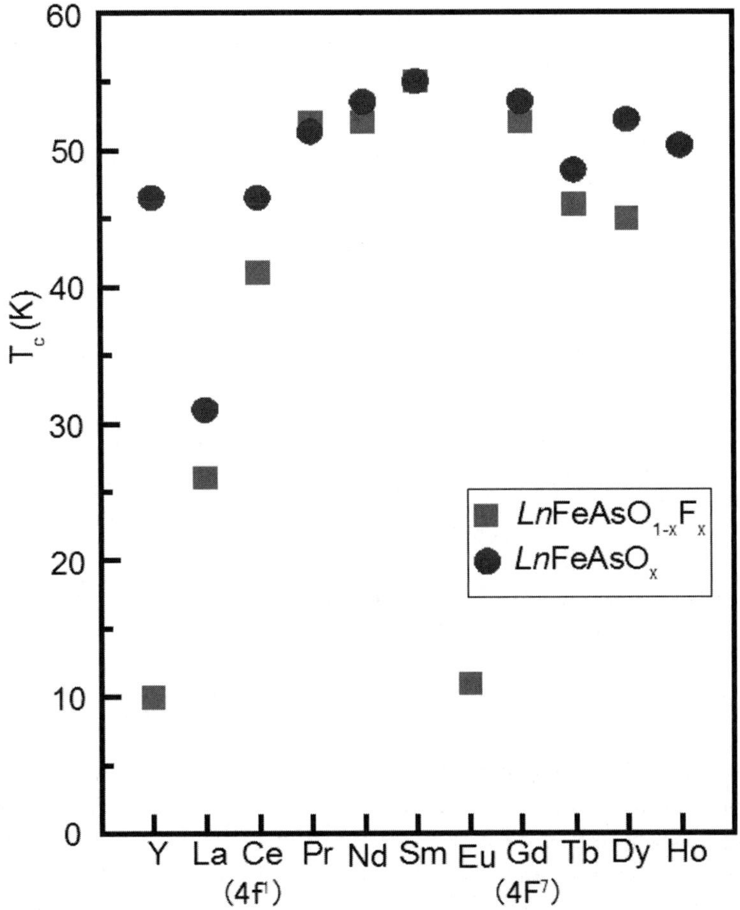

Figure 1.9 Optimal T_c's reported for Ln-1111 superconductors prepared by doping with an oxygen vacancy or F.[87]

$x = 0.4$), which utilizes the indirect hole doping, has the highest T_c.[21] No report on indirect electron-doping has been reported such as replacement of Ae sites with La^{3+} except the following case. This is due probably to the instability of La-doped 122 phase as suggested by the DFT calculations.[69] Muraba *et al.* synthesized $Sr_{1-x}La_xFe_2As_2$ using a high-pressure technique and obtained $T_c = 22$ K at $x = 0.4$.[70] Very recently, Katase *et al.*[71] clarified the comprehensive phase diagram in $Ba_{1-x}RE_xFe_2As_2$ by metastable cation doping in

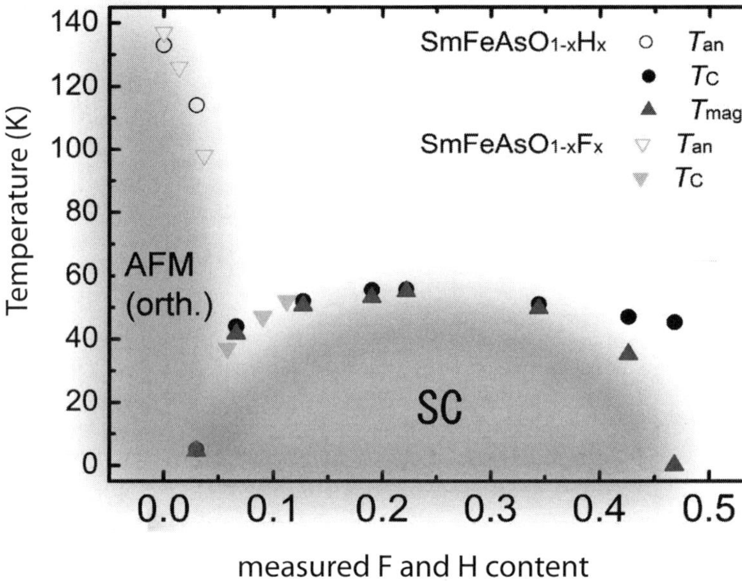

Figure 1.10 Phase diagram of $SmFeAsO_{1-x}H_x$.[49] Data on $SmFeAsO_{1-x}F_x$ are also indicated.

epitaxial thin film growth. Direct electron doping is performed by substitution of Fe^{2+} sites with Co^{2+} with an excess 3d electron such as $Ba(Fe_{1-x}Co_x)_2As_2$(T_c = 23 K at x = 0.2).[72] Similar to that for 1111-type materials, the optimum carrier concentration is 0.1–0.2/Fe.

Another feature of the 122-type is the emergence of superconductivity by substituting isoelectronic elements such as Ru (for Fe)[73,74] or P (for As)[75] as shown in Fig. 1.11. Such an isovalent substitution into the 1111-type does not lead to a superconducting state.[76–78] However, the superconducting state is induced by replacing As with P in $BaFe_2(As_{0.68}P_{0.32})_2$, ($T_c$ = 30 K) and by substituting Rh for Fe in $Ba(Fe_{1-x}Ru_x)_2As_2$,(T_c = 22 K at x = 0.44). Due to a decrease in lattice parameter upon the isovalent element replacement, the induction of superconductivity was initially explained as a chemical pressure effect. However, this explanation is not applicable to the case of $Ba(Fe_{1-x}Ru_x)_2As_2$, where

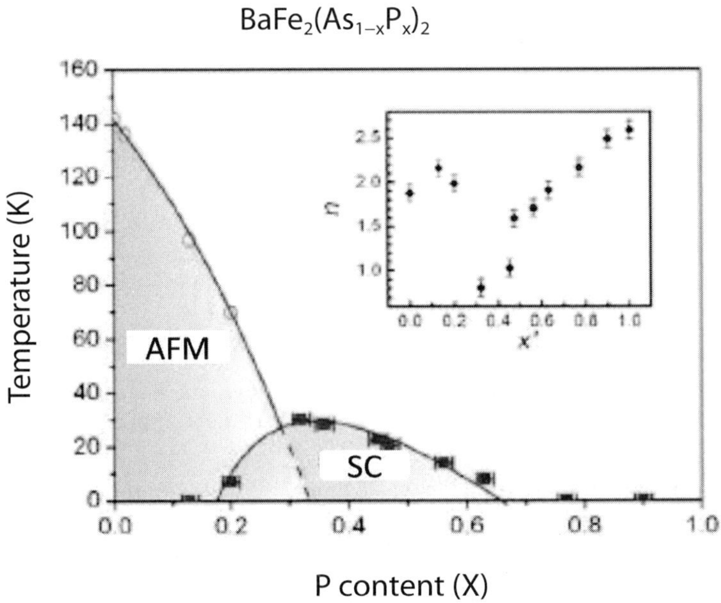

Figure 1.11 Phase diagram in $BaFe_2(As_{1-x}P_x)_2$. Inset show the exponent n in $\rho(T) = \rho_0 + \alpha T^n$.[75]

the lattice parameter increases upon replacement. Brouet *et al.*[79] investigated $Ba(Fe_{0.65}Ru_{0.35})_2As_2$ by transport measurement and angle-resolved photoemission spectroscopy (ARPES), and deduced a two-fold increase in the Fermi surface pocket compared to $BaFe_2As_2$, which results in increased carrier concentration of both electrons and holes, and significantly enhancement of the Fermi velocity. Furthermore, they suggested that Ru substitution weakens the electron correlation through the covalent effect of Ru compared to Fe. These considerations are consistent with a smaller effective mass and larger Fermi surface revealed from the measurement of the de Haas-van Alphen effect.[80] Unlike cuprate superconductors whose parent materials are Mott-type insulators, the parent materials of FeSCs are antiferromagnetic metals with sufficient conduction carriers. Hence, it is plausible for a part of parent materials that the primary effect of doping to FeSCs affects the band structure instead of controlling the carrier density.

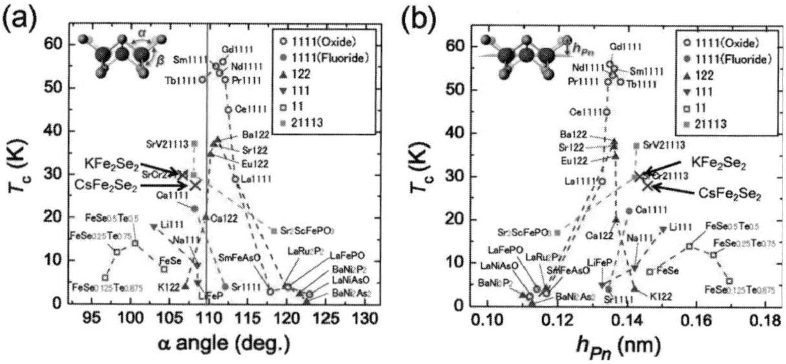

Figure 1.12 Correlation between local structure and T_c. (a) T_c vs. bond angle (α) around Fe, (b) T_c vs. Pn(Ch) height from the iron-plane.

The phase diagram (Fig. 1.7b) of 122-type materials clearly differs from that of the 1111-type. In the 1111-type, the regions showing anti-ferromagnetism and superconductivity are separated or barely overlap, whereas the 122-type materials have antiferromagnetic regions with a high T_c. Because the composition showing the highest T_c corresponds to that with a magnetic transition temperature of 0 K, this phenomenon is discussed in terms of the quantum critical point.[81]

1.4.2 *Local Structure and T_c*

Several models to explain the relationship between the structure and superconducting properties have been reported from the earlier stage of FeSC research. Lee *et al.*[82] first reported a good correlation, which is called Lee's plot hereafter, between T_c and the angle (α) of As–Fe–As in the FeAs layer. Figure 1.12a summarizes the data on various iron-based superconductors on the Lee's plot. The highest T_c in the oxide 1111-type series (LnFePnO) is observed in materials with α corresponding to a regular tetrahedron (109.47°) such as Gd-, Sm- and Nd-1111. In this figure, the highest T_c's reported and the α values mainly for the parent materials at room temperature are plotted.

Intuitively, we can imagine an increase in the density of state on the Fermi surface by the degeneracy of the orbitals resulting from the high symmetry around the Fe atom. This model has been successfully applied to fluoride 1111-type (*Ae*FeAsF), by comparing CaFFeAs ($\alpha = 108^\circ$, $T_c = 22$ K) and SrFeAsF ($\alpha = 112^\circ$, $T_c = 7$ K) where Co is doped into both parent materials to introduce superconductivity.[26] The regular tetrahedron rule can be also applied to the 122-type; one material ($AeFe_2As_2$) with an α value close to 109.47° shows a high T_c (e.g., $Ba_{1-x}K_xFe_2As_2$ shows the highest T_c at $x = 0.3$–0.45). According to the structural analysis of $BaFe_2As_2$ under high pressure by Kimber *et al.*,[83] the changes in the structure in terms of lattice parameters and As-Fe-As bond angle are similar to those by the chemical doping with K. They emphasized that a structural change due to the chemical pressure effect is more important to drive superconductivity than carrier doping.

Kuroki *et al.*,[84] who were inspired by Lee's plot, proposed a theoretical model to explain the relationship between the structure and T_c. They extracted an important parameter, h_{Pn}, which indicates the height of the pnictogen atom measured from the Fe plane, then calculated the changes in the electronic structure and magnetic properties with changes in h_{Pn}. Their calculations indicate the appearance of the electron pocket γ on the Fermi surface around (π, π) in the unfolded Brillouin zone is sensitive to h_{Pn} and this Fermi surface controls the strength of the nesting effect, T_c, and the symmetry of the superconducting gap. They concluded that T_c increases with increases in h_{Pn}. Figure 1.12b plots this relationship where T_c's and h_{Pn}'s are employed in same manner as in Fig. 1.12a. In Fig. 1.12a, materials showing high T_c's are roughly localized around $\alpha = 109.5^\circ$ for the 1111-, 122-and homologous-types, but the deviation from the regular tetrahedron rule is found in the 111-type where LiFeAs ($\alpha \approx 103^\circ$) shows a higher T_c ($\sim$18 K) than NaFeAs ($T_c \sim$9 K, $\alpha \sim 108.5^\circ$) or LiFeP ($T_c \sim$6 K, $\alpha \sim 108.5^\circ$). Although rules cannot be deduced from the h_{Pn}–T_c relationship in Fig. 1.11, a positive correlation for each material type is observed and the discrepancy for the 111-type is at least resolved. However, several discrepancies remain. h_{Pn} of KFe_2As_2 is highest in superconductors belonging to the 122-type, but its T_c is low ($\sim$4 K).[85] For the 11-type, h_{Pn} increases with the Te content, but the T_c shows a maximum at $x = 0.5$ in $FeSe_{1-x}Te_x$.[86]

Here we would like to discuss models for each type. Figure 1.13 shows the relationship between h_{Pn} and T_c for the doped Ln-1111-

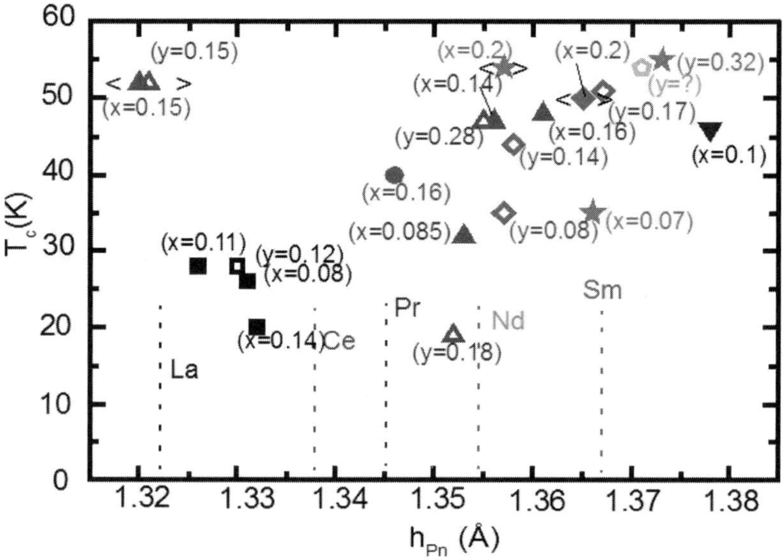

Figure 1.13 Correlation between h_{Pn} and T_c of 1111-type superconductors of $LnFeAsO_{1-x}F_x$ (closed marks) and $LnFeAsO_{1-y}$ (open marks).[87]

type superconductor.[87] Closed and open marks indicate F doped and O vacancy doped samples, respectively, and dotted lines denote the h_{Pn}'s of the parent materials. Except for two points of Pr-1111 where h_{Pn}'s were measured at low temperature (5 K), a positive correlation is roughly shown, suggesting the h_{Pn}–T_c rule is applicable to the superconductors. However, it is impossible to apply this rule to an individual doping operation where h_{Pn} changes monotonically as the amount of dopant increases and the highest T_c emerges at the optimal composition due to such doping. For example, $(Ba_{1-x}K_x)Fe_2As_2$ is a complete solid solution system[88] where h_{Pn} of $BaFe_2As_2$ (1.358 Å) increase monotonically with an increase in x toward that of KFe_2As_2 (1.429 Å) and the highest T_c emerges at $x = 0.37$, $h_{Pn} = 1.383$ Å (Fig. 1.14). The regular tetrahedron rule is effective on this material series. The T_c's of alkali-intercalated FeSe compounds also are located at a good position on the Lee's plot (See Fig. 1.12).

Regardless of the high correlation between each structural parameter and T_c (α–T_c, h_{Pn}–T_c), several distinct deviations still

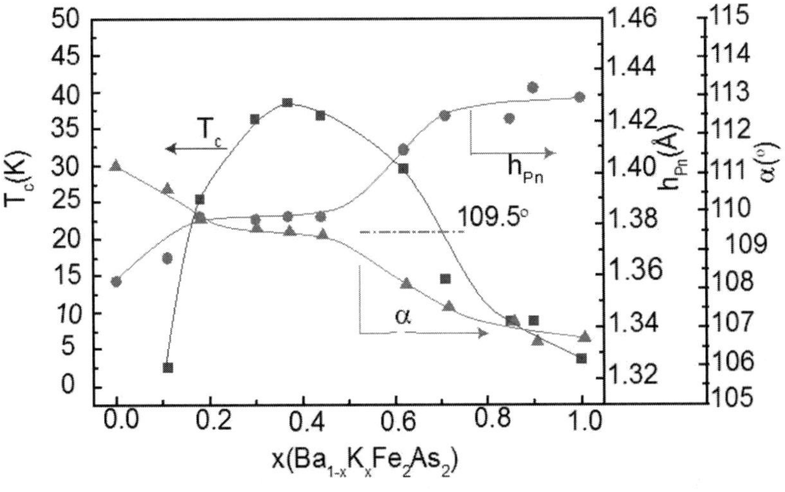

Figure 1.14 Correlation between T_c and local structural parameters (α and h_{Pn}) in $Ba_{1-x}K_xFe_2As_2$. Data were taken from ref. 88.

remain. Consequently, a structural parameter that is applicable to FeSCs as a whole has yet to be found. On the other hand, the unique characteristics of each type of FeSC have been determined as FeSC research has progressed. Hence, it is unclear whether a universal parameter to describe the characteristics of FeSCs as a whole exists.

1.5 Unique Characteristics of FeSCs

1.5.1 *Multi-Band Nature of Fe3d*

Figure 1.15 shows the calculated and observed density of state in LaOFeAs.[89] The electronic state of the Fermi level (E_F) controlling the transport property is formed by a complex tangle of five Fe3d orbitals, due to the small contribution of As, which is unlike cuprate superconductors where only Cu d_{x2-y2} contributes to the E_F. With five bands comprising E_F, multi-pockets appear on the Fermi surface. The energy levels of d_{x2-y2}, d_{xy}, and d_{yz} are sensitive to both the changes in the symmetry of the $FeAs_4$ tetrahedron and the carrier density. These energy level differences influence the shape and volume of the pockets on the Fermi surface. Consequently,

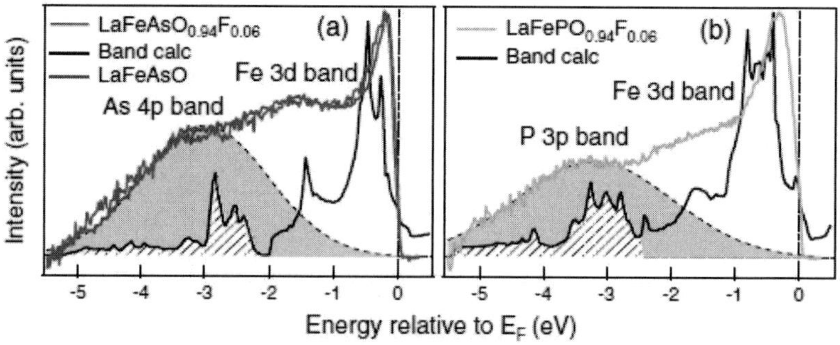

Figure 1.15 Calculated and measured density of states in LaFeAsO and LaFePO.[89] The contribution of pnicogen to the Fermi level is small.

these differences affect the strength of Cooper pair formation due to spin fluctuations and subsequently mediate T_c.[84,90,91] As described in Section 4.2, the model proposed by Kuroki *et al.*[84] classifies T_c by the height of pnictogen (h_{Pn}) from the Fe plane. This model is attributed to the fact that h_{Pn} reflects the relationships between the three bands of the 3d orbitals (d_{x2-y2}, d_{xy} and d_{yz}) and E_F. Recently, the orbital fluctuation mechanism was proposed[92,93] based on an experimental finding of a large reduction of share elastic modulus near the structural/magnetic transition.[94]

1.5.2 *Parent Material: Antiferromagnetic Metal*

Metal compound superconductors such as MgB_2 do not exhibit magnetism and their conduction electrons induce superconductivity without doping. The parent materials of cuprate superconductors are antiferromagnetic insulators with a Cu magnetic moment of $\approx 1\mu_B$, which corresponds to nine electrons localized on the 3d orbitals. The parent materials of the 1111-or 122-type FeSCs are antiferromagnetic metals, and their magnetic moment per Fe atom is 0.4–0.9μ_B which is much smaller than that due to the six electrons localized on Fe^{2+} with a high spin state ($4\mu_B$). The repulsive energy between 3d electrons in FeSC is approximately half that in cuprate. These results demonstrate the strong itinerancy of electrons in FeSCs.

In the parent compounds, 245 phase is totally different from others;[95] it is an insulator with a distinct band gap and high spin magnetic moment and high Need temperature (~500 K) such as $3\mu_B$. Apparent coexistence of AM and superconductivity is attributed to the nano-scale phase separation.[95] The antiferromagnetic spin fluctuation and the multi-band effect discussed in the above section are two factors to drive the system into the superconducting state. A plausible pairing is thought via interpocket scattering of electrons between the hole pockets (around the Γ point) and electron pockets (around the M point), and leads to a pairing situation where an isotropic gap with opposite signs on each pocket (so-called $S^{\pm}$ symmetry). However, recent finding of robust impurity effect and modestly high T_c of the 245 phase (in which the hole pocket almost disappears).[96] The symmetry of the superconducting gap is an exciting topic in research on the theory of superconductivity.

1.5.3 *Impurity Robust T_c*

Generally, T_c's of superconductors decrease drastically upon doping with magnetic impurities such as Fe, Ni, and Co. However, compared to cuprate superconductors, an analogous amount of a magnetic impurity doped into FeSCs causes only a small decrease in T_c. Additionally, superconductivity often emerges in FeSCs upon magnetic ion doping. That is, the substitution of Fe^{2+} ($3d^6$) by Co^{2+} ($3d^7$) or Ni^{2+} ($3d^8$) in the non-superconducting parent material induces superconductivity, similar to the result of electron doping into the FeAs layer. A similar result can be obtained by doping the 4d transition elements as shown in Fig. 1.16.[97]

1.5.4 *Large Critical Field and Small Anisotropy*

The superconducting performance is estimated from the critical temperature (T_c), critical current (J_c), upper critical field (H_{c2}), and anisotropy. A large H_{c2} and small anisotropy are unique properties of FeSCs. It has been reported that $H_{c2}(0)$ is 80 T for the 122-type (200 T for the 1111-type), and the anisotropic ratio of the resistivity (γ) is four for the 122-type. $H_{c2}(0)$ is defined as the upper critical

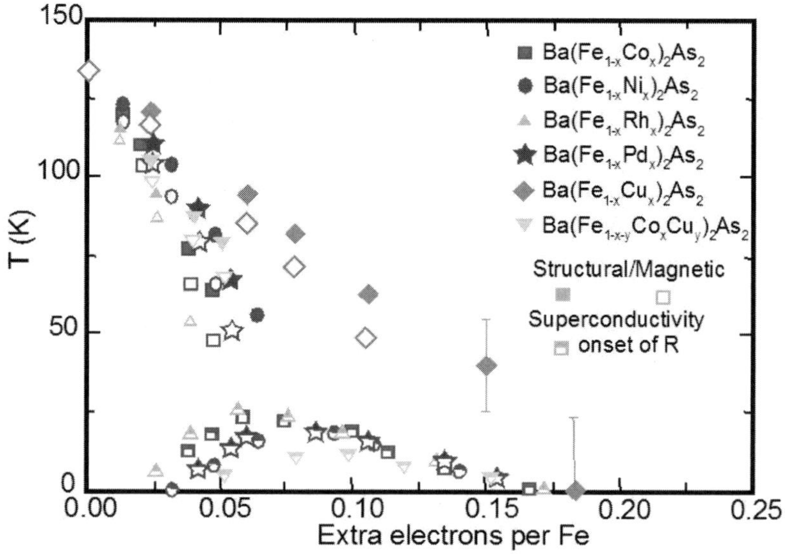

Figure 1.16 Phase diagram of $Ba(Fe_{1-x}T_{Mx})_2As_2$.[97] T_M = Transition metal.

field at 0 K, and γ is defined as the ratio of the resistivity along the crystal axes directions, a and c, measured just above T_c. $H_{c2}(0)$ of FeSC is higher than that of MgB_2 ($\approx$40 T), but is smaller than that of a typical cuprate ($\sim$350 T for $YBa_2Cu_3O_{7-\delta}$). The γ value for FeSC is compatible with that of MgB_2 ($\approx$3.5), but is lower than that of a typical cuprate (30–270 for $YBa_2Cu_3O_{7-\delta}$). These properties are favorable for practical applications such as electrical wires. Table 1.1 listed the properties of FeSCs, MgB_2 and cuprates. The values of γ, J_c and H_{c2}, for FeSCs are discussed in the next section.

1.5.5 *Advantageous Grain Boundary Nature*

High-T_c superconductors have zero power consumption and could be used to produce ideal electric power lines. The major obstacle in fabricating superconducting wires and tapes is grain boundaries—the misalignment of crystalline orientations at grain boundaries, which is unavoidable for polycrystals, largely deteriorates critical current density for each superconductors. Recently advantageous

Table 1.1 Comparison between three representative superconductors

Materials	Fe pnictides	MgB_2	Cuprates
Parent material	antiferromagnetic metal (T_N ~150 K)	non-magnetic metal	antiferromagnetic insulator (T_N ~400 K)
Fermi level	five Fe 3d orbitals	two B 2p orbitals	single Cu 3d orbital
T_c max.	56 K	40 K	~ 130K
Superconducting gap symmetry	Extended s-wave	Simple s-wave	d-wave
$H_2^c(0)$	100–200 T	~ 40T	~ 350T (YBCO)
Magnetic impurity effect	robust	sensitive	sensitive
Anisotropy of conductivity	2–4 (122) 8–10 (1111)	~ 3.5	30–250 (YBCO) ~ 1000(Bi2112)
Critical tilt angle for J_c	~°(Ba 122)		~ 5°(YBCO)

grain boundary nature was reported for Co-doped Ba122 with a small anisotropy. Katase *et al.*[98] examined the transport properties through well-defined bicrystal grain boundary junctions with various misorientation angles (θ_{GB}) for cobalt-doped $BaFe_2As_2$ epitaxial films fabricated on bicrystal substrates and found that iron pnictide superconductors have advantages over cuprates with respect to these grain boundary issues. Figure 1.17 shows the J_c of Ba122:Co epitaxial thin films as a function of θ_{GB}. The J_c through bicrystal grain boundary, J_c (BGB) remained high (>1MAcm^{-2}) and nearly constant up to a critical angle θ_c of ~9°, which is substantially larger than the θ_c of ~ 5° for $YBa_2Cu_3O_{7-\delta}$. Even at $\theta_{GB} > \theta c$, the decay of J_c (BGB) was much slower than that of $YBa_2Cu_3O_{7-\delta}$. Such grain boundary natures together with the high $H_{c2}(0)$ make the iron pnictides to be more promising materials for application to high J_c superconducting tapes.

1.6 Single Crystal

1.6.1 *Growth of 1111-Type Crystals*

After superconductivity of LaFeAsO was initially discovered, single crystal for Nd-1111-type and Sm-1111-type materials have been reported by Wen's group[99] and Kalpinski's group,[100] respectively.

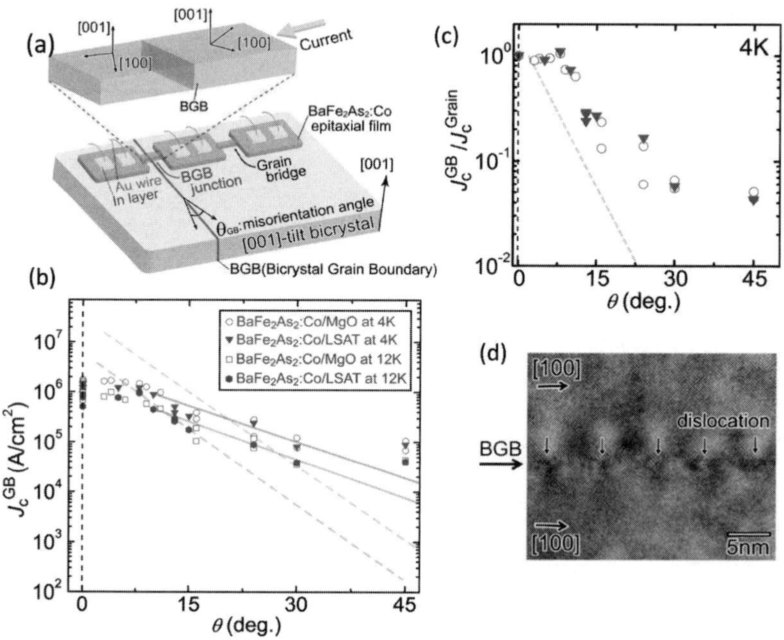

Figure 1.17 Critical current density (J_c) of Co-doped $BaFe_2As_2$ epitaxial thin films as a function of bicrystal grain boundary (BGB) angle θ. (a) sample shape for J_c measurement, (b) J_c as a function of BGB. Data of YBCO are also drawn for comparison, (c) bicrystal grain boundary J_c normalized by In grain bridge J_c vs. θ, and (d) TEM photo around BGB. Dislocations with a periodic separation (l) is seen and the length continuously decreases with θ and l at the θ_{cri} is close to the coherent length.[98]

For the 1111-type, the crystal growth was carried out using a flux of alkali halide under either atmospheric or high pressure. Although the high-pressure method can grow larger crystals within a short time,[100–102] the instrument is large and expensive. Here, we introduce in the atmospheric pressure method reported by Jia *et al.*[99,103] Powders of NdAs, NdF_3, Fe_2O_3, and Fe are mixed in a stoichiometric composition with the formula of $NdFeAsO_{0.82}F_{0.18}$ and the mixture is pressed into a pellet, placed together with NaCl powder at a weight ratio of (NaCl:NdFeAsOF = 10:1), sealed in an evacuated quartz tube, and sintered at around 1050°C for 10 days. Afterwards, it is cooled at a rate of 3°C/h to 850°C, and then the furnace is turned off for faster cooling. The resulting product contains mainly plate-like small crystals with lateral sizes ranging from 5 to 30 μm and thicknesses of 1–5 μm. Although larger plate-like crystals with a lateral size of 100 μm have been obtained under high-pressure growth (3 GPa) for 1–2 days, the crystals are too small to operate by hand. Yan *et al.*[104] prepared doped LaFeAsO superconducting single crystals with a lateral size of several millimeters under atmospheric pressure in NaAs flux, but the dopant content has yet to be controlled. Because the scale of 1111-type single crystals is generally small, the focused ion beam (FIB) technique has been employed to shape specimens and contact electrodes with sub-micrometer dimensions.[99–104]

1.6.2 *Growth of the 122-Type Crystals*

Single crystals of 122-type with a lateral size on the "millimeter" order can be prepared using a flux method with Sn or FeAs. Immediately after Johrendt's group[21] reported superconductivity of (Ba,K)Fe_2As_2, several groups prepared single crystals of $BaFe_2As_2$ doped with K[105,106] and Co[107] Plate shape single crystals measuring 1–2 × 1–2 × 0.05–0.1 mm^3 have prepared in the following manner.[105] Elemental Ba, K, Fe and As are added to Sn in the ratio of [(Ba,K)Fe_2As_2]:Sn = 1:48 as a starting material, and the mixture is placed in an MgO crucible. Then a second catch crucible containing quartz wool is set on top of this growth crucible. This assembly is sealed in a silica glass tube under approximately 1/3 atmosphere of argon gas. The sealed tube is heated to 850°C and cooled to 500°C over 36 h. Once the furnace reached at 500°C, the Sn flux was decanted from the (Ba,K)Fe_2As_2 crystals using a centrifuge.

Use of the FeAs flux is effective to obtain large crystals. According to Ni *et al.*[108] a $Ba(Fe_{1-x}Co_x)_2As_2$ single crystal with dimension of 12 × 8 × 0.8 mm^3 can be prepared by mixing Ba, FeAs and CoAs with a ratio of 1:4(1 – x):4x. The mixture is placed into an alumina crucible, and then a second crucible is attached on the growth crucible and sealed in the same manner as the Sn case. This assembly is heated at 1180°C for 2 h and cooled over 36 h to 1000°C. Once the furnace reached at 1000°C the fluxis decanted from the $Ba(Fe_{1-x}Co_x)_2As_2$ crystals using a centrifuge.

Recently, the growth of a $Ba_{0.6}K_{0.4}Fe_2As_2$ crystal without a flux agent has been reported.[109] Excess K and As are mixed with starting materials of Ba, K, Fe, and As to compensate for evaporation loss. The mixture with a nominal composition of $Ba_{0.6}K_{0.44}Fe_2As_{2.1}$ is placed in an alumina crucible and heated above 1200°C to melt completely in a tightly sealed stainless tube. It is subsequently cooled below 1050°C prior to turning off the furnace. Single crystals with a lateral size on the "millimeter" order are obtained using this method.

1.6.3 *Characteristics of a Single Crystal*

Similar to cuprate superconductors, FeSCs have a layered structure, but FeSCs are unique due to their small anisotropy. For instance, the resistivity of cuprate superconductors along the *a*- or *b*-axis direction at T_c (parallel to the CuO_2 conduction plane) is smaller than that along the *c*-axis direction with a ratio ($\gamma_\rho = \rho_c/\rho_{ab}$) of 30–250 for $YBa_2Cu_3O_{7-\delta}$[110] and over 1000 for $Bi_2SrCaCu_2O_{10}$.[111] This high anisotropy is a reflection of a thick blocking layer in cuprates. High anisotropies are also observed for the upper critical fields (H_{c2})[112] and the critical current density (J_c)[113] in cuprate superconductors.

For FeSCs, early band-structure calculations suggest a two-dimensional electronicstructure.[114] A high anisotropy in the resistivity with $\gamma_\rho \approx 100$ is reported for the non-superconducting parent compounds $BaFe_2As_2$[115] and $SrFe_2As_2$[116] as well as for superconducting Co-doped $BaFe_2As_2$.[103] In contrast, Tanatar *et al.*[104] reported that the ratio $\gamma_\rho = \rho_c/\rho_{ab}$ for $Ba(Fe_{0.936}Co_{0.074})_2As_2$ is about 3–5 at T_c, and this value is widely accepted. Moll *et al.*[119] reported a higher γ_ρ(10 at T_c) for the 1111-type $SmFeAsO_{0.7}F_{0.25}$ single crystal shaped by the FIB technique with thicker blocking layer than that of the 122-type.

For polycrystalline materials, two different critical current densities, J_c^{local} and J_c^{global}, should be examined. The former is J_c of the intra-grain and the latter is the total J_c of the specimen, which includes contributions from the inter-grain and surface. Because the inter-grain suppresses J_c in many cases, J_c^{global} of polycrystalline specimen is considerably smaller than that of J_c^{local}. Factors responsible for the suppression at the inter-grain region are voids, cracks, impurity phases, and weak-links of grains, which are remarkable in materials with high anisotropy. Such obstacles can be neglected ($J_c^{\mathrm{global}} = J_c^{\mathrm{local}}$) in single crystals. The anisotropy of J_c is observed in materials with structural anisotropy such as FeSCs. To evaluate them precisely, measurements using a single crystal are essential. Generally J_c attains a high current value on the order of $> 10^5$ A/cm^2 at 4 K for FeSCs, and direct measurement by the I-V relation is limited to the immediate vicinity of T_c. Thus, J_c is usually estimated from magnetization measurements (M-H curves) by applying Bean's model.[120] Figure 1.18 shows the J_c's estimated by Tanatar *et al.*[118] for $Ba(Fe_{0.936}Co_{0.074})_2As_2$($T_c = 23$ K) and by Moll *et al.*[119] for $SmFeAsO_{0.7}F_{0.25}$($T_c = 48$ K), where J_c is the critical current density along the *ab*-plane direction measured by applying a magnetic field to the *c*-axis direction (H//*c*) and the *ab*-plane direction (H//*ab*), respectively. The J_c's are 2.8×10^5 A/cm^2 and 2.2×10^6 A/cm^2 at 5 K for the 122 and 1111- types, respectively. In a broad range of temperature from T_c to 5 K, the magnetically determined in-plane and inter-plane critical current densities differ by a factor of three in the 122-type. Such a difference for the 1111-type is small at low temperatures, but increases as the temperature increases.

Beside single crystals, the reported critical current densities include 1×10^4 A/cm^2 at 4 K of polycrystalline $Ba_{0.6}K_{0.4}Fe_2As_2$ prepared by the PIT method [121] and 4×10^6 A/cm^2 at 4 K of $BaFe_2As_2$:Co epitaxial film grown by pulsed laser deposition.[122]

Figure 1.19 shows H_{c2}'s estimated by Ni *et al.*[108] for $(Ba_{0.55}K_{0.45})Fe_2As_2$ ($T_c = 32$ K) and by Moll *et al.*[119] for $SmFeAsO_{0.7}F_{0.25}$ ($T_c = 48$ K) where H_{c2} is the upper critical field measured by applying a magnetic field to the *c*-axis direction (H//*c*) and the *ab*-plane direction (H//*ab*) respectively. The measurement for a $(Ba_{0.55}K_{0.45})Fe_2As_2$($T_c = 32$ K) single crystal indicates $H_{c2}^{//c} =$ 55 T and $H_{c2}^{//ab} = 66$T at 15 K, where the superscripts //*c* and //*ab* indicate magnetic fields applied parallel to the *c*-axis and the *ab*-plane, respectively. From the extrapolation of their H_{c2}–T plots, estimates for both $H_{c2}^{//ab}$ and $H_{c2}^{//c}$ readily exceed 80 T at 4 K. The anisotropy ($\gamma_H = H_{c2}^{//ab}/H_{c2}^{//c}$) is small, 1.2 at 15 K, and decreases with temperature in the temperature region of 32–14 K.[123] For a $SmFeAsO_{0.7}F_{0.25}$ single crystal, $H_{c2}^{//ab}$ (which exceeds the measurement range up to 65 T at 36 K) is higher than that of the

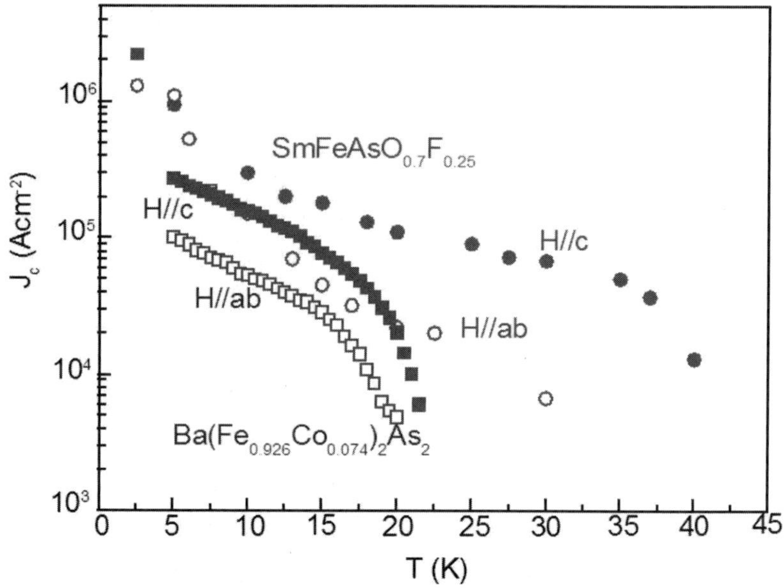

Figure 1.18 J_c's (along the *ab*-plane direction) of $Ba(Fe_{0.926}Co_{0.074})_2As_2$ and $SmFeAsO_{0.7}F_{0.25}$ single crystals for an applied field parallel to the *c*-axis and perpendicular to the *c*-axis (parallel to the *ab*-plane) calculated using Bean's model. [118,119]

$H_{c2}^{//c}$ (60T at 15 K) in the entire temperature range. Additionally, the anisotropy of H_{c2} (γ_H) decreases monotonically from 6 at 48 K to 3.5 at 37 K as temperature decreases. From the extrapolation of this result, the estimated $H_{c2}^{//}$ exceeds 150 T and $H_{c2}^{//i}$s greater than 100 T at 4 K. The 1111-type shows a higher H_{c2} and higher γ_H than those for the 122-type. The higher γ_H and γ_ρ of the 1111-type originate from the higher two-dimensionality, which is attributed to a thicker blocking layer than that in the 122-type.

1.7 Thin Film

1.7.1 *1111-Type Compounds*

Hiramatsu *et al.*[124] first succeeded in fabricating *in situ* epitaxial films of 1111-type compounds, LaFeAsO (La1111) films, on MgO

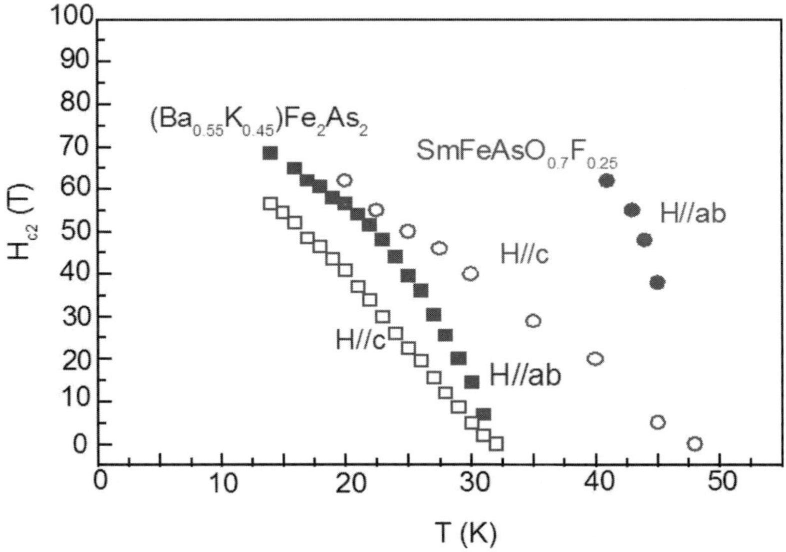

Figure 1.19 Anisotropic $H_{c2}(T)$ of $(Ba_{0.55}K_{0.45})Fe_2As_2$ and SmFeAs $O_{0.7}F_{0.25}$ single crystals for an applied field parallel to the *c*-axis and perpendicular to the *c*-axis (parallel to the *ab*-plane).[108,119]

and LSAT substrates by PLD using a second-harmonic Nd:YAG laser (533 nm). The XRD patterns shown in Fig. 1.20 clearly indicated perfect *c*-axis and in-plane alignment of the as-grown film deposited using a F-doped LaFeAsO target with a nominal content of 10 at.%. The film showed a temperature dependence of the resistivity curve resembling that of undoped La1111, suggesting that F was not effectively incorporated in the film deposited at a substrate temperature of around 780°C. Since the 1111-type compounds also become superconducting by electron doping via the introduction of oxygen deficiency, they attempted post-annealing in a H_2 atmosphere. However, the H_2 annealing resulted in only a slight change in the lattice parameters and no superconductivity was induced. In contrast, Backen *et al.*[125] obtained superconducting LaFeAs(O,F) (F-La1111) thin films by an *ex situ* PLD method.[125] The precursor films deposited on $LaAlO_3$ substrates at room temperature from a $LaFeAsO_{1-x}F_x$ ($x = 0.25$) target were annealed in a silica glass tube at approximately 1030°C. The annealed films

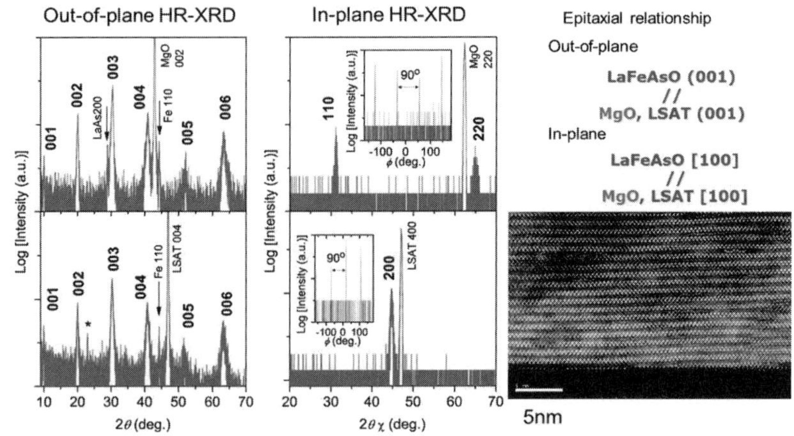

Figure 1.20 Epitaxial thin films of LaFeAsO grown on LSAT substrate by PLD.[124] XRD patterns and TEM photo of the resulting films are shown along the epitaxial relationship.

exhibited *c*-axis and in-plane orientations and a resistive T_c onset of 11 K, although zero resistance was not obtained. By optimizing the conditions of post-annealing with the bulk material having the target composition in a silica glass tube, they raised the T_c onset and T_c zero up to 28 and approximately 20 K, respectively.[126] However, the films showed weak-link behavior and a rather low self-field J_c on the order of 103 A/cm^2 at low temperatures.

1111-type compound thin films with higher T_c values have been realized by means of an MBE method. Kawaguchi *et al.*[127] first reported the successful *in situ* epitaxial growth of Nd1111 thin films on GaAs substrates, which are lattice-matched with Nd1111, though the window of the optimum growth conditions, especially for O_2 and As flux. They found to be narrow. In spite of the use of an F-containing source, NdF_3, the obtained films exhibited semiconducting behavior of the resistivity, indicating that the F content in the films was too low. Subsequently, they found that a long growth time of over 5 h resulted in a clear superconducting transition with a T_c onset and a T_c zero of up to 48 and 42 K, respectively. They also noticed the existence of an NdOF layer near the surfaces of the superconducting films and considered that this layer might supply F into the films. Moreover, by artificially depositing a NdOF/Nd1111 bilayer structure to accelerate F incorporation, they obtained superconducting F-Nd1111 films on MgO and CaF_2 substrates in addition to GaAs substrates with the highest T_c zero of 50 K on CaF_2, which is comparable to that of bulk F-Nd1111.[128,129]

1.7.2 *122-Type Compounds*

The 122-type compounds have small anisotropy and high T_c values close to that of MgB_2 for the K-doped compounds. Carrier doping by the substitution of K or Co is easier than F doping in the 1111-type compounds, particularly, for thin film growth, because F tends to evaporate from film surfaces. Thus, the largest number of works on thin films has been focused on the 122-type compounds among the iron-based superconductors.

The first superconducting epitaxial films of the 122-type compound $Sr(Fe,Co)_2As_2$ (Co-Sr122) were synthesized on (LaSr)(Al,Ta)O_3(LSAT) single-crystal substrates by Hiramatsu *et al.*[130] by pulsed laser deposition (PLD) using a second-harmonic Nd:YAG laser ($\lambda = 532$ nm). The film deposited at a higher substrate temperature of 700°C clearly exhibits in-plane orientation and the epitaxial relationship (001)[100] $SrFe_2As_2$|(001)[100] LSAT. The films exhibited a T_c onset and a T_c zero of approximately 20 and 15 K, respectively. Although the transport critical current density (J_c) for the films was rather low, 20 kA/cm^2 at 4.2 K, the measurements in a high magnetic field of up to 50 T using a pulse magnet revealed that the films have a nearly isotropic upper critical field and γ is close to 1 at low temperatures.[131] Co-Sr122 films were also found to have rather high sensitivity to humidity. They underwent rapid degradation in water vapor, and, interestingly, undoped Sr122 epitaxial films were found to exhibit superconductivity below 25 K after exposure to humid air for 2–6 hours, which can possibly be attributed to the doping of water-related species such as O^{2-} or OH^- in the I9 site between the Fe-As tetrahedron and the Sr atom in Sr122.[132]

Compared with Sr122, $BaFe_2As_2$ (Ba122) bulk polycrystal was found to be more stable in air. Katase *et al.*[133] confirmed that Co-doped Ba122 (Co-Ba122) epitaxial films prepared by PLD showed

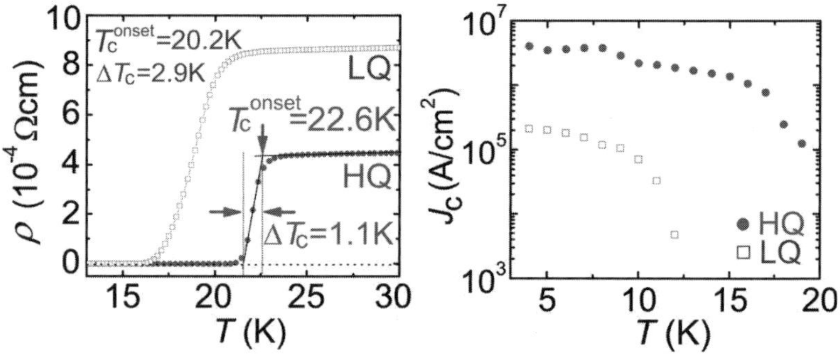

Figure 1.21 Temperature dependence of resistivity and J_c in Co-doped $BaFe_2As_2$ epitaxial thin films grown on LSAT substrates. HQ denotes the films deposited under the improved conditions.

no appreciable change in the resistive T_c even after exposure to water vapor (dew point of 20–25°C) for 48 h, and thus had much higher stability against water vapor than Sr122 films. Such high stability was considered to be due to the smallest free space of the I9 site in Ba122 among the Ae122 compounds. The Co-Ba122 epitaxial films were also found to have higher crystallinity with an atomically flat surface and exhibit a sharper resistive transition than the Co-Sr122 films. Moreover, by increasing the purity of the Co-Ba122 target and optimizing the PLD conditions, they succeeded in directly fabricating Co-Ba122 epitaxial films with a narrow resistive transition width of ΔT_c = 1.1–1.3 K and a self-field J_c of 1–4 MA/cm^2 at 4 K, as shown in Fig. 1.21, on both LSAT and MgO substrates.[134] The high-J_c film on LSAT was found to show a J_c peak for a field parallel to the *c*-axis that was larger than the peak for a field perpendicular to the *c*-axis even at a high field of 12 T at a low temperature of 4 K. This finding suggested the existence of *c*-axis correlated defects or pinning centers in the film,[135] although the origin of the pinning centers is not yet clear. Iida *et al.* also fabricated Co-Ba122 epitaxial thin films on $SrTiO_3$ (STO), LSAT, and $LaAlO_3$ (LAO) substrates by PLD using a KrF excimer laser (λ = 248 nm) and found a strong correlation between the c/a lattice constant ratio and T_c onset,[136] though a systematic correlation between a and the lattice mismatch was not observed.

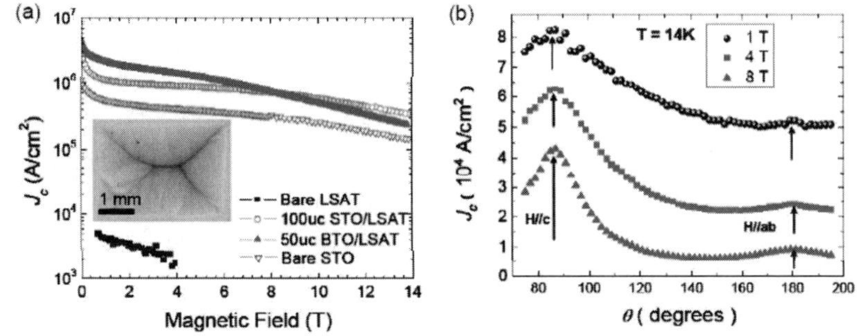

Figure 1.22 (a) Dependence of J_c on magnetic field parallel to the c-axis for Co-Ba122 thin films prepared on bare LSAT and $SrTiO_3$ (STO) substrates, a 50-unit-cell (u.c.)-thick $BaTiO_3$ buffer layer, and a 100-u.c.-thick STO buffer layer on LSAT. (b) Magnetic field angular dependence of J_c at 14 K and 1, 4, and 8 T for a Co-Ba122 thin film on STO/LSAT.[137]

High-quality Co-Ba122 epitaxial thin films were also fabricated by using a buffer layer technique. Lee *at al.*[137] reported that the epitaxial quality of Co-Ba122 thin films was significantly improved by employing an STO or $BaTiO_3$ (BTO) template layer with a thickness of 50–100 unit cells on LSAT or other substrates as compared with the films directly prepared on the substrates by PLD. More interestingly, the epitaxial films on the template exhibited a very high J_c in magnetic fields parallel to the c-axis of up to 14 T, as shown in Fig. 1.22, indicating strong c-axis flux pinning. Co-Ba122 films on an optimized template (100 unit cells of STO) were found to exhibit a J_c peak for a field along the c-axis exceeding J_c for H//ab up to a high field of 12 T at 12 K and a large maximum pinning force of above 5 GN/m^3. Recently, Katase *et al.*[138] reported high critical current densities J_c >1 MA/cm^2 were realized in cobalt-dopedBaFe2As2:Co films on flexible metal substrates with biaxially textured MgO base-layers fabricated by an ion-beam assisted deposition technique. The films showed small in-plane crystalline misorientations of 3° regardless of twice larger misorientations (7.3°) of the MgO base-layers, and exhibited high self-field J_c up to 3.5 MA/cm^2 at 2 K. These values are comparable to that on MgO single crystals and the highest J_c among iron pnictide superconducting tapes and wires ever reported. High in-field J_c suggests the existence of c-axis correlated vortex pinning centers.

K-doped Ae122 (Ae = Sr, Ba) has the highest T_c of 38 K among the 122-type compounds. The fabrication of such thin films is difficult because of the volatile and reactive nature of K. Lee *et al.*[139] first reported the synthesis of K-doped $BaFe_2As_2$ (K-Ba122) thin films by an *ex situ* PLD method in which an undoped Ba122 film deposited at room temperature was post-annealed with K shots in a silica glass tube. The films prepared on Al_2O_3 (0001) and $LaAlO_3$ (001) substrates exhibited a preferential *c*-axis orientation, and the former film showed a T_c onset and T_c zero of as high as 40 and 37.5 K, respectively. The *in situ* growth of epitaxial K-doped Ae122 films by a molecular beam epitaxy (MBE) method has recently been reported by Takeda *et al.*[140] They employed a chemically stable In-K alloy (In_8K_5) as a K source and found that the key to incorporating K into films is low-temperature growth (300–350°C) in a reduced As flux. The obtained K-Ba122 film exhibited a T_c onset and T_c end of 38.3 and 35.5 K, respectively, and substantially lower resistivity than that of the *ex situ* films. 8. However, because the K-doped films degraded immediately in an ambient atmosphere. Passivation is needed for further handling.

1.7.3 *11-Type Compounds*

The 11-type compounds have the lowest T_c among the iron-based superconductors. However, their simplest crystal structure and the pressure-induced significant increase in T_c straightforwardly lead to motivation for fabricating epitaxial films on lattice-mismatched substrates, which may show strain-induced T_c enhancement. Fabrication of the first 11-type compound thin film, an $Fe(Se_{0.5}Te_{0.5})$ film with a T_c zero of approximately 12 K, by a PLD method using a KrF excimer laser was reported by Wu *et al.*[141] Subsequently, Wang *et al.*[142] of the same group reported the properties of $FeSe_{1-x}$ thin films deposited on MgO substrates by a PLD method. They found that thin *c*-axis-oriented films did not undergo a structural transformation to a low-temperature orthorhombic phase and superconductivity, while (101) -oriented films deposited at a higher substrate temperature underwent such a structural transformation and had a T_c of approximately 8 K as observed in bulk $FeSe_{1-x}$. Similarly $FeSe_{1-x}$ superconducting thin films with (101) preferential

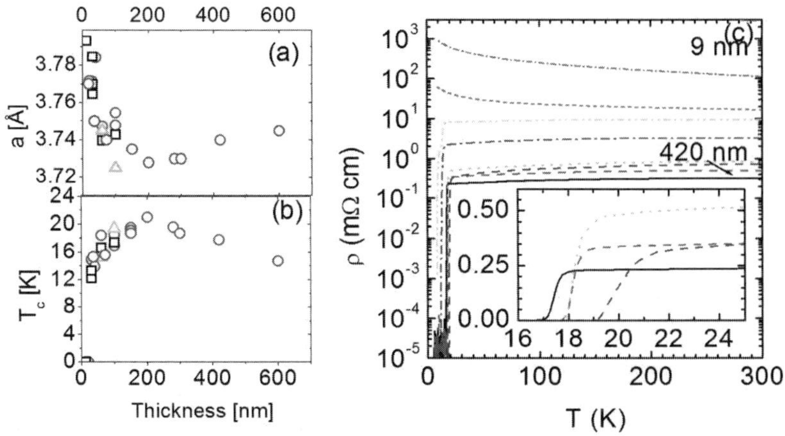

Figure 1.23 Epitaxial thin films of Fe(Se,Te). (a) *a*-axis and (b) T_c values vs film thickness on LAO (○), STO (□), and Y:ZrO (□). (c) ρ–T curves for several films on LAO with different thickness (9, 18, 36, 72, 150, 200, 280, and 420 nm). The inset shows the detail of the superconducting transition for some films.[146]

orientation and T_c values close to the bulk value were prepared on various substrates including sapphire (0001), STO, and LAO by a PLD method.[143] However, no sign of strain-induced T_c enhancement was observed in these studies. Imai *et al.*[144] investigated the properties of $Fe(Se_{0.5}Te_{0.5})$ thin films deposited on eight different substrates and found no systematic *a*-axis contraction associated with the lattice mismatch. The film T_c was found to be rather strongly correlated with the degree of in-plane orientation and the lattice parameter ratio c/a.

On the other hand, Si *et al.*[145] observed an enhancement of T_c onset to 17 K in some of their $Fe(Se_{0.5}Te_{0.5})$ films prepared on STO substrates by a PLD method and suggested that the significantly reduced lattice constant *c* was related to the T_c enhancement. Bellingeri *et al.*[146] reported a more significant T_c enhancement up to 21 K in high-quality epitaxial $Fe(Se_{0.5}Te_{0.5})$ films on LAO and yttrium-stabilized zirconia (ZrO_2:Y) substrates with lattice constants slightly smaller than the *a*-axis constant for the bulk $Fe(Se_{0.5}Te_{0.5})$. The films exhibited a significant decrease in the *a*-axis lattice constant with increasing thickness, as shown in Fig. 1.23, and a linear correlation between the *a*-axis lattice constant and T_c, indicating the important role of compressive strain along the direction parallel to the plane in enhancing T_c. The film T_c was also found to increase as the bond angle of the tetrahedron approached the ideal value of 109.28°, which is similar to the trend observed in other iron-based superconductors.[77]

1.8 Summary and Relevant New Superconductors

FeSCs typically show high J_c and H_{c2} with a low anisotropy and relatively high T_c, which are favorable features for practical applications. On the other hand, many problems remain; the content of toxic elements such as As, P or Se in FeSCs with a high T_c is a serious issue, and is an obstacle to join for researchers in industry. Although steady progress is continuing, under these conditions, the record-highest T_c (56 K) has not been broken for three years. The discoveries of $K_xFe_2Se_2$[31] with a T_c at 30 K and $Ca(Fe_{1-x}Pt_x)_{2-\delta}As_2$[41] with a T_c at 38 K are typical examples. For $K_xFe_2Se_2$, band structure calculations and preliminary angle resolved photo-emission spectroscopy (ARPES) measurements indicate that the band near the Γ-point seems to dive below the Fermi energy, leading to the absence of hole pockets, which are prerequisites for the pairing. This discrepancy with the model previously accepted for FeSCs has received attention from physicists attempting to understand the phenomenon as well as those exploring new materials.

Very recently, the new type of the FeSC with relatively high T_c, $Ca_{1-x}Ln_xFe_2As_2$, has been reported.[147] The highest T_c is 47 K for $Ca_{0.8}Pr_{0.2}Fe_2As_2$[148] or 42.7 K for $Ca_{0.8}La_{0.2}Fe_2As_2$.[149] Because the 122-type is stable and relatively easy to prepare single crystals, many researchers are undertaking theoretical evaluations and are beginning to investigate the physical properties. First thin film Josephson Junction[150] and SQUID[151] were fabricated using Co-doped Ba122.

In the 60s to early 80s when the T_c appeared to be reached the limit at 23 K of $Nb_3Al_{1-x}Ge_x$, "Matthias's rules,"[152] which are shown below, were widely accepted as a guide to finding a new superconductor with high T_c:

(1) Transition metals are better than simple metals.
(2) There are favorable electron/atom ratios [$N(E_F)$ peaks].
(3) High symmetry is good; cubic symmetry is best.
(4) Stay away from oxygen.
(5) Stay away from magnetism.
(6) Stay away from insulating phases.

Now, we know no researchers who adhere to these rules. When FeSCs were discovered after 22 years of focusing on cuprate superconductors, Mazin,[153] who first reported a FS nesting mechanism for FeSCs,[154] proposed a new set of rules to find high-T_c superconductors:

(1′) Layered structures are good.
(2′) The carrier density should not be too high (compared with, say, conventional metals).
(3′) Transition metals of the fourth period (V, Cr, Mn, Fe, Co, Ni and Cu) are good.
(4′) Magnetism is essential.
(5′) Proper Fermi surface geometry is essential (It must match the structure of the spin excitations).
(6′) Enlist theorists, at least to compute the Fermi surfaces Note that there is one corollary to these rules. Materials of interest are likely to be complex chemical compounds – work closely with solid-state chemists.

The development of cuprate superconductors emphasized the importance of a layered structure but the discovery of FeSCs eliminated elements forbidden for superconductor, expanding frontier in superconducting materials. Recently, a series of non-iron-based superconductors with the same crystal structures, a-SnO[155] with the 11-type, $LaCo_2B_2$[156] with 122-type and $CeNi_{0.8}Bi_2$[157] with the 1111-type (two bismuth have difference valence). Figure 1.24 shows crystal structures ogf these new superconductors. Although T_c of these materials is not high, we expect many discoveries of new superconductors in the parent compounds and relevant materials. *Strike while iron is hot!*

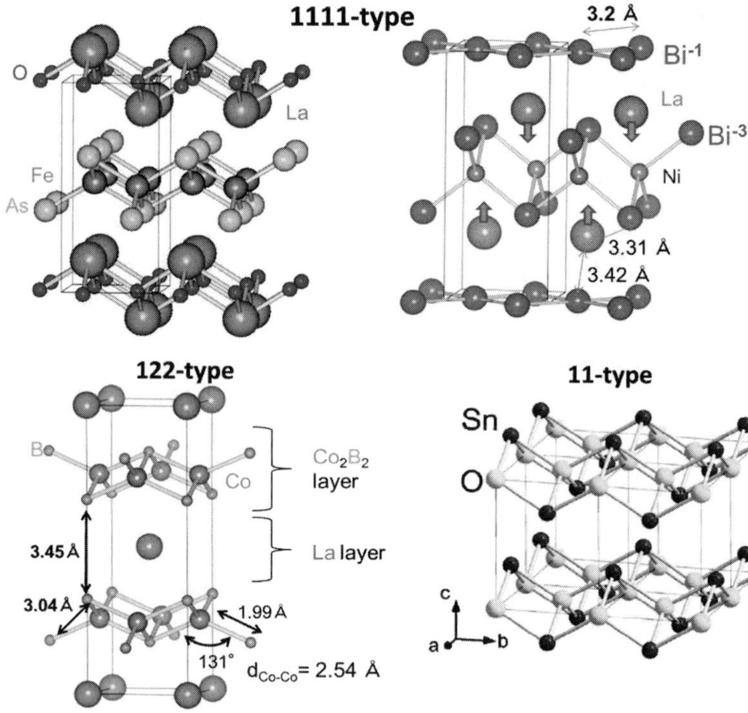

Figure 1.24 Crystal structure of newly discovered superconductors (parent materials) with 1111, 122 and 11-type lattice. Note that $CeNiBi_2$ has two kinds of Bi anion with different valence state. There coexist heavy electrons and light electrons at the Fermi level and the former is associated with Ce and the latter associated with Bi square net is responsible for superconductivity.[154] Superconductivity in $LaCo_2B_2$ is induced by substitution of Co sites with Fe, which is reverse to the case of $BaFe_2As_2$.[155] SnO is a wide-gap semiconductor but is converted to superconductor under a high pressure of >5GPa.[153] See also Color Insert.

Acknowledgment

This work is supported by a grant from the Japan Society for the Promotion of Science (JSPS) through the "Funding Program for World-Leading Innovative R&D on Science and Technology (FIRST Program)," initiated by the Council for Science and Technology Policy (CSTP).

References

1. J. G. Bednorz, and K. A. Muller, *Z. Phys. B*, 64, 189 (1986).
2. M. K. Wu, J. R. Ashburn, and C. J. Torng, P. H. Hor, R. L. Meng, L. Gao, Z. J. Huang, Y. Q. Wang, and C. W. Chu, *Phys. Rev. Lett.*, 58, 908 (1987); J. Nagamatsu, N. Nakagawa, T. Muranaka, Y. Zenitani, and J. Akimitsu, *Nature*, 410, 63 (2001).
3. J. Nagamatsu, N. Nakagawa, T. Muranaka, Y. Zenitani, and J. Akimitsu, *Nature*, 401, 63 (2001).
4. Y. Kamihara, T. Watanabe, M. Hirano, and H. Hosono, *J. Am. Chem. Soc.*, 130, 3296 (2008).
5. Y. Maeno, H. Hashimoto, K. Yoshida, S. Nishizaki, T. Fujita, J. G. Bednorz, and F. Lichtenberg, *Nature*, 372, 532 (1994).
6. D. Ginley, H. Hosono, and D. Ginley, *Handbook of Transparent Conductors,* Springer, 2010.
7. H. Kawazoe, M. Yasukawa, H. Hyodo, M. Kurita, H. Yanagi, H. Hosono, *Nature*, 389, 939 (1997).
8. H. Hosono, *Thin Solid Films*, 515, 6000 (2007).
9. S. Matsuishi, Y. Toda, M. Miyakawa, K. Hayashi, T. Kamiya, M. Hirano, I. Tanaka, H. Hosono, *Science*, 301, 626 (2003).
10. M. Miyakawa, S. W. Kim, M. Hirano, Y. Kohama, H. Kawaji, T. Atake, H. Ikegami, K. Kono, H. Hosono, *J. Am. Chem. Soc.*, 129, 7270 (2007).
11. Y. Kamihara, H. Hiramatsu, M. Hirano R. Kawamura, H. Yanagi, T. Kamiya, and H. Hosono, *J. Am. Chem. Soc.*, 128, 10012 (2006).
12. T. Watanabe, H. Yanagi, T. Kamiya, Y. Kamihara, H. Hiramatsu, M. Hirano, and H. Hosono, *Inorg. Chem.*, 46, 7719 (2007).
13. T. Watanabe, H. Yanagi, Y. Kamihara, T. Kamiya, M. Hirano, and H. Hosono, *J. Sol. State. Chem.*, 181, 2117 (2008).
14. H. Takahashi, K. Igawa, K. Arii, Y. Kamihara, M. Hirano, and H. Hosono, *Nature*, 453, 376 (2008).
15. X. H. Chen, T. Wu, G. Wu, R. H. Liu, H. Chen, and D. F. Fang, *Nature*, 453, 761 (2008).
16. Z-A. Ren, W. Lu, J. Yang, W. Yi, X-L. Shen, Z-C. Li, G-C. Che, X-L. Dong, L-L. Sun, F. Zhou, and Z-X. Zhao, *Chin. Phys. Lett.*, 25, 2215 (2008).
17. Y. Kamihara, M. Hirano, and H. Hosono, Presented at the International Conference on Intercalation Compounds, Seoul, Korea, May 2007.
18. Y. Kamihara, H. Hiramatsua, M. Hirano, H. Yanagi, T. Kamiya, and H. Hosono, *J. Phys. Chem. Sol.*, 69, 2916 (2008).

19. V. Vildosola, L. Pourovskii, R. Arita, S. Biermann, and A. Georges, *Phys. Rev. B*, 78, 064518 (2008).
20. E. M. Brüning, C. Krellner, M. Baenitz, A. Jesche, F. Steglich, and C. Geibel, *Phys. Rev. Lett.*, 101, 117206 (2008).
21. M. Rotter, M. Tegel, and D. Johrendt, *Phys. Rev. Lett.*, 101, 107006 (2008).
22. T. Mine, H. Yanagi, T. Kamiya, Y. Kamihara, M. Hirano, and Hideo Hosono, *Solid State Commun.*, 147, 111 (2008).
23. A. T. Nientiedt, W. Jeitschko, P. G. Pollmeier, and M. Brylak, *Z. Naturforsch. B*, 52, 560 (1997).
24. B. I. Zimmer, W. Jeitschko, J. H. Albering, R. Glaum, and M. Reehuis, *J. Alloys Compd.*, 229, 238 (1995).
25. P. Quebe, L. J. Terbüchte, and W. Jeitschko, *J. Alloys Compd.*, 302, 70 (2000).
26. S. Matsuishi, Y. Inoue, T. Nomura, H Yanagi, M. Hirano, and H. Hosono, *J. Am. Chem. Soc.*, 130, 14428 (2008).
27. T. C. Ozawa and S. M. Kauzlarich, *Sci. Technol. Adv. Mater.*, 9, 033003 (2008).
28. D. Johrendt, H. Hosono, R.-D. Hoffmann, and R. Rottgen, *Z. Kristallogr.*, 226, 435 (2011).
29. G. Just and P. Paufler, *J. Alloys Compd.*, 232, 1 (1996).
30. D. Mandrus, A. S. Sefat, M. A. McGuire, and B. C. Sales, *Chem. Mater.*, 22, 715 (2010).
31. J. Guo, S. Jin, G. Wang, S. Wang, K. Zhu, T. Zhou, M. He, and X. Chen, *Phys. Rev. B*, 82, 180520 (2010).
32. A. K-Maziopa, Z. Shermadini, E. Pomjakushina, V. Pomjakushin, M. Bendele, A. Amato, R. Khasanov, H. Luetkens, and K. Conder, *J. Phys.: Condens. Matter*, 23, 052203 (2011).
33. Nan Lin Wang, Presented at International Conference on Strongly Correlated Electron Systems 29 August to 3 September 2011, Cambridge, UK.: see chap. 5.
34. X. C. Wang, Q. Q. Liu, Y. X. Lv, W. B. Gao, L. X. Yang, R. C. Yu, F. Y. Li, and C. Q. Jin, *Solid State Commun.*, 148, 538 (2008).
35. J. H. Tapps, Z. Tans, B. Li, K. Sazmal, B. Lorenz, P. C. W. Chu, and A. M. Guly, *Phys. Rev. B*, 75, 060505 (R) (2008).
36. F-C. Hsu, J-Y. Luo, K-W. Yeh, T-K. Chen, T-W. Huang, P. M. Wu, Y-C. Lee, Y-L. Huang, Y-Y. Chu, D-C. Yan, and M-K. Wu, *Proc. Natl. Acad. Sci. USA*, 105, 14262 (2008).

37. S. Medvedev, T. M. McQueen, I. A. Troyan, T. Palasyuk, M. I. Eremets, R. J. Cava, S. Naghavi, F. Casper, V. Ksenofontov, G. Wortmann, and C. Felser, *Nat. Mater.*, 8, 630 (2009).
38. H. Ogino, Y. Matsumura, Y. Katsura, K. Ushiyama, S. Horii, K. Kishio, and J. Shimoyama, *Supercond. Sci. Technol.*, 22, 075008 (2009).
39. H. Ogino, Y. Katsura, S. Horii, K. Kishio, and J. Shimoyama, *Supercond. Sci. Technol.*, 22, 085001 (2009).
40. H. Ogino, S. Sato, K. Kishio, J. Shimoyama, T. Tohei, and Y. Ikuhara, *Appl. Phys. Lett.*, 97, 072506 (2010).
41. S. Kakiya, K. Kudo, Y. Nishikubo, K. Oku, E. Nishibori, H. Sawa, T. Yamamoto, T. Nozaka, and M. Nohara, *J. Phys. Soc. Jpn.*, 80, 093704 (2011).
42. C. Löhnert, T. Stürzer, M. Tegel, R. Frankovsky, G. Friederichs, and D. Johrendt, *Angew. Chem.*, 123, 9362 (2011).
43. T. Nomura, S. W. Kim, Y. Kamihara, M. Hirano, P. V. Sushko, K. Kato, M. Takata, A. L. Shluger, and H. Hosono, *Supercond. Sci. Technol.*, 21, 125028 (2008).
44. C. de la Cruz, Q. Huang, J. W. Lynn, J. Li, W. Ratcliff II, J. L. Zarestky, H. A. Mook, G. F. Chen, J. L. Luo, N. L. Wang, and P. Dai, *Nature*, 453, 899 (2008).
45. Y. Xiao, Y. Su, R. Mittal, T. Chotterju, T. Hansen, C. M. N. Kumar, S. Matsuishi, H. Hosono, and Th. Brueckel, *Phys. Rev. B*, 79, 060504R (2009).
46. T. Nomura, Y. Inoue, S. Matsuishi, M. Hirano, J-E. Kim, K. Kato, M. Takata, and H. Hosono, *Supercond. Sci. Technol.*, 22, 055016 (2009).
47. L. Malavasi, G. A. Artioli, C. Ritter, M. C. Mozzati, B. Maroni, B. Pahari, and A. Canesch, *J. Am. Chem. Soc.*, 132, 2417 (2010).
48. C. Hess, A. Kondrat, A. Narduzzo, J. E. Hamann-Borrero, R. Klingeler, J. Werner, G. Behr, and B. Büchner, *Europhys. Lett.*, 87, 17005 (2009).
49. T. Hanna, Y. Muraba, S. Matsuishi, N. Igawa, K. Kodama, S. Shamoto, and H. Hosono, *Phys. Rev. B*, 84, 024521 (2011).
50. Y. Kamihara, T. Nomura, M. Hirano, J. E. Kim, K. Kato, M. Takata, Y. Kobayashi, S. Kitao, S. Higashitaniguchi, Y. Yoda, M. Seto, and Hideo Hosono, *New J. Phys.*, 12, 033005 (2010).
51. S. Margadonna, Y. Takabayashi, M. T. McDonald, M. Brunelli, G. Wu, R. H. Liu, X. H. Chen, and K. Prassides, *Phys. Rev. B*, 79, 014503 (2009).
52. A. J. Drew, Ch. Niedermayer, P. J. Baker, F. L. Pratt, S. J. Blundell, T. Lancaster, R. H. Liu, G. Wu, X. H. Chen, I. Watanabe, V. K. Malik, A.

Dubroka, M. Rössle, K. W. Kim, C. Baines, and C. Bernhard, *Nat. Mater.*, 8, 310 (2009).

53. A. Martinelli, A. Palenzona, M. Tropeano, M. Putti, C. Ferdeghini, G. Profeta, and E. Emerich, *Phys. Rev. Lett.*, 106, 227001 (2011).
54. G. A. Artioli, L. Malavasi, M. C. Mozzati, and Y. D. Fernandez, *J. Am. Chem. Soc.*, 131, 12044 (2009).
55. L. Wang, Y. Qi, D. Wang, Z. Gao, X. Zhang, Z. Zhang, C. Wang, and Y. Ma, *Supercond. Sci. Technol.*, 23, 075005 (2010).
56. A. Kohler and G. Behr, *J. Supercond. Nov. Magn.*, 22, 565 (2009).
57. J. Karpinski, N. D. Zhigadlo, S. Katrych, Z. Bukowski, P. Moll, S. Weyeneth, H. Keller, R. Puzniak, M. Tortello, D. Daghero, R. Gonnelli, I. Maggio-Aprile, Y. Fasano, O. Fischer, K. Rogacki, B. Batlogg, *Phisica C*, 469, 370 (2009).
58. T. Tohei, T. Mizoguchi, H. Hiramatsu, Y. Kamihara, H. Hosono, and Y. Ikuhara, *Appl. Phys. Lett.*, 95, 193107 (2009).
59. H. Kito, H. Eisaki, and A. Iyo, *J. Phys. Soc. Jpn.*, 2008, 77, 063707 (2008).
60. Z-A. Ren, G-C. Che, X-L. Dong, J. Yang, W. Lu, W. Yi, X-L. Shen, Z-C. Li, L-L. Sun, F. Zhou, and Z-X. Zhao, *Europhys. Lett.*, 83, 17002 (2008).
61. C. Wang, L. Li, S. Chi, Z. Zhu, Z. Ren, Y. Li, Y. Wang, X. Lin, Y. Luo, S. Jiang, X. Xu, G. Cao, and Z. Xu, *Europhys. Lett.*, 83, 67006 (2008).
62. N. D. Zhigadlo, S. Katrych, S. Weyeneth, R. Puzniak, P. J. W. Moll, Z. Bukowski, J. Karpinski, H. Keller, and B. Batlogg, *Phys. Rev. B*, 82, 064517 (2010).
63. X. L. Dong, W. Lu, J. Yang, W. Yi, Z. C. Li, C. Zhang, Z. A. Ren, G. C. Che, L. L. Sun, F. Zhou, X. J. Zhou, and Z. X. Zhao, *Phys. Rev. B*, 82, 212506 (2010).
64. G. Cao, S. Jiang, X. Lin, C. Wang, Y. Li, Z. Ren, Q. Tao, C. Feng, J. Dai, Z. Xu, and F. Zhang, *Phys. Rev. B*, 79, 174505 (2009).
65. B. Maroni, L. Malavasi, M. Cristina Mozzati, M. S. Grandi, A. H. Hill, D. Chermisi, P. Dore, and P. Postorino, *Phys. Rev. B*, 82, 1045003 (2010).
66. Z-A. Ren and Z-X. Zhao, *Adv. Mater.*, 21, 4584 (2009).
67. H-H. Wen, G. Mu, L. Fang, H. Yang, and X. Zhu, *Europhys. Lett.*, 82, 17009 (2008).
68. K. Kasperkiewicz, J. G. Bos, A. N. Fitch, K. Prassides, and S. Margadonna, *Chem. Comm.*, 2009, 707.
69. D. Kasinatyhan, A. Ormeci, K. Koch, U. Burickhardt, W. Schnelle, A. L.-Jaspa, and H. Rosner, *New J. Phys.*, 11, 025023 (2008).
70. Y. Muraba, S. Matsuishi, S-W. Kim, T. Atou, O. Fukunaga, and H. Hosono, *Phys. Rev. B*, 82, 180512(R) (2010).

71. T. Katase, H. Hiramatsu, T. Kamiya, and H. Hosono, ArXiv:1110.0045.
72. A. S. Sefat, R. Jin, M. A. McGuire, B. C. Sales, D. J. Singh, and D. Mandrus, *Phys. Rev. Lett.*, 101, 117004 (2008).
73. Y. Qi, L. Wang, Z. Gao, D. Wang, X. Zhang, C. Wang, C. Yao, and Y. Ma, *New J. Phys.*, 13, 033020 (2011).
74. S. Sharma, A. Bharathi, S. Chandra, V. R. Reddy, S. Paulraj, A. T. Satya, V. S. Sastry, A. Gupta, and C. S. Sundar, *Phys. Rev. B*, 81, 174512 (2010).
75. S. Jiang, H. Xing, G. Xuan, C. Wang, Z. Ren, C. Feng, J. Dai, Z. Xu, and G. Cao, *J. Phys.: Condens. Matter*, 21, 382203 (2009).
76. M. A. McGuire, D. J. Singh, A. S. Sefat, B. C. Sales, and D. Mandrus, *J. Solid. State. Chem.*, 182, 2326 (2009).
77. S. C. Lee, E. Satomi, Y. Kobayashi, and M. Sato, *J. Phys. Soc. Jpn.*, 79, 023702 (2010).
78. M. Tropeano, M. R. Cimberle, C. Ferdeghini, G. Lamura, A. Martinelli, A. Palenzona, I. Pallecchi, A. Sala, I. Sheikin, F. Bernardini, M. Monni, S. Massidda, and M. Putti, *Phys. Rev. B*, 81, 184504 (2010).
79. V. Brouet, F. Rullier-Albenque, M. Marsi, B. Mansart, M. Aichhorn, S. Biermann, J. Faure, L. Perfetti, A. Taleb-Ibrahimi, P. Le Fe'vre, F. Bertran, A. Forget, and D. Colson, *Phys. Rev. Lett.*, 105, 087001 (2010).
80. H. Shishido, A. F. Bangura, A. I. Coldea, S. Tonegawa, K. Hashimoto, S. Kasahara, P. M. C. Rourke, H. Ikeda, T. Terashima, R. Settai, Y. O nuki, D. Vignolles, C. Proust, B. Vignolle, A. McCollam, Y. Matsuda, T. Shibauchi, and A. Carrington, *Phys. Rev. Lett.*, 2010, 104, 057008 (1–4).
81. Y. Nakai, T. Iye, S. Kitagawa, K. Ishida, H. Ikeda, S. Kasahara, H. Shishido, T. Shibauchi, Y. Matsuda, and T. Terashima, *Phys. Rev. Lett.*, 105, 107003 (2010).
82. C-H Lee, A. Iyo, H. Eisaki, H. Kito, M. T. Fernandez-Diazi, T. Ito, K. Kohou, H. Matsuhata, M. Brenden, and K. Yamada, *J. Phys. Soc. Jpn.*, 77, 083704 (2008).
83. S. A. J. Kimber, A. Kreyssig, Y-Z. Zhang, H. O. Jeschke, R. Valentí, F. Yokaichiya, E. Colombier, J. Yan, T. C. Hansen, and T. Chatterji, R. J. McQueeney, P. C. Canfield, A. I. Goldman, and D. N. Argyriou, *Nat. Mater.*, 8, 471–475 (2009).
84. K. Kuroki, H. Usui, S. Onari, R. Arita, and H. Aoki, *Phys. Rev. B*, 79, 224511 (2009).
85. K. Kihou, T. Saito, S. Ishida, M. Nakajima, Y. Tomioka, H. Fukazawa, Y. Kohori, T. Ito, S. Uchida, A. Iyo, C-H. Lee, and H. Eisaki, *J. Phys. Soc. Jpn.*, 79, 124713 (2010).

86. K. Horigane, H. Hiraka, and K. Ohoyama, *J. Phys. Soc. Jpn.*, 78, 074718 (2009).
87. K. Kodama, M. Ishikado, F. Esaka, A. Iyo, H. Eisaki, and S. Shamoto, *J. Phys. Soc. Jpn.*, 80, 034601 (2011).
88. M. Rotter, M. Pangerl, M. Tegel, and D. Johrendt, *Angew. Chem. Int. Ed.*, 47, 7947 (2008).
89. W. Malaeb, T. Yoshida, T. Kataoka, A. Fujimori, M. Kubota, K. Ono, H. Usui, K. Kuroki, R. Arita, H. Aoki, Y. Kamihara, M. Hirano, and Hideo Hosono, *J. Phys. Soc. Jpn.*, 77, 093714 (2008).
90. I. I. Mazin, D. J. Singh, M. D. Johannes, and M. H. Du, *Phys. Rev. Lett.*, 101, 057003 (2008).
91. H. Ikeda, R. Arita, and J. Kuneš, *Phys. Rev. B*, 81, 054502 (2010).
92. Y. Yanagi, Y. Yamakawa, N. Adachi, and Y. Ono, *J. Phys. Soc. Jpn.*, 79, 1237007 (2010).
93. H. Kontani T. Saito, and S. Onari, *Phys. Rev. B*, 84, 024528 (2011).
94. M. Yoshizawa, S. Simayi, K. Sakano, Y. Nakanishi, K. Kihou, C.-H. Lee, A. Iyo, H. Eisaki, M. Nakajima, and S. Uchida, *Solid State Commun.*, 152, 680 (2012).
95. F. Ye *et al. Phys. Rev. Lett.* 107, 137003 (2011); Z. Shermadini *et al. Phys. Rev. Lett.*, 106, 117602 (2011).
96. Y. Zhang *et al. Nat. Mater.*, 10, 273 (2011); L. Zhao *et al. Phys. Rev. B*, 83, 140508R (2011).
97. N. Ni *et al.*, *Phys. Rev. B*, 82, 024519 (2010); N. Ni, and S. L. Bud'ko, *MRS Bull.*, 36, 620 (2011).
98. T. Katase, Y. Ishimaru, A. Tsukamoto, H. Hiramatsu, T. Kamiya, K. Tanabe, and H. Hosono, *Nat. Commun.*, 2, 409 (2011).
99. Y. Jia, P. Cheng, L. Fang, H. Luo, H. Yang, C. Ren, L. Shan, C. Gu, and H-H. Wen, *Appl. Phys. Lett.*, 93, 032503 (2008).
100. N. D. Zhigadlo, S. Katrych, Z. Bukowski, S. Weyeneth, R. Puzniak, and J. Karpinski, *J. Phys.: Condens Matter*, 20, 342202 (2008).
101. C. Martin, M. E. Tillman, H. Kim, M. A. Tanatar, S. K. Kim, A. Kreyssig, R. T. Gordon, M. D. Vannette, S. Nandi, V. G. Kogan, S. L. Bud'ko, P. C. Canfield, A. I. Goldman, and R. Prozorov, *Phys. Rev. Lett.*, 102, 247002 (2009).
102. H-S. Lee, J-H. Park, J-Y. Lee, J-Y. Kim, N-H. Sung, T-Y. Koo, B. K. Cho, C-U. Jung, S. Saini, S-J. Kim, and H-J. Lee, *Supercond. Sci. Technol.*, 22, 075023 (2009).
103. L. Fang, P. Cheng, Y. Jia, X. Zhu, H. Luo, G. Mu, C. Gu, and H.-H. Wen, *J. Cryst. Growth*, 311, 358 (2009): H. Q. Luo, P. Cheng, Z. S. Wang, H. Yang,

Y. Jia, L. Fang, C. Ren, L. Shan, and H. H. Wen, *Physica C*, 469, 477 (2009); F. Nitsche, A. Jesche, E. Hieckmann, Th. Doert, and M. Ruck, *Phys. Rev. B*, 82, 134514 (2010); S. K. Kim, M. E. Tillman, H. Kim, A. Kracher, S. L. Bud'ko, R. Prozorov, and P. C. Canfield, *Supercond. Sci. Technol.*, 23, 054008 (2010).

104. J-Q. Yan, S. Nandi, J. L. Zarestky, W. Tian, A. Kreyssig, B. Jensen, A. Kracher, K. W. Dennis, R. J. McQueeney, A. I. Goldman, R. W. McCallum, and T. A. Lograsso, *Appl. Phys. Lett.*, 95, 222504 (2009); J-Q. Yan, B. Jensen, K. W. Dennis, R. W. McCallum, and T. A. Lograsso, *Appl. Phys. Lett.*, 98, 072504 (2011).
105. N. Ni, S. L. Bud'ko, A. Kreyssig, S. Nandi, G. E. Rustan, A. I. Goldman, S. Gupta, J. D. Corbett, A. Kracher, and P. C. Canfield, *Phys. Rev. B*, 78, 014507 (2008).
106. H. Luo, Z. Wang, H. Yang, P. Cheng, X. Zhu, and H-H. Wen, *Supercond. Sci. Technol.*, 21, 125014 (2008).
107. A. S. Sefat, R. Jin, M. A. McGuire, B. C. Sales, D. J. Singh, and D. Mandrus, *Phys. Rev. Lett.*, 101, 117004 (2008).
108. N. Ni, M. E. Tillman, J.-Q. Yan, A. Kracher, S. T. Hannahs, S. L. Bud'ko, and P. C. Canfield, *Phys. Rev. B*, 78, 214515 (2008).
109. C. Wang, Z. Gao, C. Yao, L. Wang, Y. Qi, D. Wang, X. Zhang, and Y. Ma, *Supercond. Sci. Technol.*, 24, 065002 (2011).
110. S. W. Tozer, A. W. Kleinsasser, T. Penney, D. Kaiser, and F. Holtzberg, *Phys. Rev. Lett.*, 59, 1768 (1987); Y. Iye, T. Tamegai, T. Sakakibara, T. Goto, N. Miura, H. Takeya, and H. Takei, *Physica C*, 153–155, 26 (1988); S. J. Hagen, T. W. Jing, Z. Z. Wang, J. Horvath, and N. P. Ong, *Phys. Rev. B*, 37, 7928 (1988).
111. D. E. Farrell, S. Bonham, J. Foster, Y. C. Chang, P. Z. Jiang, K. G. Vandervoort, D. J. Lam, and V. G. Kogan, *Phys. Rev. Lett.*, 63, 782 (1989); B. D. Biggs, M. N. Kunchur, J. J. Lin, S. J. Poon, T. R. Askew, R. B. Flippen, M. A. Subramanian, J. Gopalakrishnan, and A. W. Sleight, *Phys. Rev. B*, 39, 7309 (1989).
112. Y. Hidaka, Y. Enomoto, M. Suzuki, M. Oda, A. Katsui, and T. Murakama, *Jpn. J. Appl. Phys.*, 26, L377 (1987).
113. T. R. Dinger, T. K. Worthington, W. J. Gallagher, and R. L. Sandstrom, *Phys. Rev. Lett.*, 58, 2687 (1987).
114. F. Ma and Z-Y. Lu, *Phys. Rev. B*, 78, 033111 (2008).
115. X. F. Wang, T. Wu, G. Wu, H. Chen, Y. L. Xie, J. J. Ying, Y. J. Yan, R. H. Liu, and X. H. Chen, arXiv:0806.2452.

116. G. F. Chen, Z. Li, J. Dong, G. Li, W. Z. Hu, X. D. Zhang, X. H. Song, P. Zheng, N. L. Wang, and J. L. Luo, *Phys. Rev. B*, 78, 224512 (2008).
117. X. F. Wang, T. Wu, G. Wu, R. H. Liu, H. Chen, Y. L. Xie, and X. H. Chen, arXiv:0811.2920.
118. M. A. Tanatar, N. Ni, C. Martin, R. T. Gordon, H. Kim, V. G. Kogan, G. D. Samolyuk, S. L. Bud'ko, P. C. Canfield, and R. Prozorov, *Phys. Rev. B*, 79, 094507 (2009).
119. P. J. W. Moll, R. Puzniak, F. Balakirev, K. Rogacki, J. Karpinski, N. D. Zhigadlo, and B. Batlogg, *Nat. Mater.*, 9, 628 (2010).
120. C. P. Bean, *Phys. Rev. Lett.*, 8, 250–253 (1962).
121. K. Togano, A. Matsumoto, and H. Kumakura, *Appl. Phys. Express*, 4, 043101 (2011).
122. T. Katase, H. Hiramatsu, T. Kamiya, and H. Hosono, *Appl. Phys. Express*, 3, 063101 (2010).
123. M. M. Altarawneh, K. Collar, and C. H. Mielke, N. Ni, S. L. Bud'ko, and P. C. Canfield, *Phys. Rev. B*, 78, 220505 (2008).
124. H. Hiramatsu, T. Katase, T. Kamiya, M. Hirano, and H. Hosono, *Appl. Phys. Lett.*, 93, 162504 (2008).
125. E. Backen, S. Haindl, T. Niemeier, R. Hühne, T. Freudenberg, J. Werner, G. Behr, L. Schultz, and B. Holzapfel, *Supercond. Sci. Technol.*, 21, 122001 (2008).
126. S. Haindl, M. Kidszun, A. Kauffmann, K. Nenkov, N. Kozlova, J. Freudenberger, T. Thersleff, J. Hänisch, J. Werner, E. Reich, L. Schultz, and B. Holzapfel, *Phys. Rev. Lett.*, 104, 077001 (2010).
127. T. Kawaguchi, H. Uemura, T. Ohno, R. Watanabe, M. Tabuchi, T. Ujihara, K. Takenaka, Y. Takeda, and H. Ikuta, *Appl. Phys. Express*, 2, 093002 (2009).
128. T. Kawaguchi, H. Uemura, T. Ohno, M. Tabuchi, T. Ujihara, Y. Takeda, and H. Ikuta, *Appl. Phys. Express*, 4, 063102 (2011).
129. H. Ikuta, T. Kawaguchi, H. Uemura, T. Ohno, M. Tabuchi, T. Ujihara, and Y. Takeda: presented at International Workshop on Novel Superconductors and Super Materials, 2011 (N2S2011).
130. H. Hiramatsu, T. Katase, T. Kamiya, M. Hirano, and H. Hosono, *Appl. Phys. Express*, 1, 101702 (2008).
131. S. A. Baily, Y. Kohama, H. Hiramatsu, B. Maiorov, F. F. Balakirev, M. Hirano, and H. Hosono, *Phys. Rev. Lett.*, 102, 117004 (2009).
132. H. Hiramatsu, T. Katase, T. Kamiya, M. Hirano, and H. Hosono, *Phys. Rev. B*, 80, 052501 (2009); T. Kamiyaa, H. Hiramatsub, T. Katasea, M. Hirano, and H. Hosono, *Mat. Sci. Eng. B*, 173, 244 (2010).

133. T. Katase, H. Hiramatsu, H. Yanagi, T. Kamiya, M. Hirano, and H. Hosono, *Solid State Commun.*, 149, 2121 (2009).

134. T. Katase, H. Hiramatsu, T. Kamiya, and H. Hosono, *Appl. Phys. Express*, 3, 063101 (2010).

135. B. Maiorov, T. Katase, S. A. Baily, H. Hiramatsu, T. G. Holesinger, H. Hosono, and L. Civale, *Supercond. Sci. Technol.*, 24, 055007 (2011).

136. K. Iida, J. Hänisch, R. Hühne, F. Kurth, M. Kidszun, S. Haindl, J. Werner, L. Schultz, and B. Holzapfel, *Appl. Phys. Lett.*, 95, 192501 (2009).

137. S. Lee, J. Jiang, C. T. Nelson, C. W. Bark, J. D. Weiss, C. Tarantini, H. W. Jang, C. M. Folkman, S. H. Baek, A. Polyanskii, D. Abraimov, A. Yamamoto, Y. Zhang, X. Q. Pan, E. E. Hellstrom, D. C. Larbalestier, and C. B. Eom, *Nat. Mater.*, 9, 397 (2010).

138. T. Katase, H. Hiramatsu, T. Kamiya, and H. Hosono, *Appl. Phys. Express*, 3, 063101 (2010).

139. N. H. Lee, S. G. Jung, D. H. Kim, and W. N. Kang, *Appl. Phys. Lett.*, 96, 202505 (2010).

140. S. Takeda, S. Ueda, T. Yamagishi, S. Agatsuma, S. Takano, A. Mitsuda, and M. Naito, *Appl. Phys. Express*, 3, 093101 (2010).

141. M. K. Wu, F. C. Hsu, K. W. Yeh, T. W. Huang, J. Y. Luo, M. J. Wang, H. H. Chang, T. K. Chen, S. M. Rao, B. H. Mok, C. L. Chen, Y. L. Huang, C. T. Ke, P. M. Wu, A. M. Chang, C. T. Wu, and T. P. Perng, *Physica C*, 469, 340 (2009).

142. M. J. Wang, J. Y. Luo, T. W. Huang, H. H. Chang, T. K. Chen, F. C. Hsu, C. T. Wu, P. M. Wu, A. M. Chang, and M. K. Wu, *Phys. Rev. Lett.*, 103, 117002 (2009).

143. S. G. Jung, N. H. Lee, E. M. Choi, W. N. Kang, S. I. Lee, T. J. Hwang, and D. H. Kim, *Physica C*, 470, 1977 (2010).

144. Y. Imai, T. Akiike, M. Hanawa, I. Tsukada, A. Ichinose, A. Maeda, T. Hikage, T. Kawaguchi, and H. Ikuta, *Appl. Phys. Express*, 3, 043102 (2010).

145. W. Si, Z. W. Lin, Q. Jie, W. G. Yin, J. Zhou, G. Gu, P. D. Johnson, and Q. Li, *Appl. Phys. Lett.*, 95, 052504 (2009).

146. E. Bellingeri, I. Pallecchi, R. Buzio, A. Gerbi, D. Marrè, M. R. Cimberle, M. Tropeano, M. Putti, A. Palenzona, and C. Ferdeghini, *Appl. Phys. Lett.*, 96, 102512 (2010).

147. S. B. Lv, L. Z. Deng, M. Gooch, F. Y. Wei, Y. Y. Sun, J. Meen, Y. Y. Xue, B. Lorenz, and C. W. Chu, *Proc. Nat. Acad. Sci.*, 108, 15705 (2011).

148. S. R. Saha, N. P. Butch, T. Drye, J. Magill, S. Ziemak, K. Kirshenbaum, P. Y. Zavalij, J. W. Lynn, and J. Paglione, *Phys. Rev. B*, 85, 024525 (2012).

149. Z. Gao, Y. Qi, L. Wang, D. Wang, X. Zhang, C. Yao, C. Wang, and Y. Ma, *Europhys. Lett.*, 95, 67002 (2011).

150. T. Katase, Y. Ishimaru, A. Tsukamoto, H. Hiramatsu, T. Kamiya, K. Tanabe, and H. Hosono, *Appl. Phys. Lett.*, 96, 142507 (2010).

151. T. Katase, Y. Ishimaru, A. Tsukamoto, H. Hiramatsu, T. Kamiya, K. Tanabe, and H. Hosono, *Supercond. Sci. Technol.*, 23, 082001 (2010).

152. W. E. Picket, *Physica B*, 296, 112 (2001).

153. I. I. Mazin, *Nature*, 464, 183 (2010).

154. I. I. Mazin, D. J. Singh, M. D. Johannes, and M. H. Du, *Phys. Rev. Lett.*, 101, 057003 (2008).

155. M. K. Forthaus, K. Sengupta, O. Heyer, N. E. Christensen, A. Svane, K. Syassen, D. I. Khomskii, T. Lorenz, and M. M. Abd-Elmeguid, *Phys. Rev. Lett.*, 105, 157001 (2010).

156. H. Mizoguchi, S. Matsuishi, M. Hirano, M. Tachibana, E. Takayama-Muromachi, H. Kawaji, and H. Hosono, *Phys. Rev. Lett.*, 106, 057002 (2011).

157. H. Mizoguchi, T. Kuroda, T. Kamiya, and H. Hosono, *Phys. Rev. Lett.*, 106, 237001 (2011).

Chapter 2

Synthesis and Physical Properties of the New Potassium Iron Selenide Superconductor $K_{0.80}Fe_{1.76}Se_2$

R. Hu,[a,b] E. D. Mun,[c] D. H. Ryan,[d] K. Cho,[b] H. Kim,[a,b] H. Hodovanets,[a,b] W. E. Straszheim,[b] M. A. Tanatar,[b] R. Prozorov,[a,b] W. N. Rowan-Weetaluktuk,[d] J. M. Cadogan,[e] M. M. Altarawneh,[c] C. H. Mielke,[c] V. S. Zapf,[c] S. L. Bud'ko,[a,b] and P. C. Canfield[a,b,*]

[a]*Department of Physics and Astronomy, Iowa State University, Ames, IA 50011, USA*
[b]*Ames Laboratory, U.S. DOE, Ames, IA 50011, USA*
[c]*National High Magnetic Field Laboratory, Los Alamos National Laboratory, Los Alamos, New Mexico 87544, USA*
[d]*Physics Department and Centre for the Physics of Materials, McGill University, Montreal, H3A 2T8, Canada*
[e]*Department of Physics and Astronomy, University of Manitoba, Winnipeg, Manitoba, R3T 2N2, Canada*
*canfield@ameslab.gov

In this chapter, we review our studies of the $K_{0.80}Fe_{1.76}Se_2$ superconductor, with an attempt to elucidate the crystal growth details and basic physical properties over a wide range of temperatures and applied magnetic field, including anisotropic magnetic and electrical transport properties, thermodynamic, London penetration depth, magneto-optical imaging and Mössbauer measurements. We find that: (i) Single crystals of similar stoichiometry can be grown both

Iron-Based Superconductors: Materials, Properties, and Mechanisms
Edited by Nan-Lin Wang, Hideo Hosono, and Pengcheng Dai

ISBN 978-981-4303-22-4 (Hardcover), 978-981-4303-23-1 (eBook)
www.panstanford.com

by furnace-cooled and decanted methods; (ii) Single crystalline $K_{0.80}Fe_{1.76}Se_2$ shows moderate anisotropy in both magnetic susceptibility and electrical resistivity and a small modulation of stoichiometry of the crystal, which gives rise to broadened transitions; (iii) The upper critical field, H_{c2}(T) is ~55 T at 2 K for **H** ∥ **c**, manifesting a temperature dependent anisotropy that peaks near 3.6 at 27 K and drops to 2.5 by 18 K; (iv) Mössbauer measurements reveal that the iron sublattice in $K_{0.80}Fe_{1.76}Se_2$ clearly exhibits magnetic order, probably of the first order, from well below T_c to its Néel temperature of $T_N = 532 \pm 2$ K. It is very important to note that, although, at first glance there is an apparent dilemma posed by these data: high T_c superconductivity in a near insulating, large ordered moment material, analysis indicates that the sample may well consist of two phases with the minority superconducting phase (that does not exhibit magnetic order) being finely distributed, but connected with in an antiferromagnetic, poorly conducting, matrix, essentially making a superconducting aerogel.

2.1 Introduction

The iron-based superconductors have attracted intense research attention because of their high transition temperature and their possibly unconventional pairing mechanism, correlated to magnetism [1–4]. Similar to cuprate superconductors, iron-based superconductors have layered structures; the planar Fe layers tetrahedrally coordinbated by As or chalcogen anions (Se or Te) are believed to be responsible for superconductivity. Stacking of the FeAs building blocks with alkali, alkaline earth or rare earth oxygen spacer layers forms the basic classes of iron arsenic superconductors in these compounds: 111-type AFeAs [5], 122-type AFe_2As_2 [6–9], 1111-type ROFeAs [10, 11] and more complex block containing phases, e.g. Sr_2VO_3FeAs [12], $Sr_3Sc_2Fe_2As_2O_5$ [13], $Sr_4Sc_2Fe_2As_2O_6$ [14]. The simple binary 11-type iron chalcogenide has no spacer layers and superconductivity can be induced by doping FeTe with S[15] or Se [16]. Different from the other iron-based superconductors, FeSe is a superconductor [17], $T_c \sim 8$ K, with no static magnetic order and its transition temperature can be increased up to 37 K by applying pressure [18] or 15 K in $FeSe_{0.5}Te_{0.5}$ [16]. More recently, superconductivity above 30 K has been reported in $A_xFe_{2-y}Se_2$ (A = K, Cs, Rb or Tl) [19–23], by adding A between the Fe_2Se_2 layers, a compound with the same unit cell structure as the AFe_2As_2 compounds.

μSR measurements showed that magnetic order co-exists with bulk superconductivity in $Cs_{0.8}Fe_{1.6}Se_2$ [24], and neutron diffraction measurements on $K_{0.8}Fe_{1.76}Se_2$ [25] have suggested that not only do magnetic order and superconductivity co-exist, but that the iron moments are remarkably large (3.3 μ_B/Fe) and are ordered in a relatively complex antiferromagnetic structure that places all of the iron moments parallel to the c-axis. The magnetic ordering temperatures are quite high in both compounds: T_N(Cs) = 480 K [24], T_N(K) = 560 K [25]. The development of a paramagnetic component near T_N [24] and the unusual temperature dependence of the magnetic intensity [25] suggest that the magnetic transition may be first order in nature rather than being a more conventional second order transition. First order magnetic transitions are commonly associated with changes in crystal structure, and both synchrotron x-ray diffraction [26] and neutron diffraction [25, 27] have now shown evidence for a structural change from I4/m to I4/mmm associated with a disordering of iron vacancies that occurs in the vicinity of the magnetic transition. In this review we summarize our basic understanding of this material [28–30]. First we clarify the growth details and present elemental analysis and physical properties of $K_{0.80}Fe_{1.76}Se_2$ single crystals [28]. Then the H_{c2}-T phase diagram for the $K_{0.8}Fe_{1.76}Se_2$ is constructed and discussed [29]. At the end we present the study of the magnetic ordering of $K_{0.80}Fe_{1.76}Se_2$ using ^{57}Fe Mössbauer spectroscopy [30].

2.2 Experimental Methods

Crystals were characterized by powder x-ray diffraction using a Rigaku Miniflex x-ray diffractometer. The actual chemical composition was determined by wavelength dispersive x-ray spectroscopy (WDS) in a JEOL JXA-8200 electron microscope. Magnetic susceptibility was measured in a Quantum Design MPMS, SQUID magnetometer. In plane AC resistivity ρ_{ab} was measured by a standard four-probe configuration. Measurement of ρ_c was made in the two-probe configuration. Contacts were made by using a silver alloy. For ρ_c, contacts were covering the whole *ab*-plane area [31]. Thermoelectric power measurements were carried out by a dc, alternating temperature gradient (two heaters and two thermometers) technique [32]. Specific heat data were collected using a Quantum Design PPMS. The in-plane London penetration depth was measured by using a tunnel-diode resonator (TDR) oscillating at 14 MHz and at temperature down to 0.5 K [33]. Magneto-optical imaging was conducted by utilizing the Faraday effect in bismuth-doped iron garnet indicators with in-plane magnetization [34]. A flow-type liquid ^{4}He cryostat with sample in vacuum was used. The sample was positioned on top of a copper cold finger and an indicator was placed on top of the sample. The cryostat was positioned under polarized-light reflection microscope and the color images could be recorded on video and high-resolution CCD cameras. When linearly polarized light passes through the indicator and reflects off the mirror sputtered on its bottom, it picks up a double Faraday rotation proportional to the magnetic field intensity at a given location on the sample surface. Observed through the (almost) crossed analyzer, we recover a 2D image [35].

To investigate the upper critical field anisotropy to higher fields ($H \leq 60$ T), the magnetic field dependence of radio frequency (rf) contactless penetration depth was measured for applied field both parallel ($\mathbf{H} \parallel \mathbf{c}$) and perpendicular ($\mathbf{H} \parallel \mathbf{ab}$) to the tetragonal *c*-axis. The rf contactless penetration depth measurements were performed in a 60 T short pulse magnet with a 10 ms rise and 40 ms decay time. The rf technique has proven to be a sensitive and accurate method for determining the H_{c2} of superconductors [36]. This technique is highly sensitive to small changes in the rf penetration depth ($\sim$ 1–5 nm) in the mixed state. As the magnetic field is applied, the probe detects the transition to the normal state by tracking the shift in resonant frequency, which is proportional to the change in penetration depth as $\Delta\lambda \propto \Delta F/F_0$, where F_0 is 25 MHz in the current setup. Because of the eddy current heating caused by the pulsed field, small single crystals were chosen, where the sample was placed in a circular detection coil for $\mathbf{H} \parallel \mathbf{ab}$ and was located on the top surface of one side of the counterwound coil pair for **H** $\parallel$ **c** [37, 38]. For the **H** $\parallel$ **c** configuration, the coupling between sample and coil is weaker than that for **H** $\parallel$ **ab**, resulting in a smaller frequency shift that is still sufficient to resolve $H_{c2}(T)$. Details about this technique can be found in Refs. [37–39].

For Mössbauer measurements, two cleaved single crystal mosaic samples were prepared from the same batch of crystals. The first, for low-temperature work, was prepared by attaching several single crystal plates to a 12 mm diameter disc of 100 μm thick Kapton foil using Apiezon N grease. Care was taken to ensure that there were no gaps, but rather minimal overlap between the crystals. This sample was transferred promptly to a vibration-isolated closed-cycle refrigerator with the sample held in vacuum. The second sample, for the high-temperature work, was attached to a $\frac{1}{2}$-inch diameter 10-mil beryllium disc using diluted GE-7031 varnish before being mounted in a resistively heated oven, again with the sample in vacuum. While we operated somewhat above the maximum service temperature of the varnish, the sample was cycled above 250°C three times without any evidence of degradation.

The Mössbauer spectra were collected on conventional spectrometers using 50 mCi ^{57}Co**Rh** sources mounted on electro-mechanical drives operated in constant acceleration mode (on the high-temperature system) and sine-mode (on the low-temperature system). The spectrometers were calibrated against $\alpha-$Fe metal at room temperature. The closed-cycle refrigerator cools to 10 K, with temperature sensing and control using a calibrated silicon diode mounted on the copper sample stage. Measured gradients (centre to edge of sample) in the oven are less than 1 K up to 750 K. Control and sensing rely on four, type-K, thermocouples. Temperature stability in both cases is better than 0.2 K. Spectra were fitted using a conventional non-linear least-squares minimisation routine to a sum of equal-width Lorentzian lines. Magnetic patterns were fitted assuming first-order perturbation in order to combine the effects of the magnetic hyperfine field (B_{hf}) and the electric field gradient.

2.3 Crystal Growth and Stoichiometry

Although single crystals of $K_xFe_{2-y}Se_2$ could be grown readily from a melt, various stoichiometries of the single crystals were reported in literature, with wide ranges of the values of x, y ($0.6 \leq x < 1$ and $0 \leq y \leq 0.59$) [20–22, 40–46]. There is concensus that $K_xFe_{2-y}Se_2$ is of off-stoichiometric nature and the deficiency of K and Fe strongly influences their electrical transport properties, tuning the material from insulating to superconducting state [22, 40]. Different techniques were claimed to be suscessful in growing single crystals: self-flux growth [20], Bridgeman method [40]. In order to understand the crystal growth and obtain well controlled samples, two different ways were tried for growing single crystals of $K_xFe_{2-y}Se_2$. As-grown crystals were compared and checked for homogeneity.

The first batch of single crystals of $K_xFe_{2-y}Se_2$ were grown from $K_{0.8}Fe_2Se_2$ melt, as described in Ref. 20. The starting material was slowly furnace-cooled from 1050°C and dark shiny crystals could be mechanically separated from the solidified melt, which was consisted of crystals and fine polycrystalline material. The different stoichiometry between the starting material and resultant single crystal clearly implies that this is not simply the cooling of a stoichiometric melt to form a congruently melting, line compound. A second batch was grown from a starting composition of KFe_3Se_3. The sample was decanted [47, 48] at 850°C after cooled from 1050°C. This procedure resulted in similar but smaller crystals as the furnace-cooled samples. It clearly shows that $K_xFe_{2-y}Se_2$ crystals are grown out of a ternary high temperature solution.

The lattice parameters refined from powder x-ray diffraction pattern of the crystals for both I4/mmm and I4/m space groups were $a = 3.8897(8)$Å and $c = 14.141(3)$Å. They are in good agreement with the previous reported values in Ref. 20 ($a = 3.8912$Å, $c = 14.139$Å). Wavelength dispersive x-ray spectroscopy (WDS) analysis was performed on both types of crystals to give a better determination of stoichiometry than the semi-quantitative Energy Dispersive X-ray (EDX) spectroscopy [20–22, 40–46]. The average composition are K:Fe:Se = 0.80(2):1.76(2):2.00(3) for the furnace-cooled sample and K:Fe:Se = 0.79(2):1.85(4):2.00(4) for the decanted sample, where the atomic numbers of K and Fe are normalized to two Se per formula unit and the standard deviation σ is taken as the compositional error and shown in parentheses after value. We found there is a spread of composition, the difference between the maximum and minimum values of the measurements, 0.07, 0.06 and 0.10 for K, Fe and Se respectively, for furnace-cooled crystals and 0.04, 0.12 and 0.09 for decanted crystals, roughly within 3σ of a normal distribution of random variable. It could be associated with the broadened superconducting transition, microstructure as seen in scanning electron microscope and the paramagnetic phase observed in Mössbauer spectroscopy (discussed below). The crystals grown from solution have very similar composition to the furnace cooled samples, with only a little higher concentration of Fe, reasonable for a crystal grown out of solution with a greater excess of Fe-Se.

2.4 Physical Properties of Single Crystals of $K_{0.80}Fe_{1.76}Se_2$

2.4.1 *Transport and Thermodynamic Properties*

We compare the temperature dependent electrical resistivity and magnetization measurements of crystals grown by both the furnace cooled and decanted methods in Fig. 2.1. The in-plane resistivity of the furnace cooled sample is very similar to that of earlier reports [20, 40]. Although the superconducting transition temperature inferred from resistivity are similar ($T_c^{\mathrm{offset}} = 30.9$ K for furnace-cooled sample and $T_c^{\mathrm{offset}} = 29$ K for decanted sample), the broad resistive maxima is shifted from 160 K for furnace-cooled sample to 280 K for decanted sample. Wang *et al.* showed that the position of the hump is sensitive to Fe deficiency [40]. With decreasing Fe deficiency, the hump shifts to higher temperature. This observation agrees well with the WDS result, which shows smaller Fe deficiency in the decanted samples. Given the small difference of both types of single crystals and the similarity to samples from earlier reports of the furnace-cooled samples, for the rest of this paper we will focus their fuller characterization.

Anisotropic resistivity as a function of temperature is shown in Fig. 2.2a. It is clear that there is a broad maximum peak around 160 K for ρ_{ab} and 180 K for ρ_c. The difference of maximum positions suggest that they result from a crossover rather than transition. The anisotropy is probably due to the layered structure

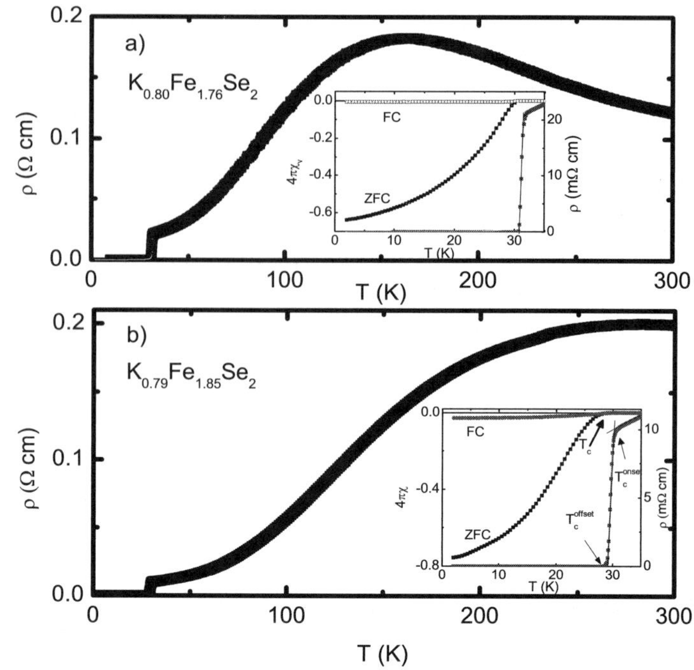

Figure 2.1 Comparison of the in-plane resistivity, and low temperature magnetic susceptibility of two types of $K_xFe_{2-y}Se_2$ single crystals, a) furnace cooled; b) decanted sample. Inset shows the low temperature region of the resistivity (to the right axis) together with zero-field-cooled and field-cooled magnetic susceptibility in a field of 50 Oe.

of $K_{0.80}Fe_{1.76}Se_2$. Figure 2.2b shows the anisotropy ρ_c/ρ_{ab}, reaches the maximum of 6 around 180 K and decreases to 4 around 300 K. It is comparable to the anisotropy of AFe_2As_2 [49]. But a much larger resistivity anisotropy of 30–45 was reported in $(Tl,K)Fe_xSe_2$ [50], this may imply that the specific composition influences carrier tunneling significantly. An expanded view around the superconducting transition is shown in the inset to Fig. 2.2a. For both of the current directions, the transition width is about 0.7 K, but the inferred T_c value from ρ_c is slightly higher than that of ρ_{ab}.

Figure 2.3a shows the magnetic susceptibility of $K_{0.80}Fe_{1.76}Se_2$ for two directions of an applied field of 50 Oe. For **H** ∥ **ab**, the zero-field-cooled (ZFC) curve decreases slowly with temperature and for

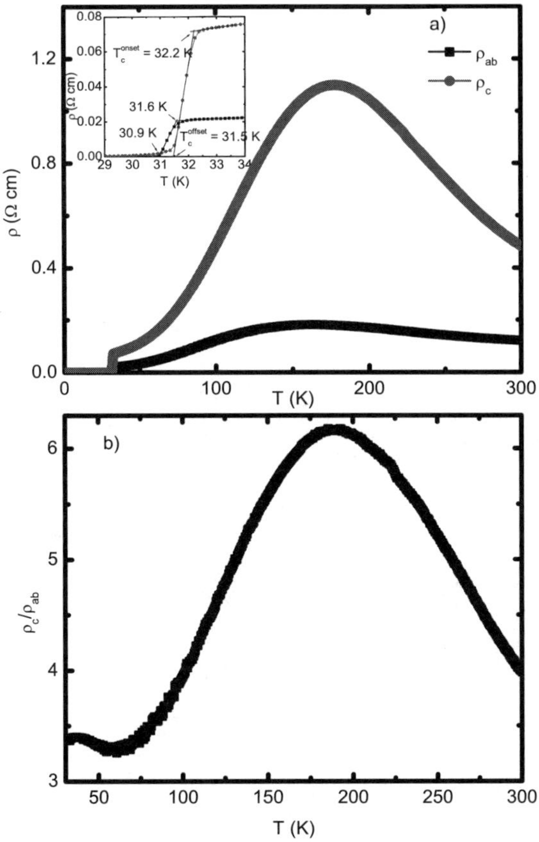

Figure 2.2 (a) Anisotropic resistivity as a function of temperature. Inset is an expanded view around the transition. (b) Anisotropy of resistivity vs temperature.

H || **c** the transition becomes sharper. Similar behavior can be seen in $Tl_{0.58}Rb_{0.42}Fe_{1.72}Se_2$ [50]. This temperature dependence of the ZFC curve is similar to an inhomogeneous superconductor with a range of transition temperatures and may be related to the small spread of stoichiometry found in WDS data. Both of the ZFC curves in Fig. 2.3a approach −0.6 consistent with substantial shielding and T_c inferred from both curves is the same, $T_c = 30.1 \pm 0.1$ K, within experimental error. The magnetic susceptibility M/H (H = 50 kOe) as a function of temperature for both field directions is shown in Fig. 2.3b. Similar

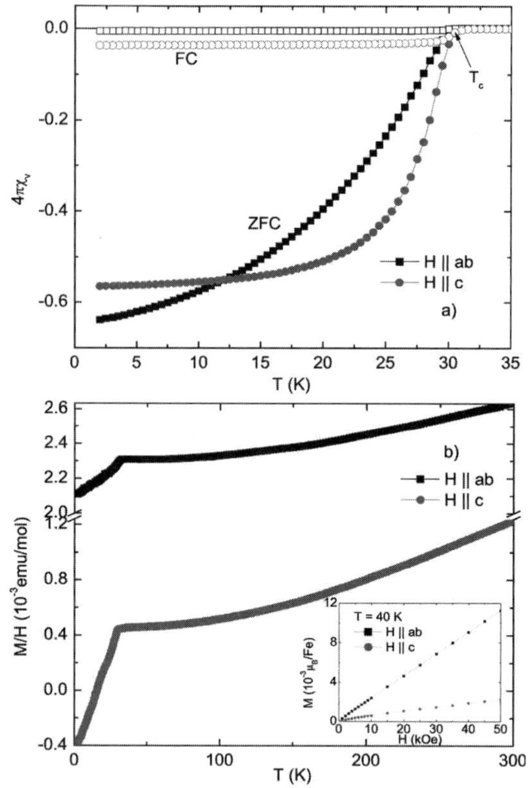

Figure 2.3 (a) Temperature dependence of low field (H = 50 Oe) magnetic susceptibility for H$\|ab$ and H$\|c$; (b) Magnetic susceptibility M/H, measured in 50 kOe for two field directions. Inset shows field dependence of magnetization at 40 K for both field directions.

temperature dependence is observed for both field directions, i.e. M/H decreases almost linearly with decreasing temperature above 150 K and shows a sudden drop below 30 K associated with superconductivity. χ_{ab} is clearly larger than χ_c over the whole temperature range. No anomalies in magnetic susceptibility can be correlated with the broad maxima in resistivity. The linear field dependence of magnetization at 40 K for both directions (Fig. 2.3b inset) indicates that there are no ferromagnetic impurities, and the non-Curie-Weiss like temperature of the susceptibility indicates that the system might

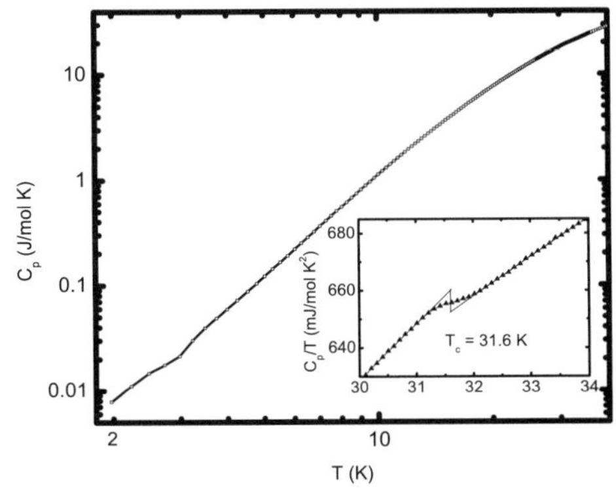

Figure 2.4 Specific heat as a function of temperature on a log-log plot. Inset shows the heat capacity jump at the superconducting transition. The solid line is an isoentropic estimate of T_c and ΔC_p.

be deep in an antiferromagnetic state, consistent with what was suggested for $Cs_{0.8}Fe_2Se_{1.96}$ [24] and $K_{0.8}Fe_{1.6}Se_2$ [25].

Specific heat data was collected to verify the bulk thermodynamic nature of the superconducting transition. C_p vs T at low temperature is shown in Fig. 2.4 on a log-log plot. In the superconducting state, below 15 K, C_p roughly follows a T^3 power law. This implies a dominant phonon contribution and a very small electronic term. C_p/T vs T is plotted in the inset for $T \sim T_c$ and a clear jump of specific heat associated with the superconducting transition at 31.6 K is seen and $\Delta C_p/T_c = 7.7$ mJ/mol K^2, can be identified. The jump is substantially less than jump seen for K-doped Ba122 samples; in comparison to the $\Delta C_p/T_c$ versus T_c presented by Bud'ko *et al.* [51], this jump is $\sim$ 15% of what would be expected from a doped 122 material [52].

The thermoelectric power (TEP) as a function of temperature is shown in Fig. 2.5. Three different samples with different eletrical and thermal contact were shown to have consistent $T_c = 31.6$ K inferred from $S(T) = 0$. The data for three samples are similar over the whole temperature range. The origins of local minimum and maximum found between $100 - 200$ K are not clear, but it is very likely that they are associated with the multiband structure of $K_{0.80}Fe_{1.76}Se_2$ and the crossover (metal-like at low temperature)

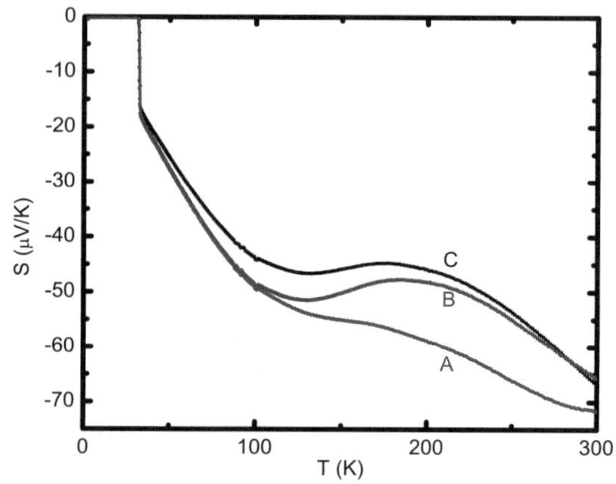

Figure 2.5 Thermoelectric power as a function of temperature. Samples A and B use silver paste as contact (contact resistance $\sim 1-3$ kΩ). Sample C uses silver wires attached by In-Sn solder as contact (contact resistance ~ 200 Ω).

observed in resistivity. The negative sign of the thermopower indicates that electron like carriers are dominant, thus in agreement with the observation of electron only pockets at the Fermi surface by Angle Resolved Photoemission Spectroscopy (ARPES) [43]. The large absolute value of S above 50 K is similar to TEP data observed for Co-doped $BaFe_2As_2$ [53] and has been reproduced by other, recent TEP measurements [54, 55].

2.4.2 *London Penetration Depth and Magneto-Optical Imaging*

London penetration depth measurements with good reproducibility were performed on several single crystal samples. Figure 2.6 shows the normalized frequency shift, proportional to differential magnetic susceptibility, $\delta f_{norm} = (f(T) - f(T_c))/(f(T_c) - f(T_{min}))$, where $f(T_{min})$ is the resonant frequency at the lowest temperature $\simeq 0.5$ K and $f(T_c)$ is the frequency in the normal state right above T_c. Consistent measurements on several samples indicate little or no variation within the batch. The transition itself is quite unusual - it shows quite a sharp onset, but then is smeared almost over the entire temperature interval. This also might be due to the small variation of the stoichiometry or impurities. It is also possible

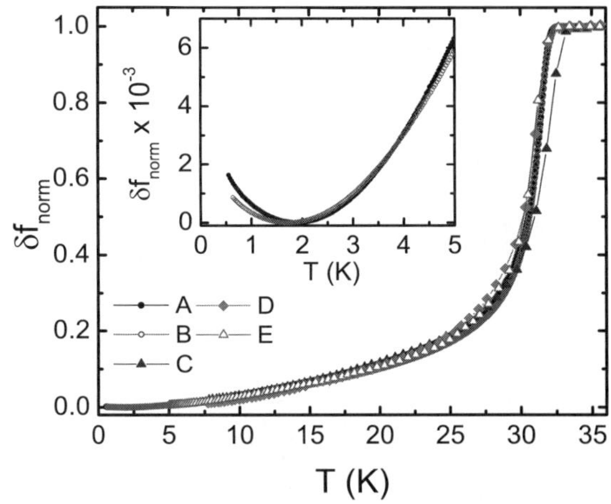

Figure 2.6 Normalized London penetration depth expressed via resonant frequency shift, $\Delta f_{norm} = (f(T) - f(T_c))/(f(T_c) - f(T_{min}))$ proportional to magnetic susceptibility. $f(T_{min})$ is the resonant frequency at the lowest temperature $\simeq 0.5$ K. $f(T_c)$ is the frequency in the normal state right above T_c. Inset shows an upturn, presumably due to paramagnetic ions and/or impurities below 2 K from two samples A and B.

that the observed behavior is indicative of strongly anisotropic gap function or even nodes. In addition, there is a clear upturn at low temperatures. It has been shown in both, high-T_c cuprates [56] and 1111 pnictides [57] that this upturn can be caused by the paramagnetic ions.

Magneto-optical imaging can shed more light on the homogeneity of the superconducting state (at least for length scales larger than the wavelength of optical light) and gives a rough estimate critical current density. A magneto-optical image of a trapped flux in a field-cooled sample is shown in Fig. 2.7. We did not observe any noticeable Meissner expulsion, similar to other 122 pnictides [58]. When magnetic field was turned off, it revealed a typical "Bean" roof, again similar to other pnictide superconductors [59, 60]. As can be seen in Fig. 2.7, the magnetic flux distribution is relatively uniform; however, some macroscopic variations (upper left corner) might indicate some smooth variation of stoichiometry across the sample and may help to explain the broadened transition curves. In order to quantify the critical state, Fig. 2.7 also shows profiles of the magnetic induction taken along two lines (shown in the figure).

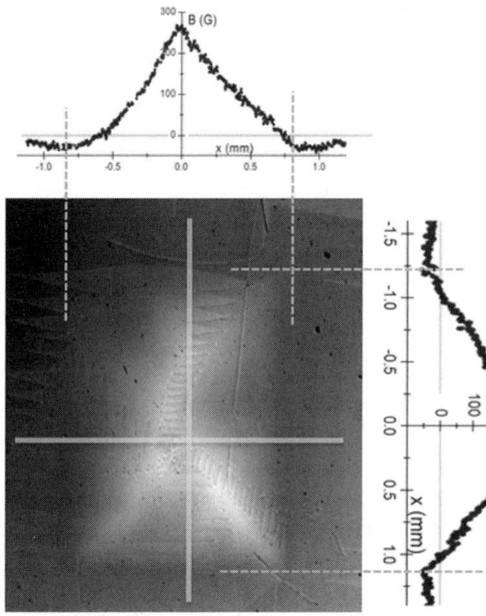

Figure 2.7 Magneto-optical image of single crystal $K_{0.80}Fe_{1.76}Se_2$. Grey lines show the cuts along which the magnetic induction was measured and shown on the side panels.

The remanence reaches about 250 Oe. A simple one-dimensional estimate, using

$$\frac{4\pi}{c}j_c = \frac{dB}{dx}$$

gives:

$$j_c = \frac{250}{0.77}\frac{10}{4\pi} \approx 2.6 \times 10^3 \text{ A/cm}^2$$

This shows that the current samples cannot support large critical current density even at low temperatures. Similar numbers are estimated from the magnetization measurements [61].

2.4.3 *Anisotropic $H_{c2}(T)$*

The anisotropic H_{c2} curves for $K_{0.8}Fe_{1.76}Se_2$ are inferred from measurements of magnetoresistance (for $\mathbf{H} \leq 14$ T) and from high magnetic field measurements of radio frequency (rf) contactless

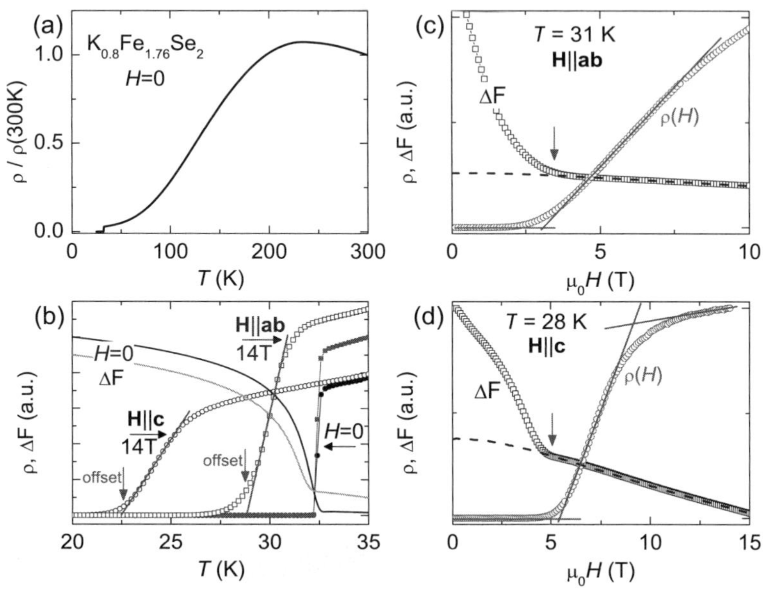

Figure 2.8 (a) Temperature dependence of the normalized *ab*-plane resistivity $\rho(T)$ of the $K_{0.8}Fe_{1.76}Se_2$ single crystal at $H = 0$, where $\rho(300\ K) = 0.12\ \Omega$ cm. (b) Low temperature region of the resistance for two samples at $H = 0$ (closed symbols) and 14 T (open symbols) and the warming curves of rf shift (ΔF) for two samples (solid lines). Vertical arrows indicate T_c^{offset} and lines on the top of 14 T data are guide to the eye. (c) Comparison of the *ab*-plane resistance $R(H)$ and ΔF for **H** ∥ **ab** at $T = 31$ K. (d) Comparison of the *ab*-plane resistance $R(H)$ and ΔF for **H** ∥ **c** at $T = 28$ K. The dashed lines in (c) and (d) are the Δ F taken at $T = 35$ K as a normal state, background signal. The solid lines in (c) and (d) are guides to the eye for offset and onset criteria of H_c and vertical arrows indicate the deviation of ΔF from the background signal (see text).

penetration depth for applied field both parallel (**H** ∥ **c**) and perpendicular (**H** ∥ **ab**) to the tetragonal *c*-axis. Figure 2.8a shows the temperature dependence of the normalized resistivity for the $K_{0.8}Fe_{1.76}Se_2$ sample. The offset and zero-resistance ($R < 3 \times 10^{-5}\Omega$) temperatures were estimated to be $T_c^{\mathrm{offset}} \simeq 32.2$ K and $T_c^{\mathrm{zero}} \simeq 32$ K, respectively, as shown in Fig. 2.8b. The solid lines in Fig. 2.8b are warming curves of the rf shift (ΔF) at $H = 0$ for two different samples. As the temperature decreases, the rf shift

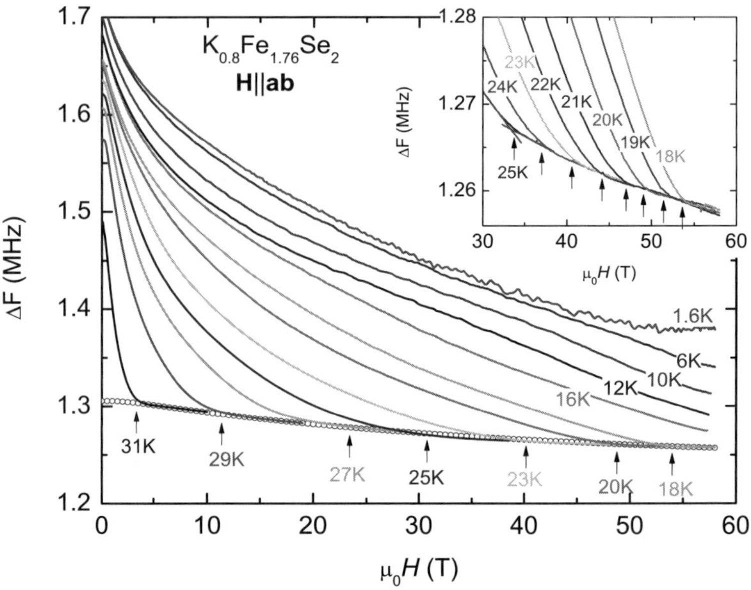

Figure 2.9 Frequency shift (ΔF) as a function of magnetic field for **H** ∥ **ab** at selected temperatures. Open symbols are ΔF taken at $T = 35$ K as a normal state, background signal. The arrows indicate H_{c2} determined from the point deviating from background signal. Inset shows the low temperature data close to H_c. The straight lines on the $T = 25$ K curve are guides to the eye for determining the point at which the rf signal intercepts the slope of the normal state background.

suddenly increases at T_c, where $T_c = 32$ and 32.4 K for two samples were determined from $d\Delta F/dT$. A clear anisotropy in the response of the superconductivity under applied fields was observed between **H** ∥ **ab** and **H** ∥ **c** as shown in Fig. 2.8b for $H = 14$ T curves. To compare the superconducting transition between resistance and ΔF measurement, resistance data measured in a superconducting magnet and ΔF taken in pulsed magnetic fields at $T = 31$ K for **H** ∥ **ab** and at $T = 28$ K for **H** ∥ **c** are plotted in Figs. 2.8c and d, respectively. As shown in the figures, the deviation from the background signal of ΔF is close to the H_c^{offset} criterion of the resistance curves.

The ΔF vs H plots shown in Figs. 2.9 and 2.10 can be used to infer the temperature dependence of the upper critical field $H_{c2}(T)$ by simply taking the first point deviating from the normal state

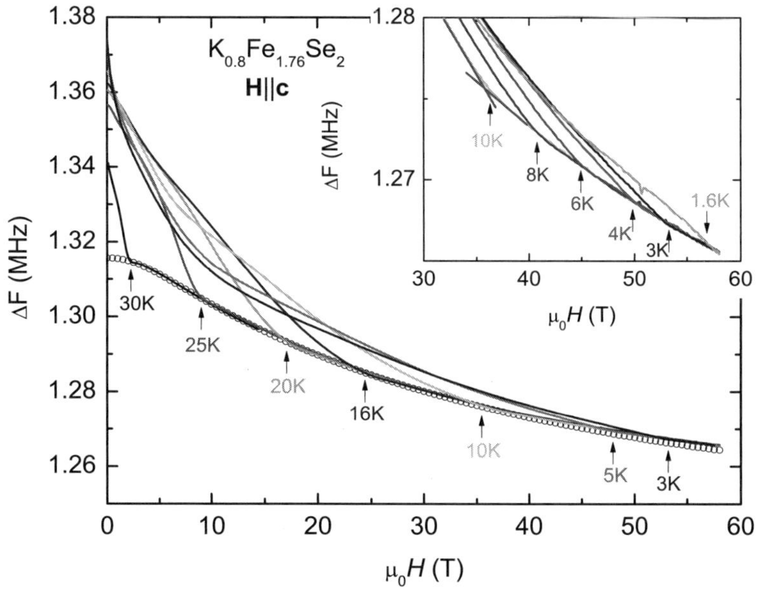

Figure 2.10 Frequency shift (ΔF) as a function of magnetic field for **H** ∥ **c** at selected temperatures. Open symbols are ΔF taken at $T = 35$ K as a normal state, background signal. The arrows indicate H_{c2} determined from the point deviating from background signal. Inset shows the low temperature data close to H_c. The straight lines on $T = 10$ K curve are guides to the eye for determining the point at which the rf signal intercepts the slope of the normal state background.

background. Arrows in Figs. 2.9 and 2.10 indicate the determined H_{c2}. The difference between the H_{c2} values determined by the first deviation and slope change point criteria was used to determine the H_{c2} error bar size.

The $H_{c2}(T)$ curves for both **H** ∥ **ab** (H_{c2}^{ab}) and **H** ∥ **c** (H_{c2}^{c}) in $K_{0.8}Fe_{1.76}Se_2$ are plotted in Fig. 2.11, as determined from the $H \leq 14$ T resistance and from the $H \leq 60$ T data taken from the down sweep of pulsed field magnetic field rf measurements. The curvature of $H_{c2}(T)$ has been reported to vary depending on the criteria used to determine H_{c2}, for example in the case of highly two dimensional, high-T_c cuprate superconductors [62]. In this study, the shape of H_{c2} curves does not change qualitatively when H_{c2} is defined by different criteria or even different measurements. On the other hand,

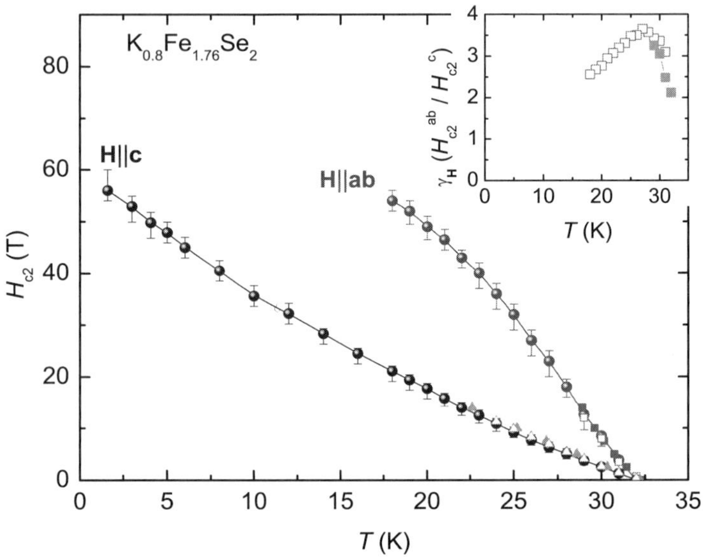

Figure 2.11 Anisotropic $H_{c2}(T)$ for $K_{0.8}Fe_{1.76}Se_2$ single crystals. Solid circles are obtained from the pulsed field rf shift measurements and closed (open) square and triangle symbols are taken from temperature (magnetic field)-dependent resistance measurements. Inset shows the temperature dependence of the anisotropy $\gamma_H = H_{c2}^{ab}/H_{c2}^{c}$ as determined from pulsed field rf shift (open squares) and resistance (solid squares) measurements.

the shapes of the upper critical field curves for **H** $\parallel$ **ab** and **H** $\parallel$ **c** clearly do not manifest the same temperature dependence. As is evidenced from Fig. 2.11, a conventional linear field dependence of H_{c2} is observed close to the T_c, with clearly different slopes for the two field orientations. In the low field region the H_{c2} curves are consistent with earlier studies. [40, 44] Towards higher fields, $H_{c2}^{c}(T)$ presents an almost linear temperature dependence down to 1.5 K, whereas the curve of $H_{c2}^{ab}(T)$ has a tendency to saturate. The anisotropy parameter, $\gamma_H \equiv H_{c2}^{ab}/H_{c2}^{c}$, is about $\sim$ 2 near T_c, but shows a maximum around 27 K with $\gamma_H \sim 3.6$, and decreases considerably for lower temperatures. In all known examples so far, the temperature dependence of γ_H was opposite to that of $\gamma_\lambda \equiv \lambda_c/\lambda_{ab}$. It would be interesting to examine $\gamma_\lambda(T)$ in this material, in particular to see if it goes through a minimum at $\sim$ 27 K.

The zero temperature limit of H_{c2} can be estimated by using the Werthamer–Helfand–Hohenberg (WHH) theory [63], which gives $H_{c2} = 0.69T_c(dH_{c2}/dT)|_{T_c}$. The value of H_{c2} (0) for **H** $\|$ **ab** and **H** $\|$ **c** were estimated to be $\sim$ 102 T and $\sim$ 31 T respectively, where $T_c = 32$ K, $dH_{c2}^{ab}/dT \sim -4.6$ T/K and $dH_{c2}^{c}/dT \sim -1.4$ T/K were used. Clearly these values do not capture the salient physics for this compound. On the other hand, in the simplest approximation, the Pauli limit (H_P) is given by $1.84T_c$ [64–66], giving $H_P \sim 59$ T. This low temperature value of H_{c2} may indeed capture some of the basic physics associated with $K_{0.8}Fe_{1.76}Se_2$. To explain the observed H_{c2} curves in detail, a more complete theoretical treatment is needed, one that does not exclude the strong electron-phonon coupling and multiband nature of Fe-based compounds. Anisotropic superconducting coherence length can be calculated using $H_{c2}^{ab} = \frac{\phi_0}{2\pi\xi_{ab}\xi_c}$ and $H_{c2}^{c} = \frac{\phi_0}{2\pi\xi_{ab}^2}$. [67] If $H_{c2}^{c} = 60$ T and H_{c2}^{ab} is assumed to be between 60 and 100 T, then $\xi_{ab} \sim 2.3$ nm, and 1.4 nm $\lesssim \xi_c \lesssim$ 2.3 nm.

The behavior of $H_{c2}(T)$ for $K_{0.8}Fe_{1.76}Se_2$ is found to be very similar to that of several 122 systems as well as doped FeSe [39, 68–70]. It should be noted that the H_{c2} curves for two orientations in K-doped $BaFe_2As_2$ system seem to cross at low temperature due to the flattening of $H_{c2}^{ab}(T)$ curve [39, 69], additionally, the H_{c2} curves for $FeTe_{0.6}Se_{0.4}$ shows a crossing between **H** $\|$ **ab** and **H** $\|$ **c** curves below 4.5 K because of the subsequent flattening of the H_{c2}^{ab} curve at low temperatures [71, 72]. However in the Co-doped system, the anisotropic $H_{c2}(T)$ curves do not show such crossing [68, 70], a result similar to what was found in this study. Thus, an intriguing feature of $H_{c2}(T)$ curves for Co- and K-doped $BaFe_2As_2$, $FeTe_{0.6}Se_{0.4}$ and $K_{0.8}$ $Fe_{1.76}Se_2$ systems is that the anisotropy near T_c is as large as 3 but drops towards $\sim$ 1 as $T \rightarrow 0$ K. The $H_{c2}(T)$ anisotropy in $K_{0.8}Fe_{1.76}Se_2$ is particularly noteworthy given that it exists deep within an antiferromagnetically ordered state [27, 30]. In the case of Co-doped Ba122, $\gamma_H(T) \sim 1$ when $T_c < T_N$ with clear anisotropy only emerging when the antiferromagnetic state is suppressed [68].

2.4.4 ^{57}*Fe Mössbauer Spectroscopy*

Room temperature neutron diffraction studies of $Cs_yFe_{2-x}Se_2$ [26] and $A_yFe_{2-x}Se_2$ (A = Rb, K) [73] have suggested that the iron moments may be much smaller ($\sim$2.5 μ_B/Fe) and also that the magnetic structure may be far more complex than initially suggested, with the iron atoms being distributed among two (magnetically) inequivalent sublattices and carrying very different magnetic moments. Moreover, even the ordering *direction* has been questioned and it is possible that the iron moments may lie in the *ab*-plane, at least for $Cs_yFe_{2-x}Se_2$ [26], rather than parallel to the *c*-axis as initially suggested [25]. Given the these questions surrounding the magnetic ordering of the iron moments in the $A_yFe_{2-x}Se_2$ system, we have undertaken a ^{57}Fe Mössbauer study $K_{0.8}Fe_{1.76}Se_2$. Whereas Mössbauer spectroscopy cannot be used to determine magnetic structures directly, it is a quantitative local probe that can be used to set hard limits on possible structures. As we will show below, the observation of a single, well-split magnetic component allows us to rule out any structure in which the iron sub-lattice is further subdivided into multiple, inequivalent sites, and the scale of the splitting ($\sim$29 T at 10 K) is consistent with the 3.31 μ_B moment derived from neutron scattering [25].

Several conclusions can be reached simply by inspection of the spectrum taken at 10 K (Fig. 2.12). The spectrum is dominated by a single, well-split, magnetic component. This confirms that $K_{0.8}Fe_{1.76}Se_2$ is indeed magnetically ordered in the superconducting state (recall $T_c \sim 30$ K for this sample). A small quadrupole splitting of 0.33$\pm$0.02 mm/s is present and the linewidth (full width at half maximum) is 0.200$\pm$0.007 mm/s, slightly larger than our typical instrumental width of 0.15 mm/s. The single magnetic component allows us to rule out any magnetic structures involving multiple iron sub-sites with moments that differ by more than a few percent. As we will show below, the large hyperfine field (B_{hf} $\sim$29 T) is inconsistent with a small iron moment and so places further limits on possible magnetic structures. Finally, two of the lines normally present in a magnetically split ^{57}Fe Mössbauer spectra, are essentially absent from the 10 K pattern.

A magnetic field at the ^{57}Fe nucleus, either externally applied or transferred from an ordered moment on the iron atom, lifts the degeneracy of the nuclear states and, in combination with the selection rules for the $\frac{3}{2} \rightarrow \frac{1}{2}$ transition, leads to a six-line pattern with intensities of 3:R:1:1:R:3 (counting from left to right in

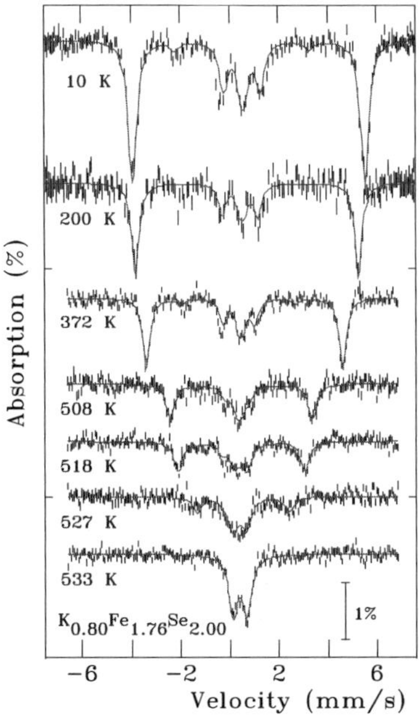

Figure 2.12 (color online) ^{57}Fe Mössbauer spectra of $K_{0.80}Fe_{1.76}Se_{2.00}$ showing the evolution of the magnetic ordering on heating from 10 K (well below $T_c \sim 30$ K) to 533 K where the material is paramagnetic. The extreme weakness of the $\Delta m_I = 0$ transitions in the ordered state indicates that the moments are almost parallel to the crystal *c*-axis (see text), while the growth of a central paramagnetic component above 500 K is characteristic of a first order magnetic transition. Solid lines are fits as described in the text. See also Color Insert.

Fig. 2.12). For a powder sample, $R = 2$, however if there is a unique angle, θ, between the magnetic field and the direction of the γ−beam used to record the spectrum, then the intensity, R, of the $\Delta m_I = 0$ transitions is given by:

$$R = \frac{4\sin^2\theta}{1+\cos^2\theta}$$

$R = 0$ implies that θ is also zero so that the magnetic field, and by extension, the moments that lead to it, are parallel to the γ-beam. Since the sample consists of an *ab*-plane mosaic of single crystals, setting $\theta = 0$ means that the magnetic ordering direction is parallel to the *c*-axis, ruling out any magnetic structures that involve planar ordering of the iron moments. We note that R is a relatively soft function of θ near zero, and a free fit to the intensity of the $\Delta m_I = 0$ transitions is consistent with an angle of $18\pm4^\circ$, and leads to a slight improvement in the least square fit error, χ^2, for the fit. Such an angle would not be consistent with a purely planar ordering of the iron moments (indeed, if the ordering were planar, then R would be 4, and the $\Delta m_I = 0$ transitions would provide the strongest features in the spectrum) but it is too large to be dismissed as being due to a simple mis-alignment of the mosaic. This suggests that there is a small canting of the antiferromagnetic structure away from the c-axis.

Estimating the iron moment from the observed hyperfine field requires some care as the scaling is imperfect at best [74]. However, some data exist on binary iron–chalcogenides that can be used as a guide (Table 2.1). If we use the factor of 6.2 T/μ_B for Fe_7Se_8 with our measured B_{hf} of 29.4 T we obtain a rather large estimate of 4.7 μ_B/Fe for the iron moment in this system. This is significantly larger than the 3.31 μ_B/Fe reported on the basis of neutron diffraction [25], however it does suggest that the iron moment is indeed substantial as even the larger conversion factor for the sulphide yields 3.5 μ_B/Fe. If we assume that B_{hf} is at least proportional to the iron moment, then we can use the observed change in B_{hf} between 10 K and 293 K to scale the 3.31 μ_B/Fe observed at 11 K [25] to get an estimate of 3.0 μ_B/Fe for the moment at room temperature for comparison with the much smaller value of 2.55 μ_B/Fe reported by Pomjakushin *et al.* [73]. However, the strong temperature dependence of magnetic signal noted by Bao *et al.* [25] suggests a very rapid decline in ordered moment to about 2.8 μ_B/Fe by room temperature. It is possible that much of the variation may be intrinsic to the material and its variable stoichiometry, so that combined measurements on a well characterised sample will be needed to settle this.

Impurities may provide a possible origin for the variation in measured moments. Mössbauer spectroscopy, while sensitive to the presence of impurity phases, does not rely on normalisation to the total sample in order to determine moments, they come rather from the observed line splitting, and not the intensity. Neutron diffraction, by contrast, while providing far more information on the magnetic ordering, ultimately relies on peak intensities, normalized

Table 2.1 Average hyperfine fields (B_{hf}) derived from ^{57}Fe Mössbauer spectroscopy and average iron moments derived from neutron diffraction for approximately equi-atomic iron–chalcogenide compounds with estimated field–moment conversion factors. The Fe–Te system exhibits significant variability and measurements have yet to be made on common samples making the conversion factor unreliable. There is however a clear trend to lower values in the sequence S→Se→Te

Compound	Average	Average	Conversion
	B_{hf}	moment	Factor
	(T)	μ_B/Fe	T/μ_B
Sulphides			
Fe_7Se_8	26.8 [75]	3.16 [76]	8.5
Selenides			
Fe_7Se_8	24.1 [77]	3.86 [78]	6.2
Tellurides			
$Fe_{1.125}Te$	—	2.07 [79]	
$Fe_{1+x}Te$	—	1.96–2.03 [80]	
$0.076 \leq x \leq 0.141$			
$Fe_{1.068}Te$	—	2.25 [81]	
$Fe_{1.05}Te$	—	2.54 [82]	
$Fe_{1.11}Te$	11 [83]	—	
$Fe_{1.08}Te$	10.34 [84]		4.3–5.2

to the total nuclear scattering, to determine the magnetic moments. It is clear from the 10 K spectrum shown in Fig. 2.12 that there is a central paramagnetic component present that involves about 12±2% of the iron in the sample. Such high apparent impurity levels in single crystal samples that had no impurities detected by powder x-ray diffraction, deserves further attention. If the paramagnetic component is not an "impurity" then it must either be intrinsic to the structure or a property of the material.

At the temperatures of interest here, $K_{0.8}Fe_{1.76}Se_2$ adopts a vacancy-ordered I4/m modification of the parent $ThCr_2Si_2$−type I4/mmm structure with iron essentially filling a 16*i* site and

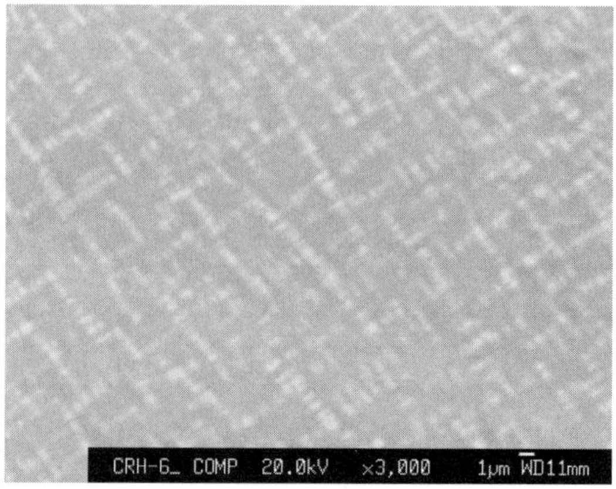

Figure 2.13 Backscattered electron analysis (BSE) image of a cleaved crystal surface of $K_{0.80}Fe_{1.76}Se_{2.00}$ taken at an accelerating voltage of 20 kV. The lighter regions have lower potassium concentrations than the darker background area.

leaving ordered vacancies on the (almost) empty 4*d* site [26, 46]. Occupations of $\sim$8% for the Fe-4*d* site have been reported [46]. If we assume full occupation of the Fe-16*i* site in our sample, this leaves 9% of the iron in the 4*d* site. Partial occupation of the Fe-16*i* site would leave more iron to be accommodated in the 4*d* site. As we see no evidence for a second magnetic component that could be associated with iron in the 4*d* site, it is possible that the iron in these more isolated sites does not order, in which case our estimate of $\gtrsim$9% in the 4*d* site is fully consistent with the 12$\pm$2% paramagnetic component observed in the Mössbauer spectrum.

Another possible origin of the 12$\pm$2% non-magnetic Fe component in the low temperature (including room temperature) state can be seen in the backscattered electron analysis (BSE) image shown in Fig. 2.13. This image reveals that there is, at the micron scale, a modulation in the surface composition that can be correlated, through a preliminary line-scan analysis of the WDS data, with reductions of K content or increase of Fe content in the lighter regions. Auger electron analysis further confirmed this observation and gave a rough estimate of a composition of $K_{0.9}Fe_{1.7}Se_2$ for the dark region and $K_{0.6}Fe_{1.9}Se_2$ for the light region. It should be noted,

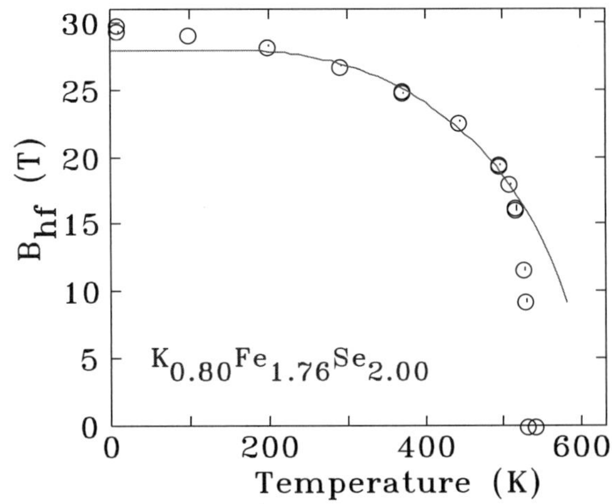

Figure 2.14 (color online) Temperature dependence of the magnetic hyperfine field (B_{hf}) in $K_{0.80}Fe_{1.76}Se_{2.00}$. The solid line is a fit to a $J = \frac{1}{2}$ Brillouin function between 200 K and 500 K that yields an expected transition of 600±30 K, well above the observed value of 532±2 K. Fitted errors on B_{hf} are less than 0.1 T, much smaller than the plotting symbols. The rapid collapse above 500 K is accompanied by the growth of a paramagnetic component (see Fig. 2.15).

though, that such patterns appear in samples grown by furnace cooling as well as samples decanted from a liquid melt [28].

Raising the temperature leads to the expected decline in B_{hf}, however it is clear from Fig. 2.12 that magnetic order persists up to 530 K, confirming that $K_{0.8}Fe_{1.76}Se_2$ has a remarkably high ordering temperature. The temperature dependence of B_{hf} shown in Fig. 2.14 yields an ordering temperature of $T_N = 532 \pm 2$ K. However this is not the result of the fit to a $J = \frac{1}{2}$ Brillouin function shown in Fig. 2.14 as this predicts a transition temperature of 600±30 K and the observed behaviour departs from this curve above 500 K. The two points that bracket the transition are at 530 K, where a clear magnetic signal is seen, and at 533 K where the sample is no longer magnetic, setting the transition at 532 ± 2 K.

A neutron diffraction study of $K_{0.8}Fe_{1.6}Se_2$ found two regions in which the temperature dependence of the magnetic parameter was unusual [25]. From 50 K to 450 K they found a linear dependence of the (101) magnetic peak intensity, suggesting that μ_{Fe}^2 is a linear function of temperature. The clear curvature of B_{hf}(T) in this region, shown in Fig. 2.14, is not consistent with this form, as squaring our observed B_{hf}(T) to get something that would scale with the scattering intensity in a neutron diffraction pattern leads to *increased* curvature rather than linear behaviour.

Above 500 K, Wei Bao *et al.* reported a very rapid decrease in the (101) intensity [25] leading to an ordering temperature of $\sim$560 K. Whereas our sample composition is slightly different and our ordering temperature slightly lower, we see the *same* abrupt loss of magnetic order in Fig. 2.14. It is noteworthy that magnetic susceptibility measurements show somewhat smaller T_N than that revealed by neutron diffraction but very similar T_N to that indicated by Mössbauer spectrum in this article [27, 55, 89]. Inspection of the spectra above 500 K shown in Fig. 2.12 reveals that the intensity of the magnetic peaks decreases visibly as their splitting falls. The ability to uniquely separate the amount of a magnetic phase (seen through line intensities) from the magnitude of the magnetic order (seen independently through line splittings) is an important strength of Mössbauer spectroscopy. Tracking the fraction of the iron that is present as a magnetically ordered form (Fig. 2.15) confirms that the magnetic phase is disappearing even faster than the splitting that marks the order. This strongly suggests that the magnetic phase is transforming before it reaches its true ordering temperature (which we estimate to be about 600 K) and that the observed transition is being driven by a first order structural event. This view is supported by the neutron diffraction work of Wei Bao *et al.* [25] where they also tracked the intensity of the (110) structural peak that is associated with the I4/m vacancy-ordered structure of *b* below 580 K. This peak starts to lose intensity at the same temperature at which the (101) magnetic peak starts its sudden decline. As we see both a weakening of the magnetic order and a reduction in the magnetic fraction above 500 K, it is possible that the break-up of the vacancy-ordered magnetic form reduces the magnetic connectivity of the ordered phase until it forms a non-percolating network of finite clusters. The magnetic order is then lost at a temperature below both its intrinsic ordering temperature, and the temperature at which the vacancy-ordered I4/m structure fully transforms to the high-temperature I4/mmm form.

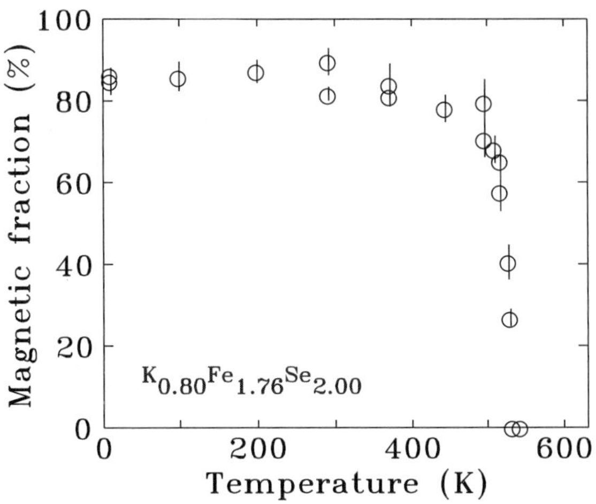

Figure 2.15 (color online) Temperature dependence of the magnetic fraction in $K_{0.80}Fe_{1.76}Se_{2.00}$. The rapid collapse above 500 K indicates that the magnetic transition has first order character and may be associated with a structural transition.

2.4.5 *Phase Separation and Possible Superconducting Aerogel*

The data presented so far offer a rather contradictory set of observations. On one hand $K_{0.8}Fe_{1.76}Se_2$ appears to have a high value of T_c, a fair-sized shielding fraction, and $H_{c2}(T)$ anisotropy that is consistent with many of the other Fe-based superconductors. On the other hand, the electrical resistivity of $K_{0.8}Fe_{1.76}Se_2$ is anomalously high, with increasingly non-metallic temperature dependence depending on precise Fe stoichiometry, the specific heat jump, $\Delta C_p/T_c$ is relatively small, and there is large, local moment-like antiferromagnetic order of the Fe sublattice with a first order transition near 530 K. If the sample were to be considered homogeneous, with all of these features being associated with the same, single phase, then we would need to consider $K_{0.8}Fe_{1.76}Se_2$ to be an anomalous and very different type of Fe-based superconductivity.

As it stands, though, there are several indications, in the data presented, that $K_{0.8}Fe_{1.76}Se_2$ is not homogeneous, but rather is phase separated into a non-magnetically ordered, minority phase that is superconducting and a majority phase that manifest high temperature, large moment, antiferromagnetic order and is probably near insulating.

The Mössbauer and electron microscopy data shown in Figs. 2.12–2.15 indicate that there may well be a mesoscopic phase separation into a majority phase with a large hyperfine field on the Fe site and a minority phase with essentially no hyperfine field on the Fe site. This, combined with the reduced jump in $\Delta C_p/T_c$ and the apparently high electrical resistivity, point toward a scaffold-like network (or aerogel-like pattern) of conducting (and below 30 K, superconducting) phase that exists on a sub-micron length scale. Such a network would be consistent with the moderate shielding seen in the ZFC low field magnetization data as well as the magneto optical results. Such a phase separated, minority, superconducting phase would allow the majority phase to host the large moment anti-ferromagnetism and also be a poor conductor, explaining the curious composite resistivity data that is so sensitive on Fe stoichiometry. It is important to note, though, that given the clear anisotropy in $H_{c2}(T)$ data, this minority phase has to remain, at least partially, oriented with the host matrix. Given the similarity in $H_{c2}(T)$ anisotropy between $K_{0.8}Fe_{1.76}Se_2$ and other Fe-based superconductors, it is tempting to assume that the orientation of the minority phase is closer to complete than to random, but this will need to be tested directly. After our initial reports of phase separation [30], the phase separation scenario in $K_xFe_{2-y}Se_2$ is proposed by more and more recent reports, from Transmission Electron Microscopy [85, 86], single-crystal XRD [87] and magnetic hysteresis loops [88].

2.5 Summary

$K_xFe_{2-y}Se_2$ superconductor has been extensively studied since its discovery. But the non-stoichiometry present a complex material problem that hinders better understanding of its properties. Our single crystal growth from the furnace-cooled and decanted methods implies that it is grown from excess Fe-Se flux and crystals with similar stoichiometry can be obtained from both methods with $T_c \sim 30$ K. Single crystals of $K_{0.8}Fe_{1.76}Se_2$ exhibit moderate anisotropy in both magnetic susceptibility and electrical resistivity with $\chi_{ab}/\chi_c \sim 2$ and $\rho_c/\rho_{ab} \sim 4$ at 300 K. Broadened superconducting transitions seen in several measurements may be associated with a small variation of stoichiometry of the crystal, consistent with what was shown by WDS analysis. The upper critical field of $K_{0.8}Fe_{1.76}Se_2$ is determined as $H_{c2}^{ab}(18\text{ K}) \simeq 54$ T and $H_{c2}^{c}(1.6\text{ K}) \simeq 56$ T. The anisotropy parameter γ_H initially increases with decreasing temperature, passed through a maximum of ~ 3.6 near 27 K, then decreases to $\sim$2.5 at 18 K. The observed γ_H values show a weakening anisotropic effect at low temperatures. Although the Fe-based superconductors have a layered crystal structure, a weak anisotropy of H_{c2} may be a common feature, suggesting that the inter-layer coupling and the three dimensional Fermi surface may play an important role in the superconductivity of this family. Our ^{57}Fe Mössbauer spectroscopy study confirms the presence of magnetic order from well below $T_c \sim 30$ K to $T_N = 532 \pm 2$ K. The large magnetic splitting of 29.4$\pm$0.1 T at 10 K indicates that the iron moments are large, consistent with values of 3.3 μ_B/Fe observed by neutron diffraction at 11 K[25], while the line intensities indicate that the ordering is almost parallel to the c-axis. An apparent paramagnetic impurity phase can be attributed to iron atoms in the $4d$ site or the stoichiometry variation of the microstructure seen in BES image. Analysis of the spectra taken in the vicinity of T_N shows that the magnetic fraction decreases rapidly above 500 K and that the loss of order is driven by a first order structural transition.

In addition to the above, conspicuous observations, there is growing evidence that $K_{0.8}Fe_{1.76}Se_2$ is a phase separated sample, with a metallic (and at low temperature, superconducting) minority phase that does not manifest long range magnetic order and a majority phase that undergoes a first order, antiferromagnetic phase transition near 530 K and may well be either non-conducting or very poorly conducting. As such, $K_{0.8}Fe_{1.76}Se_2$ would essentially be a superconducting aerogel embedded in a matrix of antiferromagnetic (near) insulator. If there is indeed such a phase separation, then $K_{0.8}Fe_{1.76}Se_2$ can be understood, or at least categorized, as another example of Fe-based superconductivity similar at a qualitative level to other, better understood, and single phase, examples.

Acknowledgements

This work was carried out at the Iowa State University and supported by the AFOSR-MURI grant #FA9550-09-1-0603 (R.H. and P.C.C.). Part of this work was performed at Ames Laboratory, US DOE, under contract # DE-AC02-07CH 11358 (K.C., H.K., H.H., W.E.S., M.A.T., R.P., S.L.B. and P.C.C.). S.L.B. also acknowledges partial support from the State of Iowa through Iowa State University. R.P. acknowledges support from the Alfred P. Sloan Foundation.Work at the NHMFL-PFF is supported by the NSF, the DOE and the State of Florida. Mössbauer work was supported by the Natural Sciences and Engineering Research Council of Canada and Fonds Québécois de la Recherche sur la Nature et les Technologies. J.M.C. acknowledges support from the Canada Research Chairs programme.

References

1. Kenji Ishida, Yusuke Nakai, and Hideo Hosono, J. Phys. Soc. Jpn., 78 062001 (2009).
2. M. D. Lumsden and A. D. Christianson, J. Phys.: Condens. Matter, 22, 203203 (2010).
3. Paul C. Canfield and Sergey L. Bud'ko, Annu. Rev. Condens. Matter Phys., 1, 27 (2010).
4. Johnpierre Paglione and Richard L. Greene, Nat. Phys., 6, 645 (2010).
5. X. C. Wang, Q. Q. Liu, Y. X. Lv, W. B. Gao, L. X. Yang, R. C. Yu, F. Y. Li, and C. Q. Jin, Solid State Commun., 148, 538 (2008).
6. M. Rotter, M. Tegel, and D. Johrendt, Phys. Rev. Lett., 101, 107006 (2008).
7. Athena S. Sefat, Rongying Jin, Michael A. McGuire, Brian C. Sales, David J. Singh, and David Mandrus, Phys. Rev. Lett., 101, 117004 (2008).
8. N. Ni, M. E. Tillman, J.-Q. Yan, A. Kracher, S. T. Hannahs, S. L. Bud'ko, and P. C. Canfield, Phys. Rev. B, 78, 214515 (2008).
9. A. Leithe-Jasper, W. Schnelle, C. Geibel, and H. Rosner, Phys. Rev. Lett., 101, 207004 (2008).
10. Y. Kamihara, T. Watanabe, M. Hirano, and H. Hosono, J. Am. Chem. Soc., 130, 3296 (2008).
11. X. H. Chen, T. Wu, G. Wu, R. H. Liu, H. Chen, and D. F. Fang, Nature, 453, 761 (2008).

12. X. Zhu, F. Han, G. Mu, P. Cheng, B. Shen, B. Zeng, and H. H. Wen, Phys. Rev. B, 79, 220512 (2009).
13. X. Zhu, F. Han, G. Mu, B. Zeng, P. Cheng, B. Shen, and H. H. Wen, Phys. Rev. B, 79 024516 (2009).
14. G. F. Chen, T. L. Xia, H. X. Yang, J. Q. Li, P. Zheng, J. L. Luo, and N. L. Wang, Supercond. Sci. Tech., 22, 072001 (2009).
15. Rongwei Hu, Emil S. Bozin, J. B. Warren, and C. Petrovic, Phys. Rev. B, 80, 214514 (2009).
16. Y. Mizuguchi, Y. Hara, K. Deguchi, S. Tsuda, T. Yamaguchi, K. Takeda, H. Kotegawa, H. Tou, and Y. Takano, Supercond. Sci. Tech., 23, 054013 (2010).
17. F. C. Hsu, J. Y. Luo, K. W. The, T. K. Chen, T. W. Huang, P. M. Wu, Y. C. Lee, Y. L. Huang, Y. Y. Chu, D. C. Yan, and M. K. Wu, Proc. Nat. Acad. Sci., 105, 14262 (2008).
18. S. Medvedev, T. M. McQueen, I. Trojan, T. Palasyuk, M. I. Eremets, R. J. Cava, S. Naghavi, F. Casper, V. Ksenofontov, G. Wortmann, and C. Felser, Nature Mater., 8, 630 (2009).
19. Jiangang Guo, Shifeng Jin, Gang Wang, Shunchong Wang, Kaixing Zhu, Tingting Zhou, Meng He, and Xiaolong Chen, Phys. Rev. B, 82, 180520 (2010).
20. J. J. Ying, X. F. Wang, X. G. Luo, A. F. Wang, M. Zhang, Y. J. Yan, Z. J. Xiang, R. H. Liu, P. Cheng, G. J. Ye, and X. H. Chen, Phys. Rev. B, 83, 212502 (2011).
21. Chun-Hong Li, Bing Shen, Fei Han, Xiyu Zhu, and Hai-Hu Wen, Phys. Rev. B, 83, 184521 (2011).
22. Minghu Fang, Hangdong Wang, Chiheng Dong, Zujuan Li, Chunmu Feng, Jian Chen, and H. Q. Yuan, EPL, 94, 27009 (2011).
23. A. F. Wang, J. J. Ying, Y. J. Yan, R. H. Liu, X. G. Luo, Z. Y. Li, X. F. Wang, M. Zhang, G. J. Ye, P. Cheng, Z. J. Xiang, and X. H. Chen, Phys. Rev. B, 83, 060512(R) (2011).
24. Z. Shermadini, A. Krzton-Maziopa, M. Bendele, R. Khasanov, H. Luetkens, K. Conder, E. Pomjakushina, S. Weyeneth, V. Pomjakushin, O. Bossen, and A. Amato, Phys. Rev. Lett., 106, 117602 (2011).
25. Wei Bao, Q. Huang, G. F. Chen, M. A. Green, D. M. Wang, J. B. He, X. Q. Wang, and Y. Qiu, Chinese Phys. Lett., 28, 086104 (2011).
26. V. Yu. Pomjakushin, D. V. Sheptyakov, E. V. Pomjakushina, A. Krzton-Maziopa, K. Conder, D. Chernyshov, V. Svitlyk, and Z. Shermadini, Phys. Rev. B, 83, 144410 (2011).

27. Wei Bao, G. N. Li, Q. Huang, G. F. Chen, J. B. He, M. A. Green, Y. Qiu, D. M. Wang, and J. L. Luo, arXiv:1102.3674 (2011).
28. R. Hu, K. Cho, H. Kim, H. Hodovanets, W. E. Straszheim, M. A. Tanatar, R. Prozorov, S. L. Bud'ko, and P. C. Canfield, Supercond. Sci. Technol., 24, 065006 (2011).
29. E. D. Mun, M. M. Altarawneh, C. H. Mielke, V. S. Zapf, R. Hu, S. L. Bud'ko, and P. C. Canfield, Phys. Rev. B, 83, 100514(R) (2011).
30. D. H. Ryan, W. N. Rowan-Weetaluktuk, J. M. Cadogan, R. Hu, W. E. Straszheim, S. L. Bud'ko, and P. C. Canfield, Phys. Rev. B, 83, 104526 (2011).
31. Jeffrey E. Marchese, Matteo Cirillo, and Niels Grøbech-Jensen, Phys. Rev. B, 79, 094517 (2009).
32. Eundeok Mun, Sergey L. Bud'ko, Milton S. Torikachvili, and Paul C. Canfield, Meas. Sci. Technol, 21, 21055104 (2010).
33. R. Prozorov and R. W. Giannetta, Supercond. Sci. Tech., 19, R41 (2006).
34. L. A. Dorosinskii, M. V. Indenbom, V. I. Nikitenko, Yu. A. Ossip'yan, A. A. Polyanskii, and V. K. Vlasko-Vlasov, Physica C, 203, 149 (1992).
35. Ch. Jooss, J. Albrecht, H. Kuhn, S. Leonhardt, and H. Kronmuller, Rep. Prog. Phys., 65, 651 (2002).
36. C. Mielke, J. Singleton, M.-S. Nam, N. Harrison, C. C. Agosta, B. Fravel, and L. K. Montgomery, J. Phys.: Condens. Matter, 13, 8325 (2001).
37. T. Coffey, Z. Bayindir, J. F. DeCarolis, M. Bennett, G. Esper, and C. C. Agosta, Rev. Sci. Instrum., 71, 4600 (2000).
38. M. M. Altarawneh, C. H. Mielke, and J. S. Brooks, Rev. Sci. Instrum., 80, 066104 (2009).
39. M. M. Altarawneh, K. Collar, C. H. Mielke, N. Ni, S. L. Bud'ko, and P. C. Canfield, Phy. Rev. B, 78, 220505(R) (2008).
40. D. M. Wang, J. B. He, T.-L. Xia, and G. F. Chen, Phys. Rev. B, 83, 132502 (2011).
41. Y. Zhang, L. X. Yang, M. Xu, Z. R. Ye, F. Chen, C. He, J. Jiang, B. P. Xie, J. J. Ying, X. F. Wang, X. H. Chen, J. P. Hu, and D. L. Feng, Nat. Mater., 10, 273 (2011).
42. I. R. Shein and A. L. Ivanovskii, Phys. Lett. A, 375(6), 1028–1031 (2011).
43. T. Qian, X.-P. Wang, W.-C. Jin, P. Zhang, P. Richard, G. Xu, X. Dai, Z. Fang, J.-G. Guo, X.-L. Chen, and H. Ding, Phys. Rev. Lett., 106, 187001 (2011).
44. Yoshikazu Mizuguchi, Hiroyuki Takeya, Yasuna Kawasaki, Toshinori Ozaki, Shunsuke Tsuda, Takahide Yamaguchi, and Yoshihiko Takano, Appl. Phys. Lett., 98, 042511 (2011).

45. D. A. Torchetti, M. Fu, D. C. Christensen, K. J. Nelson, T. Imai, H. C. Lei, and C. Petrovic, Phys. Rev. B, 83, 104508 (2011).
46. P. Zavalij, W. Bao, X. F. Wang, J. J. Ying, X. H. Chen, D. M. Wang, J. B. He, X. Q. Wang, G. F Chen, P-Y Hsieh, Q. Huang, and M. A. Green, Phys. Rev. B, 83, 132509 (2011).
47. Z. Fisk, J. P. Remeika, in: K. A. Gschneider, J. Eyring (Eds.), Handbook on the Physics and Chemistry of Rare Earths, Vol. 12, Elsevier, Amsterdam, (1989).
48. P. C. Canfield and Z. Fisk, Phil. Magaz. B, 65, 1117 (1992).
49. M. A. Tanatar, N. Ni, G. D. Samolyuk, S. L. Bud'ko, P. C. Canfield, and R. Prozorov, Phys. Rev. B, 79, 134528 (2009).
50. Hangdong Wang, Chihen Dong, Zujuan Li, Shasha Zhu, Qianhui Mao, Chunmu Feng, H. Q. Yuan, and Minghu Fang, Europhys. Lett., 93, 47004 (2011).
51. Sergey L. Bud'ko, Ni Ni, and Paul C. Canfield, Phys. Rev. B, 79, 220516(R) (2009).
52. N. Ni, S. L. Bud'ko, A. Kreyssig, S. Nandi, G. E. Rustan, A. I. Goldman, S. Gupta, J. D. Corbett, A. Kracher, and P. C. Canfield, Phys. Rev. B, 78, 014507 (2008).
53. Eun Deok Mun, Sergey L. Bud'ko, Ni Ni, Alex N. Thaler, and Paul C. Canfield, Phys. Rev. B, 80, 054517 (2009).
54. Kefeng Wang, Hechang Lei, and C. Petrovic, Phys. Rev. B, 83, 174503 (2011).
55. Y. J. Yan, M. Zhang, A. F. Wang, J. J. Ying, Z. Y. Li, W. Qin, X. G. Luo, J. Q. Li, Jiangping Hu, and X. H. Chen, Sci. Rep., 2, 212 (2012).
56. R. Prozorov and R. W. Giannetta, P. Fournier, and R. L. Greene, Phys. Rev. Lett., 85, 3700 (2000).
57. C. Martin, M. E. Tillman, H. Kim, M. A. Tanatar, S. K. Kim, A. Kreyssig, R. T. Gordon, M. D. Vannette, S. Nandi, V. G. Kogan, S. L. Bud'ko, P. C. Canfield, A. I. Goldman, and R. Prozorov, Phys. Rev. Lett., 102, 247002 (2009).
58. R. Prozorov, M. A. Tanatar, Bing Shen, Peng Cheng, Hai-Hu Wen, S. L. Bud'ko, and P. C. Canfield, Phys. Rev. B, 82, 180513(R) (2010).
59. R. Prozorov, M. E. Tillman, E. D. Mun, and P. C. Canfield, New J. Phys., 11, 035004 (2009).
60. R. Prozorov, N. Ni, M. A. Tanatar, V. G. Kogan, R. T. Gordon, C. Martin, E. C. Blomberg, P. Prommapan, J. Q. Yan, S. L. Bud'ko, and P. C. Canfield, Phys. Rev. B, 78, 224506 (2008).
61. Hechang Lei and C. Petrovic, Phys. Rev. B, 83, 184504 (2011).

62. Y. Ando, G. S. Boebinger, A. Passner, L. F. Schneemeyer, T. Kimura, M. Okuya, S. Watauchi, J. Shimoyama, K. Kishio, K. Tamasaku, N. Ichikawa, and S. Uchida, Phys. Rev. B, 60, 12475 (1999).
63. N. R. Werthamer, E. Helfand, and P. C. Hohemberg, Phys. Rev., 147, 295 (1966).
64. A. M. Clogston, Phys. Rev. Lett., 9, 266 (1962).
65. B. S. Chandrasekhar, Appl. Phys. Lett., 1, 7 (1962).
66. K. Maki, Phys. Rev., 148, 362 (1966).
67. Charles P. Poole, Jr. (editor) *Handbook of Superconductivity* (Academic Press, San Diego) (2000).
68. N. Ni, M. E. Tillman, J.-Q. Yan, A. Kracher, S. T. Hannahs, S. L. Bud'ko, and P. C. Canfield, Phys. Rev. B, 78, 214515 (2008).
69. H. Q. Yuan, J. Singleton, F. F. Balakirev, S. A. Baily, G. F. Chen, J. L. Luo, and N. L. Wang, Nature (London), 457, 565 (2009).
70. M. Kano, Y. Kohama, D. Graf, F. Balakirev, A. S. Sefat, M. A. McGuire, B. C. Sales, D. Mandrus, and S. W. Tozer, J. Phys. Soc. Jpn., 78, 084719 (2009).
71. S. Khim, J. W. Kim, E. S. Choi, Y. Bang, M. Nohara, H. Takagi, and K. H. Kim, Phys. Rev. B, 81, 184511 (2010).
72. M. H. Fang, J. H. Yang, F. F. Balakirev, Y. Kohama, J Singleton, B. Qian, Z. Q. Mao, H. D. Wang, and H. Q. Yuan, Phys. Rev. B, 81, 020509(R) (2010).
73. V. Yu. Pomjakushin, E. V. Pomjakushina, A. Krzton-Maziopa, K. Conder, and Z. Shermadini, J. Phys.: Condens. Matter, 23, 156003 (2011).
74. S. M. Dubiel, J. Alloys Compd., 488, 18 (2010).
75. H. Kobayashi, M. Sato, T. Kamimura, M. Sakai, H. Onodera, N. Kuroda, and Y. Yamaguchi, J. Phys.: Condens. Matter, 9, 515 (1997).
76. A. V. Powell, P. Vaqueiro, K. S. Knight, L. C. Chapon, and R. D. Sánchez, Phys. Rev. B, 70, 014415 (2004).
77. H. N. Ok and S. W. Lee, Phys. Rev. B, 8, 4267 (1973).
78. A. F. Andresen and J. Leciejewicz, J. de Phys., 25, 574 (1964).
79. D. Fruchart, P. Convert, P. Wolfers, R. Madar, J. P. Senateur, and R. Fruchart, Mater. Res. Bull., 10, 169 (1975).
80. Wei Bao, Y. Qiu, Q. Huang, M. A. Green, P. Zajdel, M. R. Fitzsimmons, M. Zhernenkov, S. Chang, M. Fang, B. Qian, E. K. Vehstedt, J. Yang, H. M. Pham, L. Spinu, and Z. Q. Mao, Phys. Rev. Lett., 102, 247001 (2009).
81. S. Li, C. de la Cruz, Q. Huang, Y. Chen, J. W. Lynn, J. Hu, Y.-L. Huang, F.-C. Hsu, K.-W. Yeh, M.-K. Wu, and P. Dai, Phys. Rev. B, 79, 054503 (2009).

82. A. Martinelli, A. Palenzona, M. Tropeano, C. Ferdeghini, M. Putti, M. R. Cimberle, T. D. Nguyen, M. Affronte, and C. Ritter, Phys. Rev. B, 81, 094115 (2010).
83. E. Hermon, W. B. Muir, J. Quaroni, and R. C. Sweet, Can. J. Phys., 52, 1800 (1974).
84. Y. Mizuguchi, T. Furubayashi, K. Deguchi, S. Tsuda, T. Yamaguchi, and Y. Takano, Physica C, 470, S338 (2010).
85. J. Q. Li, Y. J. Song, H. X. Yang, Z. Wang, H. L. Shi, G. F. Chen, Z. W. Wang, Z. Chen, and H. F. Tian, arXiv:1104.5340 (2011).
86. Z. Wang, Y. J. Song, H. L. Shi, Z. W. Wang, Z. Chen, H. F. Tian, G. F. Chen, J. G. Guo, H. X. Yang, and J. Q. Li, Phys. Rev. B, 83, 140505(R) (2011).
87. Alessandro Ricci, Nicola Poccia, Boby Joseph, Gianmichele Arrighetti, Luisa Barba, Jasper Plaisier, Gaetano Campi, Yoshikazu Mizuguchi, Hiroyuki Takeya, Yoshihiko Takano, Naurang Lal Saini and Antonio Bianconi, Supercond. Sci. Technol., 24, 082002 (2011).
88. B. Shen, B. Zeng, G. F. Chen, J. B. He, D. M. Wang, H. Yang, and H. H. Wen, EPL, 96, 37010 (2011).
89. R. H. Liu, X. G. Luo, M. Zhang, A. F. Wang, J. J. Ying, X. F. Wang, Y. J. Yan, Z. J. Xiang, P. Cheng, G. J. Ye, Z. Y. Li, and X. H. Chen, EPL, 94, 27008 (2011).

Chapter 3

Angle-Resolved Photoemission Spectroscopy of Iron Pnictides

Takafumi Sato,[a,b,*] **Pierre Richard,**[c,d] **Kosuke Nakayama,**[a] **Takashi Takahashi,**[a,d] **and Hong Ding**[c]

[a] *Department of Physics, Tohoku University, Sendai, 980-8578, Japan*
[b] *TRiP, Japan Science and Technology Agency (JST), Kawaguchi 332-0012, Japan*
[c] *Beijing National Laboratory for Condensed Matter Physics, and Institute of Physics, Chinese Academy of Sciences, Beijing 100190, China*
[d] *WPI Research Center, Advanced Institute for Materials Research, Tohoku University, Sendai 980-8577, Japan*
*t-sato@arpes.phys.tohoku.ac.jp

In this chapter, we review recent progress in angle-resolved photoemission spectroscopy (ARPES) studies of iron-pnictide superconductors, by particularly focusing on the experimental results on the $BaFe_2As_2$ system where the ARPES investigations have been most intensively performed. We discuss (i) the evolution of the Fermi surface topology upon electron- and hole-carrier doping, (ii) the superconducting-gap symmetry, (iii) the many-body interactions, and (iv) the electronic states in the antiferromagnetic phase of a parent compound, as well as their implications to the mechanism of superconductivity.

Iron-Based Superconductors: Materials, Properties, and Mechanisms
Edited by Nan-Lin Wang, Hideo Hosono, and Pengcheng Dai

ISBN 978-981-4303-22-4 (Hardcover), 978-981-4303-23-1 (eBook)
www.panstanford.com

3.1 Introduction

The discovery of superconductivity in $LaFeAs(O_{1-x}F_x)$[1,2] has triggered intensive research activities on the high-temperature (T_c) superconductivity of iron (Fe) pnictides, and it also opened new avenues for research in both basic science and technological applications. Variety of samples with different crystal structures have been discovered in the past four years, and they commonly share a FeAs plane (or FeX_c plane where X_c is a chalcogen atom) in a unit cell. Remarkable aspects of the FeAs-based superconductors are that the electrons in the Fe orbitals, generally believed to be the foe to superconductivity due to its magnetic tendency, play a determining role in the superconductivity of Fe-pnictides.[1,3,4]

Iron pnictides also share some commonalities with high-T_c copper-oxide (cuprate) superconductors. First, both have a layered crystal structure. Second, most parent compounds of both cuprates and pnictides exhibit antiferromagnetic (AF) order or spin-density-wave (SDW),[5] and superconductivity emerges either by hole or electron doping into the parent compounds.[1,6] There are also some characteristic differences between cuprates and pnictides. For example, the parent compound of cuprates is a Mott insulator, whereas that of pnictides is a semimetal. Cuprates are well described by a single-band model while the presence of multiple orbitals plays an essential role in pnictides.

To understand the origin of such characteristic commonalities and dissimilarities, it is essential to investigate the electronic structure of these materials. In this sense, angle-resolved photoemission spectroscopy (ARPES) plays a crucial role, since it has a unique capability to directly determine the momentum (k) and energy of electrons in solids. In fact, in cuprates, the ARPES technique greatly contributed in elucidating anomalous physical properties and the mechanism of superconductivity by observing a large holelike Fermi surface around the optimally-doped region,[7,8] a pseudogap above T_c in the underdoped region,[9–11] and $d_{x^2-y^2}$-like anisotropic superconducting gap.[12,13] Recent remarkable progress in the energy and momentum resolutions of ARPES spectrometers further enables the determination of fine electronic structure in the close vicinity of the Fermi level (E_F), and indeed played an important role in determining the superconducting pairing symmetry of novel superconductors such as MgB_2 ($T_c = 37$ K)[14,15] and graphite intercalated compound C_6Ca ($T_c = 11.5$ K).[16] It is thus of particular importance to investigate the electronic states of iron pnictides in terms of high-resolution ARPES to obtain insight into the origin of anomalous physical properties and the mechanism of superconductivity.

In this chapter, we review the present status of ARPES studies on iron pnictides, by mainly focusing on the results of the $BaFe_2As_2$ system (*122* system) where the ARPES studies have been most intensively performed. First we explain the basic principle of the ARPES technique, and then the Fermi surface and pairing symmetry of hole- and electron-doped *122* systems. Then we describe many-body interactions and their relationship to the pairing mechanism, and finally the anomalous electronic states in the AF phase of the parent compound.

3.1.1 *Principle of ARPES*

ARPES is the most sophisticated method among a variety of photoemission spectroscopies (PES), enabling us to determine experimentally the electronic band structure and Fermi surface of single crystalline materials. In ARPES experiments, as shown in Fig. 3.1a, we irradiate the crystal surface with vacuum ultraviolet (VUV) light, and measure the kinetic energy of photoelectrons emitted from the crystal surface as a function of emission angle (θ: polar angle). In comparison with angle-integrated PES, which provides us the information only on the electronic density of states (DOS), ARPES provides detailed information on the electronic structure as a function of momentum.

When we measure the polar angle θ and the kinetic energy (E_k) of photoelectrons, we are able to map out the band dispersions of electrons in a crystal as a function of the momentum parallel to the surface ($k_\parallel$) by using the simple formula $k_\parallel = \sqrt{2mE_k}\sin(\theta)/\hbar$, where the momentum parallel to the surface is conserved during the photoemission process as shown in Fig. 3.1b. This method is very powerful when the measured crystal is a quasi-two-dimensional since it completely determines the band dispersion. For three-

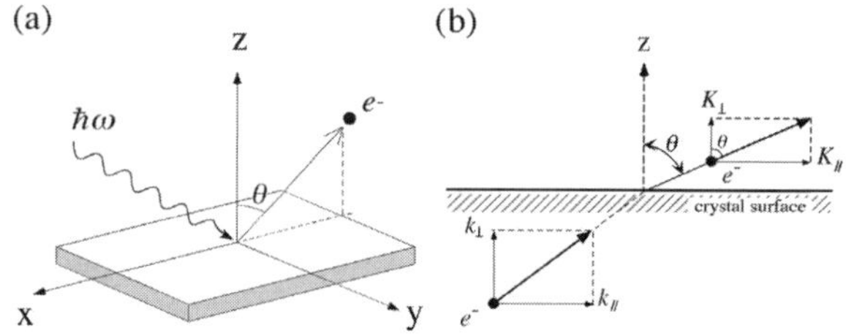

Figure 3.1 Principle of (a) photoemission spectroscopy and (b) ARPES.

dimensional materials, many photon energies are needed in order to probe different values of $k_{\perp}$ which are not conserved during the photoemission process, and some information of final states is needed as well.

A modern ARPES spectrometer usually consists of (i) a large hemispherical analyzer capable of the simultaneous energy and angle detection of photoelectrons, (ii) a high-intensity VUV source typically based on synchrotron radiation source, high-flux helium/xenon discharge lamps,[17] or VUV laser,[18–20] (iii) a low-temperature cryostat with a sample holder, (iv) an ultrahigh-vacuum pumping system, (v) a sample transfer system, and (vi) a sample preparation and characterization system. A typical example of an ARPES spectrometer (at Tohoku University) is displayed in Fig. 3.2.

Figure 3.2 Photograph of a high-resolution ARPES spectrometer. See also Color Insert.

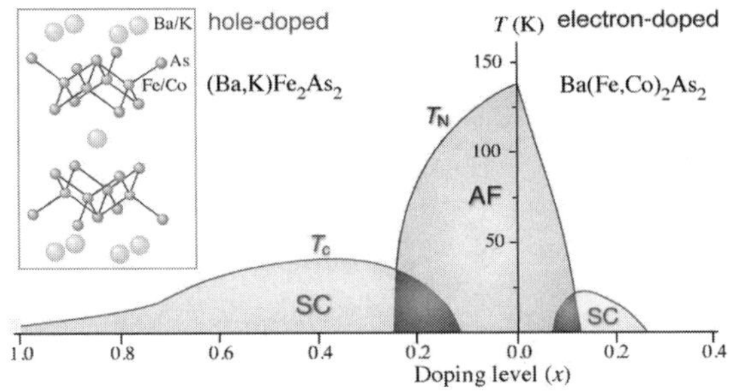

Figure 3.3 Electronic phase diagram of electron- and hole-doped $BaFe_2As_2$ system.[6,21]

3.2 Experimental Results

Before going into the details of the ARPES results, it is convenient to show the typical electronic phase diagram of the *122* system. As shown in Fig. 3.3, the parent compound $BaFe_2As_2$ undergoes an AF transition at $T_N \sim 140$ K. Doping of hole and electron carriers by replacing Ba with K or by replacing Fe with Co leads to the reduction of T_N, and eventually superconductivity appears and the AF order vanishes in both hole and electron-doped systems. There are also some characteristic differences between the two systems, such as the maximum T_c value (37 K and 25 K for the hole- and electron-doped systems, respectively) and the doping width of the superconducting dome: superconductivity does not vanish even in the extreme over-hole doped (KFe_2As_2) whereas it disappears by much smaller amount of electron doping. The relatively easy control of the carrier signs and concentrations is a characteristic of the *122* system and thus provides a precious opportunity to perform systematic ARPES studies in a wide doping range.

3.2.1 *Fermi Surface and Pairing Symmetry*

3.2.1.1 Hole-doped system

An ARPES measurement on optimally hole-doped $Ba_{0.6}K_{0.4}Fe_2As_2$ single crystals ($T_c = 37$ K) reveals the presence of four Fermi surface sheets in the normal state,[22] as shown in Fig. 3.4a. They consist

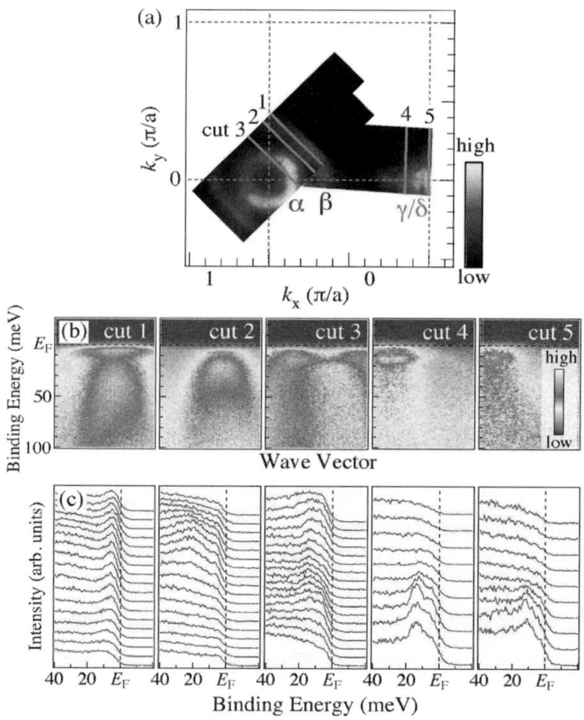

Figure 3.4 (a) Fermi surface of optimally-hole-doped $Ba_{0.6}K_{0.4}Fe_2As_2$. (b), (c) Representative ARPES intensity plots in the superconducting state (T = 15 K) for several cuts shown in (a), and corresponding EDCs, respectively.

of an inner holelike Fermi-surface pocket (from now on we refer to it as the α Fermi surface) and an outer holelike Fermi-surface sheet (the β Fermi surface), both centered at the zone center Γ, and two electron-like Fermi surfaces (the inner γ and the outer δ Fermi surface) centered at M, or (π, 0) defined in the unreconstructed Brillouin zone. The α Fermi surface is nearly circular with an enclosed area of $\sim$ 4% of the reconstructed Brillouin-zone area. The β Fermi surface is more square-like with an area of $\sim$ 18%, and the electron-like γ and δ Fermi surfaces occupy an area of $\sim$ 2% and $\sim$ 4%, respectively. Thus, according to the Luttinger theorem, the total hole concentration is 40% by taking into account the possible degeneracy of the α Fermi surface as in band calculations.[23] This value is similar to the nominal bulk hole concentration of 40% per two Fe atoms in a doubled unit cell, confirming that the ARPES results on the *122* system likely reflect the bulk properties.

When the samples are cooled down below T_c, as shown in the energy distribution curves (EDCs) in Fig. 3.4c at $T = 15$ K, the leading edge of the spectra on all Fermi-surface sheets shifts away from E_F, indicating the opening of an energy gap. In addition, sharp quasiparticle peaks are observed in the vicinity of the Fermi vector (k_F) points, enabling us to estimate the gap values from the quasiparticle peak positions. As can be seen in Fig. 3.4c, the gap opened on the β Fermi surface is clearly smaller than those on the α and γ/δ Fermi surfaces. The spectral linewidth near the β Fermi surface is sharper than the ones near the α and γ/δ Fermi surfaces, indicating a longer lifetime or reduced scattering rate for the low-energy β band excitations.

Opening of the superconducting gap is also demonstrated by the temperature-dependent measurements. Figure 3.5 shows the temperature evolution of EDCs and the symmetrized counterparts for representative k_F points on (a), (b) the α Fermi surface, and (c), (d) the β Fermi surface. As better illustrated by the symmetrized EDCs, the quasiparticle peak gradually reduces its weight upon increasing temperature and it disappears near T_c. Simultaneously, the spectral weight around E_F gradually recovers. Such a characteristic behavior is similar to the temperature evolution of the quasiparticle peak in optimally- and over-doped high-T_c cuprates,[24] indicating that the observed temperature-induced spectral change in $Ba_{0.6}K_{0.4}Fe_2As_2$ indeed originates from the opening of the superconducting gap.

The most important issue for clarifying the superconducting pairing symmetry is the k dependence of the superconducting gap. Figure 3.6 displays the k dependence of symmetrized EDCs obtained at various locations on all the measured Fermi surfaces (α-δ). The energy position of the quasiparticle peak, corresponding to the superconducing-gap size $|\Delta|$, has no meaningful k dependence on each Fermi-surface sheet, demonstrating an isotropic s-wave character with no indication of gap node. Such a s-wave feature for the superconducting gap has been confirmed by several ARPES experiments.[25–27] The energy position of the quasiparticle peak is about 12 meV on the α, γ, and δ Fermi surfaces, while that on the β Fermi surface is about 6 meV.

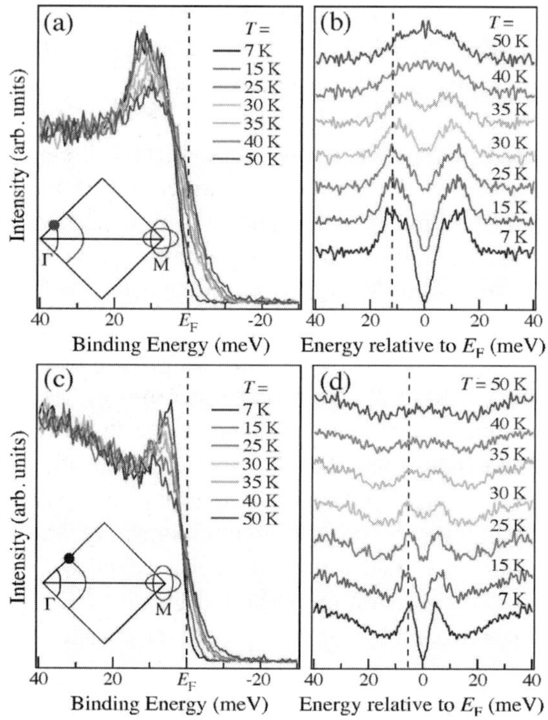

Figure 3.5 Temperature dependence of EDCs and symmetrized EDCs in $Ba_{0.6}K_{0.4}Fe_2As_2$ at the representative k_F points on (a)–(b) the α band and (c)–(d) the β band.

Such k dependence of the superconducting gap is better illustrated in the polar plot of Fig. 3.7 where the gap size is plotted against the Fermi-surface angle θ defined in Fig. 3.6a. The averaged gap size for α, β, γ, and δ Fermi surfaces are 12.3, 5.8, 12.2, and 11.4 meV, respectively, again demonstrating that the superconducting gap of the β Fermi surface is strongly suppressed.[28] Estimated $2|\Delta|/k_B T_c$ values of ~7.5 for the α, γ, and δ Fermi surfaces indicate a strong coupling regime, whereas that for the β Fermi surface (~3.5) is in a weak coupling regime. Interestingly, the α and γ/δ Fermi surfaces are well connected by the AF wave vector of the parent compound $Q = (\pi, 0)$, satisfying the interband nesting condition which can enhance the kinetic process where a zero

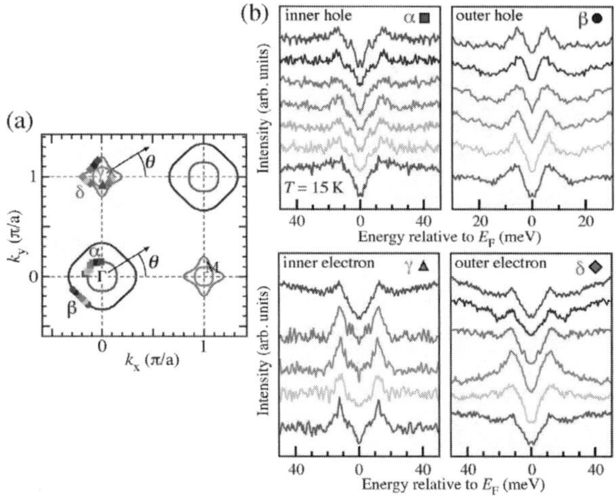

Figure 3.6 k dependence of symmetrized EDCs in $Ba_{0.6}K_{0.4}Fe_2As_2$ for different Fermi surfaces (α-δ) in the superconducting state ($T = 15$ K).

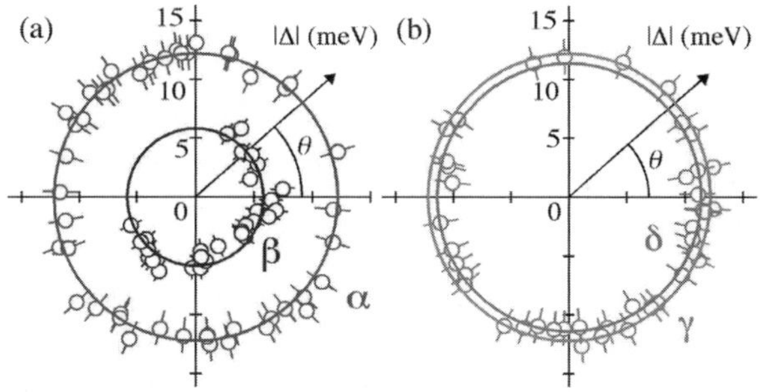

Figure 3.7 Superconducting gap size of $Ba_{0.6}K_{0.4}Fe_2As_2$ at 15 K extracted from symmetrized EDCs shown on polar plots for the (a) α, β, and (b) γ, δ Fermi surfaces as a function of θ.

momentum pair formed on the α Fermi surface is scattered onto the γ/δ Fermi surface by fluctuations near the wave vector Q, whereby increasing the pairing amplitude. A pairing mediated by $Q \sim (\pi, 0)$ spin fluctuations would lead to the opening of an unconventional sign-changing s-wave ($s\pm$-wave) gap on the hole and electron pockets.[29–31]

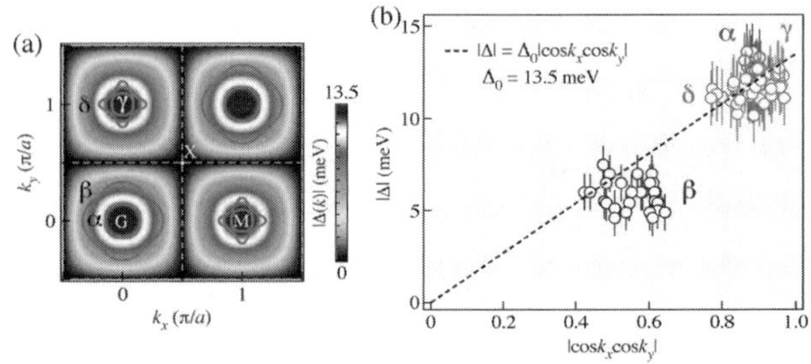

Figure 3.8 Theoretical superconducting gap value $|\Delta(k)| = \Delta_0|\cos k_x \cos k_y|$ with $\Delta_0 = 13.5$ meV as a function of the 2D wave vector. Dashed lines represent zero-gap (nodal) lines. (b) Plot of the superconducting-gap values at 15 K in $Ba_{0.6}K_{0.4}Fe_2As_2$ as a function of $|\cos k_x \cos k_y|$. The dashed line shows the best fit using the gap function $|\Delta(k)| = \Delta_0|\cos k_x \cos k_y|$.

Since the full set of superconducting gaps as a function of momentum is available, it is tempting to quantitatively compare the experimental results with theoretical models. As an example of the superconducting gap form for the $s\pm$-wave symmetry,[32,33] a simple formula with $\Delta(k) = \Delta_0 \cos k_x \cos k_y$ is adopted, and its magnitude as a function of the 2D wave vector is shown in Fig. 3.8a. This formula predicts (i) the opening of larger (smaller) superconducting gaps on smaller (larger) Fermi surfaces, (ii) the sign change between the Fermi surfaces at Γ and the ones at M. We note that, although the $\cos k_x \cos k_y$ value becomes zero between the Γ and M line (orange lines in Fig. 3.8a), it does not create a gap node in the case of $Ba_{0.6}K_{0.4}Fe_2As_2$ since observed Fermi surfaces do not intersect this nodal line. In Fig. 3.8b, we plot the experimentally determined gap values as a function of $|\cos k_x \cos k_y|$. The observed gap sizes among four Fermi surface sheets basically follow the $\Delta_0 \cos k_x \cos k_y$ function with $\Delta_0 = 13.5$ meV. However, closer look reveals a finite deviation between the experiment and the model. For example, according to the model, the gap size on the β Fermi surface varies from 6 (along Γ-M) to 8.5 meV (along Γ-X), whereas the experimentally observed nearly isotropic gap value (5.8 ± 0.8 meV) does not follow this trend. Similar deviation is also seen in the δ Fermi surface. These results indicate that the observed Fermi-surface-dependent superconducting gaps are not simply explained by the $\cos k_x \cos k_y$

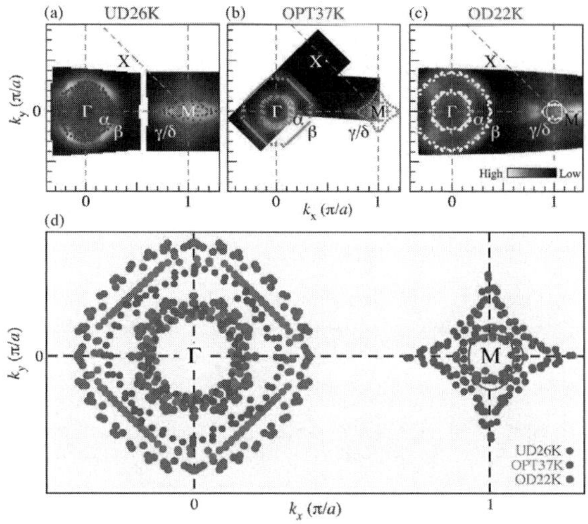

Figure 3.9 ARPES intensity plots at E_F as a function of the 2D wave vector for (a) underdoped $Ba_{0.75}K_{0.25}Fe_2As_2$, (b) optimally-doped $Ba_{0.6}K_{0.4}Fe_2As_2$, and (c) overdoped $Ba_{0.3}K_{0.7}Fe_2As_2$. (d) Comparison of Fermi surface among the samples with three different doping levels. See also Color Insert.

order parameter with a single energy scale Δ_0, indicating possible k_z dependence and/or orbital dependence for this material.

The next important issue is the evolution of the Fermi surface and the superconducting gap as a function of carrier doping, which is crucial in establishing the universality of the superconducting mechanism. Figures 3.9a-c show the comparison of ARPES intensity at E_F plotted as a function of the 2D wave vector for superconducting samples with different hole doping levels covering the (a) underdoped (T_c = 26 K)[34] and (c) overdoped (T_c = 22 K)[35] regions. The observed Fermi surface topology is qualitatively similar in these compounds: the α and the β hole pockets are clearly observed in the under- and over-doped regions. Also, the γ/δ electron pockets appear to be still present. As visible in Fig. 3.9d, both the α and β hole pockets monotonically expand upon hole doping, while the γ and δ electron pockets systematically shrink, as naturally expected from the chemical potential shift induced by carrier change.

Figure 3.10 displays the symmetrized EDCs in the superconducting state measured at various locations on the Fermi surface for

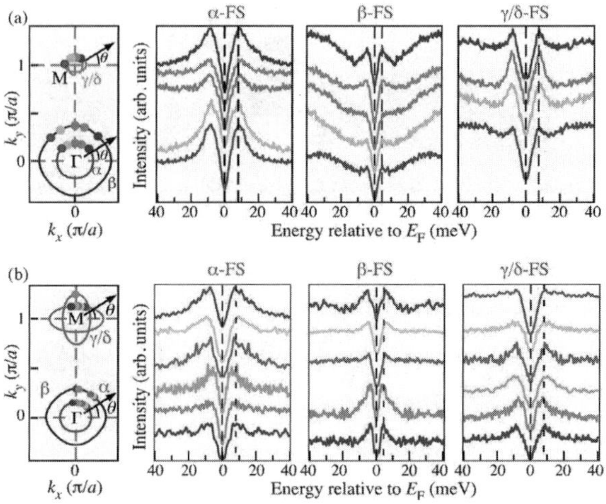

Figure 3.10 k dependence of symmetrized EDCs in the superconducting state for (a) overdoped $Ba_{0.3}K_{0.7}Fe_2As_2$ and (b) underdoped $Ba_{0.75}K_{0.25}Fe_2As_2$. EDCs for the underdoped sample were divided by that at 35 K to remove the influence of gradually decreasing DOS in the normal state.

the (a) overdoped and (b) underdoped samples. In both samples, one can identify a quasiparticle peak away from E_F which shows no discernible variation as a function of k on each Fermi surface, indicative of the isotropic s-wave character. The gap size of the β band is always smaller than that of the other Fermi surfaces, as in the case of the optimally-doped sample.

Figure 3.11a compares the Fermi-surface-angle dependence of the superconducting gap size among underdoped, optimally-doped, and overdoped samples. The superconducting gap commonly shows an isotropic character on each Fermi surface sheets, indicating that the Fermi-surface-dependent nodeless superconducting gap is a common feature of the *122* system in a wide doping range around the optimally-doped region. Interestingly, there are quantitative differences in the magnitude of the superconducting gap. For example, for the α band, the overall gap magnitude is apparently larger in the optimally-doped sample than that in the underdoped and overdoped samples. Similar trend is also recognized for the

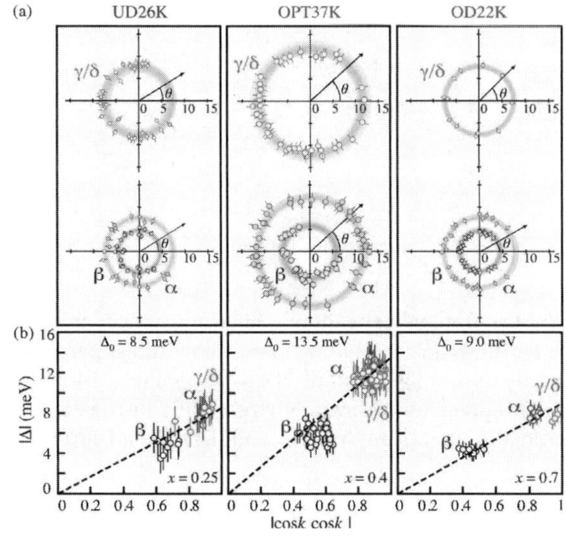

Figure 3.11 (a) Comparison of the Fermi-surface-angle dependence of the superconducting gap value of underdoped, optimally-doped, and overdoped $Ba_{1-x}K_xFe_2As_2$. (b) Same as (a) but plotted as a function of $|\cos k_x \cos k_y|$.

other bands (β, γ, and δ). These results suggest that the magnitude of the superconducting gap is closely related to the value of T_c.

Figure 3.11b displays a comparison of experimentally determined gap values as a function of $|\cos k_x \cos k_y|$ for three doping levels. As in the case of the optimally-doped sample (see Fig. 3.8b), the observed gap sizes roughly follow the $\Delta_0 \cos k_x \cos k_y$ function irrespective of the doping level, whereas the values of Δ_0 are significantly different for different doping levels, suggesting the basic validity of the $\Delta_0 \cos k_x \cos k_y$-like gap function.

Figure 3.12a shows the comparison of momentum-averaged superconducting gap values for each Fermi surface sheet. It is again confirmed that the gap size is largest in the optimally-doped sample. More significantly, the ratio of $2|\Delta|/k_B T_c$ are similar for the same Fermi surface sheets irrespective of the doping levels ($\sim$7–9 for the α, γ, and δ Fermi surfaces, and $\sim$3–5 for the β Fermi surface), strongly suggesting that the transition temperature of pnictide superconductors is determined by pairing strength.

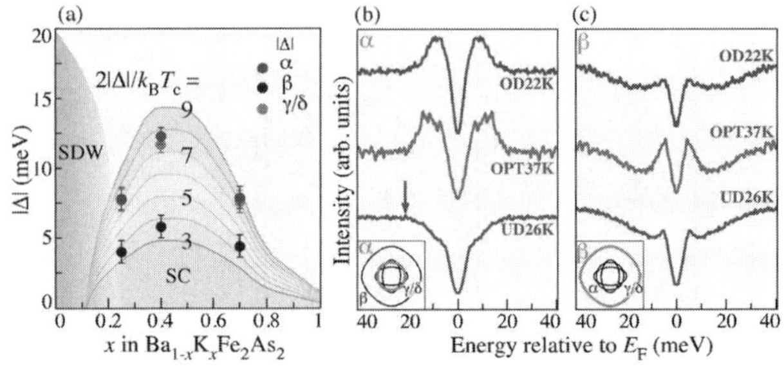

Figure 3.12 (a) Doping dependence of averaged superconducting gap size on each Fermi surface. Colored curves show the superconducting gap size obtained by assuming constant $2|\Delta|/k_B T_c$ values. (b) and (c) Doping dependence of representative symmetrized EDCs in the superconducting state measured at k_F points on the (a) α and (b) β Fermi surfaces.

Figure 3.12b displays the comparison of symmetrized EDCs in the superconducting state at representative momentum points on the α band. In the optimally and overdoped samples, a quasiparticle peak is clearly identified away from E_F, whereas the peak is significantly suppressed in the underdoped sample, with a shoulder-like feature below 10 meV. Such a suppression of the peak weight is related to the opening of a possible pseudogap as indicated by an arrow in the EDC of the underdoped sample. The magnitude of this pseudogap ($\sim$20 meV) is about twice as large as that of the superconducting gap. Moreover, there is no clear evidence for the pseudogap in the β band as shown in Fig. 3.12c where the nesting of Fermi surface is ill-defined. These results, together with the experimental fact that the pseudogap survives well above T_c,[34] suggest that the observed pseudogap is not directly related to the superconducting pairing and likely originates from the AF fluctuations. Such fluctuations would remarkably suppress the superconducting quasiparticle weight in the specific momentum region, like in the case of the antinodal region of high-T_c cuprate superconductors.[9–11]

Another important issue is the electronic states in heavily overdoped region. Figure 3.13a shows the ARPES intensity plot of KFe_2As_2 ($T_c \sim 3$ K) at E_F as a function of the 2D wave vector measured at 15 K. Two holelike Fermi surfaces centered at the Γ

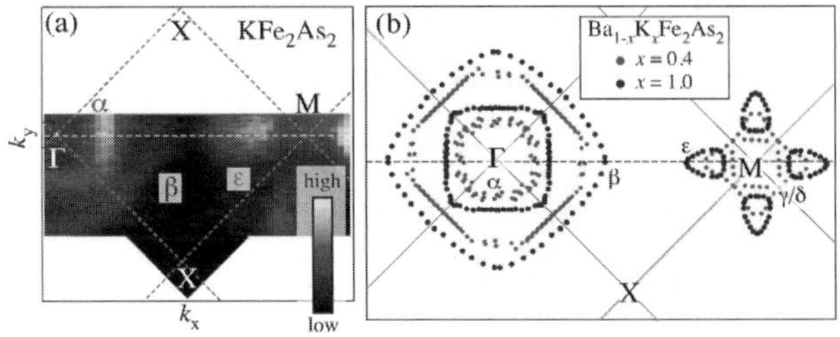

Figure 3.13 (a) ARPES intensity plot as a function of the 2D wave vector of extremely overdoped KFe_2As_2. (b) Comparison of k_F points between overdoped KFe_2As_2 ($x = 1.0$) and optimally-doped $Ba_{0.6}K_{0.4}Fe_2As_2$ ($x = 0.4$).

point are ascribed to the α (inner) and β (outer) hole pockets. One can also find small bright spots slightly away from the M point, called here the ϵ pockets, which have a holelike character judging from the band dispersion near E_F.[36] No evidence for the two electron pockets (the γ and δ Fermi surface)[4,23,37–40] has been found at the M point, likely due to the excess of hole doping which lifts the band bottom of the electron pockets above E_F.

Figure 3.13b displays a direct comparison of the momentum location of k_F points between the optimally-doped sample ($x = 0.4$) and extremely overdoped sample ($x = 1.0$). As clearly visible, the volume of the α and β Fermi surfaces expands due to hole doping. The Fermi surface topology at the M point of these two doping levels is drastically different since the small γ and δ electron pockets observed at $x = 0.4$ are completely absent at $x = 1.0$. This marked difference is reasonably explained by a simple chemical potential shift of ~20 meV due to hole doping. This chemical potential shift is much smaller than the expected value from the band calculations (~120 meV), possibly due to the strong mass renormalization of the near-E_F bands.

As seen in the schematic band structure shown in Fig. 3.14, the α holelike and γ electronlike Fermi surfaces of the optimally-doped sample are well connected by the AF wave vector $Q = (\pi, 0)$. Therefore, the enhanced inter-band scattering via the Q vector would promote the kinetic process of pair scattering between these two Fermi surfaces by low-energy fluctuations, leading to

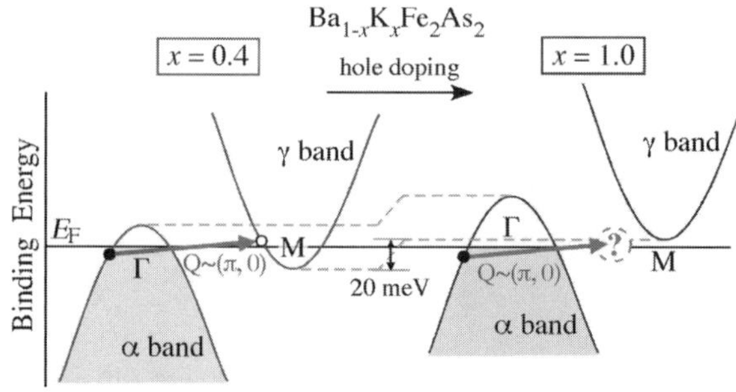

Figure 3.14 Schematic view of the interband scattering by the AF wave vector between the hole and electron bands centered at the Γ and M points, respectively. The interband scattering is markedly suppressed in the extremely overdoped region due to the absence of the electron pockets at the M point.

an increase of the pairing amplitude.[29-31] On the other hand, in the overdoped region ($x = 1.0$), this interband scattering would diminish since the γ band is in the unoccupied side and the size of the α band expands significantly. This suggests that the interband scattering between the holelike and electronlike bands plays an important role in achieving high-T_c values in this iron pnictide.

3.2.1.2 Electron-doped system

One characteristic feature of the electronic phase diagram of the *122*-based iron pnictides is that the superconductivity emerges in both the hole-doped and electron-doped regions (see Fig. 3.3). Since the Fermi-surface-dependent nodeless superconducting gap is established in the hole-doped region, a next important question is whether or not a similar superconducting gap exists in the electron-doped region. Clarifying a possible electron-hole symmetry or asymmetry is essential for a deeper understanding of the superconducting mechanism of iron pnictides.

Figure 3.15a shows the ARPES intensity plot of optimally electron-doped $BaFe_{1.85}Co_{0.15}As_2$ ($T_c = 25$ K) at E_F as a function of the 2D wave vector.[41] A single Fermi surface centered at the Γ point and two elongated pockets centered at the M point are observed. These Fermi surface sheets are ascribed to the hole β

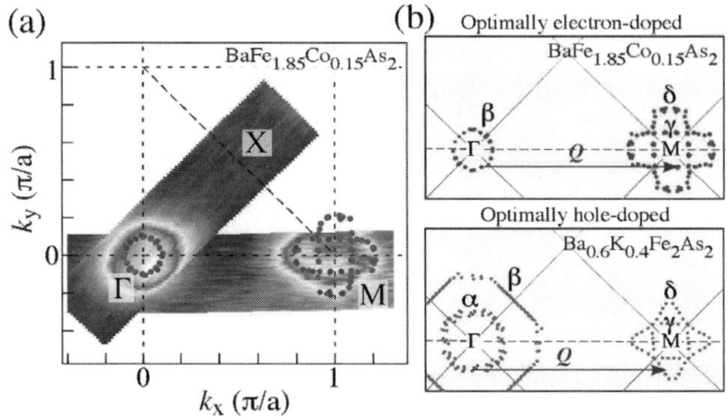

Figure 3.15 (a) ARPES intensity plot as a function of the 2D wave vector of optimally electron-doped $BaFe_{1.85}Co_{0.15}As_2$. (b) Fermi surface comparison between optimally electron-doped $BaFe_{1.85}Co_{0.15}As_2$ and optimally hole-doped $Ba_{0.6}K_{0.4}Fe_2As_2$.

pocket and the electron γ/δ pockets, respectively. Figure 3.15b directly compares experimentally determined k_F points for electron-doped (top) and hole-doped (bottom) samples in the optimally-doped region. Upon electron doping into the hole-doped sample, the α Fermi surface disappears and the β Fermi surface significantly shrinks. Simultaneously, the γ/δ pockets expand. These features are naturally understood by the rigid-band-like chemical potential shift where the α band sinks below E_F in an electron-doped sample due to the excess of electron doping. Interestingly, the nesting condition still holds in an electron-doped sample while the involved Fermi surfaces (β and γ) are different from the hole-doped sample (α and δ).

Next we turn our attention to the k-dependence of the superconducting gap. Figures 3.16a and b show the symmetrized EDCs measured at 8 K at various k_F points on the β hole pocket and the ellipsoidal electron pocket (γ/δ). The superconducting gap of each Fermi surface is nearly constant, as in the case of the hole-doped counterpart. As seen in the polar plot of the superconducting gap in Fig. 3.16d, the estimated gap size is nearly independent of momentum on each Fermi surface, indicating an isotropic s-wave nature, while the average gap values of the hole and electron pockets are slightly different (6.6 and 5.0 meV, respectively). This

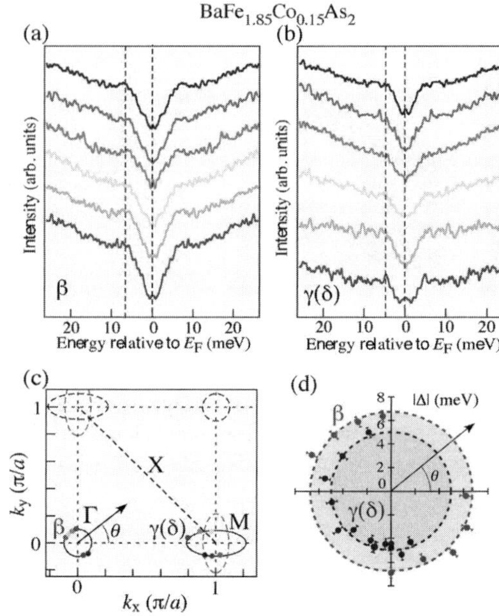

Figure 3.16 (a), (b) Symmetrized EDCs in the superconducting state of $BaFe_{1.85}Co_{0.15}As_2$ measured at various k_F locations shown in (c). (d) Superconducting gap size extracted from symmetrized EDCs shown on polar plots.

result unambiguously establishes that the Fermi-surface-dependent isotropic nature of the superconducting gap is an essential feature of the *122* compound in both hole- and electron-doped systems.

As in the case of the hole-doped system, the investigation of the electronic states in the heavily overdoped region is particularly important in the electron-doped sample since the superconductivity completely vanishes in the heavily electron-doped system.

Figure 3.17 displays a direct comparison of the momentum location of k_F points between the non-superconducting heavily overdoped $BaFe_{1.7}Co_{0.3}As_2$ and the superconducting optimally-doped $BaFe_{1.85}Co_{0.15}As_2$ samples.[47] Upon electron doping, the holelike β Fermi surface, as observed in the optimally-doped sample,[41] disappears or becomes indistinguishably small in the overdoped sample. Simultaneously, the size of the two electron pockets at the M point significantly expands with electron doping.

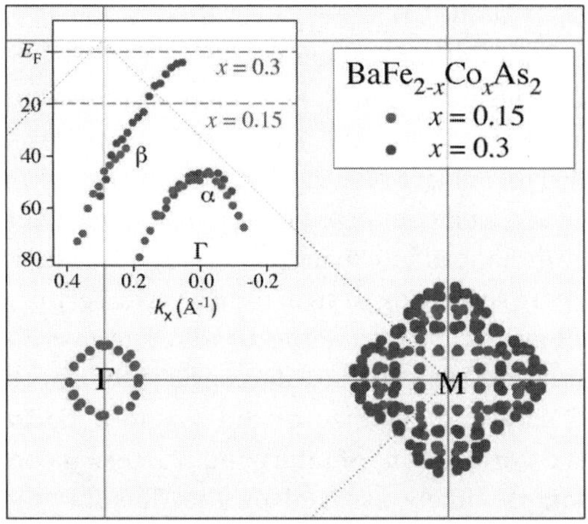

Figure 3.17 Comparison between the Fermi surface of optimally-doped and overdoped samples of electron-doped $BaFe_{2-x}Co_xAs_2$. Inset shows the comparison of band dispersion near the Γ point.

Indeed, the estimated volume of the β Fermi surface with respect to the first unfolded Brillouin zone for the overdoped sample is 0-0.5 %, much smaller than that for the optimally doped sample (1.5±0.5 %), and the volume of the electronlike Fermi surface is larger in the overdoped sample (6.5±0.5 %) as compared to the optimally-doped sample (3.5±0.5 %). It is emphasized here that the Fermi surfaces at the Γ point of the superconducting and non-superconducting samples are drastically different since the β hole pocket observed in the optimally-doped sample is absent or fairly small in the overdoped sample. This marked difference is reasonably explained by a chemical potential shift due to electron doping. As shown in the inset of Fig. 3.17, the energy position of bands in the vicinity of E_F near the Γ point for the overdoped and optimally-doped samples quantitatively matches when we shift down the bands of the optimally-doped sample by 20 meV, suggesting the basic applicability of a rigid-band model.

A simple extrapolation using a tight-binding formula suggests that the holelike β band in the optimally-doped sample tops at less than 20 meV above E_F.[41] The β band for the overdoped sample would be nearly completely occupied if we assume the observed chemical potential shift of 20 meV.

What is the mechanism of T_c reduction and resultant disappearance of superconductivity in the overdoped region? As shown in Fig. 3.15b, the β Fermi surface and the γ Fermi surface of the optimally-doped sample are well connected by the AF wave vector $Q = (\pi, 0)$. Therefore, the enhanced interband scattering *via* the Q vector would promote the pair scattering between these two Fermi surfaces.[29–31] On the other hand, in the non-superconducting overdoped sample, this interband scattering would decrease since the β band is in the occupied side and the size of the γ and δ pockets expands. The absence (deterioration) of interband scattering condition in the overdoped sample, likely assisted by the Co-substitution-induced disorder, would completely kill superconductivity. It is noted here that similar mechanism is at work in the hole-doped samples as discussed in Fig. 3.13. Also, a finite difference in the electronic phase diagram in the overdoped region between the hole- and electron-doped systems is well explained by taking into account the electron-hole asymmetry of the band structure as revealed by the core-level and valence-band photoemission spectroscopy.[42]

It is remarked here that another type of 122 compounds which is isostructural to the Ba122 system but contains FeSe planes in a unit cell has been recently discovered.[43,44] Interestingly, ARPES studies have revealed that holelike bands at the Γ point sink below E_F and do not form the Fermi surface[45,46] due to heavily electron-doped nature. This observation suggests that the simple weak-coupling interband scattering scenario may be insufficient for the superconductivity, instead the strong-coupling local pairing may be more important.

3.2.2 *Many-Body Interactions*

The many-body interactions responsible for superconductivity can sometimes be reflected from fine features in the electronic energy dispersion in the vicinity of E_F. ARPES is a powerful technique to directly access the electronic dispersion of materials and its interactions with bosonic excitations. Indeed, a direct signature of an electron-mode coupling is an anomaly in the electronic energy

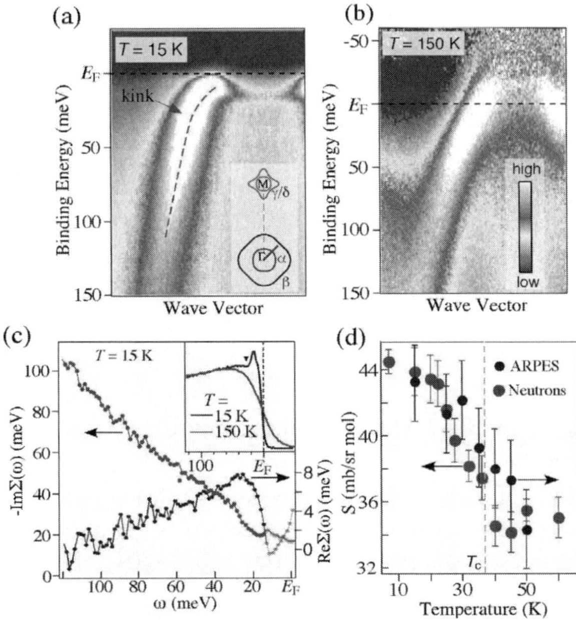

Figure 3.18 (a) ARPES intensity plot in the superconducting state (15 K) along a cut crossing the α band in optimally hole-doped $Ba_{0.6}K_{0.4}Fe_2As_2$. (b) ARPES intensity at 150 K divided by a FD function, recorded along the same cut as in (a). (c) Real and imaginary parts of $\Sigma(\omega)$. The inset compares the partial density of states along the cuts measured at 15 K and 150 K, respectively. (d) Comparison of the temperature dependence of the neutron magnetic intensity and the strength of the dispersion kink $\mathrm{Re}\Sigma(\omega)$.

dispersion (kink). For example, previous ARPES studies revealed a kink in cuprates,[48–51] which is believed to be linked to the pairing. It is thus very important to clarify whether or not a similar kink exists in iron pnictides.

Figure 3.18a shows an ARPES intensity plot of optimally-doped $Ba_{0.6}K_{0.4}Fe_2As_2$ as a function of wave vector and binding energy at $T = 15$ K measured along a cut which crosses the α Fermi surface as indicated in the inset.[55] The ARPES intensity plot shows a clear discontinuity or kink in the electronic dispersion around 25 meV. On the other hand, as shown in Fig. 3.18b which plots the ARPES intensity at $T = 150$ K divided by the Fermi-Dirac distribution (FD) function, such a kink-like feature is not clearly observed.

As shown in the inset of Fig. 3.18c, the emergence of a dispersion anomaly is also reflected in the partial density of states (DOS) obtained by integrating the EDCs over a wide momentum range along the measured cut, as a characteristic dip in EDC (marked by a dot).

In the presence of many-body interactions, an electron-mode coupling is usually described by a complex self-energy $\Sigma(\omega)$. The real part of $\Sigma(\omega)$ is obtained by the difference between the peak position of momentum distribution curves (MDCs) at 15 K and 150 K, where the MDC dispersion at 150 K is used as an approximation of the bare band dispersion without electron-mode coupling. The imaginary part of $\Sigma(\omega)$ is approximated by the half-width at half-maximum of the Lorenztian functions used to fit the MDCs, multiplied by the velocity of the bare band at each energy ω. The results of the self-energy analysis are shown in Fig. 3.18c. Both the real and imaginary parts of $\Sigma(\omega)$ exhibit a well defined anomaly at a position similar to the dip in the EDC, as expected in the presence of an electron-mode coupling. Particularly, the drop in the absolute value of the imaginary part of $\Sigma(\omega)$ indicates a reduced quasiparticle scattering rate. In simple models, a coupling with an Einstein mode of energy Ω in the presence of a superconducting gap Δ will be observed at a binding energy $\Omega + \Delta$.[24] Subtracting the 12 meV superconducting gap of the α band from the kink energy, the mode energy is assigned to be $\Omega = 13\pm2$ meV.

An inelastic neutron scattering study of $Ba_{0.6}K_{0.4}Fe_2As_2$ revealed a spin resonance centered around 14 meV at the AF wavevector.[56] As shown in Fig. 3.18d, the temperature dependence of the corresponding neutron scattering intensity agrees remarkably well with the ARPES result for the maximum value of Re $\Sigma(\omega)$, pointing towards a same origin. Indeed, the opening of a superconducting gap Δ in the strong coupling regime is naturally accompanied by the opening of an energy gap in the electronic spin excitations, which can give rise to a resonance mode due to a divergence in the magnetic spin susceptibility (resonance mode).[33,57,58] In the presence of such a feedback process, induced kinks in the dispersion are expected around 2Δ, in agreement with the result for the α bands.

We now discuss the orbital dependence of the dispersion anomaly. Figure 3.19a displays the ARPES intensity plot for the *k*

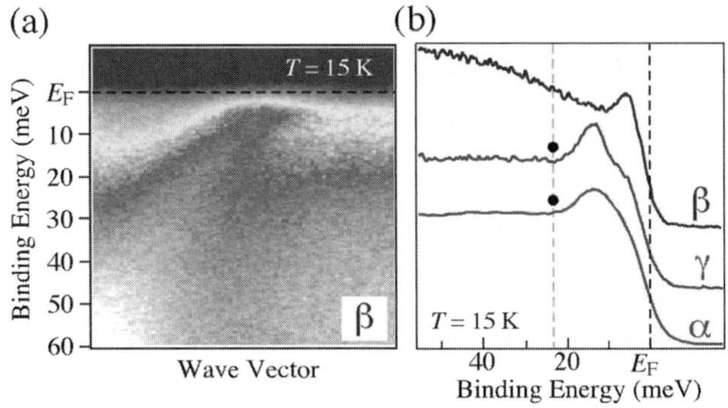

Figure 3.19 ARPES intensity plots of $Ba_{0.6}K_{0.4}Fe_2As_2$ in the superconducting state (15 K) along a cut crossing the β band. Comparison of EDCs in the superconducting state around k_F points of the α, β, and γ bands.

cut crossing the β Fermi surface. Surprisingly, the β band exhibits a kink neither at 25 meV nor at $\Delta_\beta+\Omega = 19$ meV, where $\Delta_\beta = 6$ meV. The partial DOS around the k_F points of the β band supports this observation (see Fig. 3.19b), in contrast to the α and γ bands, which both show a dip around 25 meV in the partial DOS. This behavior is not simply explained by a conventional isotropic electron-phonon coupling and suggests the band selective nature of the mode. The mode is preferentially coupled to the electrons in the α and γ bands, which are well connected via the AF momentum transfer $Q = (\pi, 0)$.

It is interesting to compare the kink in pnictides with the kink observed in cuprates. While phonons have been proposed to explain this feature,[51] some experiments and calculations contradict this hypothesis[48,52–54] and the nature of the kink in cuprates remains an open question. Although a one-to-one correspondence between the Fe-based and Cu-based superconductors is not expected, the kinks observed in the two families of compounds behave similarly: both are observed at an energy close to 2Δ at locations that can be linked by an AF wavevector, in addition to having a strong temperature dependence and disappearing above T_c. It is therefore tempting to assume that they also share the same origin. More significantly, the emergence of an AF resonance mode has been intimately associated with the opposite pairing phases between the connected Fermi surface sections in both cuprates[57] and Fe-based superconductors.[33,58]

Thus, the observation of a band-selective mode coupling supports the idea of antiphase s-wave pairing in iron pnictides.

3.2.3 *Parent Compound*

It is possible that high-T_c superconductivity in pnictides emerges by tuning interactions already present in the parent compounds. While it is admitted that the parent compounds $BaFe_2As_2$, $SrFe_2As_2$, and $CaFe_2As_2$ exhibit a magnetic ordering below a transition temperature T_{SDW} that is accompanied by an orthorhombic distortion,[59–61] the nature of the force driving this transition remains controversial. Earlier ARPES studies to elucidate the nature of the low-temperature electronic states of the pnictide parent compounds support competing scenarios, including spin-density-wave (SDW)[62] and exchange splitting[63] models. Even more challenging is to answer whether the signatures of tiny Fermi surfaces inferred from quantum oscillation experiments,[64,65] which cannot be experimentally located in the k space, are analogous to the Fermi-surface reconstructions claimed in cuprates[66] or consequences of more fundamental topological anomalies in the electronic structure, such as the Dirac cone in graphene.[67] It is thus imperative to characterize directly and more precisely the electronic states that trigger the low-temperature properties of the parent compounds of pnictides.

Figure 3.20a shows a Fermi-surface mapping[68] of $BaFe_2As_2$ ($T_{SDW} = 138$ K) obtained at 25 K. Two regions of similar size with high intensity are centered around the Γ point and the M point, respectively. Interestingly, there is a very bright spot observed around $(0.75\pi, 0)$ on the Γ-M symmetry line. Here we define this point as the Λ point. The intensity plot of the band dispersion along the Γ-M cut is given in Fig. 3.20b. Even after renormalization by a factor of 3 for a better correspondence with the gross experimental features, the LDA calculations in the paramagnetic state cannot be successfully folded across the 2D low-temperature Brillouin-zone boundary to account for a perfect description of all bands and especially for the small intensity spot at the Λ point. In addition to the α and β_1 bands expected from a comparison with optimally-doped samples, an extra holelike dispersion crossing E_F (β_2) is observed around the Γ point. This band is also folded across the

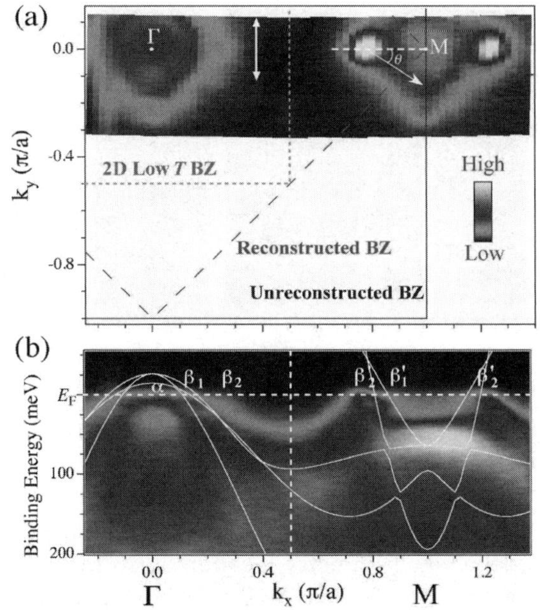

Figure 3.20 (a) ARPES intensity plot of $BaFe_2As_2$ at $T = 25$ K as a function of 2D wave vector. (b) Band dispersion at $T = 25$ K along the ΓM cut, together with the LDA calculated bands (white curves) renormalized by a factor of 3. See also Color Insert.

2D low-temperature Brillouin-zone boundary. It crosses E_F at $\Lambda = (0.75\pi, 0)$, which corresponds to the bright spot. Interestingly, an electronlike band crosses E_F at the same point.

Top panels of Fig. 3.21 show the ARPES intensity at $T = 25$ K measured across various cuts going through the Λ point and indexed according to the θ angle defined in Fig. 3.20a. The corresponding second energy derivative intensity plots are displayed in the bottom panels. They exhibit a complicated hybridization pattern. For instance, the hybridization between the electron-like band and the β_1' at $\theta = 0$ manifests itself by a gap in the energy derivative intensity plot in the ~20–30 meV binding energy range. Surprisingly, the portion of the band structure around the Λ point, from E_F to a binding energy of about 20 meV, seems not to change much with θ: an electronlike band and a holelike band intersect around E_F and their angular symmetry defines a cone structure. All these results

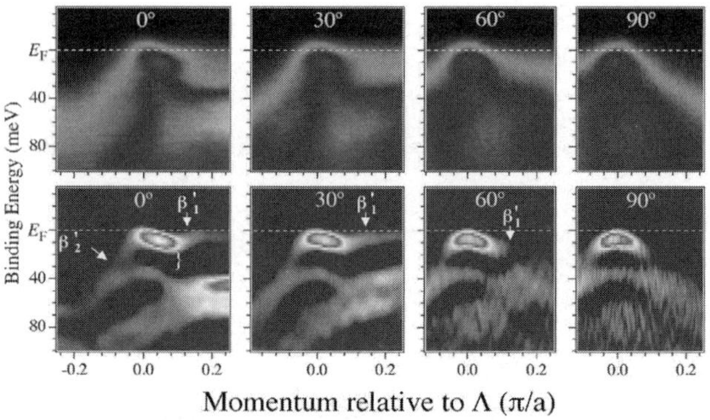

Figure 3.21 (Top) ARPES intensity plots of $BaFe_2As_2$ at $T = 25$ K across the Λ point along cuts with different Fermi-surface angles. (Bottom) Same as the top panels but plotted with the second derivatives of the EDCs.

indicate the presence of a Dirac cone in the band dispersion below T_{SDW}. Unlike the high symmetry location of isotropic Dirac cone in graphene, the Λ point in $BaFe_2As_2$ is not a high-symmetry point.

Figure 3.22a shows the polar plot of extracted Fermi velocity (v_F) as a function of θ by assuming a linear dispersion in the vicinity of E_F. The size of the plain circle represents the averaged v_F values of 330±60 meV·Å. Although it is consistent with the data within uncertainties, a better agreement is achieved if a small anisotropy is assumed. The analysis reveals that the apex of the cone is located 1±5 meV above E_F, which corresponds to E_F within uncertainties. In Figure 3.22b, a 3D representation of the anisotropic Dirac cone is schematically shown.

Within the framework of SDW mean-field theory of both two-band and five-band tight-binding models, Ran *et al.* demonstrated that nodes in the SDW gap function of undoped FeAs-based superconductors must exist due to the occurrence of Dirac cones forming as a consequence of the symmetry-enforced degeneracy at the Γ and M high-symmetry points, even in the presence of perfect nesting.[69] These nodes are located in the same region as the ARPES experiment, strongly suggesting that the Dirac cone is created by the nodal SDW. It is also noted that a recent theory based on the first-principles calculations nicely reproduces the experimentally-determined Dirac-cone feature as well as the band dispersion in the AF phase.[70]

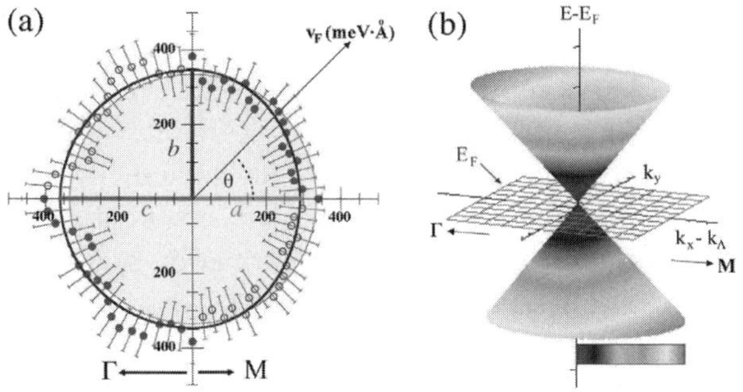

Figure 3.22 (a) Fermi-surface-angle dependence of Fermi velocities v_F's. (b) 2D band dispersion of the Dirac cone.

It is noted here that k_z dispersion[71–73] should be also taken into account to reach complete understanding of the Dirac cone and the quantum oscillation behavior. Although the α band has a strong k_z dependence, Liu *et al.* reported only little variations for the β_1 and β_2 bands.[71] The latter, which coincides with the Λ point, is found around both the Γ and M points, confirming the band folding. Due to the sensitivity of the Dirac cone apex position to the precise electronlike and holelike dispersions generating the cone, it is likely that even a very small wiggling in the k_z dispersion can lead to a shift of the cone apex and produce the series of very small Fermi surfaces suggested from quantum oscillation experiments.[64,65] Even though the failure of quantum oscillation experiments to detect the large holelike β_1 Fermi surface pocket observed by ARPES and predicted by calculations remains puzzling, the observation of tiny Fermi surfaces in the parent compound of an iron-pnictide superconductor derived from a fundamental topological anomaly, i.e. a Dirac cone dispersion, is of crucial importance to understand the unconventional electronic properties of these materials and allows us to describe them as nodal SDW materials.

3.3 Concluding Remarks and Summary

In this chapter, we reviewed representative ARPES results of iron pnictides by focusing on the $BaFe_2As_2$ system. The results clearly indicate the multi-orbital character of the Fermi surface, and the universality of Fermi-surface-dependent nodeless superconducting gaps in electron- and hole-doped systems. The disappearance/suppression of superconductivity in heavily overdoped region has been interpreted in terms of the degradation of the interband scattering condition. The nature of the many-body interactions and the electronic states of the parent compound were also discussed in relation to the AF/SDW formation and fluctuations. The ARPES results presented here strongly suggests the importance of AF fluctuations to the mechanism of superconductivity in iron pnictides.

Acknowledgments

The ARPES study presented in this chapter has been performed in collaboration with Y. Sekiba, T. Kawahara, K. Umezawa, S. Souma, K. Sugawara, T. Arakane, A. Takayama, K. Terashima, T. Qian, Y.-M. Xu, J. H. Bowen, M. Neupane, Z.-H. Pan, S.-C. Wang, Z. Wang, X. Dai, Z. Fang, G. F. Chen, J. L. Luo, N. L. Wang, L. J. Li, G. H. Cao, Z.-A. Xu, and H.-H. Wen.

References

1. Kamihara, Y., Watanabe, T., Hirano, M., and Hosono, H. (2008). Iron-based layered superconductor La[$O_{1-x}F_x$]FeAs (x = 0.05–0.12) with T_c = 26 K, *J. Am. Chem. Soc.* **130**, pp. 3296–3297.
2. Takahashi, H., Igawa, K., Arii, K., Kamihara, Y., Hirano, M., and Hosono, H. (2008). Superconductivity at 43 K in an iron-based layered compound $LaO_{1-x}F_xFeAs$, *Nature (London)* **453**, pp. 376–378.
3. Sato, T., Souma, S., Nakayama, K., Terashima, K., Sugawara, K., Takahashi, T., Kamihara, Y., Hirano, M., and Hosono, H. (2008). Superconducting gap and pseudogap in iron-based layered superconductor La($O_{1-x}F_x$)FeAs, *J. Phys. Soc. Jpn.* **77**, 063708-1–3.
4. Ding, H., Richard, P., Nakayama, K., Sugawara, K., Arakane, T., Sekiba, Y., Takayama, A., Souma, S., Sato, T., Takahashi, T., Wang, Z., Dai, X., Fang, Z., Chen, G. F., Luo, J. L., and Wang, N. L. (2008). Observation of Fermi-surface-dependent nodeless superconducting gaps in $Ba_{0.6}K_{0.4}Fe_2As_2$, *Europhys. Lett.* **83**, 47001-1–4.

5. Cruz, C. de la, Huang, Q., Lynn, J. W., Li, J., Ratcliff II, W., Zarestky, J. L., Mook, H. A., Chen, G. F., Luo, J. L., Wang, N. L., Dai, P. (2008). Magnetic order close to superconductivity in the iron-based layered $LaO_{1-x}F_xFeAs$ systems, *Nature (London)* **453**, pp. 899–902.
6. Rotter, M., Tegel, M., and Johrendt, D. (2008). Superconductivity at 38 K in the iron arsenide $(Ba_{1-x}K_x)Fe_2As_2$, *Phys. Rev. Lett.* **101**, 107006-1–4.
7. Aebi, P., Osterwalder, J., Schwaller, P., Schlapbach, L., Shimoda, M., Mochiku, T., and Kadowaki, K. (1994). Complete Fermi surface mapping of $Bi_2Sr_2CaCu_2O_{8+x}$ (001): Coexistence of short range antiferromagnetic correlations and metallicity in the same phase, *Phys. Rev. Lett.* **72**, pp. 2757–2760.
8. Ding, H., Bellman, A. F., Campuzano, J. C., Randeria, M., Norman, M. R., Yokoya, T., Takahashi, T., Katayama-Yoshida, H., Mochiku, T., Kadowaki, K., Jennings, G., and Brivio, G. P. (1996). Electronic excitations in $Bi_2Sr_2CaCu_2O_8$: Fermi surface, dispersion, and absence of bilayer splitting, *Phys. Rev. Lett.* **76**, pp. 1533–1536.
9. Ding, H., Yokoya, T., Campuzano, J. C., Takahashi, T., Randeria, M., Norman, M. R., Mochiku, T., Kadowaki, K., and Giapintzakis, J. (1996). Spectroscopic evidence for a pseudogap in the normal state of underdoped high-T_c superconductors, *Nature (London)* **382**, pp. 51–54.
10. Norman, M. R., Ding, H., Randeria, M., Campuzano, J. C., Yokoya, T., Takeuchi, T., Takahashi, T., Mochiku, T., Kadowaki, K., Guptasarma, P., and Hinks, D. G. (1998). Destruction of the Fermi surface in underdoped high-T_c superconductors, *Nature (London)* **392**, pp. 157–160.
11. Loeser, A. G., Shen, Z.-X., Dessau, D. S., Marshall, D. S., Park, C. H., Fournier, P., and Kapitulnik, A. (1996). Excitation gap in the normal state of underdoped $Bi_2Sr_2CaCu_2O_{8+\delta}$, *Science* **273**, pp. 325–329.
12. Ding, H., Norman, M. R., Campuzano, J. C., Randeria, M., Bellman, A. F., Yokoya, T., Takahashi, T., Mochiku, T., and Kadowaki, K. (1996). Angle-resolved photoemission spectroscopy study of the superconducting gap anisotropy in $Bi_2Sr_2CaCu_2O_{8+x}$, *Phys. Rev. B* **54**, pp. R9678–R9681.
13. Shen, Z.-X., Dessau, D. S., Wells, B. O., King, D. M., Spicer, W. E., Arko, A. J., Marshall, D., Lombardo, L. W., Kapitulnik, A., Dickinson, P., Doniach, S., DiCarlo, J., Loeser, A. G., and Park, C. H. (1993). Anomalously large gap anisotropy in the *a-b* plane of $Bi_2Sr_2CaCu_2O_{8+\delta}$, *Phys. Rev. Lett.* **70**, pp. 1553–1556.
14. Souma, S., Machida, Y., Sato, T., Takahashi, T., Matsui, H., Wang, S.-C., Ding, H., Kaminski, A., Campuzano, J. C., Sasaki, S., and Kadowaki, K. (2003). The origin of multiple superconducting gaps in MgB_2, *Nature (London)* **423**, pp. 65–67.

15. Tsuda, S., Yokoya, T., Takano, Y., Kito, H., Matsushita, A., Yin, F., Itoh, J., Harima, H., and Shin, S. (2003). Definitive experimental evidence for two-band superconductivity in MgB_2, *Phys. Rev. Lett.* **91**, 127001-1–4.
16. Sugawara, K., Sato, T., and Takahashi, T. (2008). Fermi-surface-dependent superconducting gap in C_6Ca, *Nat. Phys.* **5**, pp. 40–43.
17. Souma, S., Sato, T., Takahashi, T., and Baltzer, P. (2007). High-intensity xenon plasma discharge lamp for bulk-sensitive high-resolution photoemission spectroscopy, *Rev. Sci. Instrum.* **78**, 123104-1–4.
18. Koralek, J. D., Douglas, J. F., Plumb, N. C., Griffith, J. D., Cundiff, S. T., Kapteyn, H. C., Murnane, M. M., and Dessau, D. S. (2007). Experimental setup for low-energy laser-based angle resolved photoemission spectroscopy, *Rev. Sci. Instrum.* **78**, 053905-1–5.
19. Liu G., Wang G., Zhu Y., Zhang H., Zhang G., Wang X., Zhou Y., Zhang W., Liu H., Zhao L., Meng J., Dong X., Chen C., Xu Z., and Zhou X. J. (2008). Development of a vacuum ultraviolet laser-based angle-resolved photoemission system with a superhigh energy resolution better than 1 meV, *Rev. Sci. Instrum.* **79**, 023105.
20. Kiss, T., Shimojima, T., Ishizaka, K., Chainani, A., Togashi, T., Kanai, T., Wang, X.-Y., Chen, C.-T., Watanabe, S., and Shin, S. (2008). A versatile system for ultrahigh resolution, low temperature, and polarization dependent laser-angle-resolved photoemission spectroscopy, *Rev. Sci. Instrum.* **79**, 023106-1–7.
21. Pratt, D. K., Tian, W., Kreyssig, A., Zarestky, J. L., Nandi, S., Ni, N., Budfko, S. L., Canfield, P. C., Goldman, A. I., and McQueeney, R. J. (2009). Coexistence of competing antiferromagnetic and superconducting phases in the underdoped $Ba(Fe_{0.953}Co_{0.047})_2As_2$ compound using x-ray and neutron scattering techniques, *Phys. Rev. Lett.* **103**, 087001-1–4.
22. Ding, H., Nakayama, K., Richard, P., Souma, S., Sato, T., Takahashi, T., Neupane, M., Xu, Y.-M., Pan, Z.-H., Fedorov, A. V., Wang, Z., Dai, X., Fang, Z., Chen, G. F., Luo, J. L., and Wang, N. L. (2011). Electronic structure of optimally doped pnictide $Ba_{0.6}K_{0.4}Fe_2As_2$: a comprehensive angle-resolved photoemission spectroscopy investigation, *J. Phys.: Condens. Matter* **23**, 135701-1–6.
23. Xu, G., Zhang, H., Dai, X., and Fang, Z. (2008). Electron-hole asymmetry and quantum critical point in hole-doped $BaFe_2As_2$, *Europhys. Lett.* **84**, 67015-1–5.
24. Norman, M. R., Randeria, M., Ding, H., and Campuzano, J. C. (1998). Phenomenology of the low-energy spectral function in high-T_c superconductors, *Phys. Rev. B* **57**, R11093–R11096.

25. Zhao, L., Liu, H., Zhang, W., Meng, J., Jia, X., Liu, G., Dong, X., Chen, G. F., Luo, J. L., Wang, N. L., Wang, G., Zhou, Y., Zhu, Y., Wang, X., Zhao, Z., Xu, Z., Chen, C., Zhou, X. J. (2008). Multiple nodeless superconducting gaps in $(Ba_{0.6}K_{0.4})Fe_2As_2$ superconductor from angle-resolved photoemission spectroscopy, *Chin. Phys. Lett.* **25**, pp. 4402–4405.

26. D. V. Evtushinsky, D. S. Inosov, V. B. Zabolotnyy, A. Koitzsch, M. Knupfer, B. Buchner, M. S. Viazovska, G. L. Sun, V. Hinkov, A. V. Boris, C. T. Lin, B. Keimer, A. Varykhalov, A. A. Kordyuk, S. V. Borisenko (2009). Momentum dependence of the superconducting gap in $Ba_{1-x}K_xFe_2As_2$, *Phys. Rev. B* **79**, 054517-1–12.

27. Wray, L., Qian, D., Hsieh, D., Xia, Y., Li, L., Checkelsky, J. G., Pasupathy, A., Gomes, K. K., Parker, C. V., Fedorov, A. V., Chen, G. F., Luo, J. L., Yazdani, A., Ong, N. P., Wang, N. L., and Hasan, M. Z. (2008). Momentum-dependence of superconducting gap, strong-coupling dispersion kink, and tightly bound Cooper pairs in the high-T_c $(Sr,Ba)_{1-x}(K,Na)_xFe_2As_2$ superconductors, *Phys. Rev. B* **78**, 184508-1–5.

28. Nakayama, K., Sato, T., Richard, P., Xu, Y.-M., Sekiba, Y., Souma, S., Chen, G. F., Luo, J. L., Wang, N. L., Ding, H., and Takahashi, T. (2009). Superconducting gap symmetry of $Ba_{0.6}K_{0.4}Fe_2As_2$ studied by angle-resolved photoemission spectroscopy, *Europhys. Lett.* **85**, 67002-1–5.

29. Mazin, I. I., Singh, D. J., Johannes, M. D., and Du, M. H. (2008). Unconventional superconductivity with a sign reversal in the order parameter of $LaFeAsO_{1-x}F_x$, *Phys. Rev. Lett.* **101**, 057003-1–4.

30. Kuroki, K., Onari, S., Arita, R., Usui, H., Tanaka, Y., Kontani, H., and Aoki, H. (2008). Unconventional pairing originating from the disconnected fermi surfaces of superconducting $LaFeAsO_{1-x}F_x$, *Phys. Rev. Lett.* **101**, 087004-1–4.

31. Wang, F., Zhai, H., Ran, Y., Vishwanath, A., and Lee, D.-H. (2009). Functional renormalization-group study of the pairing symmetry and pairing mechanism of the FeAs-based high-temperature superconductor, *Phys. Rev. Lett.* **102**, 047005-1–4.

32. Seo, K., Bernevig, B. A., and Hu, J. P. (2008). Pairing symmetry in a two-orbital exchange coupling model of oxypnictides, *Phys. Rev. Lett.* **101**, 206404-1–4.

33. Korshunov, M. M., and Eremin, I. (2008). Theory of magnetic excitations in iron-based layered superconductors, *Phys. Rev. B* **78**, 140509(R)-1–4.

34. Xu, Y.-M., Richard, P., Nakayama, K., Kawahara, T., Sekiba, Y., Qian, T., Neupane, M., Souma, S., Sato, T., Takahashi, T., Luo, H., Wen, H.-H.,

Chen, G.-F., Wang, N.-L., Wang, Z., Fang, Z., Dai, X., and Ding, H. (2011). Fermi surface dichotomy of superconducting gap and pseudogap in underdoped pnictides, *Nature Commun.* **2**, 392-1–5.

35. Nakayama, K., Sato, T., Richard, P., Xu, Y.-M., Kawahara, T., Umezawa, K., Qian, T., Neupane, M., Chen, G. F., Ding, H., Takahashi, T. (2011). Universality of superconducting gaps in overdoped $Ba_{0.3}K_{0.7}Fe_2As_2$ observed by angle-resolved photoemission spectroscopy, *Phys. Rev. B* **83**, 020501(R)-1–4.

36. Sato, T., Nakayama, K., Sekiba, Y., Richard, P., Xu, Y.-M., Souma, S., Takahashi, T., Chen, G. F., Luo, J. L., Wang, N. L., and Ding, H. (2009). Band structure and Fermi surface of an extremely overdoped iron-based superconductor KFe_2As_2, *Phys. Rev. Lett.* **103**, 047002-1–4.

37. D. J. Singh (2008). Electronic structure and doping in $BaFe_2As_2$ and LiFeAs: Density functional calculations, *Phys. Rev. B* **78**, 094511-1–7.

38. Dai, J., Li, Z., Yang, J., and Hou, J. G. (2008). A first principles study on the electronic and magnetic properties of $Ba_{1-x}K_xFe_2As_2$, *ArXiv:0808.0065v1*.

39. Ma, F., Lu, Z.-Y., and Xiang, T. (2010). Electronic structures of ternary iron arsenides AFe_2As_2 (A=Ba, Ca, or Sr), *Front. Phys. China* **5(2)**, 150.

40. Nekrasov, I. A., Pchelkina, Z. V., and Sadovskiil, M. V. (2008). Electronic structure of prototype AFe_2As_2 and ReOFeAs high-temperature superconductors: a comparison, *JETP Letters* **88**, pp. 144–149.

41. Terashima, K., Sekiba, Y., Bowen, J. H., Nakayama, K., Kawahara, T., Sato, T., Richard, P., Xu, Y.-M., Li, L. J., Cao, G. H., Xu, Z.-A., Ding, H., and Takahashi, T. (2009). Fermi surface nesting induced strong pairing in iron-based superconductors, *Proc. Natl. Acad. Sci.* **106**, pp. 7330–7333.

42. Neupane, M., Richard, P., Xu, Y.-M., Nakayama, K., Sato, T., Takahashi, T., Fedrov, A. V., Xu, G., Dai, X., Fang, Z., Wang, Z., Chen, G.-F., Wang, N.-L., Wen, H.-H., and Ding, H. (2011). Electron-hole asymmetry in superconductivity of pnictides originated from the observed rigid chemical potential shift, *Phys. Rev. B* **83**, 094522-1–6.

43. Guo, J., Jin, S., Wang, G., Wang, S., Zhu, K., Zhou, T., He, M., and Chen, X. (2010). Superconductivity in the iron selenide $K_xFe_2Se_2$ ($0 \leq x \leq 1.0$) *Phys. Rev. B* **82**, 180520(R)-1–4.

44. Fang, M.-H., Wang, H.-D., Dong, C.-H., Li, Z.-J., Feng, M.-C., Chen, J., and Yuan, H. Q. (2011). Fe-based superconductivity with $T_c = 31$ K bordering an antiferromagnetic insulator in $(TI,K)Fe_xSe_2$, *Europhys. Lett.* **94**, 27009-1–4.

45. Qian, T., Wang, X.-P., Jin, W.-C., Zhang, P., Richard, P., Xu, G., Dai, X., Fang, Z., Guo, J.-G., Chen, X.-L., and Ding, H. (2011). Absence of a holelike Fermi surface for the iron-based $K_{0.8}Fe_{1.7}Se_2$ superconductor revealed by angle-resolved photoemission spectroscopy, *Phys. Rev. Lett.* **106**, 187001-1–4.
46. Wang, X.-P., Qian, T., Richard, P., Zhang, P., Dong, J., Wang, H.-D., Dong, C.-H., Fang, M.-H., Ding, H. (2011). Strong nodeless pairing on separate electron Fermi surface sheets in (Tl, K)$Fe_{1.78}Se_2$ probed by ARPES, *Europhys. Lett.* **93**, 57001-1–4.
47. Sekiba, Y., Sato, T., Nakayama, K., Terashima, K., Richard, P., Bowen, J. H., Ding, H., Xu, Y.-M., Li, L. J., Cao, G. H., Xu, Z.-A., and Takahashi, T. (2009). Electronic structure of heavily electron-doped $BaFe_{1.7}Co_{0.3}As_2$ studied by angle-resolved photoemission, *New J. Phys.* **11**, 025020-1–8.
48. Norman, M. R., Ding, H., Campuzano, J. C., Takeuchi, T., Randeria, M., Yokoya, T., Takahashi, T., Mochiku, T., and Kadowaki, K. (1997). Unusual dispersion and line shape of the superconducting state spectra of $Bi_2Sr_2CaCu_2O_{8+\delta}$, *Phys. Rev. Lett.* **79**, pp. 3506–3509.
49. Valla, T., Fedorov, A. V., Johnson, P. D., Wells, B. O., Hulbert, S. L., Li, Q., Gu, G. D., and Koshizuka, N. (1999). Evidence for quantum critical behavior in the optimally doped cuprate $Bi_2Sr_2CaCu_2O_{8+\delta}$, *Science* **285**, pp. 2110–2113.
50. Kaminski, A., Randeria, M., Campuzano, J. C., Norman, M. R., Fretwell, H., Mesot, J., Sato, T., Takahashi, T., and Kadowaki, K. (2001). Renormalization of spectral line shape and dispersion below T_c in $Bi_2Sr_2CaCu_2O_{8+\delta}$, *Phys. Rev. Lett.* **86**, 1070–1073.
51. Lanzara, A., Bogdanov, P. V., Zhou, X. J., Kellar, S. A., Feng, D. L., Lu, E. D., Yoshida, T., Eisaki, H., Fujimori, A., Kishio, K., Shimoyama, J.-I., Noda, T., Uchida, S., Hussain, Z., and Shen, Z.-X. (2001). Evidence for ubiquitous strong electron-phonon coupling in high-temperature superconductors, *Nature (London)* **412**, pp. 510–514.
52. Johnson, P. D., Valla, T., Fedorov, A. V., Yusof, Z., Wells, B. O., Li, Q., Moodenbaugh, A. R., Gu, G. D., Koshizuka, N., Kendziora, C., Jian, S., and Hinks, D. G. (2001). Doping and temperature dependence of the mass enhancement observed in the cuprate $Bi_2Sr_2CaCu_2O_{8+\delta}$, *Phys. Rev. Lett.* **87**, 177007-1–4.
53. Terashima, K., Matsui, H., Hashimoto, D., Sato, T., Takahashi, T., Ding, H., Yamamoto, T., and Kadowaki, K. (2006). Impurity effects on electron-mode coupling in high-temperature superconductors, *Nat. Phys.* **2**, pp. 27–31.

54. Giustino, F., Cohen, M. L., Louie, S. G. (2008). Small phonon contribution to the photoemission kink in the copper oxide superconductors, *Nature (London)* **452**, pp. 975–978.
55. Richard, P., Sato, T., Nakayama, K., Souma, S., Takahashi, T., Xu, Y.-M., Chen, G. F., Luo, J. L., Wang, N. L., and Ding, H. (2009). Angle-resolved photoemission spectroscopy of the Fe-based $Ba_{0.6}K_{0.4}Fe_2As_2$ high temperature superconductor: evidence for an orbital selective electron-mode coupling, *Phys. Rev. Lett.* **102**, 047003-1–4.
56. Christianson, A. D., Goremychkin, E. A., Osborn, R., Rosenkranz, S., Lumsden, M. D., Malliakas, C. D., Todorov, l. S., Claus, H., Chung, D. Y., Kanatzidis, M. G., Bewley, R. I., Guidi, T. (2008). Unconventional superconductivity in $Ba_{0.6}K_{0.4}Fe_2As_2$ from inelastic neutron scattering, *Nature (London)* **456**, pp. 930–932.
57. Chubukov, A. V., Pines, D., and Schmalin, J. A. (2008). in *Superconductivity*, edited by Bennemann K. H., and Ketterson J. B. (Springer-Verlag, Berlin), p. 1349.
58. Maier, T. A., and Scalapino D. J. (2008). Theory of neutron scattering as a probe of the superconducting gap in the iron pnictides, *Phys. Rev. B* **78**, 020514(R)-1–4.
59. Huang, Q., Qiu, Y., Bao, W., Green, M. A., Lynn, J. W., Gasparovic, Y. C., Wu, T., Wu, G., and Chen, X. H. (2008). Neutron-diffraction measurements of magnetic order and a structural transition in the parent $BaFe_2As_2$ compound of FeAs-based high-temperature superconductors, *Phys. Rev. Lett.* **101**, 257003-1–4.
60. Zhao, J., Yao, D.-X., Li, S., Hong, T., Chen, Y., Chang, S., Ratcliff II, W., Lynn, J. W., Mook, H. A., Chen, G. F., Luo, J. L., Wang, N. L., Carlson, E. W., Hu, J., and Dai, P. (2008). Low energy spin waves and magnetic interactions in $SrFe_2As_2$, *Phys. Rev. Lett.* **101**, 167203-1–4.
61. Wilson, S. D., Yamani, Z., Rotundu, C. R., Freelon, B., Bourret-Courchesne, E., and Birgeneau, R. J. (2010). Neutron diffraction study of the magnetic and structural phase transitions in $BaFe_2As_2$, *Phys. Rev. B* **79**, 184519-1–10.
62. Hsieh, D., Xia, Y., Wray, L., Qian, D., Gomes, K., Yazdani, A., Chen, G. F., Luo, J. L., Wang, N. L., Hasan, M. Z. (2008). Experimental determination of the microscopic origin of magnetism in parent iron pnictides, *ArXiv:0812.2289v1*.
63. Yang, L. X., Zhang, Y., Ou, H. W., Zhao, J. F., Shen, D. W., Zhou, B., Wei, J., Chen, F., Xu, M., He, C., Chen, Y., Wang, Z. D., Wang, X. F., Wu, T., Wu, G., Chen, X. H., Arita, M., Shimada, K.,Taniguchi, M., Lu, Z. Y., Xiang, T., and

Feng, D. L. (2009). Electronic structure and unusual exchange splitting in the spin-density-wave state of the $BaFe_2As_2$ parent compound of iron-based superconductors, *Phys. Rev. Lett.* **102**, 107002-1–4.

64. Sebastian, S. E., Gillett, J., Harrison, N., Lau, P. H. C., Singh, D. J., Mielke, C. H., and Lonzarich, G. G. (2008). Quantum oscillations in the parent magnetic phase of an iron arsenide high temperature superconductor, *J. Phys.: Condens. Matter* **20**, 422203-1–5.

65. Analytis, J. G., McDonald, R. D., Chu, J.-H., Riggs, S. C., Bangura, A. F., Kucharczyk, C., Johannes, M., and Fisher, I. R. (2009). Quantum oscillations in the parent pnictide $BaFe_2As_2$: Itinerant electrons in the reconstructed state, *Phys. Rev. B* **80**, 064507-1–5.

66. Doiron-Leyraud, N., Proust, C., LeBoeuf, D., Levallois, J., Bonnemaison, J.-B., Liang, R., Bonn, D. A., Hardy, W. N., and Taillefer, L. (2007). Quantum oscillations and the Fermi surface in an underdoped high-T_c superconductor, *Nature (London)* **447**, pp. 565–568.

67. Novoselov, K. S., Geim, A. K., Morozov, S. V., Jiang, D., Katsnelson, M. I., Grigorieva, I. V., Dubonos, S. V., and Firsov, A. A. (2005). Two-dimensional gas of massless Dirac fermions in graphene, *Nature (London)* **438**, pp. 197–200.

68. Richard, P., Nakayama, K., Sato, T., Neupane, M., Xu, Y.-M., Bowen, J. H., Chen, G. F., Luo, J. L., Wang, N. L., Dai, X., Fang, Z., Ding, H., and Takahashi, T. (2010). Observation of Dirac cone electronic dispersion in $BaFe_2As_2$, *Phys. Rev. Lett.* **104**, 137001-1–4.

69. Ran, Y., Wang, F., Zhai, H., Vishwanath, A., and Lee, D.-H. (2009). Nodal spin density wave and band topology of the FeAs-based materials, *Phys. Rev. B* **79**, 014505-1–9.

70. Yin, Z. P., Haule K., and Kotliar, G. (2011). Magnetism and charge dynamics in iron pnictides, *Nature Phys.* **7**, pp. 294–297.

71. Liu, C., Kondo, T., Ni, N., Palczewski, A. D., Bostwick, A., Samolyuk, G. D., Khasanov, R., Shi, M., Rotenberg, E., Budfko, S. L., Canfield, P. C., and Kaminski, A. (2009). Three- to two-dimensional transition of the electronic structure in $CaFe_2As_2$: a parent compound for an iron arsenic high-temperature superconductor, *Phys. Rev. Lett.* **102**, 167004-1–4.

72. Vilmercati, P., Fedorov, A., Vobornik, I., Manju, U., Panaccione, G., Goldoni, A., Sefat, A. S., McGuire, M. A., Sales, B. C., Jin, R., Mandrus, D., Singh, D. J., and Mannella, N. (2009). Evidence for three-dimensional Fermi-surface topology of the layered electron-doped iron superconductor $Ba(Fe_{1-x}Co_x)_2As_2$, *Phys. Rev. B* **79**, 220503(R)-1–4.

73. Malaeb, W., Yoshida, T., Fujimori, A., Kubota, M., Ono, K., Kihou, K., Shirage, P. M., Kito, H., Iyo, A., Eisaki, H., Nakajima, Y., Tamegai, T., and Arita, R. (2009). Three-dimensional electronic structure of superconducting iron pnictides observed by angle-resolved photoemission spectroscopy, *J. Phys. Soc. Jpn.* **78**, pp. 123706-1–4.

Chapter 4

Quantum Oscillations in Iron Pnictide Superconductors

Suchitra E. Sebastian
Cavendish Laboratory, Cambridge University,
JJ Thomson Avenue, Cambridge CB3 OHE, UK
suchitra@phy.cam.ac.uk

Insight into the electronic structure of the pnictide family of superconductors is obtained from quantum oscillation measurements. Here we review experimental quantum oscillation data that reveal a transformation from large quasi-two-dimensional electron and hole cylinders in the paramagnetic overdoped members of the pnictide family to significantly smaller three-dimensional Fermi surface sections in the antiferromagnetic parent members, via a potential quantum critical point at which an effective mass enhancement is observed. Similarities with the Fermi surface evolution from the overdoped to the underdoped normal state of the cuprate superconducting family are discussed, along with the enhancement in antiferromagnetic correlations in both these classes of materials, and the potential implications for superconductivity.

Iron-Based Superconductors: Materials, Properties, and Mechanisms
Edited by Nan-Lin Wang, Hideo Hosono, and Pengcheng Dai

ISBN 978-981-4303-22-4 (Hardcover), 978-981-4303-23-1 (eBook)
www.panstanford.com

4.1 Quantum Oscillations

Quantum oscillation measurements are a vital tool to characterise the electronic structure of metals. Originally conceived as a tool to probe elementary metals, quantum oscillations have grown in scope and have now been observed in complex intermetallics and oxides, including heavy fermions,[1,2] other unconventional superconductors,[3,4] and even doped Mott insulating cuprate superconductors.[5–13] The particular strength of quantum oscillations is their utility as a bulk probe of electronic structure, providing high momentum space resolution of constant energy surfaces at the Fermi energy.

In the presence of an applied magnetic field, a free electron gas whose surface of constant energy is spherical is arranged into a series of concentric cylinders known as 'Landau tubes' (Fig. 4.1). Electron motion is now confined to the surface of these 'Landau tubes', resulting in quantised motion along 'cyclotron orbits'. The quantisation of Landau tube areas arises from the quantum

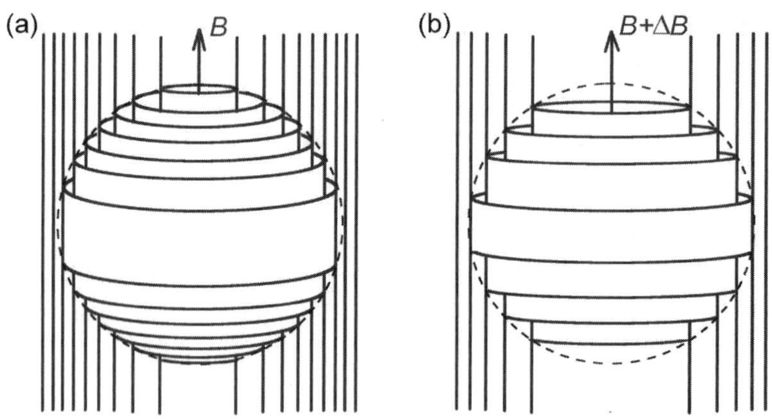

Figure 4.1 (a) Schematic of Landau levels for spherical Fermi surface as in a free electron gas in a magnetic field B. (b) Schematic of Landau levels, when the applied magnetic field is increased to $B + \Delta B$. The Landau levels increase in area with an increase in magnetic field, causing fewer of them to be accommodated within the Fermi surface as the magnetic field is increased. [Modified from ref. 14]

condition

$$\Delta E = \hbar\omega_c \tag{4.1}$$

where ω_c is the angular frequency of an electron executing cyclotron motion, given by

$$\omega_c = \frac{eB}{m^*} = \frac{2\pi eB/\hbar^2}{\partial\Lambda/\partial\epsilon} \tag{4.2}$$

where m^* is the effective cyclotron mass, and Λ is the cross-sectional area in reciprocal space of a constant energy Landau tube intersected by a plane normal to the magnetic field B. From equations 4.1 and 4.2, the condition for Landau tube area quantisation is yielded as

$$\Lambda_r = \left(r + \frac{1}{2}\right)\frac{2\pi eB}{\hbar} \tag{4.3}$$

When the magnetic field is increased, each Landau tube expands in area (shown in Fig. 4.1), causing its height accommodated within the Fermi surface to shrink, until it spans the extremity of the Fermi surface (with extremal area $\Lambda = A$) just before it 'pops' out of the Fermi surface entirely. As each Landau tube pops out of the Fermi surface, there is a sudden discontinuous change in the total energy of the occupied states. Such an event happens periodically as tubes of successively small quantum numbers r pass through the Fermi surface at equal intervals of $1/B$ (from eqn. 4.3). The frequency of Landau tubes popping out is given by

$$F = \frac{\phi_0 A}{2\pi^2} = 2\frac{A}{A_{BZ}}\frac{\phi_0}{ab} \tag{4.4}$$

where $\phi_0 = \frac{h}{e}$ is the magnetic flux quantum, A_{BZ} is the area of the Brillouin zone in reciprocal space, and a and b are the unit cell dimensions in real space assuming a tetragonal lattice.[14] We label the frequency harmonics as pF, where p represents the pth harmonic. All derivatives of the free energy, such as magnetisation, consequently display oscillatory behaviour. These quantum oscillations in magnetisation are referred to as 'de Haas–van Alphen' oscillations, and in electrical transport are referred to as 'Shubnikov de Haas' oscillations.[14] The frequency of oscillations reflects the area of the extremal Fermi surface orbit. A complex Fermi surface geometry with multiple orbits, or a Fermi surface comprising multiple sections would yield an oscillatory signal comprising multiple frequencies, which can be analysed by filtering or Fourier transform methods.

4.1.1 *Angular Dependence*

4.1.1.1 Fermi surface geometry

Quantum oscillations have an important role in characterising Fermi surface geometry, with the potential to reveal significant information about the physics of the systems studied, especially in strongly correlated systems. For instance, in heavy fermion systems, the geometry of the Fermi surface can reveal information about the participation or otherwise of f-electrons in the electronic conduction of the system.[1,2] In underdoped high temperature cuprate superconductors, quantum oscillations measurements of the Fermi surface geometry have proved an important indicator of likely Fermi surface reconstruction by a long range ordering wavevector.[10,11,13]

The area of extremal orbits of the Fermi surface can be studied with great precision from the measured quantum oscillation frequencies. To further identify the momentum space location of the Fermi surface sections and map out the precise Fermi surface geometry, comparison with calculated theoretical band structures is typically required. Measurement of the quantum oscillation frequency as a function of the magnetic field direction specifies the extremal areas of the Fermi surface normal to all directions relative to the crystalline axes. This provides geometrical information about the Fermi surface that can be compared with band structure calculations. For a quasi-two-dimensional Fermi surface, the angular dependence provides information relating to the c-axis dispersion of the Fermi surface.[15]

4.1.1.2 Spin splitting

The quantum oscillation amplitude as a function of angle is also modulated by an effect known as spin splitting: separate sets of Landau levels arise from spin-up and spin-down electrons, leading to a phase difference between oscillations from spin-up and spin-down electrons, and hence a modulation of the oscillation amplitude due to interference. Whereas the Zeeman splitting between spin up and spin down electrons is invariant with angle in the absence of significant spin-orbit effects, the Landau level separation is a function of the projected magnetic field $B\cos\theta$ for quasi-two-dimensional systems, where θ is the angle between the applied magnetic field and the crystalline axis. The quantum oscillation amplitude as a function of angle is consequently modulated by a prefactor

$$R_s = \cos\left(\frac{1}{2}\frac{\pi g^*_\theta m^*_\theta}{m_e}\right) \tag{4.5}$$

where g^*_θ is the effective spin-splitting factor, and m^*_θ is the effective quasiparticle mass corresponding to a magnetic field applied at an angle θ to the crystalline c-axis. In cases where the spins are ordered in such a way that the spin moment is suppressed, the interference term is not expected to appear – potentially proving a diagnostic tool in detecting forms of spin order.[11,13]

4.2 Magnetic Field Dependence

An important source of quantum oscillation amplitude modulation is 'phase smearing', i.e. effects occur that are equivalent to the superposition of oscillations over a finite phase window, resulting in an amplitude modulation of the net oscillations. One such amplitude reduction factor (R_{Dingle}, denoted by R_D) arises from a finite energy spread due to quasiparticle scattering time, and is given by

$$R_D = e^{\frac{-B_c}{B}} \tag{4.6}$$

where $\frac{B_c}{B} = \frac{p\pi}{\omega_c\tau} = \frac{p\pi r_c}{l}$; $\hbar\omega_c$ represents the Landau level spacing, τ is the scattering time, l is the mean free path, and r_c is the cyclotron radius. As a consequence, crystal quality is paramount in observing sufficiently large quantum oscillations. The exponential rate of growth of the experimentally observed quantum oscillation amplitude with magnetic field yields a measure of the purity of the measured single crystal, with a faster rate of growth indicating a higher impurity concentration.

4.3 Temperature Dependence

Another source of phase smearing is finite temperature. For quasiparticles with Fermi–Dirac statistics, the temperature dependence of the quantum oscillation amplitude is governed by the sharp Fermi–Dirac distribution that grows more rounded with increasing temperature. A reduction at finite temperatures in the abruptness with which the particle occupation as a function of energy vanishes causes phase smearing, and a consequent reduction in the amplitude of the measured quantum oscillations. In this case, the magnitude of amplitude reduction as a function of temperature is given by a derivative of the Fermi–Dirac distribution

$$R_{\mathrm{T}} = \frac{\pi\lambda}{\sinh \pi\lambda} \tag{4.7}$$

where $\lambda = \frac{2p\pi m^* k_{\mathrm{B}} T}{e\hbar B}$, k_{B} is the Boltzmann constant. For $\pi\lambda \gtrapprox 1$,

$$R_T \approx 2\pi\lambda e^{-\pi\lambda} = \frac{4\pi^2 p k_{\mathrm{B}} T}{\hbar\omega_{\mathrm{c}}} e^{-\frac{2\pi^2 p k_{\mathrm{B}} T}{\hbar\omega_{\mathrm{c}}}}$$

where λ is expressed in terms of the Landau level spacing $\hbar\omega_{\mathrm{c}}$, analogous to the magnetic field dependent damping. A measure of quantum oscillation amplitude as a function of temperature is used to directly verify Fermi–Dirac statistics,[8] and to measure the effective cyclotron mass of elementary quasiparticles.

The enhancement in effective quasiparticle mass over the band mass is a direct measure of the strength of correlations, of particular importance in strongly correlated materials. A value of effective cyclotron mass m^* is extracted by a fit to the temperature dependence of the experimentally measured quantum oscillation amplitude. The effective mass can furthermore, serve as a diagnostic of a quantum critical point at which a relevant susceptibility diverges. Quantum oscillations have proven an effective way to detect a quantum critical point, especially underlying an unconventional superconducting phase by measuring a divergent effective mass (i.e. a collapse in Fermi velocity) on the suppression of superconductivity by an applied magnetic field.[1,16] Furthermore, the vicinity of a quantum critical point may be a propitious location in phase space to look for a breakdown of Fermi–Dirac statistics, given that a divergence in effective mass signals a breakdown in the quasiparticle concept.[17]

4.4 Iron Pnictide Superconductors

We now turn to the iron pnictide family of superconductors, and the measurements of quantum oscillations in these materials. A landmark advance in the field of superconductivity was made with the discovery of the family of superconducting iron pnictides.[18,19] These tetragonal structure-type materials that contain tetrahedrally coordinated Fe^{2+} on a square lattice superconduct at temperatures as high as 56 K.[20] Proving next only to the copper oxide superconductors in superconducting temperatures, iron pnictide materials are of particular interest due to the unconventional nature of superconductivity they exhibit.[21] Materials include the '1111' family with the ZrCuSiAs-structure type, containing *R*FeAsO members (*R*=rare earth),[18–20] the '122' family with the $ThCr_2Si_2$-structure type, containing AFe_2As_2 (*A*=alkaline earth, alkali metal)[22–24] members, the '111' family with the Ce_2Sb-structure type,[25–27] containing *A*FeAs members, and the '11' family with the PbO-structure type, containing Fe*Ch* (*Ch*=chalcogen)[28] members, among others. Nearly all these materials such as the '1111' and '122' families are antiferromagnetic in their groundstate,[22,23,29] and become superconducting on the application of either charge or isovalent chemical doping[19,20,30–35] or the application of pressure.[36,37] Intriguingly, the phase diagram of the pnictides has some broad similarities with other unconventional superconductors such as heavy fermions, organics, and high temperature cuprate superconductors[21] (shown in Fig. 4.2).

4.5 Quantum Oscillations in Antiferromagnetic Parent Iron Pnictides

Soon after the discovery of superconductivity in the iron pnictide family of materials, quantum oscillations were discovered in the antiferromagnetic parent iron arsenide $SrFe_2As_2$ using the resonant oscillatory technique in a pulsed field magnet and torque magnetometry in a DC hybrid magnet,[39] and the superconducting iron phosphide LaFePO using torque and resistivity techniques in superconducting and DC hybrid magnets.[40] Quantum oscillations

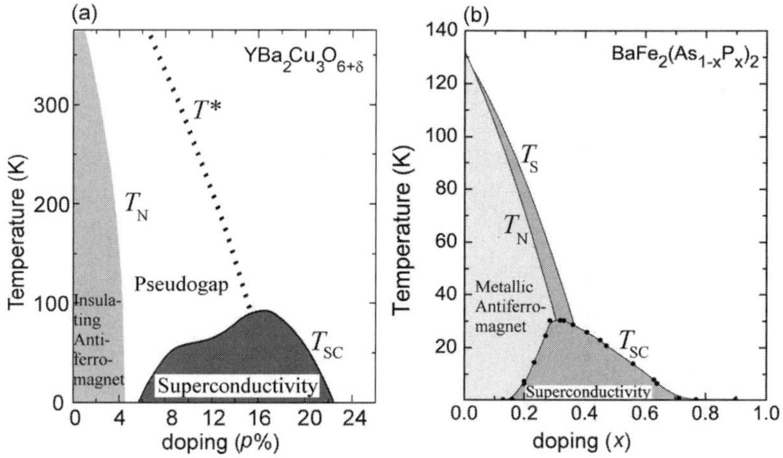

Figure 4.2 (a) Schematic phase diagram for a copper oxide high temperature superconductor, the example of $YBa_2Cu_3O_{6+\delta}$ is shown in the figure (for hole doping $p \approx 18\delta\%$), from ref. 38. At zero to low dopings, an insulating antiferromagnetic phase with Néel temperature T_N is exhibited, evolving with higher doping to a d-wave superconductor with superconducting temperature T_{sc} and a 'pseudogap' normal state that onsets at T^*. (b) Schematic phase diagram for an iron pnictide superconductor (the example of $BeFe_2(As_{1-x}P_x)_2$ from refs. 31,35 shown here). At zero to low dopings, a metallic antiferromagnetic phase with Néel temperature T_N preceded by a structural phase transition at T_s is exhibited, evolving at higher dopings to a superconductor with superconducting temperature T_{sc}.

have now been measured in other members of the pnictide family of superconductors including other '122' and '111' materials by a variety of experimental techniques including magnetic torque measurements, resonant oscillatory measurements, and electrical transport measurements in a range of magnets — from pulsed, to DC resistive and hybrid, to superconducting magnets.[40–51] Here we focus on quantum oscillations in the parent and overdoped '122' family and the evolution of quantum oscillations as a function of doping in these materials.

Figures 4.3a,c show a sample of quantum oscillations measured over different magnetic field ranges in single crystals of $SrFe_2As_2$ and detwinned single crystals of $BaFe_2As_2$, which are

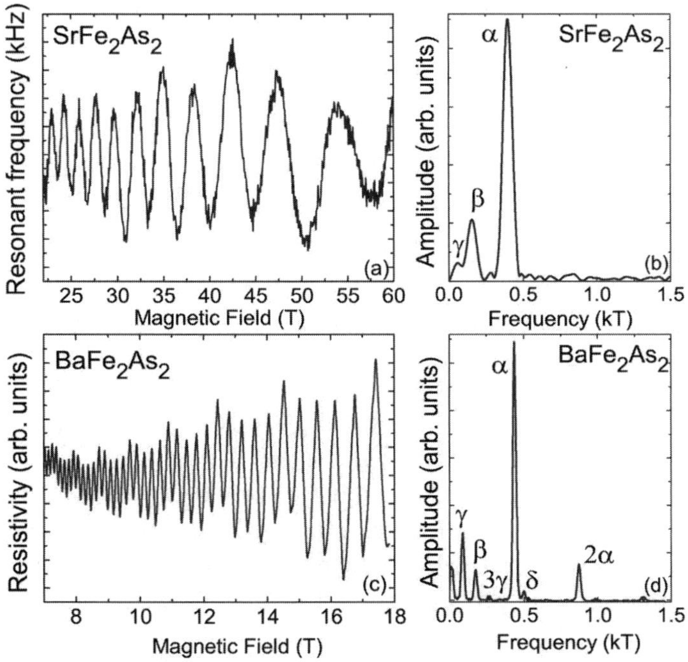

Figure 4.3 (a) Quantum oscillations measured by the tunnel diode oscillator technique on a single crystal of $SrFe_2As_2$ between 22 T and 60 T at an applied magnetic field angle of 10° to the crystalline *c*-axis and a temperature of 0.5 K. (b) Fourier transform of the quantum oscillations in $SrFe_2As_2$ shown in (a), with peaks identifying quantum oscillation frequencies that correspond to $F_\alpha = 370(20)$ T, $F_\beta = 140(20)$ T, and $F_\gamma = 70(20)$ T for an applied magnetic field along the crystalline *c*-axis. From refs. 39,44. (c) Quantum oscillations measured in the resistivity of a single detwinned crystal of $BaFe_2As_2$ from 7 T to 18 T, with the magnetic field parallel to the crystalline *c*-axis, at a temperature of 0.17 K. (d) Fourier transform of the quantum oscillations in $BaFe_2As_2$ shown in (c), with peaks identifying quantum oscillation frequencies that correspond to $F_\alpha = 440$ T, $F_\beta = 170$ T, $F_\gamma = 90$ T, and $F_\delta = 500$ T, for an applied magnetic field along the crystalline *c*-axis. From ref. 43.

antiferromagnetic parent pnictides with ordering temperatures of 205 K and 140 K, respectively.[22,23] Prominent oscillations periodic in inverse magnetic field are measured in the field range 7 T to 60 T. Figures 4.3b,d show Fourier transforms of measured quantum oscillations in $SrFe_2As_2$ and $BaFe_2As_2$ over different field ranges.

Low frequency Fourier peaks are observed at $F_\alpha = 370(20)$ T, $F_\beta = 140(20)$ T, $F_\gamma = 70(20)$ T in $SrFe_2As_2$,[39] and at $F_\delta = 500$ T, $F_\alpha = 440$ T, $F_\beta = 170$ T, $F_\gamma = 90$ T in $BaFe_2As_2$.[41,43]

4.5.1 *Fermi Surface Geometry: Nonmagnetic and Antiferromagnetic Band Structure Calculations*

The crystal structure of the '122' iron pnictide is shown in Fig. 4.4a. Shown in green in Figs. 4.4a,b is the basal plane unit cell containing a single Fe ion, of side a. However, the alternating positioning of the As ions above (marked as 'u') and below (marked as 'd') the Fe-ion plane (shown in Figs. 4.4a,b) mean that the primitive unit cell contains two Fe ions. The basal plane unit cell containing two Fe ions of side $\tilde{a}$ (where $\tilde{a} = \sqrt{2}a$) is shown in pink in Figs. 4.4a,b. For the remainder of this chapter, we adopt this two-Fe ion notation in describing the unit cell and the Brillouin zone, and the ordering wavevector within the Brillouin zone. An alternate convention arises on noting that the low energy part of the electronic structure is equivalent to a Brillouin zone that is twice as large, and corresponds to a unit cell axa in dimension, containing a single Fe ion. The first convention can be related to the second, by noting that the $X = (\pi/a, 0)$ and $Y = (0, \pi/a)$ points in the one-Fe unit cell correspond to the $\tilde{M} = (\pi/\tilde{a}, \pi/\tilde{a})$ point in the two-Fe unit cell. In the nonmagnetic state, local density approximation (LDA) calculations yield an electronic structure of AFe_2As_2 that comprises two electron-like surfaces near the $\tilde{M}$ point, and two hole surfaces near the Γ point (shown in Fig. 4.5).[39,53]

It is clear that the experimentally measured Fermi surface sections, each occupying $\lessapprox 2\%$ of nonmagnetic Brillouin zone (% of Brillouin zone area occupied by each Fermi surface section shown in Table 4.1) are vastly smaller in size than the quasi-two-dimensional sections occupying up to $\approx 25\%$ of the nonmagnetic Brillouin zone expected in the calculated nonmagnetic band structure. A Fermi surface transformation into sections of much smaller size is expected from reconstruction due to an additional superstructure, which would increase the size of the unit cell in real space, thereby reducing the size of the Brillouin zone in momentum space, and causing the Fermi surface to break up into smaller sections. Such

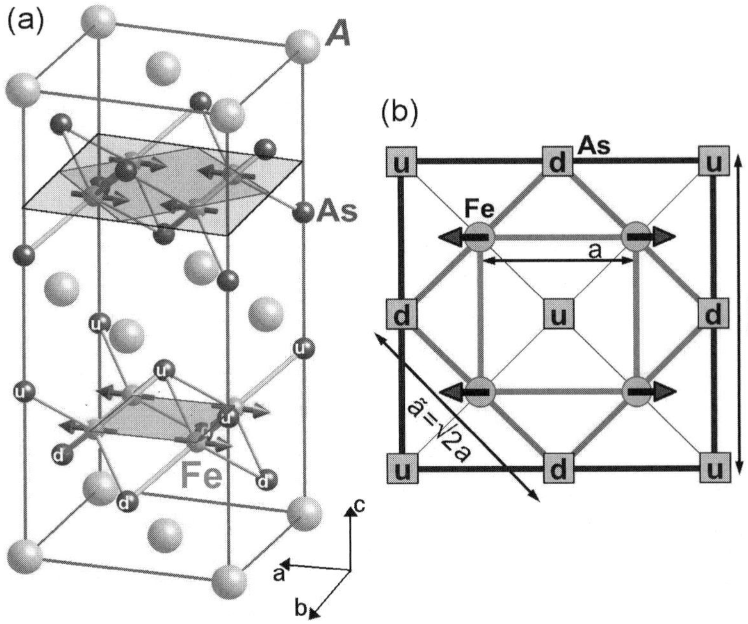

Figure 4.4 (a) Unit cell of the antiferromagnetic '122' family of AFe_2As_2 pnictides: A = Sr,Ba,Ca. A atoms shown in blue, Fe ions in orange, As ions in purple. Ordering of Fe moments (shown in red) arranged parallel to each other along the b direction, and antiparallel toward each other along the orthogonal a direction. The one-Fe ion unit cell is shown in green, with side a (top-view shown in (b)). However, the arrangement of As ions located above (labelled 'u') and below (labelled 'd') the plane of Fe-ions causes the minimal unit cell to be defined as a larger unit cell containing two Fe ions, shown in pink, with side $\tilde{a} = \sqrt{2}a$ (top-view shown in b). The antiferromagnetic unit cell is obtained by doubling the two-Fe ion unit cell, and is shown in blue. The side of the antiferromagnetic unit cell is given by $\sqrt{2}a$, the top view is shown in b. Modified from ref. 52. (b) top view of each of the one-Fe ion, two-Fe ion, and doubled antiferromagnetic unit cells accompanied by dimensions, and the alignment of Fe-spins due to antiferromagnetic order. See also Color Insert.

a superstructure in antiferromagnetic AFe_2As_2 might be expected to occur at the quasi-nesting wavevector $Q = (\pi/\tilde{a}, \pi/\tilde{a})$ between the electron and hole pockets which are separated in momentum space in the nonmagnetic electronic structure, given the enhanced Lindhard function in the vicinity of this wavevector.[54–57] Indeed,

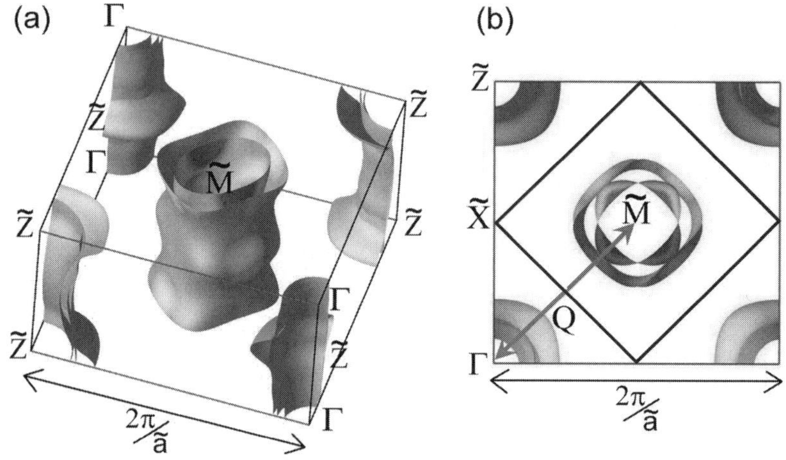

Figure 4.5 (a) Calculated paramagnetic Fermi surface of $SrFe_2As_2$ in the original nonmagnetic two-Fe ion Brillouin zone, calculated using using the local density approximation (LDA) method, comprising two hole cylinders located at the Γ point and two electron cylinders at the $\tilde{M}$ point of the Brillouin zone. High symmetry points in the two-Fe ion Brillouin zone are denoted as Γ, $\tilde{M}$, $\tilde{X}$, $\tilde{Z}$. Shading of the surfaces is according to the calculated Fermi velocity. From ref. 39. (b) A top down view of the nonmagnetic two-Fe ion unit cell accompanied by the antiferromagnetic ordering wavevector $Q = (\pi/\tilde{a}, \pi/\tilde{a})$[29] to yield an antiferromagnetic unit cell (shown in blue). From ref. 8. See also Color Insert.

the onset of an antiferromagnetic wavevector $Q = (\pi/\tilde{a}, \pi/\tilde{a})$ at the Néel temperature has been measured by inelastic neutron scattering in the parent pnictide materials, in which neighbouring Fe moments are arranged parallel to each other along one direction, and antiparallel to each other along the orthogonal direction (shown in Fig. 4.4).[29] In $SrFe_2As_2$ and $BaFe_2As_2$, a spin moment of $\approx 0.9\,\mu_B$ per Fe ion is measured,[59,60] and indications from the breadth of the measured high energy dispersion are that the spin density wave is itinerant in character.[61]

The anticipated electronic structure in this parent antiferromagnetic state as a consequence of the $Q = (\pi/\tilde{a}, \pi/\tilde{a})$ ordering wavevector is yielded by folding the non-magnetic Brillouin zone shown in Fig. 4.5a along the magnetic wavevector Q, such that the $\tilde{M}$ point is folded onto the Γ point. The unit cell in the basal plane is therefore doubled, yielding the antiferromagnetic unit cell shown in dark blue in Figs. 4.4a,b. The resulting Brillouin zone is reduced in size from the black square shown in Fig. 4.5a,b to the blue diamond shown in Fig. 4.5b. The reduction in size of the Brillouin zone results in the superposition of the hole pocket at Γ, and the electron pocket at $\tilde{M}$, causing a gapping of large sections of the original Fermi surface, and leaving behind small three-dimensional hole and electron pockets (shown in Fig. 4.6).[39] Figure 4.6 shows the result of a local density approximation calculation of the antiferromagnetic state of $SrFe_2As_2$ in which magnetic ordering is included; a value of negative on-site Coulomb repulsion U is used that suppresses the magnetic moment to the experimentally observed value of $\approx 0.9\ \mu_B$.[41,45] The calculated antiferromagnetic Fermi surface (Fig. 4.6a,d) is seen to comprise small three-dimensional Fermi surface pockets in contrast to the larger quasi-two-dimensional Fermi surface sections expected in the nonmagnetic electronic structure (Fig. 4.5).

4.5.2 *Experimental Comparison with Band Structure*

Table 4.1 shows the measured quantum oscillation frequencies and effective quasiparticle masses in the antiferromagnetic parent pnictide $SrFe_2As_2$ and $BaFe_2As_2$, along with the effective mass enhancement of each Fermi surface section, and fraction of the original two-Fe ion Brillouin zone area occupied by each Fermi surface section. The size and angular dependence of the measured quantum oscillation frequencies that correspond to the extremal orbits of the Fermi surface pockets are compared with expectations from the calculated antiferromagnetic band structure in Fig. 4.7. Figure 4.7a shows a comparison of the angular dependence of extremal orbits expected from the calculated band structure, with an angular dependence of the measured quantum oscillation frequencies in $BaFe_2As_2$.[43] The observed frequencies extend over the entire angular range, and are weakly angle-dependent, indicating largely three-dimensional Fermi surface sections. Good agreement between experimental and calculated frequencies is seen for the α and δ frequencies for a band shift

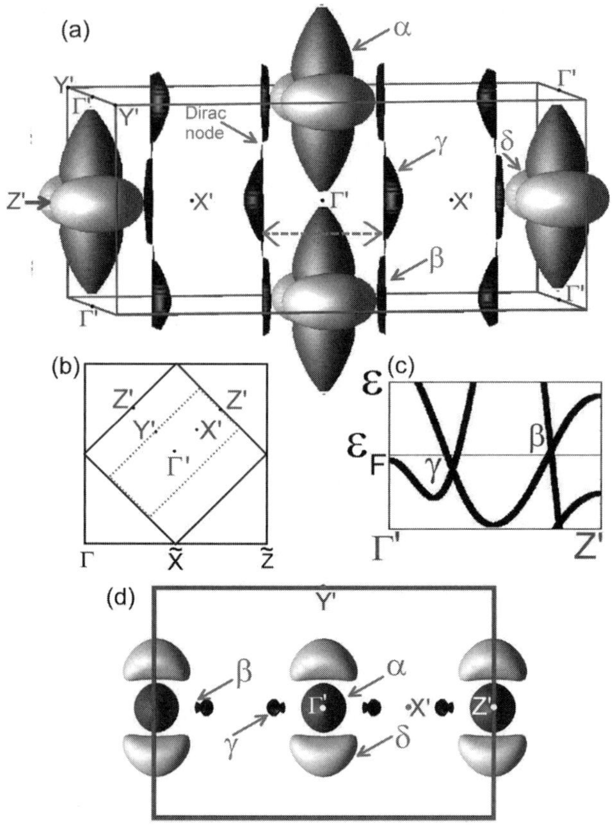

Figure 4.6 (a) Calculated antiferromagnetic Fermi surface of $SrFe_2As_2$ obtained using the local density approximation (LDA) method in which magnetic ordering is included.[41,45] High symmetry points in the new antiferromagnetic Brillouin zone are denoted by Γ', X', Y', Z'. (b) The relative dimensions of the reduced antiferromagnetic Brillouin zone (in red) are compared to the conventional original nonmagnetic two-Fe ion (in black). While the α, and δ surfaces denote the larger ellipsoidal surfaces formed from parabolic bands, the γ and β pockets are smaller closed surfaces with pointed ends formed from linear band crossings at a Dirac point (shown in (c)). The dashed arrow shows a sample construction of the β and γ Dirac pockets at a Dirac point. (d) shows the top view of the Brillouin zone with Fermi surface pockets and high symmetry points labelled. A comparison of the experimentally measured extremal orbits with the band structure calculations is shown in Table 4.1. Figures provided courtesy of M. D. Johannes. See also Color Insert.

Table 4.1 Measured quantum oscillation frequencies (F) in antiferromagnetic parent '122' pnictide materials $SrFe_2As_2$ and $BaFe_2As_2$, corresponding Fermi surface cross-sectional areas as a fraction of the original two-Fe ion Brillouin zone area ($\frac{A}{A_{BZ}}$), measured effective quasiparticle mass ($\frac{m^*}{m_e}$), effective quasiparticle mass enhancement compared to the band mass ($\frac{m^*}{m_b}$)

	$SrFe_2As_2$				$BaFe_2As_2$			
Fermi surface	F (T)	$\frac{A}{A_{BZ}}$ (%)	$\frac{m^*}{m_e}$	$\frac{m^*}{m_B}$	F (T)	$\frac{A}{A_{BZ}}$ (%)	$\frac{m^*}{m_e}$	$\frac{m^*}{m_B}$
α (hole)	370(20)	1.38	2.0(1)	1.8	440	1.7	2.1(1)	2.8
β (Dirac)	140(20)	0.52	1.5(1)	2.1	170	0.7	1.8	–
γ (Dirac)	70(20)	0.26	–	–	90	0.3	0.9	–
δ (electron)	–	–	–	–	500	1.9	2.4(3)	2.0

The corresponding Fermi surface sections are illustrated in Fig. 4.6. From refs. 39,41,43.

of -44 meV and $+65$ meV, respectively for the bands corresponding to the larger electron-like δ pocket and hole-like α pocket shown in Fig. 4.6. Interestingly, this band shift corresponds to a calculated magnetic moment of $1.6\mu_B$ per Fe ion, which overestimates the experimentally measured magnetic moment of 0.9 μ_B per Fe ion, potentially indicating the relevance of fully magnetic models in describing the parent antiferromagnetic pnictide materials.

The identification of the γ and β pocket is less unambiguous, given their extremely small size, as seen in Fig. 4.6. A more detailed study of these small Fermi surface pockets is presented in Section 4.5.3. Using the observed enhancement in effective mass compared to the calculated band mass for the Fermi surface pockets measured in $SrFe_2As_2$ and $BaFe_2As_2$ (Table 4.1), the contribution to the Sommerfeld coefficient from the measured pockets is estimated to be 3.5 mJ/K^2mol for $SrFe_2As2_2$ and 5.0 mJ/K^2mol for $BaFe_2As_2$, in good agreement with measured values of heat capacity.[39,41,43] The modest effective mass enhancement places these materials in a regime of intermediate correlation strength. Quantum oscillation measurements of the Fermi surface in the parent $A\mathrm{Fe_2As_2}$ family enables experimental identification of the reconstructed Fermi surface as corresponding to the calculated antiferromagnetic band structure in this material, reconstructed from the original band structure by an ordering wavevector $Q = (\pi/\tilde{a}, \pi/\tilde{a})$.

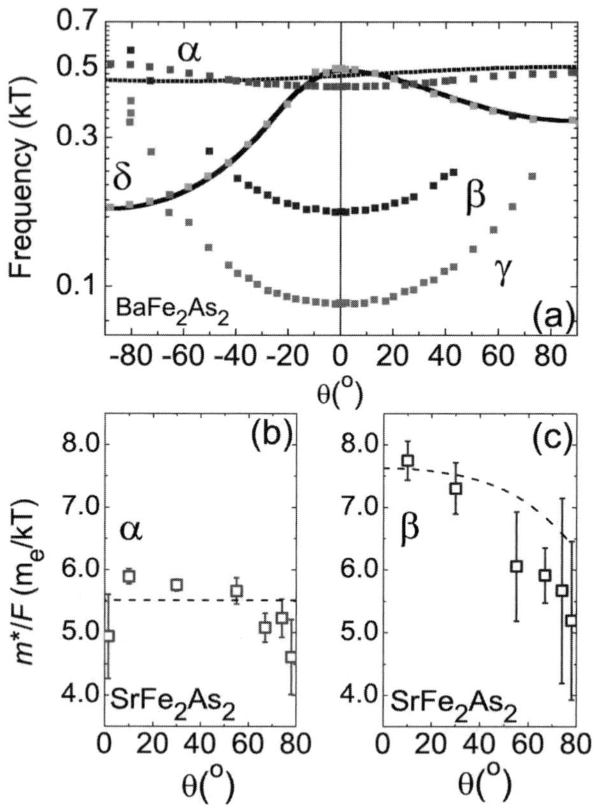

Figure 4.7 (a) Experimentally measured angular dependence of the α, β, γ, and δ frequencies in $BeFe_2As_2$ compared with expected angular dependence of α and δ from a calculated Fermi surface (similar to that shown in Fig. 4.6). Good agreement is seen for the α and δ frequencies for a band shift less than 65 meV, corresponding to a magnetic moment size of 1.65 μ_B per Fe ion. From ref. 43. (b) Angular dependence of the ratio of effective mass to quantum oscillation frequency experimentally measured for the α pocket in $SrFe_2As_2$ — the ratio is largely constant (dashed line is a guide to the eye), as expected for a parabolic band (c) angular dependence of the ratio of effective mass to frequency experimentally measured for the β pocket in $SrFe_2As_2$ — the value dips sharply for higher angles (dashed line is a guide to the eye), consistent with the expectation for linear bands crossing at a Dirac point as described in Section 4.5.3. Modified from ref. 44.

4.5.3 *Dirac Nodes*

A more in-depth comparison is required to identify the topology of the smallest γ and β pockets observed in the Fermi surface of antiferromagnetic $SrFe_2As_2$ and $BaFe_2As_2$. A theoretical analysis shows that the experimentally observed spin density wave gives rise to symmetry-enforced band degeneracy at high-symmetry points in the Brillouin zone, the experimentally observed spin density wave with ordering wavevector $Q = (\pi/\tilde{a}, \pi/\tilde{a})$ resulting in a nontrivial band topology. Even were nesting to be perfect, topological protection would preserve the Dirac nodes that result for such a band topology, thereby preventing full gapping of the Fermi surface by the spin density wave; a conclusion that has been shown to survive the introduction of strong interactions.[62] The location of Dirac nodes which are located close to the Fermi energy in the case of antiferromagnetic $SrFe_2As_2$ and $BaFe_2As_2$[41,44,45] (shown in Fig. 4.6) yields small Dirac pockets. The geometry of the small closed Dirac pockets with pointed ends is not sufficient to distinguish these from ellipsoidal pockets; a true test of such a Dirac point can only be yielded by a comparison of the angular dependence of the associated frequency and effective cyclotron mass, in this case, for the β and γ pockets. For a parabolic band, this ratio would remain constant; for a linear crossing, however, the angular-dependent ratio of the frequency and effective cyclotron mass is expected to deviate from a constant value. For a linear dispersion of the form $\epsilon = \pm\hbar v^*|k| + 2t\cos(ck_z/2) + \mu$ where c is the bilayer spacing for the body-centred-tetragonal crystal structure (12.3 Å), v^* is the characteristic Dirac velocity, t is the interlayer hopping parameter, $|k| = \sqrt{k_x^2 + k_y^2}$, and μ is the chemical potential, a striking experimental signature is expected. As θ is increased, the orbits would begin to encompass regions of the Fermi surface close to the Dirac point (Fig. 4.6 left inset), where velocities are high and masses are very light — leading to a reduction in orbitally averaged cyclotron masses, and consequently a rapid reduction in m^*/F with angle compared to the expectation for a parabolic band.[42,44]

Figures 4.7b,c show the ratio of m^*/F as a function of angle for both the α pocket and the β pocket in $SrFe_2As_2$. The ratio m^*/F remains largely constant as a function of angle for the α pocket, consistent with its parabolic band origin. However, the ratio m^*/F dips steeply with angle in the case of the β pocket, as would be expected for a pocket associated with linear band crossing at a Dirac point.[44] The linear dispersion characteristic of a Dirac cone is also directly observed via photoemission experiments.[63] The Fermi surface of the antiferromagnetic parent '122' pnictide family is therefore seen to exhibit Dirac points as a consequence of the orbital character of the bands in this material.[44,45,62]

4.6 Quantum Oscillations in Overdoped Paramagnetic Iron Pnictides

Quantum oscillations have been observed in the end $A\mathrm{Fe_2P_2}$ members of the superconducting '122' pnictide family (Fig. 4.2).[46–48] Figure 4.8 shows quantum oscillations measured on $\mathrm{SrFe_2P_2}$.[46] The first noticeable aspect of these oscillations is that the measured frequencies are considerably higher than in the parent antiferromagnetic parent member of the series (Fig. 4.3), indicating larger Fermi surface sections (measured frequencies and the size of the corresponding Fermi surface cross-sectional area as a fraction of the original two-Fe ion Brillouin zone are shown in Table 4.2). The Fourier transform in Fig. 4.8b shows the component frequencies in the measured quantum oscillations, the angular dependence of which is shown in Fig. 4.8c. A band structure calculation of the expected Fermi surface of $\mathrm{SrFe_2P_2}$ consists of two concentric electron cylinders located at the Brillouin zone corners (i.e. the $\tilde{M}$ point), and two hole cylinders centred at the Γ point (Fig. 4.9), which is very similar to that calculated for the nonmagnetic phase of $\mathrm{SrFe_2As_2}$ (Fig. 4.5a). Some band structure calculations[64] and photoemission experiments[65] experiments indicate an additional third hole pocket located at the Γ point, that is more three-dimensional in character.

Measured quantum oscillation frequencies and effective mass enhancements are shown in Table 4.2 for $\mathrm{SrFe_2P_2}$ and $\mathrm{BaFe_2P_2}$. These yield experimental results that correspond well with band structure calculations, with the largest frequency corresponding to a cross-sectional area as large as $\approx 25\%$ the size of the original two-

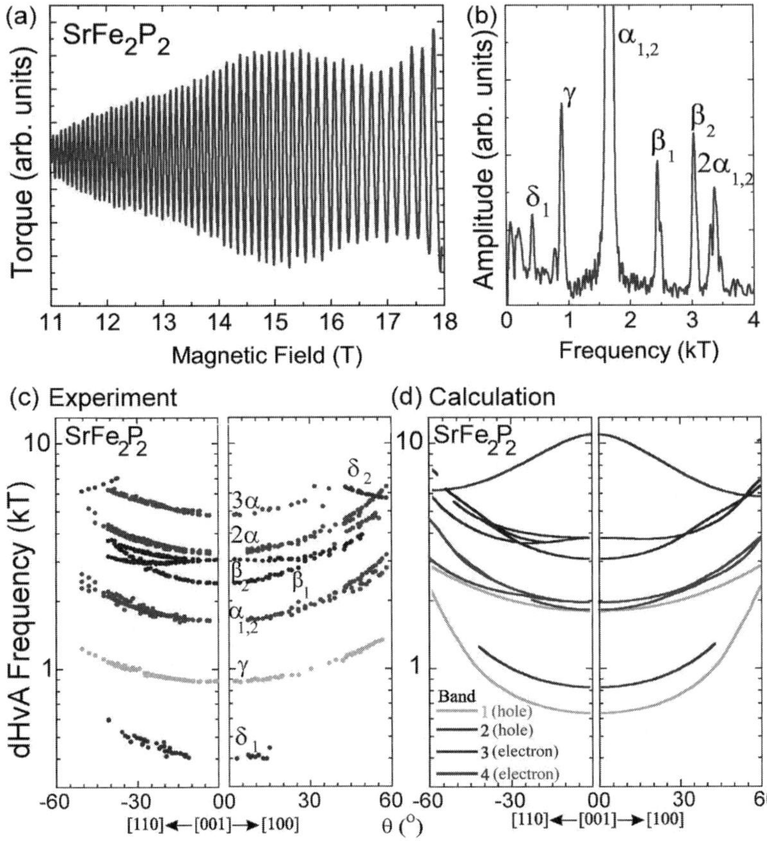

Figure 4.8 (a) Quantum oscillations measured using torque magnetometry on a single crystal of $SrFe_2P_2$ (at an angle of 9° to the crystalline *c*-axis) (b) Fourier analysis of the quantum oscillations shown in (a) to extract the constituent frequencies from Fourier peaks labelled according to the Fermi surface sections to which they correspond. (c) Angular dependence of experimentally measured quantum oscillation frequencies. (d) Calculated angular dependence of quantum oscillation frequencies expected for the band structure shown in Fig. 4.9. Excellent agreement is seen between the angular dependence of the experimentally measured quantum oscillations and extremal orbits calculated from the band structure. From ref. 46.

Fe ion Brillouin zone. Figures 4.8c,d show the angular dependence of the experimentally measured quantum oscillation frequencies compared with the predictions for the calculated band structure

Table 4.2 Measured quantum oscillation frequencies F in the end members $SrFe_2P_2$ and $BaFe_2P_2$ of the '122' pnictide family, corresponding Fermi surface cross-sectional areas as a fraction of the original two-Fe ion Brillouin zone area ($\frac{A}{A_{BZ}}$), measured effective quasiparticle mass ($\frac{m^*}{m_e}$), effective quasiparticle mass enhancement compared to the band mass ($\frac{m^*}{m_b}$)

	$SrFe_2P_2$				$BaFe_2P_2$			
Fermi surface	F (kT)	$\frac{A}{A_{BZ}}$ (%)	$\frac{m^*}{m_e}$	$\frac{m^*}{m_B}$	F (kT)	$\frac{A}{A_{BZ}}$ (%)	$\frac{m^*}{m_e}$	$\frac{m^*}{m_B}$
γ (hole)	0.89	3.4	1.49(2)	1.4	1.25	4.8	1.64	1.80
δ_1 (hole)	0.41	1.6	1.6(1)	1.3	0.79	3.1	2.30	1.59
δ_2 (hole)	6.02	23.2	3.41(5)	1.7	6.58	25.4	3.32	1.59
β_1 (electron)	2.41	9.3	1.92(2)	1.6	2.14	8.3	1.65	1.68
β_2 (electron)	3.06	12.0	2.41(3)	1.6	2.31	8.9	1.54	1.81
α_1 (electron)	1.637	6.3	1.13(1)	2.1	1.18	4.8	1.75	1.82
α_2 (electron)	1.671	6.5	1.13(1)	2.1	1.35	4.6	1.58	1.88

The corresponding Fermi surface sections are illustrated in Fig. 4.9. From refs. 46,48.

in $SrFe_2P_2$. Good agreement is obtained on shifting electron bands α and β up by 59 and 49 meV, respectively, and the hole band γ down by 110 meV[46] in the calculated band structure. Using the observed enhancement in effective mass compared to the calculated band mass for the Fermi surface pockets measured in $SrFe_2P_2$, the contribution to the Sommerfeld coefficient from the measured Fermi surface pockets is estimated to be 10.4(2) $mJ/mol/K^2$, in good agreement with the value of 11.6(2) $mJ/mol/K^2$ measured by heat capacity. The size of quasiparticle effective mass enhancement places these materials in a regime of intermediate correlation strength. The significant transformation in Fermi surface from nonmagnetic end members of the '122' series to the antiferromagnetic parent members of this family of materials is discussed more in Section 4.7.2.

4.6.1 *Quasi-Nesting of Hole and Electron Cylinders*

An important aspect of the Fermi surface geometry revealed by quantum oscillations in the AFe_2P_2 end members of the '122' family is the significant c-axis dispersion (i.e. warping) of all the Fermi surface sections, rendering them more three-dimensional instead of two-dimensional. It is also obvious that while the Lindhard function

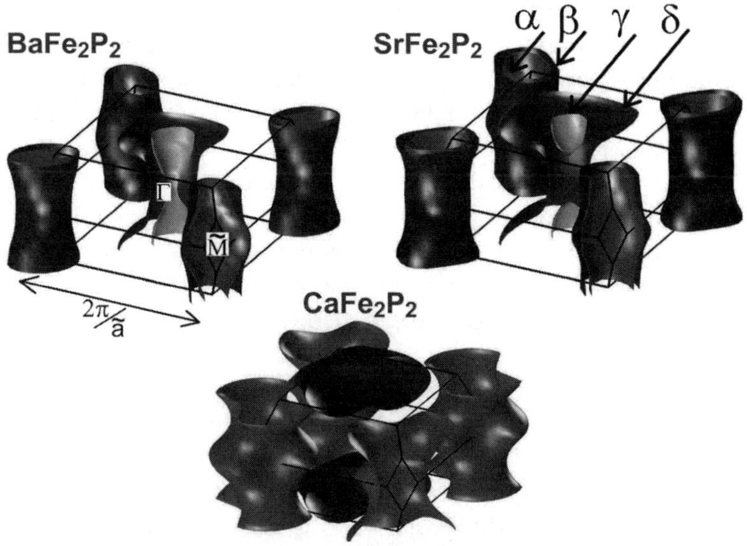

Figure 4.9 Fermi surfaces of $BaFe_2P_2$, $SrFe_2P_2$, and $CaFe_2P_2$ from Local Density Approximation calculations with bands shifted for best agreement with experimental quantum oscillation data.[46–48] Bands corresponding to extremal orbits α, β, γ, and δ indicated by arrows. The quasi-two-dimensional cylinders in $SrFe_2As_2$ and $BaFe_2As_2$ are warped, meaning that surfaces α, β, and δ each have a minimal ($orbit_1$) and maximal ($orbit_2$) extremal orbit corresponding to the 'neck' and 'belly' of the warped surface. A comparison of the experimentally measured extremal orbits with the band structure calculations is shown in Table 4.2. From ref. 66. See also Color Insert.

(representing the bare spin susceptibility) is expected to have a maximum at $Q = (\pi/\tilde{a}, \pi/\tilde{a})$, a perfect nesting condition is not satisfied by the pocket geometry. Nesting effects are far from ideal in either the unreconstructed band structure of the antiferromagnetic parent members (Fig. 4.5), or the measured Fermi surface of the end members of the pnictide family (Fig. 4.9), making it unlikely that these effects are principally responsible for antiferromagnetism.

Quantum oscillation measurements have been performed on the other end members of the '122' pnictide family: $BaFe_2P_2$, and $CaFe_2P_2$.[47–49] Figure 4.9 shows the calculated Fermi surface geometry corresponding to quantum oscillations measured in each of these families of materials.[46–48,66] On traversing the periodic table from the Ba members of the series, to the Sr and Ca members of the series, an interesting evolution is seen. The pocket sizes and geometries in $SrFe_2P_2$ and $BaFe_2P_2$ are fairly similar, with the chief difference being the better fulfilment of nesting criteria between inner hole (γ) and outer electron (α) pockets in $BaFe_2P_2$. The Fermi surface sections in $CaFe_2P_2$, however, become particularly different, with the outer electron pockets becoming more warped, and the central warped concentric hole cylinders in $BaFe_2P_2$ and $SrFe_2P_2$ transforming into a single three-dimensional surface in $CaFe_2P_2$ (Fig. 4.9). In fact, the electronic structure in $CaFe_2P_2$ corresponds to that of the collapsed tetragonal structure, where the ratio of c-axis to a-axis lattice constant is significantly reduced from the orthorhombic parent antiferromagnetic structure.[33] A potential relation of the increased quasi-three-dimensionality in $CaFe_2P_2$ to the reduced optimal-T_{sc} in this families of materials (optimal T_{sc} $\approx$30 K in $BaFe_2(As_{1-x}P_x)_2$ and $SrFe_2(As_{1-x}P_x)_2$, compared to optimal T_{sc} $\approx$15 K in $CaFe_2(As_{1-x}P_x)_2$) is consequently indicated.[32,33,35]

4.7 Cuprates and Iron Pnictides: Electronic Structure Comparison

Importantly, quantum oscillations are a key tool in making a comparison between the normal state of the cuprate and pnictide superconductors. At first glance, the cuprate and pnictide phase diagrams appear similar, both manifesting an antiferromagnetic parent groundstate that evolves to a superconducting state with elevated superconducting temperatures on doping (Fig. 4.2). However, the metallic antiferromagnetic groundstate of the pnictide superconductors persists with doping, overlapping with the superconducting region over a significant portion of the phase diagram, up to optimal dopings. In contrast, the antiferromagnetic Mott insulating groundstate of the cuprates persists over a narrow range with doping, and is separated in doping from the onset of the superconducting dome. A key question pertains to the nature of the normal state of the cuprates in the underdoped region of superconductivity. Low frequency quantum oscillations were discovered after two decades in the underdoped region of the phase diagram in the cuprate superconductor $YBa_2Cu_3O_{6+\delta}$ at hole dopings $p \approx$ 9–12% (i.e. δ = 0.49–0.61), corresponding to a Fermi surface of area ≈2% of the original Brillouin zone.[5,8–13] Quantum oscillations have subsequently been observed in other underdoped cuprate superconductors, including $YBa_2Cu_4O_8$[6,7] and $Nd_{2-x}Ce_xCuO_4$,[67] revealing similarly small Fermi surface sections.

4.7.1 *Enhancement in Lindhard Function in Pnictides and Cuprates*

An interesting comparison can be drawn between the Fermi surface in the cuprates and the pnictides with possible relevance to the origin of superconductivity in these materials.[68] At first sight, the nonmagnetic Fermi surface in the cuprate materials appears very different from the pnictide materials. The pnictide family of materials belong to the class of compensated semimetals, with an electronic structure comprising both hole and electron pockets — these are separated in momentum space, as seen in Figs. 4.5 and 4.9. In contrast, the cuprate materials are effectively single band in character. A significant Jahn–Teller distortion in the quasi-two-dimensional cuprates removes the degeneracy of the two-fold e_g orbital in copper, and leads to an effective single orbital model. The intra-orbital Coulomb interaction U causes the half filled band to be split into a filled lower Hubbard and an empty upper Hubbard band (Fig. 4.10c), yielding a Mott insulating ground state. On doping mobile carriers as shown in Fig. 4.10c, a single carrier band is yielded. The electronic structure on the overdoped hole-carrier side has been measured by angular magnetoresistance oscillations, quantum oscillations, and angle resolved photoemission, revealing a single hole surface centred at the Γ point of the Brillouin zone (schematic shown in Fig. 4.10a).[13,69] A single hole band occupying ≈65% of the original Brillouin zone has been observed by quantum oscillations on the overdoped side of the hole-doped cuprate superconductor $Tl_2Ba_2CuO_{6+\delta}$.[69]

A notable commonality, however, is seen between the electronic structure of the cuprate and pnictide families on identifying the quasi-nesting wavevector associated with a peak in the Lindhard function in both these cases. In the case of the pnictide materials,

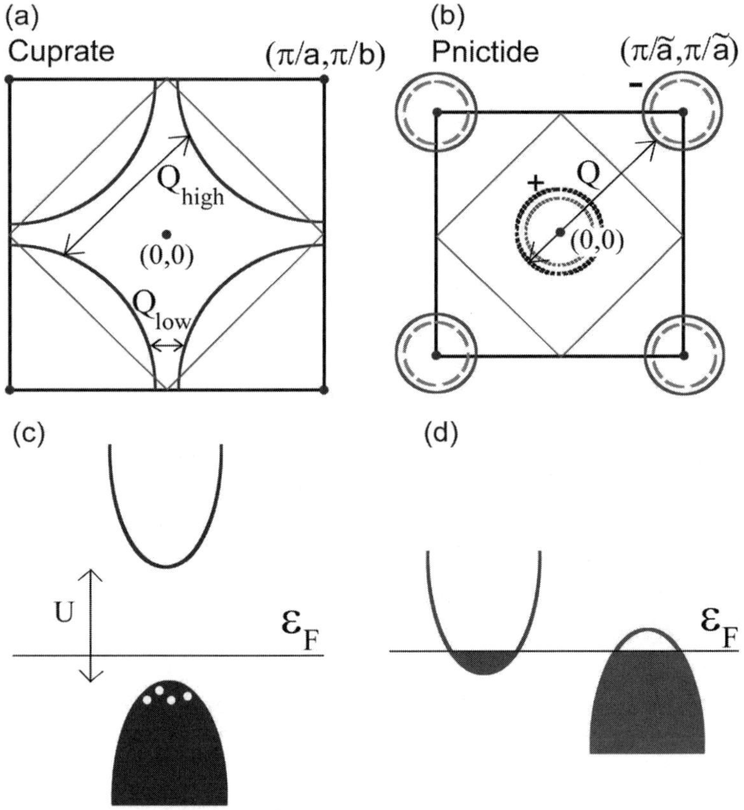

Figure 4.10 (a) Schematic of single hole band in the original Brillouin zone of a hole-doped cuprate superconductor. $Q_{\text{high}} = (\pi/\tilde{a}, \pi/\tilde{a})$ and $Q_{\text{low}} = (\Delta, 0)$ show intra-Fermi surface nesting wavevectors at which the Lindhard function is enhanced. The red diamond indicates the antiferromagnetic Brillouin zone. (b) Schematic of the hole pockets (indicated by '+') and electron pockets (indicated by '−') from multibands in the original Brillouin zone of a pnictide superconductor. $Q = (\pi/\tilde{a}, \pi/\tilde{a})$ shows the nesting wavevector between hole and electron pockets. The red diamond indicates the antiferromagnetic Brillouin zone. (c) Schematic of the half-filled band split by onsite Coulomb repulsion U to yield a Mott insulator with filled lower Hubbard band, and empty upper Hubbard band in the parent cuprate materials. The white dots show doped hole carriers that constitute the single large hole Fermi surface shown in (a). (d) Schematic of a compensated semimetal representing the pnictide materials with electron and hole bands separated in momentum space, yielding multiple Fermi surfaces as shown in (b).

a quasi-nesting wavevector $Q = (\pi/\tilde{a}, \pi/\tilde{a})$ connects the electron pockets at the Γ location with the hole pockets at the M location in the pnictides, manifesting itself as a peak in the Lindhard function.[54–58] The system then becomes unstable to an excitonic instability, i.e. a combination of electrons and holes with relative momentum wavevector Q, experimentally found to be manifested as an antiferromagnetic spin density wave.[70] Indeed, this ordering wavevector at which the Lindhard function is maximised is responsible for Fermi surface reconstruction into the observed small Fermi surface sections in the parent members of the pnictide family.

In the case of the cuprates, an intraband wavevector $Q_{\text{high}} = (\pi/\tilde{a}, \pi/\tilde{a})$ connecting opposite sides of the single large band and associated with antiferromagnetic correlations is found to enhance the Lindhard function.[71] In this case, the Lindhard function is also enhanced at small wavevectors such as $Q_{\text{low}} = (\Delta, 0)$, which is associated with charge density wave type of order, and has been related to Fermi surface reconstruction into the small Fermi surface sections observed by quantum oscillations in the underdoped regime.[13] The $Q = (\pi/\tilde{a}, \pi/\tilde{a})$ wavevector associated with antiferromagnetic correlations at which the Lindhard function is enhanced in both pnictide and cuprate families of materials could contribute to the enhancement of superconducting temperatures in these and other unconventional superconductors.[21,68,72] A possible test of the relevance of antiferromagnetic correlations in contributing to elevated superconducting temperatures is in the case of multiband systems where the location of electron and hole pockets coincide in momentum space; here the contribution of antiferromagnetic correlations to superconductivity would be expected to be suppressed.

4.7.2 *Quantum Critical Point under Superconducting Dome*

Models involving electronic mediation have been suggested to explain unconventional superconductivity in materials such as the cuprates and pnictides. In these models, an enhancement in electronic susceptibility, such as for example the density or magnetic susceptibility is anticipated to favour superconductivity.[72] While qualitative insight may be gained from the Lindhard function which represents the bare spin susceptibility, an experimentally determined electronic susceptibility is the best indication of where maximally enhanced superconductivity may be expected within such models. A magnetic instability manifested as a quantum critical point at zero temperature would yield an enhanced spin susceptibility. An attendant challenge, however, pertains to experimentally accessing the vicinity of a quantum critical point, given that a new phase such as superconductivity often conceals the notional quantum critical point. Quantum oscillations provide a good experimental probe in the vicinity of a quantum critical point, at which the effective quasiparticle mass would be expected to diverge.

In the case of pnictide superconductors, superconductivity can be suppressed sufficiently by applied magnetic fields up to 65 T, such that quantum oscillations are observed in the doping range $x = 0, 0.41-1$ in the $BaFe_2(As_{1-x}P_x)_2$ family.[73] The doping range accessed by quantum oscillations extends very close to optimal doping and reaching dopings at which the superconducting temperatures are as high as $T_{sc} \approx 25$ K ($\approx 0.8T_{sc}^{max}$). Intriguingly, an increase in effective mass, corresponding to a fall in Fermi velocity is observed in the region approaching optimal doping (shown in Fig. 4.11a). The Fermi surface geometry remains largely unchanged above optimal doping, corresponding to quasi-nested warped cylindrical electron and hole pockets at the $\tilde{M}$ and Γ points of the Brillouin zone, respectively, which reduce in size by $\approx$ 30% from $x = 1$ to $x = 0.41$.[73] A drastic change is, however, observed compared to the Fermi surface geometry in the undoped regime, in which the largest Fermi surface extremal area is approximately five times smaller than the largest Fermi surface extremal area for $x = 1$. The decrease in Fermi velocity between $x = 1$ and $x = 0.41$ in $BaFe_2(As_{1-x}P_x)_2$ is consistent with an ultimate collapse of Fermi velocity (proportional to the inverse effective quasiparticle mass) at a quantum critical point located at the doping where the antiferromagnetic phase boundary would end, were superconductivity to be absent. The collapse in Fermi velocity approaching optimal doping is consistent with a divergent magnetic susceptibility at an antiferromagnetic instability marking the notional quantum critical point underlying the superconducting maximum.

In the case of cuprate superconductors, a drastic change is observed between the Fermi surface extremal area in the overdoped

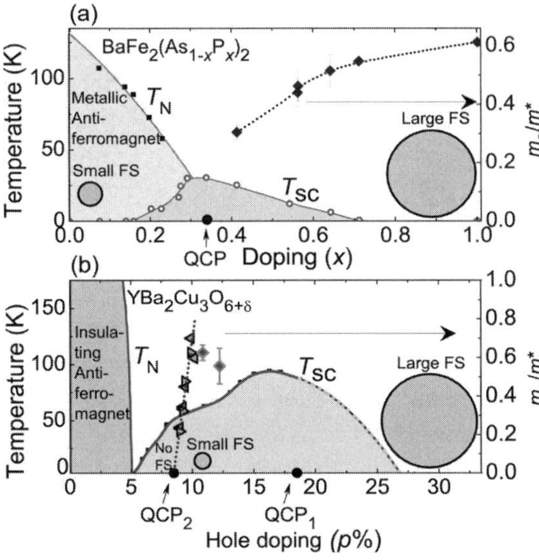

Figure 4.11 (a) Phase diagram of the doped pnictide $BaFe_2(As_{1-x}P_x)$, showing the drastic change in Fermi surface from a small area in the undoped metallic antiferromagnetic regime to a large area in the overdoped region (illustrated by gray circles with area proportional to the largest measured quantum oscillation frequency). A putative quantum critical point (labelled QCP) at the antiferromagnetic instability at the end of the metallic antiferromagnetic phase, which separates the small from the large Fermi surface is indicated by the collapse of experimentally measured Fermi velocity (proportional to the inverse effective quasiparticle mass shown on the rhs axis) on approaching optimal doping. Modified from ref. 73. (b) Phase diagram of the high temperature cuprate superconductor $YBa_2Cu_3O_{6+\delta}$. In this case the antiferromagnetic phase on the undoped side is insulating. A quantum critical point is indicated by a collapse in the experimentally measured Fermi velocity (proportional to the inverse quasiparticle mass shown on the rhs axis) at a low doping denoted as QCP_2, likely separating a charge density wave phase above QCP_2 in which the Fermi surface area is small, and a spin density wave phase below QCP_2 in which no Fermi surface is observed.[13] Left-facing and right-facing triangle symbols are from quantum oscillations measured up to 55 T and 85 T, respectively, from ref. 16; diamond symbols are from ref. 74. A notional quantum critical point (QCP_1) is expected at optimal doping separating the underdoped region in which a small Fermi surface is observed, and the overdoped region in which a large Fermi surface is observed in the $Tl_2Ba_2CuO_{6+\delta}$ family of materials.[69] Relative Fermi surface areas are illustrated by gray circles with area proportional to the measured quantum oscillation frequency. Modified from ref. 16

regime and in the underdoped regime, where the extremal area is less than thirty times smaller than in the overdoped regime. However, the use of quantum oscillations to detect evidence for a divergent effective mass at optimal doping separating these two regimes has proved challenging given the robustness of superconductivity in this region of the phase diagram, and the high magnetic fields required to suppress superconductivity. Intriguingly, quantum oscillations measured up to 85 T observe a steep increase in effective cyclotron mass, corresponding to a collapse in Fermi velocity approaching a lower doping (hole doping $p \approx 8.5\%$ (i.e. $\delta \approx 0.45$) in $YBa_2Cu_3O_{6+\delta}$) in the vicinity of a plateau (local maxima) in the superconducting dome (Fig. 4.11b).[16] In this case, the nature of the enhanced susceptibility associated with the quantum critical point signalled by the rapid collapse in Fermi velocity has not yet been precisely ascertained, but is likely associated with the onset of a spin density wave below this doping, and charge density wave above this doping.[13]

In both these families of cuprate and pnictide superconductors, and in other families of unconventional superconductors,[21] similar quantum critical points located at a magnetic or other form of instability underlying the superconducting dome maximum appear consistent with common mechanisms of unconventional superconductivity mediation.[72] Quantum oscillations that directly measure the effective quasiparticle mass are a crucial tool in revealing the divergent susceptibility signalling such a quantum critical point in diverse families of materials. A broader question pertains to whether such a potential quantum critical point underlying the superconducting dome is related to the observed enhancement in superconducting temperatures.

4.8 Conclusion

The measurement of quantum oscillations in the pnictide family of superconductors has greatly advanced our understanding of these materials. A Fermi surface comprising quasi-two-dimensional hole and electron cylinders at the Γ and $\tilde{M}$ points of the two-Fe ion Brillouin zone, respectively, in the overdoped paramagnetic side of the phase diagram evolves to a Fermi surface comprising small three-dimensional sections in the antiferromagnetic parent pnictide material via a potential quantum critical point where the quasiparticle effective mass is enhanced. The quasi-nesting wavevector $Q = \left(\pi/\tilde{a}, \pi/\tilde{a}\right)$ that connects sections of the Fermi surface, and the potential quantum critical point under the superconducting dome appear to be common features of the pnictide with the cuprate family of superconductors. While enhanced antiferromagnetic correlations appear both in pnictide and cuprate superconductors, an open question still pertains to their relation to elevated superconducting temperatures.

Acknowledgements

SES acknowledges support from the Royal Society and King's College, University of Cambridge, and thanks A. V. Chubukov, M. D. Johannes, G. G. Lonzarich, and I. I. Mazin for useful discussions.

References

1. R. Settai, H. Shishido, T. Kubo, S. Araki, T. C. Kobayashi, H. Harima, Y. Onuki, Drastic change of the electronic states at the quantum critical point in heavy fermions, *J. Magn. Mag. Matls.* **310**, 541–547, (2007).
2. S. E. Sebastian, N. Harrison, C. D. Batista, S. A. Trugman, V. Fanelli, M. Jaime, T. P. Murphy, E. C. Palm, H. Harima, and T. Ebihara, Heavy holes as a precursor to superconductivity in antiferromagnetic $CeIn_3$, *Proc. Natl. Acad. Sci.* **106**, 7741–7744, (2009).
3. C. Bergemann, S. R. Julian, A. P. Mackenzie, S. NishiZaki and Y. Maeno, Detailed topography of the Fermi surface of Sr_2RuO_4, *Phys. Rev. Lett.* **84**, 2662–2665, (2000).
4. J. Wosnitza, Fermi Surfaces of Low-Dimensional Organic Metals and Superconductors, *Springer, New York*, (1996).
5. N. Doiron-Leyraud, C. Proust, D. LeBoeuf, J. Levallois, J.-B. Bonnemaison, R. X. Liang, D. A. Bonn, W. N. Hardy, and L. Taillefer, Quantum oscillations and the Fermi surface in an underdoped high-T_c superconductor *Nature* **447**, 565–568, (2007).
6. E. A. Yelland, J. Singleton, C. H. Mielke, N. Harrison, F. F. Balakirev, B. Dabrowski, & J. R. Cooper, Quantum Oscillations in the Underdoped Cuprate $YBa_2Cu_4O_8$ *Phys. Rev. Lett.* **100**, 047003, (2008).

7. A. F. Bangura, J. D. Fletcher, A. Carrington, J. Levallois, M. Nardone, B. Vignolle, P. J. Heard, N. Doiron-Leyraud, D. LeBoeuf, L. Taillefer, S. Adachi, C. Proust, N. E. Hussey, Small Fermi surface pockets in underdoped high temperature superconductors: observation of Shubnikov de Haas oscillations in $YBa_2Cu_4O_8$ *Phys. Rev. Lett.* **100**, 047004, (2008).
8. S. E. Sebastian, N. Harrison, C. H. Altarawneh, R. X. Liang, D. A. Bonn, W. N. Hardy, & G. G. Lonzarich, Fermi-liquid behavior in an underdoped high-T_c superconductor *Phys. Rev.* B **81**, 140505, (2010).
9. S. E. Sebastian, N. Harrison, E. Palm, T. P. Murphy, C. H. Mielke, R. X. Liang, D. A. Bonn, W. N. Hardy, & G. G. Lonzarich, A multi-component Fermi surface in the vortex state of an underdoped high-T_c superconductor *Nature* **454**, 200–203, (2008).
10. A. Audouard, C. Jaudet, D. Vignolles, R. X. Liang, D. A. Bonn, W. N. Hardy, L. Taillefer, & C. Proust, Multiple quantum oscillations in the de Haas—van Alphen spectra of the underdoped high-temperature superconductor $YBa_2Cu_3O_{6.5}$, *Phys. Rev. Lett.* **103**, 157003, (2009).
11. B. J. Ramshaw, B. Vignolle, J. Day, R. X. Liang, W. N. Hardy, C. Proust, D. A. Bonn, Angle dependence of quantum oscillations in $YBa_2Cu_3O_{6.59}$ shows free-spin behaviour of quasiparticles *Nature Phys.* **7**, 234-238, (2011).
12. S. E. Sebastian, N. Harrison, R. Liang, D. A. Bonn, W. N. Hardy, C. Mielke, & G. G. Lonzarich, Quantum oscillations from nodal bilayer magnetic breakdown in the underdoped high temperature superconductor $YBa_2Cu_3O_{6+x}$ *Phys. Rev. Lett.* **108**, 196403, (2012).
13. S. E. Sebastian, N. Harrison and G. G. Lonzarich, Towards resolution of the Fermi surface in underdoped high-T_c superconductors *Rep. Prog. Phys.* in proof, (2012).
14. D. Shoenberg, *Magnetic Oscillations in Metals* (Cambridge: Cambridge University Press), (1984).
15. K. Yamaji, On the angle dependence of the magnetoresistance in quasi-two-dimensional organic superconductors, *J. Phys. Soc. Jpn.* **58**, 1520–1523, (1989).
16. S. E. Sebastian, N. Harrison, M. M. Altarawneh, C. H. Mielke, R. Liang, D. A. Bonn, W. N. Hardy, G. G. Lonzarich, Metal-insulator quantum critical point beneath the high T_c superconducting dome, *Proc. Natl. Acad. Sci.* **107**, 6175–6179, (2010).
17. P. Coleman, C. Pepin, Q. Si, and R. Ramazashvili, How do Fermi liquids get heavy and die? *J. Phys.: Condens. Matter* **13**, R723–R738, (2001).
18. Y. Kamihara, H. Hiramatsu, M. Hirano, R. Kawamura, H. Yanagi, T. Kamiya and H. Hosono, Iron-based layered superconductor: LaOFeP, *J. Am. Chem. Soc.* **128**, 10012, (2006).

19. Y. Kamihara, T. Watanabe, M. Hirano and H. Hosono, Iron-based layered superconductor $La[O_{1-x}F_x]FeAs$ ($x = 0.05 - 0.12$) with $T_c = 26$ K, *J. Am. Chem. Soc.* **130**, 3296–3297, (2008).
20. Z. Ren, W. Lu, J. Yang, W. Yi, X. Shen, Z. Li, G. Che, X. Dong, Li. Sun, F. Zhou, Z. Zhao, Superconductivity at 55 K in iron-based F-doped layered quarternary compound $SmO_{1-x}F_xFeAs$, *Chin. Phys. Lett.* **25**, 2215, (2008).
21. M. R. Norman, The challenge of unconventional superconductivity, *Science* **332**, 196–200, (2011).
22. M. Rotter, M. Tegel, D. Johrendt, I. Schellenberg, W. Hermes and R. Pöttgen, Spin-density-wave anomaly at 140 K in the ternary iron arsenide $BaFe_2As_2$, *Phys. Rev. B* **78**, 020503, (2008).
23. C. Krellner, N. Caroca-Canales, A. Jesche, H. Rosner, A. Ormeci, and C. Geibel, Magnetic and structural transitions in layered iron arsenide systems: AFe_2As_2 versus RFeAsO, *Phys. Rev. B.* **78**, 100504(R) (2008).
24. J. K. Dong, S. Y. Zhou, T. Y. Guan, H. Zhang, Y. F. Dai, X. Qiu, X. F. Wang, Y. He, X. H. Chen, and S. Y. Li, Quantum criticality and nodal superconductivity in the FeAs-based superconductor KFe_2As_2, *Phy. Rev. Lett.*, **104**, 087005, (2010).
25. X. C. Wang, Q. Q. Liu, Y. X. Lv, W. G. Gao, L. X. Yang, R. C. Yu, F. Y. Li, and C. Q. Jin, The superconductivity at 18 K in LiFeAs system, *Solid State Commun.* **148**, 538 (2008).
26. J. H. Tapp, Z. Tang, B. Lv, K. Sasmal, B. Lorenz, P. C. W. Chu, and A. M. Guloy, LiFeAs: An intrinsic FeAs-based superconductor with $T_c = 18$ K, *Phys. Rev. B* **78**, 060505 (2008).
27. M. J. Pitcher, D. R. Parker, P. Adamson, S. J. C. Herkelrath, A. T. Boothroyd, R. M. Ibberson, M. Brunelli, and S. J. Clarke, Structure and superconductivity of LiFeAs, *Chem. Commun.*, 5918–5920, (2008).
28. F.-C. Hsu, J.-Y. Luo, K.-W. Yeh, T.-K. Chen, T.-W. Huang, P. M. Wu, Y.-C. Lee, Y.-L. Huang, Y.-Y. Chu, D.-C. Yan, *et al.*, Superconductivity in the PbO-type structure α-FeSe, *Proc. Nat. Acad. Sci. (USA)* **105**, 14262, (2008).
29. C. de la Cruz, Q. Huang, J. W. Lynn, J. Y. Li, W. Ratcliff, II, J. L. Zaretsky, H. A.Mook, G. F. Chen, J. L. Luo, N. L. Wang, & P. Dai, Magnetic order close to superconductivity in the iron-based layered $LaO_{1-x}F_xFeAs$, *Nature (London)* **453**, 899 (2008).
30. M. Rotter, M. Tegel, and D. Jorendt, Superconductivity at 38 K in the iron arsenide $(Ba_{1-x}K_x)Fe_2As_2$, *Phys. Rev. Lett.* **101**, 107006, (2008).
31. S. Jiang, H. Xing, G. Xuan, C. Wang, Z. Ren, C. Feng, J. Dai, Z. Xu and G. Cao, Superconductivity up to 30 K in the vicinity of the quantum critical point in $BaFe_2(As_{1-x}P_x)_2$, *J. Phys. Cond. Mat.* **21**, 382203, (2009).

32. T. Kobayashi, S. Miyasaka, S Tajima, Single crystal growth and physical properties of $SrFe_2(As_{1-x}P_x)_2$, http://arxiv.org/abs/1112.3743v1, (2011).

33. S. Kasahara, T. Shibauchi, K. Hashimoto, Y. Nahai, H. Ikeda, T. Terashima, and Y. Matsuda, Abrupt recovery of Fermi-liquid transport by the *c*-axis collapse in $CaFe_2(As_{1-x}P_x)_2$ single crystals, *Phys. Rev. B* **83**, 060505, (2011).

34. A. S. Sefat, R. Jin, M. A. McGuire, B. C. Sales, D. J. Singh and D. Mandrus, Superconductivity at 22 K in Co-doped $BaFe_2As_2$ crystals, *Phys. Rev. Lett.* **101**, 117004, (2008).

35. S. Kasahara, T. Shibauchi, K. Hashimoto, K. Ikada, S. Tonegawa, R. Okazaki, H. Shishido, H. Ikeda, H. Takeya, K. Hirata, T. Terashima and Y. Matsuda, Evolution from non-Fermi to Fermi-liquid transport via isovalent doping in $BaFe_2(As_{1-x}P_x)_2$ superconductors, *Phys. Rev. B* **81**, 184519, (2010).

36. P. L. Alireza, Y. T. C. Ko, J. Gillett, C. M. Petrone, J. M. Cole, G. G. Lonzarich and S. E. Sebastian, Superconductivity up to 29 K in $SrFe_2As_2$ and $BaFe_2As_2$ at high pressures, *J. Phys. Cond. Mat.* **21**, 012208, (2009).

37. Y. Mizuguchi, F. Tomioka, S. Tsuda, T. Yamaguchi, and Y. Takano, Superconductivity at 27 K in tetragonal FeSe under high pressure, *Appl. Phys. Lett.* **93**, 152505, (2008).

38. R. Liang, D. A. Bonn, W. N. Hardy, Evaluation of CuO_2 plane hole doping in $YBa_2Cu_3O_{6+x}$ single crystals, *Phys. Rev. B* **73**, 180505(R), (2006).

39. S. E. Sebastian, J. Gillett, N. Harrison, P. H. C. Lau, D. J. Singh, C. H. Mielke and G. G. Lonzarich, Quantum oscillations in the parent magnetic phase of an iron arsenide high temperature superconductor *J. Phys. Cond. Matt.* **20**, 422203, (2008).

40. A. I. Coldea, J. D. Fletcher, A. Carrington, J. G. Analytis, A. F. Bangura, J. -H. Chu, A. S. Erickson, I. R. Fisher, N. E. Hussey, and R. D. McDonald, Fermi surface of superconducting LaFePO determined from quantum oscillations, *Phys. Rev. Lett.* **101**, 216402, (2008).

41. J. G. Analytis, R. D. Mcdonald, J. H. Chu, S. C. Riggs, A. F. Bangura, C. Kucharczyk, M. Johannes and I. R. Fisher, Quantum oscillations in the parent pnictide $BaFe_2As_2$: itinerant electrons in the reconstructed state, *Phys. Rev. B* **80**, 064507 (2009).

42. N. Harrison, R. D. Mcdonald, C. H. Mielke, E. D. Bauer, F. Ronning and J. D. Thompson, Quantum oscillations in antiferromagnetic $CaFe_2As_2$ on the brink of superconductivity, *J. Phys.-Condens. Matter* **21**, 322202, (2009).

43. T. Terashima, N. Kurita, M. Tomita, K. Kihou, C. H. Lee, Y. Tomioka, T. Ito, A. Iyo, H. Eisaki, T. Liang, M. Nakajima, S. Ishida, S. Uchida, H. Harima and S. Uji, Complete Fermi surface in $BaFe_2As_2$ observed via Shubnikov-de Haas oscillation measurements on detwinned single crystals, *Phys. Rev. Lett.* **107**, 176402, (2011). In this work the F_β frequency is suggested to correspond to the $2F_\gamma$ harmonic.

44. M. Sutherland, D. J. Hills, B. S. Tan, M. M. Altarawneh, N. Harrison, J. Gillett, E. C. T. O'Farrell, T. M. Benseman, I. Kokanovic, P. Syers, J. R. Cooper, and S E. Sebastian, Evidence for Dirac nodes from quantum oscillations in $SrFe_2As_2$, *Phys. Rev. B* **84**, 180506(R), (2011).

45. N. Harrison, S. E. Sebastian, Dirac nodal pockets in the antiferromagnetic parent phase of FeAs superconductors, *Phys. Rev. B* **80**, 224512, (2009).

46. J. G. Analytis, C. M. J. Andrew, A. I. Coldea, A. Mccollam, J. H. Chu, R. D. Mcdonald, I. R. Fisher and A. Carrington, Fermi surface of $SrFe_2P_2$ determined by the de Haasvan Alphen effect, *Phys. Rev. Lett.* **103**, 076401, (2009).

47. A. I. Coldea, C. M. J. Andrew, J. G. Analytis, R. D. McDonald, A. F. Bangura, J. H. Chu, I. R. Fisher and A. Carrington, Topological change of the Fermi surface in ternary iron pnictides with reduced c/a ratio: a de Haasvan Alphen study of $CaFe_2P_2$, *Phys. Rev. Lett.* **103**, 026404, (2009).

48. B. Arnold, S. Kasahara, A. Coldea, T. Terashima, Y. Matsuda, T. Shibauchi and A. Carrington, Nesting of electron and hole Fermi surfaces in nonsuperconducting $BeFe_2P_2$, *Phys. Rev. B* **83**, 220504, (2011).

49. J. G. Analytis, J.-H. Chu, R. D. McDonald, S. C. Riggs, and I. R. Fisher, Enhanced Fermi-surface nesting in superconducting $BaFe_2(As_{1-x}P_x)_2$ revealed by the de Haas–van Alphen effect, *Phys. Rev. Lett.* **105**, 207004, (2010).

50. T. Terashima, M. Kimata, N. Kurita, H. Satsukawa, A. Harada, K. Hazama, M. Imai, A. Sato, K. Kihou, C. Lee, H. Kito, H. Eisaki, A. Iyo, T. Saito, H. Fukazawa, Y. Kohori, H. Harima, and S. Uji, Fermi surface and mass enhancement in KFe_2As_2 from de Haas–van Alphen effect measurements, *J. Phys. Soc. Jpn.* **79**, 053702, (2010).

51. C. Putzke, A. I. Coldea, I. Guillamon, D. Vignolles, A. McCollam, D. LeBoeuf, M. D. Watson, I. I. Mazin, S. Kasahara, T. Terashima, T. Shibauchi, Y. Matsuda, A. Carrington, A de Haas–van Alphen study of the Fermi surfaces of superconducting LiFeP and LiFeAs, *Phys. Rev. Lett.* **108**, 047002, (2011).

52. A. I. Goldman, D. N. Argyriou, B. Ouladdiaf, T. Chatterji, A. Kreyssig, S. Nandi, N. Ni, S. L. Bud'ko, P. C. Canfield, and R. J. McQueeney,

Lattice and magnetic instabilities in $CaFe_2As_2$: a single-crystal neutron diffraction study, *Phys. Rev. B* **78**, 100506(R), (2008).

53. D. J. Singh and M. H. Du, Density functional study of $LaFeAsO_{1-x}F_x$: a Low carrier density superconductor near itinerant magnetism, *Phys. Rev. Lett.* **100**, 237003, (2008).
54. I. I. Mazin, D. J. Singh, M. D. Johannes and M. H. Du, Unconventional superconductivity with a sign reversal in the order parameter of $LaFeAsO_{1-x}$, *Phys. Rev. Lett.* **101**, 057003, (2008).
55. K. Kuroki, S. Onari, R. Arita, H. Usui, Y. Tanaka, H. Kontani and H. Aoki, Unconventional superconductivity originating from disconnected Fermi surfaces in the iron-based oxypnictide, *Phys. Rev. Lett.* **101**, 087004, (2008).
56. A. V. Chubukov, D. V. Efremov and I. Eremin, Magnetism, superconductivity, and pairing symmetry in iron-based superconductors, *Phys. Rev. B* **78**, 134512, (2008).
57. V. Cvetkovic and Z. Tesanovic, Multiband magnetism and superconductivity in Fe-based compounds, *EPL* **85**, 37002, (2009).
58. I. I. Mazin and J. Schmalian, Pairing symmetry and pairing state in ferropnictides: theoretical overview, *Physica C* **469**, 614–627, (2009).
59. Q. Huang, Y. Qiu, W. Bao, M. A. Green, J. W. Lynn, Y. C. Gasparovic, T. Wu, G. Wu, X. H. Chen, Neutron-diffraction measurements of magnetic order and a structural transition in the parent $BaFe_2As_2$ compound of FeAs-based high-temperature superconductors, *Phys. Rev. Lett.* **101**, 257003, (2008).
60. J. Zhao, D. X. Yao, S. Li, T. Hong, Y. Chen, S. Chang, W. Ratcliff, J. W. Lynn, H. A. Mook, G. F. Chen, J. L. Luo, N. L. Wang, E. W. Carlson, J. Hu, P. Dai, Low energy spin waves and magnetic interactions in $SrFe_2As_2$, *Phys. Rev. Lett.* **101**, 167203, (2008).
61. R. A. Ewings, T. G. Perring, J. Gillett, S. D. Das, S. E. Sebastian, A. E. Taylor, T. Guidi, and A. T. Boothroyd, Itinerant spin excitations in $SrFe_2As_2$ measured by inelastic neutron scattering, *Phys. Rev. B* **83**, 214519, (2011).
62. Y. Ran, F. Wang, H. Zhai, A. Vishwanath, D.-H. Lee, Nodal spin density wave and band topology of the FeAs-based materials, *Phys. Rev. B* **79**, 014505, (2009).
63. P. Richard, K. Nakayama, T. Sato, M. Neupane, Y.-M. Xu, J. H. Bowen, G. F. Chen, J. L. Luo, N. L. Wang, X. Dai, Z. Fang, H. Ding, and T. Takahashi, Observation of Dirac cone electronic dispersion in $BaFe_2As_2$, *Phys. Rev. Lett.* **104**, 137001, (2010).

64. G. Xu, H. Zhang, X. Dai and Z. Fang, Electron-hole asymmetry and quantum critical point in hole-doped $BaFe_2As_2$, *European Phys. Lett.* **84**, 67015, (2008).

65. V. Brouet, M. Marsi, B. Mansart, A. Nicolaou, A. Taleb-Ibrahimi, P. Le Fèvre, F. Bertran, F. Rullier-Albenque, A. Forget, and D. Colson, Nesting between hole and electron pockets in $Ba(Fe_{1-x}Co_x)_2As_2$ (x=0-0.3) observed with angle-resolved photoemission, *Phys. Rev. B* **80**, 165115, (2009).

66. A. Carrington, Quantum oscillation studies of the Fermi surface of iron-pnictide superconductors, *Rep. Prog. Phys.* **74**, 124507, (2011).

67. T. Helm, M. V. Kartsovnik, M. Bartkowiak, N. Bittner, M. Lambacher, A. Erb, J. Wosnitza, and R. Gross, Evolution of the Fermi surface of the electron-doped high-temperature superconductor $Nd_{2-x}Ce_xCuO_4$ revealed by Shubnikov-de Haas oscillations, *Phys. Rev. Lett.*, **103**, 157002, (2009).

68. D. N. Basov and A. V. Chubukov, Manifesto for a higher T_c, *Nature Phys.* **7**, 272–276, (2011).

69. B. Vignolle, A. Carrington, R. A. Cooper, J. M. M. French, A. P. Mackenzie, C. Jaudet, D. Vignolles, C. Proust, N. E. Hussey. Quantum oscillations in an overdoped high-T_c superconductor, *Nature* **455**, 952–955, (2008).

70. B. I. Halperin and T. M. Rice, The Excitonic State at the Semiconductor-Semimetal Transition, in Solid State Physics, edited by F. Seitz, D. Turnbull, and H. Ehrenreich (Academic Press, N. Y.) **21**, 115, (1968).

71. P. Monthoux and G. G. Lonzarich, Magnetic interactions in a single-band model for the cuprates and ruthenates, *Phys. Rev. B* **71**, 054504, (2005).

72. P. Monthoux, D. Pines & G. G. Lonzarich, Superconductivity without phonons, *Nature* **450**, 20, (2007).

73. H. Shishido, A. F. Bangura, A. I. Coldea, S. Tonegawa, K. Hashimoto, S. Kasahara, P. M. C. Rourke, H. Ikeda, T. Terashima, R. Settai, Y. Onuki, D. Vignolles, C. Proust, B. Vignolle, A. Mccollam, Y. Matsuda, T. Shibauchi and A. Carrington, Evolution of the Fermi surface of $BaFe_2(As_{1-x}P_x)$ on entering the superconducting dome, *Phys. Rev. Lett.* **104**, 057008, (2010).

74. J. Singleton, C. de la Cruz, R. D. McDonald, S. Li, M. Altarawneh, P. Goddard, I. Franke, D. Rickel, C. H. Mielke, X. Yao, and P. Dai, Magnetic quantum oscillations in $YBa_2Cu_3O_{6.61}$ and $YBa_2Cu_3O_{6.69}$ in fields of Up to 85 T: patching the hole in the roof of the superconducting dome, *Phys. Rev. Lett.* **104**, 086403, (2010).

Chapter 5

Optical Investigation on Iron-Based Superconductors

Nan-Lin Wang* **and Zhi-Guo Chen**
Beijing National Laboratory for Condensed Matter Physics,
Institute of Physics, Chinese Academy of Sciences,
Beijing 100190, China
*nlwang@aphy.iphy.ac.cn

5.1 Introduction

The discoveries of superconductivity in Fe- and Ni-based layered pnictide compounds[1–3] have generated tremendous interest in condensed matter community. For the iron-based compounds, not only could they have high superconducting transition temperatures[4–6] but also show intriguing interplay between structure, magnetism and superconductivity. The undoped compounds commonly display the structural and magnetic phase transitions, which, depending on materials, could occur either at the same temperature or separately.[7–9] Upon electron or hole doping or application of pressure, both the magnetic order and the structural transition are suppressed, and superconductivity emerges.[8,10,11] A variety of Fe-based superconductors with different structures have been discovered. Based on the magnetic structures of the parent compounds,

Iron-Based Superconductors: Materials, Properties, and Mechanisms
Edited by Nan-Lin Wang, Hideo Hosono, and Pengcheng Dai

ISBN 978-981-4303-22-4 (Hardcover), 978-981-4303-23-1 (eBook)
www.panstanford.com

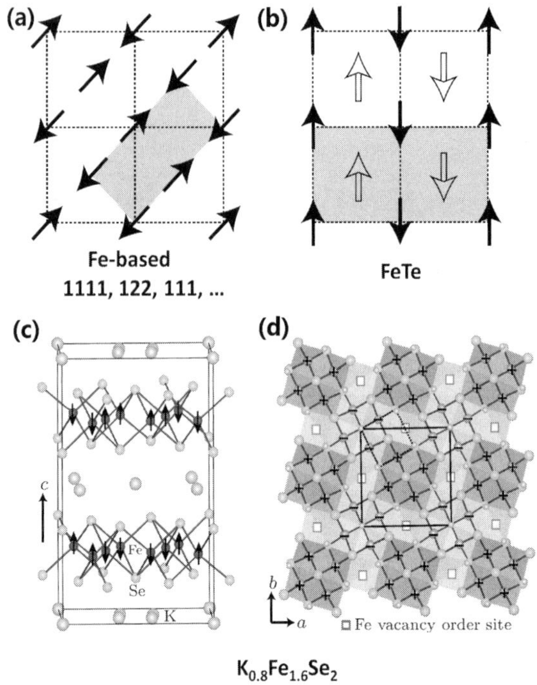

Figure 5.1 Three different types of magnetic structures for the parent compounds of Fe-based superconductors.

the family of iron-based superconductors could be classified into three different categories: (1) the Fe-arsenic compounds, including the RFeAsO (R=La, and rare-earth elements) (1111 structural type series),[7,10,12] AFe_2As_2 (A = Ca, Sr, Ba, Eu) (122-type)[13–15] and $(Li,Na)_xFeAs$ (111-type);[16] (2) the Fe-selenides Fe(Te,Se) (11-type);[17,18] and (3) the alkaline metal intercalated Fe-selenides $A_xFe_{2-y}As_2$ (A = K, Rb, Cs and Tl).[19] Their magnetic structures are shown in Fig. 5.1. The three different groups behave quite differently in a number of physical properties, although the superconductivity in those compounds must have the same origin. Comparative studies on the different Fe-based systems could help isolate key factors responsible for the superconductivity.

A large number of theoretical and experimental studies on the Fe-based superconductors have been carried out with the goal of understanding the mechanisms responsible for the high temperature superconductivity. In the context of these studies, optical spectroscopy measurements have provided much useful information about charge dynamics and band structures of the materials, since the technique probes both free carriers and interband excitations. It also plays a crucial role in identifying energy gaps in the broken symmetry states, revealing the correlation effect on the basis of the spectral weight analysis and determining the transport lifetime of quasi-particles.

In this chapter, we review the optical spectroscopy studies on the iron-based parent compounds and superconductors. The chapter is organized as follows: Section II is a very brief introduction about the basic concepts and elementary theory of the optical properties of solids. Section III focuses on the low energy excitations of parent compounds. Very different optical excitations were observed for Fe-pnictides, FeTe and Fe-selenides $K_xFe_{2-y}As_2$. The complex interplay between the Fermi surface (FS) topology and the magnetism was discussed. Section IV presents the general feature of optical response in the paramagnetic phase and the analysis based on two different approaches: the Drude-Lorentz decompositions and generalized Drude model. Section V focuses on the correlation effect in pnictides and chalcogenides. The Hund's coupling and the coexistence of the itinerant and localized Fe 3d electrons are emphasized. Section VI addresses the anisotropic charge responses with E|| *ab*-plane and E|| *c*-axis. The in-plane anisotropy in detwinned parent compound was also discussed. Section VII emphasizes the results from the superconducting response. A number of phenomena such as spectral features of pairing energy gaps, the condensed carrier density, the Josephson-coupling plasma edge in nanoscale phase-separated $K_xFe_{2-y}As_2$, the superconductivity induced high-energy spectral weight transfer, and the coherent peak just below T_c due to type-II coherent factor, were discussed.

During the last three years, a large number of the papers from the optical spectroscopy probes have appeared in the literature. A small number of the results were reviewed in early articles.[20,21] In this short review, we shall mainly focus on recent works and a few of selected issues. We apologize to many authors who have published excellent papers but not reviewed in detail here.

5.2 Introduction About Optical Properties of Solids

5.2.1 *Optical Constants*

The optical properties of a medium are often described by a number of optical constants. We shall first give a brief introduction about the relationships between different optical constants.[22,23]

Let us consider an electromagnetic wave in a medium propagating along the x axis with the electric field in the y direction, it is characterized by a plane wave,

$$E_y = E_0 e^{i(qx-\omega t)} = E_0 e^{i\omega(x/v-t)} = E_0 e^{i\omega(nx/c-t)}, \tag{5.1}$$

where $v = \omega/q = c/n$ is the phase velocity, n is the refractive index of the medium. In the presence of absorption, the intensity of the electric field would decay with the propagation of a distance of x,

$$E_y = E_0 e^{-\omega kx/c} e^{i\omega(nx/c-t)}, \tag{5.2}$$

where k is called an attenuation factor. The intensity of the electric field becomes $I \propto E_y^2 = E_0\ e^{-2\omega kx/c}$. In general, n and k are not constants but frequency-dependent, reflecting dispersive property. If one introduces a complex refractive index $N(\omega) = n(\omega) + ik(\omega)$, the electric field still has the form of simple plane wave,

$$E_y = E_0 e^{i\omega[N(\omega)x/c-t]}. \tag{5.3}$$

Now consider an electromagnetic plane wave shining on the interface between vacuum and a medium with the complex refractive index $N(\omega)$ (Fig. 5.2), then part of the radiation will be

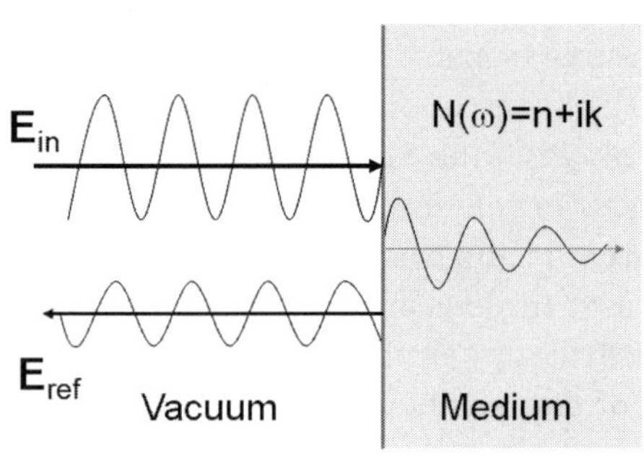

Figure 5.2 The reflection of the incident beam at the interface between vacuum and a medium.

reflected and part will transmit into the medium. In the case of normal incidence, the ratio of the electric field of the reflected beam over that of the incident beam is known to be a function of $n(\omega)$ and $k(\omega)$ of the medium,

$$\frac{E_{ref}}{E_{in}} = \frac{n+ik-1}{n+ik+1} = \sqrt{\frac{(n-1)^2+k^2}{(n+1)^2+k^2}}e^{i\theta} = r(\omega)e^{i\theta}. \quad (5.4)$$

Here, θ is phase shift of the reflected wave with respect to the incoming wave. From this equation we get following formula of reflectivity,

$$R = |\frac{E_{ref}}{E_{in}}|^2 = |r(\omega)|^2 = \frac{(n-1)^2+k^2}{(n+1)^2+k^2}, \quad (5.5)$$

and

$$\tan\theta = \frac{2k}{n^2+k^2-1}; \quad (5.6)$$

or

$$n = \frac{1-R}{1+R-2R^{1/2}\cos\theta}, k = \frac{-2R^{1/2}\sin\theta}{1+R-2R^{1/2}\cos\theta}. \quad (5.7)$$

It is easy to show that the complex refractive index is related to the complex dielectric function and the complex conductivity by following formulae: $N(\omega) \equiv n(\omega) + ik(\omega) = \sqrt{\varepsilon(\omega)}$, and $\varepsilon(\omega) = 1 + \mathrm{i}\frac{4\pi\sigma(\omega)}{\omega}$.[a] Then, the real part and imaginary part of the dielectric function and conductivity can be written as

$$\varepsilon_1 = n^2(\omega) - k^2(\omega), \varepsilon_2 = 2n(\omega)k(\omega), \quad (5.8)$$

and

$$\sigma_1 = \omega\varepsilon_2/4\pi, \sigma_2 = \omega(1-\varepsilon_1)/4\pi. \quad (5.9)$$

Now we have several pairs of optical constants: R(ω), $\theta(\omega)$; n(ω), k(ω); $\varepsilon_1(\omega)$, $\varepsilon_2(\omega)$; and $\sigma_1(\omega)$, $\sigma_2(\omega)$. If two of them are known, the rest could be easily calculated. In some measurement techniques, for example, the ultrafast laser-based time-domain Terahertz (THz) spectroscopy, or the ellipsometry measurement configuration, both the real and imagine parts of the complex refractive index or dielectric function could be directly determined. But those measurements

[a]In solids, the ions or high energy interband excitations also contribute to the dielectric constant. Then, the relation between the dielectric function and the conductivity could be generally expressed as $\varepsilon(\omega) = \varepsilon_\infty + \mathrm{i}\frac{4\pi\sigma(\omega)}{\omega}$.

usually cover only a limited range of frequencies. In many cases, reflectivity spectrum over broad frequencies of a material could be more easily measured with the use of a Fourier transform infrared spectrometer or grating-type spectrometer. Then one needs to know the phase shift θ in order to determine all other related optical constants. This was bridged by the Kramers-Kronig relation. Because of the causality condition of optical response, the real and imagine parts of a response function are not independent, but linked by an integral relation being called Kramers-Kronig transformation. In the optical reflectance measurement, the reflected light could be considered as the response of the incident light. Then, the phase shift of the reflected light relative to the incident light, $\theta(\omega)$, can be calculated through Kramers-Kronig transformation by the following relation:

$$\theta(\omega) = -\frac{\omega}{\pi} P \int_0^{+\infty} \frac{In R(s)}{s^2 - \omega^2} ds. \tag{5.10}$$

In this equation, the principle integral is from zero frequency to infinity. Since the detected frequency range is finite, appropriate extrapolations of $R(\omega)$ at low and high frequency are needed in order to work out the phase shift $\theta(\omega)$. Once the phase shift is obtained, all other optical constants could be calculated easily.

5.2.2 *Interband and Intraband Excitations*

We shall concentrate on the electronic response of a solid material to the external optical radiation. Under the periodic lattice potentials, the electrons in solids form energy bands. Then, the optical spectroscopy probes the transitions of the electrons from initial states to final states with the absorption of the photon energies. The initial states must be the occupied states of electrons and the final states must be the empty states permitted by the selection rule. The initial state and final state are allowed in either the same or different energy bands. If the initial state and the final state are in different bands, what the optical spectroscopy detects is the interband excitations (shown in Fig. 5.3). It is not difficult to derive the imaginary part of the dielectric function associated with the

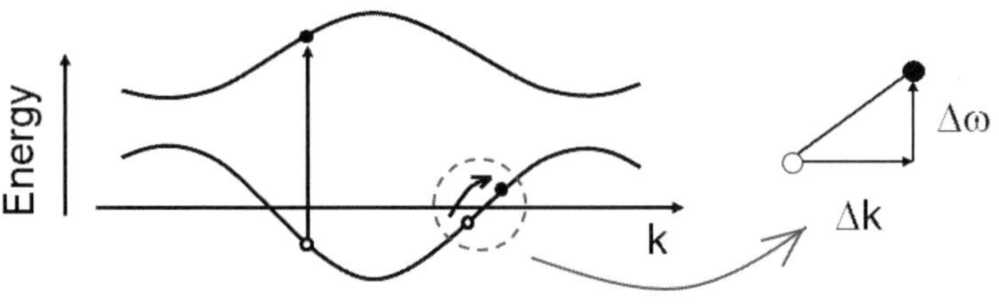

Figure 5.3 The inter- and intra-band transitions of electrons in solids. Intraband transition has an associated change of the momentum of the electron.

interband transitions–so called the Kubo-Greenwood formula,[22,23]

$$\varepsilon_2(\omega) = \frac{8\pi^2 e^2}{m^2\omega^2} J\,(\hbar\omega)|\boldsymbol{p}_{vc}(\hbar\omega)|^2, \tag{5.11}$$

where $J\,(\hbar\omega)$ is the joint density of the states, $|\boldsymbol{p}_{vc}(\hbar\omega)|$ is the dipole transition matrix element. Then, one can obtain the real part of the optical conductivity from

$$\sigma_1(\omega) = \frac{1}{4\pi}\omega\varepsilon_2(\omega). \tag{5.12}$$

$\varepsilon_2(\omega)$ or $\sigma_1(\omega)$ are closely related to the joint density of the states and the dipole transition matrix element, which could be obtained through band structure calculations. Comparing the experimental optical constants with those by theoretical calculations, one can extract useful information about the electronic band structure.

If the initial state and final state are in the same band, what the optics probes is the intraband excitations across the Fermi level. Because of the band dispersion, the intraband transition of electrons always has an associated change of momentum (shown in the enlarged part of Fig. 5.3). However, the photon momentum of infrared radiation is virtually zero, a direct intraband transition by solely absorbing the phonon energy would violate the momentum conservation, which, therefore, is forbidden. The occurrence of the intraband excitations must need the assistance of either the elastic impurities scattering or the inelastic scattering involving the electron-boson (for example, phonons, spin fluctuations, etc.)

coupling to conserve the electron momentum. Thus, the intra-band excitations are always involved with the elastic or inelastic scattering processes.

5.2.3 *Drude Model and Drude-Lorentz Model*

The simplest model of understanding the charge conduction in metal is the Drude model, in which the electrons are assumed to be a cloud of classic gas.[22,23] The electrons move freely between positive-charged ions and do not interact with each other between collisions. The central assumption of the model is the existence of an average relaxation time τ which governs the relaxation of the system to equilibrium. The conductivity for the Drude model could be easily derived as

$$\sigma(\omega) = \frac{\omega_p^2}{4\pi}\frac{1}{1/\tau - i\omega}. \tag{5.13}$$

The other expressions of optical constants in the Drude model and the schematic curve features are displayed in Fig. 5.4. For free-electron gas, the reflectivity $R(\omega)$ approaches unity at low frequencies, then drops rapidly near the plasma edge (corresponding to the screened plasma frequency in a solid material, $\omega_p/\sqrt{\varepsilon_\infty}$). The real part of the dielectric function crosses the horizontal axis at the plasma edge where there is a peak in the energy-loss function Im(-$1/\varepsilon$). The real part of the conductivity $\sigma_1(\omega)$ has a shape of half Lorentz peak centered at zero frequency. The width of the Drude peak Γ is just the scattering rate $\Gamma = 1/\tau$. The value at zero-frequency limit is the dc conductivity $\sigma_0 = \omega_p^2\tau/4\pi$, where $\omega_p^2 = 4\pi ne^2/m_e$ is the square of the plasma frequency. The integration of $\sigma_1(\omega)$ from zero to infinite frequency is $\int_0^\infty \sigma_1(\omega)d\omega = \omega_p^2/8$.

In contrast to the Drude model which describes the free carrier response, the Lorentz model is widely used to represent the response of bound-carriers or interband transitions. The Lorentz model could be easily derived by assuming that the electrons are bound by a harmonic potential and subject to a viscous damping force. The complex dielectric function can be expressed as

$$\varepsilon(\omega) = 1 + \frac{\omega_{pe}^2}{(\omega_0^2 - \omega^2) - i\omega/\tau} \tag{5.14}$$

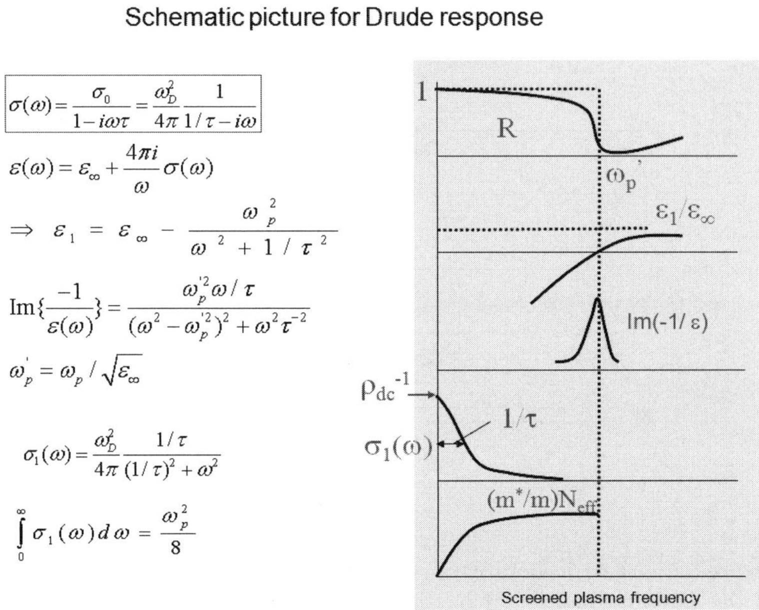

Figure 5.4 Schematic picture for the optical constants in the Drude model. Left part displays some useful formulae for the Drude model.

where ω_{pe} is the oscillator strength or plasma frequency and ω_0 is the oscillator frequency. The reflectivity $R(\omega)$, phase shift ϕ_r, complex dielectric function $\varepsilon(\omega)$, and energy loss function $-1/\varepsilon(\omega)$ are shown in Fig. 5.5. By setting the oscillator frequency $\omega_0 = 0$, namely the electrons are not bound, the Lorentz model naturally reduces to the Drude model.

The Drude and Lorentz models are classic theory of the electrons in solids. In reality, free carriers and bound electrons may coexist in a material, so a combined model, Drude-Lorentz model is often used for describing different types of excitations. the complex conductivity has the form

$$\sigma(\omega) = \frac{1}{4\pi}\left[\frac{\omega_{pD}^2\tau_D}{1-i\omega\tau_D} + \frac{\sum\omega_{pj}^2\omega}{i(\omega_j^2-\omega^2)+\omega/\tau_j}\right]. \quad (5.15)$$

Its first term represents the Drude response of the free carriers, the second term stands for the response of bound electrons or

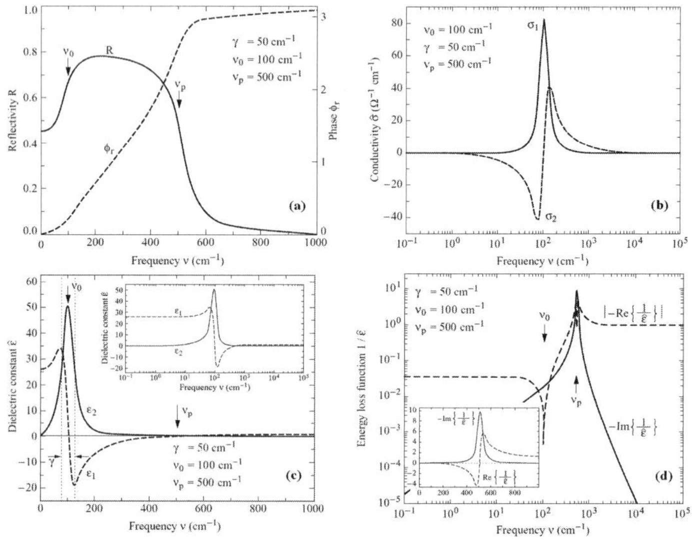

Figure 5.5 The behaviors of the physical quantities of the optical response in the Lorentz model.[23] Here, the oscillator frequency (or the center frequency) $\nu_0 = \omega_0/(2\pi c) = 100\ \mathrm{cm}^{-1}$, the width $\gamma = 1/(2\pi c\tau) = 50\ \mathrm{cm}^{-1}$, and the oscillator strength $\nu_p = \omega_{pe}/(2\pi c) = 500\ \mathrm{cm}^{-1}$.

interband transitions. The Drude-Lorentz model was widely used to decompose the different contributions in Fe-based superconductors.

5.2.4 *Extended Drude Model*

Within the Drude model, the electrons are regarded to be quasi-free and the scattering can be described by a frequency-independent quantity $1/\tau$. However, in many metals, many-body interactions or various scattering channels may exist so that the scattering rate is no longer frequency-independent, but varies with temperature and frequency. Thus, an extended Drude model was introduced to account for the inelastic scattering or the coupling of electrons to a bosonic spectrum.[24] It is shown that, because of the causality of response function, one can not solely change the constant $1/\tau$ to a frequency-dependent one $1/\tau(\omega)$ in the Drude model. An imaginary part has to be added to the damping rate, which can be interpreted as a modification of the mass of the electron that enters the expression for the plasma frequency. Introducing the complex damping function, also called as a memory function, $M(\omega) = 1/\tau(\omega) - i\omega\lambda(\omega)$, the extended Drude model could be expressed as

$$\sigma(\omega) = \frac{\omega_p^2}{4\pi} \frac{1}{1/\tau(\omega) - i\omega[1 + \lambda(\omega)]}. \tag{5.16}$$

Usually, one use the above equation to derive the frequency-dependent scattering rate or the effective mass from the real and imaginary parts of conductivity with the plasma frequency determined from the sum-rule $\omega_p^2 = 8\int_0^\infty \sigma_1(\omega)d\omega$,

$$\frac{1}{\tau(\omega)} = \frac{\omega_p^2}{4\pi} Re\left(\frac{1}{\sigma(\omega)}\right), \tag{5.17}$$

$$\frac{m^*}{m_b} = 1 + \lambda(\omega) = -\frac{\omega_p^2}{4\pi}\frac{1}{\omega} Im\left(\frac{1}{\sigma(\omega)}\right), \tag{5.18}$$

where m_b is the band mass, $\lambda(\omega)$ is the mass enhancement factor. In practice, the integration for calculating plasma frequency ω_p does not go to infinity in its upper limit but to a frequency above the intraband transition but below the interband transitions.

As mentioned above, the intraband transitions of electrons could not occur by solely absorbing the photon energy due to the requirement of momentum conservation. In general, the electron-boson coupling effect is involved in the intraband transition process. Memory function is particularly useful for the exploration of the electron-boson coupling. A well-known example is the electron-phonon interaction in metals. The frequency-dependent scattering rate at the zero temperature has the expression,[25]

$$1/\tau(\omega) = \frac{2\pi}{\omega}\int_0^\infty \alpha_{tr}^2 F(\Omega)(\omega - \Omega)d\Omega, \tag{5.19}$$

where $\alpha_{tr}^2 F(\Omega)$ is the Eliashberg function, proportional to the phonon density of state. At finite temperature, the formula should be replaced by,[26]

$$\frac{1}{\tau(\omega, T)} = \frac{\pi}{\omega}\int_0^\infty \alpha_{tr}^2 F(\Omega)\left[2\omega \coth\frac{\Omega}{2T} - (\omega + \Omega)\coth\frac{\omega + \Omega}{2T} + (\omega - \Omega)\coth\frac{\omega - \Omega}{2T}\right]d\Omega. \tag{5.20}$$

Through the inversion of the optical scattering rate as a function of frequency, usually by a maximum entropy method, the Eliashberg spectral function $\alpha_{tr}^2 F(\Omega)$ can be extracted, which may yield useful information about the coupling between the charge carriers and the bosonic modes. The extracted Eliashberg spectral function is often compared with the same quantity determined by other measurement techniques, e.g. neutron scattering or tunnelling spectroscopy. Such a theme was widely used in the study of high-T_c cuprates, and also employed by a few groups to interpret the optical data of Fe-pnictides, although the bosonic mode to which the electron couples may not be phonon.

The above electron-phonon coupling formulae are valid for metals or the normal state of superconductors. In the case of superconducting state, a pairing energy gap opens in the excitation spectrum. Then the above formulae have to be modified. For example, the expression of scattering rate at zero temperature becomes,[25]

$$1/\tau(\omega) = \frac{2\pi}{\omega}\int_0^{\omega-2\Delta} d\Omega(\omega-\Omega)\alpha_{tr}^2 F(\Omega) E\left[\sqrt{1-\frac{4\Delta^2}{(\omega-\Omega)^2}}\right], \tag{5.21}$$

where $E(x)$ is the second kind elliptic integral.

5.2.5 *Sum Rules*

Sum rules and conservation laws play an important role in metals and superconductors.[27] The conductivity sum rule is particularly useful and represents the conservation of charges. A fundamentally important sum rule is the f-sum rule:

$$\int_0^{\infty} \sigma_1(\omega)d\omega = \frac{\pi ne^2}{2m_e} = \omega_p^2/8 \tag{5.22}$$

where n is the density of the electrons in the material, and m the bare mass of the electron. The sum rule is independent on the nature of interaction, temperature, phase transitions, etc. It has the explicit implication that at energies higher than the total bandwidth of a solid, electrons behave as free particles.

As mentioned above, in practice the integration up to an infinite frequency is not possible. Usually, the summarization is limited to a finite frequency, which is referred to as partial sum rule. For the case of a single band crossing the Fermi level and being well separated from other bands, Kubo derived following sum rule:[27,28]

$$\int_0^W \sigma_1(\omega)d\omega = W_K = \frac{\pi e^2}{2N}\Sigma\nabla^2_{k_x}\epsilon_k n_k \tag{5.23}$$

where n_k is the electron distribution function and ϵ_k is the band dispersion. The upper limit of the integration is much larger than the bandwidth of a given band crossing the Fermi level but still smaller than the energy of interband transitions. For $\epsilon_k = k^2/2m_e$, the Kubo sum rule reduces to the f-sum rule. In general, however, W_K depends on T and on the state of the system because of n_k. The spectral weight changes with T is frequently called as a "violation of the conductivity sum rule," an issue being first studied in detail by Hirsch.[29] In reality, there is no true violation: the change of the spectral weight of a given band would be compensated by an appropriate change in the spectral weight in other bands, and the total spectral weight over all bands is conserved. Generally, the scale of the temperature dependence of W_K is an indicator how far in energy the change of conductivity extends. Because relevant energy scales increase with the interaction strength, the temperature dependence of W_K is also an indicator of whether a system is in a weak or strong coupling regime. If the relevant energy scale is much smaller than the Fermi energy, the band dispersion could be expanded near k_F as $\epsilon_k = v_F(k-k_F) + (k-k_F)^2/2m_B + O(k-k_F)^3$, and $\nabla^2_{k_x}\epsilon_k \approx 1/m_B$, then $W_K = n\pi e^2/(2m_B) = \omega^2_{p,B}/8$.

In the tight binding model, the above summarization of conductivity spectral weight could be shown to be equal to the kinetic energy of the system $K = 2\sum \epsilon_k n_k$. For a weak coupling case, the kinetic energy or the plasma frequency of the concerned band determined from the partial sum rule would be close to those determined from the band structure calculations. However, for the strong coupling materials, due to the strong modification of the band dispersion by correlation effect, the total intraband spectral weight would be suppressed so that $m_{exp} > m_B$, then $K_{exp}/K_{band} < 1$. The spectral weight analysis was widely used to show the strength of the correlation effect of a material,[27] including the Fe-based superconducting systems.

5.2.6 *Optical Response of Broken Symmetry States of Metals*

As a bulk measurement technique, optical spectroscopy is a powerful tool of probing energy gaps in the broken symmetry states, e.g. superconductivity and density wave states. Historically, infrared spectroscopy was the first spectroscopic technique in probing the pairing energy gap of a superconductor.[30]

The superconductivity in conventional metals was successfully explained by the BCS theory. The same theoretical frame was also used to understand the density wave instabilities. However, there are crucial differences in the electromagnetic response of the two-types broken symmetry states. The key reason is that the two phenomena have different coherent factors. Furthermore, spin density wave (SDW) arises as a consequence of the electron-electron interactions. It does not couple to the underlying lattice.

The BCS theory was extended by Mattis and Bardeen to calculate the conductivity as a function of frequency and temperature. The real part of conductivity in the broken symmetry state normalized to that in the normal state has the form[23]

$$\frac{\sigma_1(\omega, T)}{\sigma_n} = \frac{2}{\hbar\omega}\int_{\Delta}^{\infty}\frac{[f(E)-f(E+\hbar\omega)](E^2 \pm \Delta^2 + \hbar\omega E)}{(E^2-\Delta^2)^{1/2}[(E+\hbar\omega)^2-\Delta^2]^{1/2}}dE$$
$$-\frac{1}{\hbar\omega}\int_{\Delta-\hbar\omega}^{-\Delta}\frac{[1-2f(E+\hbar\omega)](E^2 \pm \Delta^2 + \hbar\omega E)}{(E^2-\Delta^2)^{1/2}[(E+\hbar\omega)^2-\Delta^2]^{1/2}}dE, \tag{5.24}$$

where $2\Delta(T)$ describes the temperature dependent single particle gap, and $f(E,T) = [1 + e^{E/k_BT}]^{-1}$ the Fermi–Dirac distribution. The positive sign in the equation is appropriate for case II, the negative sign is appropriate for case I coherence factors. For the BCS superconducting state, case II coherence factors apply. This leads to a smooth rise of the real part of conductivity for frequencies $\omega \geq 2\Delta$. In addition, at frequencies far below 2Δ, the real part of conductivity has a "coherent peak" just below T_c. The shapes of dissipations for tow different coherent factors are shown in Fig. 5.6. For density wave states, case I coherence factors appear in the electrodynamic response, leading to a characteristic square root singularity at the gap frequency.

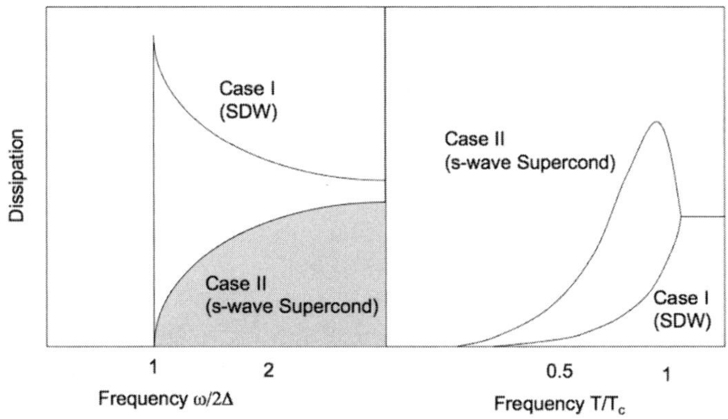

Figure 5.6 Approximate frequency- and temperature-dependence of the dissipation rate from the Mattis-Bardeen theory for Type I and Type II coherence factors (left panel $T = 0$ K; right panel $\omega = 0.1\Delta$).[31]

For a superconductor, the pairing energy gap is the minimum energy to break a Cooper pair. In weak-coupling BCS theory, this energy at zero temperature is given by $2\Delta = 3.5k_B T_c$. Due to the energy gap formation, there exists substantial suppression in the low-frequency conductivity. According to the Ferrell-Glover-Tinkham (FGT) sum rule,[32] the difference between the conductivity at $T \simeq T_c$ and $T \ll T_c$ (the so-called missing area) is related to the formation of a superconducting condensate,

$$\omega_{ps}^2 = 8\int_{0^+}^{\omega_c} [\sigma_1(\omega, T \simeq T_c) - \sigma_1(\omega, T \ll T_c)]\, d\omega. \quad (5.25)$$

where $\omega_{ps}^2 = 4\pi n_s e^2/m^*$ is the square of the superconducting plasma frequency, n_s is the condensed carrier density, and ω_c is the high-frequency cut-off frequency which should be chosen such that the ω_{ps}^2 converges smoothly. The penetration depth is related to the superconducting plasma frequency by $\lambda = c/\omega_{ps}$. The equation states that the spectral weight lost in $\sigma_1(\omega)$ in the superconducting state has been transferred to the zero frequency delta function response of the superconducting condensate.

The superconducting penetration depth can also be estimated from the imaginary part of the complex conductivity in the low-frequency limit via

$$\lambda(\omega) = c/\omega_{ps} = c/\sqrt{4\pi\omega\sigma_2(\omega)}. \quad (5.26)$$

The determination of λ from this equation at low frequency limit relies only on the imaginary part of conductivity at $T \ll T_c$. When the FGT sum rule is fulfilled, the λ value determined from the missing area in the real part of conductivity should equal to the value obtained from the imagnary part of conductivity in the low-frequency limit.

For unconventional superconductors, due to the presence of strong electrons correlations, the integral of conductivity up to a cutoff frequency higher than the band width but still lower than the interband transitions could be still temperature dependent. In such a case being usually referred to as the "violation of the sum rule" as we mentioned above, the FGT sum rule should be modified as

$$\omega_{ps}^2 = 8\int_{0^+}^{\omega_c} [\sigma_1(\omega, T \simeq T_c) - \sigma_1(\omega, T \ll T_c)]d\omega + \Delta W_K. \quad (5.27)$$

The first term is simply FGT sum rule, while the ΔW_K corresponds to the high-frequency part of the integral, which is in fact the kinetic energy contribution. In a system where kinetic energy plays a prominent role, the simple FGT sum rule may be violated. There are a number of optical works on the high-T_c cuprates whether a fraction of the condensate comes from the high energy interband transitions, but the results are controversial.[33–37] Theoretically, it is found that the spectral weight difference between the normal and superconducting states depends highly on the choice of the integral cutoff frequency (ω_c).

5.3 Optical Studies on the Parent Compounds

In 2008, the discovery of iron-based superconductors opened a new chapter in condensed matter physics.[3] Similar to the cuprates, the Fe-based compounds exhibit complex phase diagrams. The undoped parent compounds are electron-and hole-compensated semimetals. A spin-density-wave (SDW) instability or a long range antiferromagnetic (AFM) order usually develops at low temperature. Carrier (either electron or hole) doping or application of pressure could suppress the SDW or AFM order and induce superconductivity. Figure 5.7 shows the phase diagram of a prototype iron-based superconducting system: K- and Co-doped $BaFe_2As_2$. It is

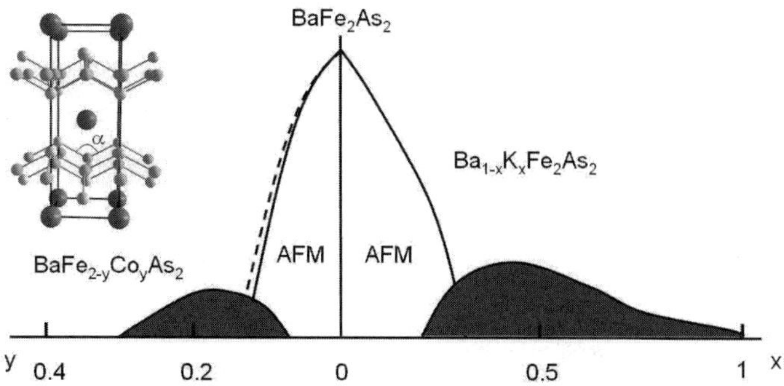

Figure 5.7 Phase diagram of a prototype iron-pnictide system: hole and electron-doped $BaFe_2As_2$. The dash curve indicates the structural phase transition. For the parent $BaFe_2As_2$, a tetragonal to orthorhombic structural phase transition occurs at the same temperature as the magnetic phase transition. With Co-doping, the structural and magnetic phase transitions separate.

widely believed that the superconductivity is intimately related to the antiferromagnetism. Therefore, identifying the nature of the antiferromagnetism in the parent compounds becomes an essential step towards understanding the mechanism of superconductivity in iron-based systems.

5.3.1 *Spin Density Wave Gap in FeAs-Based Compounds*

Identification of competing orders and SDW instability in FeAs-based compounds represents an important progress at the early stage. In an early study on the polycrystaline samples of $LaFeAsO_{1-x}F_x$, Dong *et al.*[38] revealed a partial energy gap formation for the parent compound below an abrupt phase transition. They attributed the gap formation to the development of an SDW order being driven by the nesting instability between the disconnected hole and electron Fermi surfaces (FSs) with a (π,π) wave vector, as evidenced by a strong peak in the Lindhard function calculation. Band structure calculations further indicate that the energy gain is mainly associated with the development of magnetic order rather than lattice distortion. A stripe type antiferromagentic spin structure was predicted based on the (π,π) nesting wave vector,

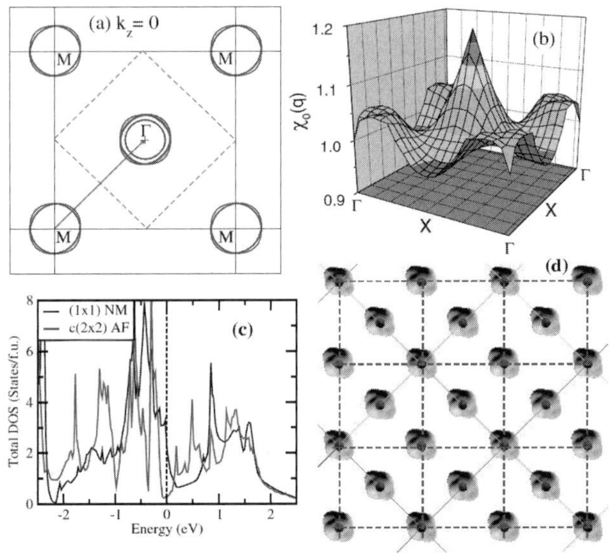

Figure 5.8 The results of theoretical calculations for LaOFeAs: (a) the Fermi surfaces at $k_z = 0$. The blue lines are original calculated FS, while red lines show the FS shifted by $q = (\pi, \pi, 0)$. (b) The calculated Lindhard response functionn $\chi_0(q)$ is strongly peaked at M point. (c) The total DOS of nonmagnetic state and antiferromagnetic state. (d) Stripe antiferromagnetic ordering pattern.[38] See also Color Insert.

as shown in Fig. 5.8. The superconductivity in the doped samples competes with the SDW order in the undoped compound. Soon after, similar competing orders and SDW instability were also found in $CeFeAsO_{1-x}F_x$ system,[5] suggesting that those phenomena are generic for those systems.

The stripe-type antiferromagnetic structure was confirmed by the neutron diffraction experiments on LaFeAsO shortly[7] and various other rare-earth based 1111 parent compounds.[10,12] The same type magnetic structure was also found in parent compounds of other 122[13–15] and 111[16] structure-type FeAs-based systems. Furthermore, neutron diffraction experiments also revealed the presence of structural phase transitions which occur either simultaneously or at temperatures above the magnetic phase transitions.

The presence of disconnected electron and hole FSs at zone center and corners was experimentally revealed by the angle resolved photoemission spectroscopy (ARPES) technique.[39] On such basis, many researchers have addressed the itineracy of electrons and important implication of FS nesting and interpocket scattering for the superconductivity in Fe-pnictides.[40–45] However, there have been much debate over the origin of the magnetic order. The stripe type antiferromagnetic spin structure can also been well explained by a Heisenberg J_1-J_2 model with J_1 and J_2 representing the nearest and the nest nearest neighbor exchange interactions between localized spin moments.[46–50] Since different antiferromagnetic structures have been found for the parent compounds of different Fe-pnictides or Fe-chalcogenides systems later, the FS nesting scenario apparently cannot explain the variation of magnetic structures. Therefore, FS nesting does not appear to be the major driving force for the formation of magnetic moment and antiferromagnetism.

Because of the multiple nature of the Fe 3d orbitals and bands, the Fe-based compounds likely contains both itinerant and localized Fe 3d electrons which contribute respectively to the FSs and local moments. The available optical measurements suggest that the Fermi surfaces formed by itinerant electrons do couple with the local moments contributed by localized electrons. When the magnetic wave vector matches with the wave vector connecting the electron and hole FSs, the FSs were strongly affected and gap formation or band reconstruction would occur. On the other hand, if no such match exists, the FSs of the compounds would not be affected.

Although the SDW gap formation in the FS was first identified in the polycrystalline samples of 1111-based compounds,[5,38] accurate and detailed information was obtained after the single crystal samples became available. Optical spectroscopic measurements on single crystals of AFe_2As_2 (A = Ba, Sr) with the electric field parallel to the *ab*-plane (i.e. $\mathbf{E} \parallel ab$-plane) were first reported by Hu *et al.*[51] The $BaFe_2As_2$ and $SrFe_2As_2$ undergo the SDW/structural transitions at about 140 K and 200 K, respectively. As shown in Fig. 5.9, AFe_2As_2 (A = Ba, Sr) exhibit metallic response. The plasma frequency ω_p estimated from the low-frequency Drude component is about 1.5 eV above the SDW transition. In the SDW order state, a

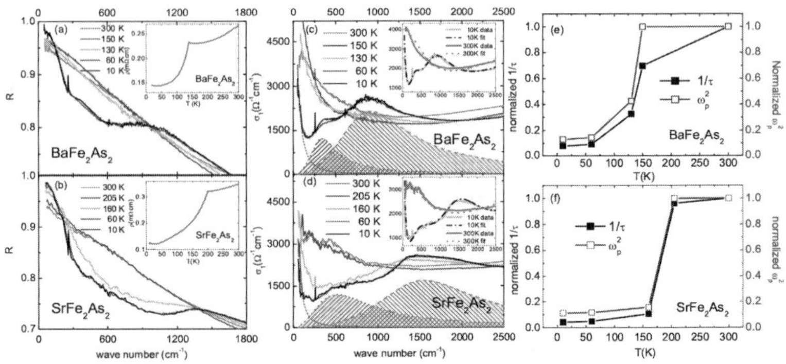

Figure 5.9 Optical spectra of single-crystalline AFe_2As_2 (A=Ba, Sr) measured within *ab*-plane. Reflectance for (a) $BaFe_2As_2$ and (b) $SrFe_2As_2$. Inset: the dc resistivity. The real part of conductivity for (c) $BaFe_2As_2$ and (d) $SrFe_2As_2$. Normalized scattering rate $1/\tau$ and plasma frequency squared ω_p^2 for (e) $BaFe_2As_2$ and (f) $SrFe_2As_2$.[51] See also Color Insert.

large part of Drude component is removed due to the formation of energy gaps. The residual Drude components become very narrow, yielding evidence for a reduction of carrier scattering rate. A Drude-Lorentz analysis of the conductivity spectra indicate that, across the transition, the reduction of the scattering rate is even faster than the suppression of plasma frequency, which can then explain why the resistivity show a sharp drop after the SDW transition. Accompanied by the suppression of the Drude component, two energy gaps with different energy scales are clearly observed in the conductivity spectra, reflecting the multi-band characteristic of the iron-based systems. Similar spectral features were also reported in subsequent work on related 122 compounds.[52–54]

NaFeAs, one of the parent compound of 111 structural type systems, was found to have two separate phase transition: magnetic transition at 41 K and structural transition at 50 K.[16,55] In-plane optical properties for $Na_{1-\delta}FeAs$ revealed similar optical response as other 122-based parent systems. An energy gap in $\sigma_1(\omega)$ is observed below the magnetic phase transition, accompanied by a spectral weight transfer from the free-carrier Drude term to energies above this gap. Correlated with the lower SDW transition temperature T_{SDW}, both the gap 2Δ and the energy scale associated

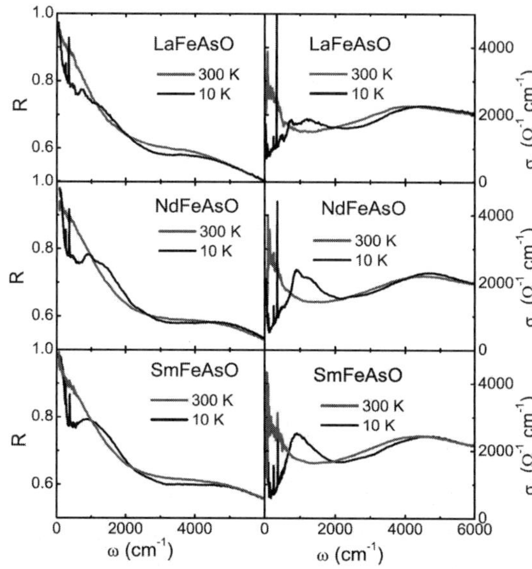

Figure 5.10 The *ab*-plane reflectance spectra optical reflectance and conductivity spectra below 6000 cm^{-1} at 300 K and 10 K for single-crystalline La-, Nd-, and Sm-based 1111-type parent compounds.[59]

with the spectral weight redistribution are smaller in comparison with 122 FeAs-based compounds.[56]

The single crystals of 1111 structural compounds are much more difficult to grow. Millimeter sized single crystals were obtained only when the NaAs solvent was used as a flux.[57] Figure 5.10 displays the optical spectra measured on single crystal samples of the parent compounds of several 1111 systems ReOFeAs (Re = La, Nd, and Sm).[58,59] Similar to 122 system, very clear SDW energy gap formations were seen in the spectra. In fact, a careful temperature-dependent optical study indicated that the spectral features corresponding to the partial energy gaps in the spin-density-wave (SDW) state are already present below the structural phase transition. This indicates that the electronic state below the structural phase transition is already very close to that in the SDW state.

Summarizing the optical spectra collected on 122, 111, and 1111 systems, we can conclude that the FSs are strongly affected by the

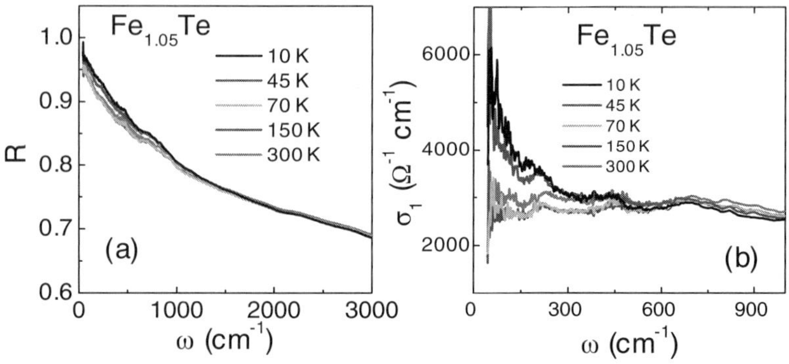

Figure 5.11 (a) Reflectivity R(ω) and the real part of conductivity $\sigma_1(\omega)$ for $Fe_{1.05}Te$.[62]

magnetic or structure phase transition. Much of the free carrier spectral weight was removed due to the opening of the SDW energy gaps. Nevertheless, as we shall see below, those are not the generic properties for all parent compounds of Fe-based superconductors.

5.3.2 *Absence of SDW Gap in* $FeTe_{1+x}$

Different from FeAs-based parent compounds, $FeTe_{1+x}$ has a bi-stripe antiferromagnetic structure below the magnetic/strctural transition at 67 K. The ordered magnetic moment is about $2\mu_B$, much larger than FeAs-based compounds.[17,18] The FSs determined by both band structural calculations and ARPES experiments are similar to other Fe-pnictides, however the magnetic wave vector corresponding to the bi-stripe AFM ordering is (π,0) in the folded Brillioun zone, which is not the wave vector connecting the hole and electron FSs.[60,61] Figure 5.11 shows the reflectivity R(ω) and the real part of conductivity $\sigma_1(\omega)$ for $Fe_{1.05}Te$.[62] It is important to note that the conductivity spectra above T_{SDW} are rather flat, without showing a clear Drude-like behavior at low frequency. As is well known, the width of Drude peak is determined by the scattering rate (or inverse of the transport lifetime) of the quasiparticle, the measurement result indicates that there is no well defined quasiparticle with sufficiently long transport life time above magnetic transition temperature T_N. The charge transport is rather incoherent. Below T_N, the low-ω R(ω) increases fast. Correspondingly, a Drude component in $\sigma_1(\omega)$ develops quickly from the incoherent background. Notably, unlike the parent compounds of LaFeAsO and AFe_2As_2 (A = Ba, Sr), there is no partial gap formation in optical spectra below T_{SDW}.[62] The absence of SDW gap formation has been confirmed by recent ARPES experiment on $Fe_{1.05}Te$.[60,61] Those results imply that the itinerant electrons experience rather strong scattering at high temperature, presumably from the disordered local moments formed by the localized Fe 3d electrons, but become less scattered in the magnetic ordered phase.

It is easy to understand the absence of the energy gap in the magnetic ordered state based on the folding effect of the energy bands induced by the formation of magnetic Brillouin zone, as shown in Fig. 5.12. When the nested Fermi surfaces are connected by the (π, π) anti-ferromagnetic wave vector, band mixing due to

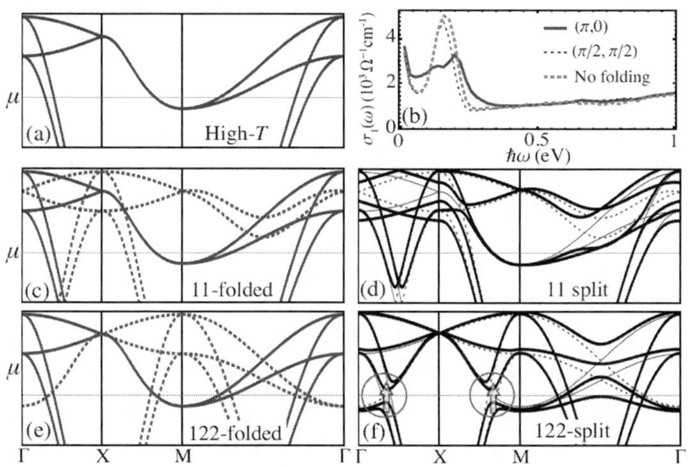

Figure 5.12 Tight-binding model band structure for (a) high-temperature unit cell, (c) folded at $\mathbf{Q}_{AFM} = (\pi, 0)$ and (e) $\mathbf{Q}_{AFM} = (\pi, \pi)$. (d) and (f), respectively, correspond to (c) and (e) with a small mean-field exchange potential included. Allowed optical transitions are indicated by arrows in (f). (b) shows *ab*-averaged conductivity of a Kubo+five-orbital tightbinding model calculation in the high temperature (dashed) and magnetic phases with 200 meV exchange potential with (π, π) (solid) and $(\pi, 0)$ (dotted) folding.[53]

the folding of the the magnetic Brillouin zone causes formation of energy gaps near the Fermi level so that the total energy of the electron system is lowered. However, there is no such band mixing near Fermi level for FeTe with the folding by the magnetic wave vector $\mathbf{Q}_{AFM} = (\pi, 0)$.[53] The absence of energy gap below T_N plus the incoherent charge dynamics above T_N make it hard to account for the bi-stripe antiferromagnetism in FeTe in terms of the SDW picture.

5.3.3 *Fully Localized Fe 3d Electrons in* $K_{0.8}Fe_{1.6}Se_2$

$A_xFe_2Se_2$ (A = K, Rb, Cs) is another interesting system. The superconducting transition temperature could be over 30 K at ambient pressure.[63] Different from all other Fe-based superconducting systems, the superconductivity in this system is found to be very close to semiconducting/insulating phases.[64] A very strong and totally new antiferromagnetic order was found by neutron diffraction experiment. The ordered magnetic moment $\sim$ 3.31 μ_B/Fe is unprecedentedly large, and the magnetic transition occurs at a record high temperature of $T_N \approx 550$ K.[19,65,66]

Actually, two semiconducting/insulating phases were identified from first principle calculations: one is the $KFe_{1.5}Se_2$ which is found to have a small energy gap of 20 meV and a stripe type AFM order;[67] another is $K_{0.8}Fe_{1.6}Se_2$ with a lager energy gap of 0.6 eV and a blocked AFM order.[68] Experimentally, only the latter one of blocked AFM order was found.

The optical properties of the insulating $K_{0.83}Fe_{1.53}Se_2$ single crystal were studied by Chen *et al.*[69] Figure 5.13(a)–(d) show the *ab*-plane reflectivity $R(\omega)$ and the real part of conductivity $\sigma_1(\omega)$. They show clearly nonmetallic optical behaviors: the reflectivity decreases with lowering temperature and the Drude-like response of free-carriers at low frequencies is absent in the conductivity spectra. The sizes of the energy gaps for the indirect and direct interband transitions are estimated to be roughly 30 meV and 0.37 eV, respectively.

In addition, there are two peculiar features in the infrared spectra of $K_{0.83}Fe_{1.53}Se_2$ which were not observed in all other iron-pnictides/chacogennides: a double peak structure between 4000–

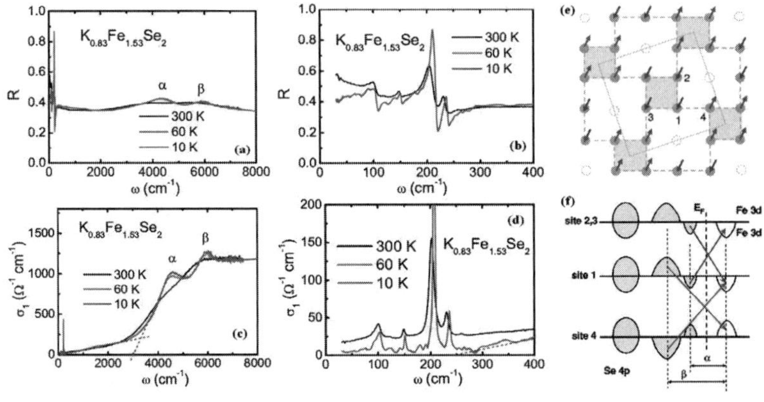

Figure 5.13 The *ab*-plane optical spectra of $K_{0.83}Fe_{1.53}Se_2$ single crystals: (a) optical reflectance $R(\omega)$ spectra at different temperatures up to 8000 cm^{-1}. (b) An expanded plot of $R(\omega)$ spectra below 400 cm^{-1}. (c) Conductivity $\sigma(\omega)$ spectra at different temperatures up to 8000 cm^{-1}. (d) An expanded plot of $\sigma(\omega)$ spectra below 400 cm^{-1}. The dashed straight lines are used for extrapolation of energy gaps. (e) The structure of blocked antiferromagnetism determined by the neutron scattering experiments. The grey solid lines represent a $\sqrt{5} \times \sqrt{5} \times 1$ unit cell. (f) A schematic picture for the band occupation for each Fe site. Site 4 has opposite spin-polarized occupations with site 1, 2, and 3. Possible d–d transitions are indicated by the arrows. The double peaks represent the coexistence of the ferro- and antiferromagnetic correlations between the neighboring sites.[69]

6000 cm^{-1} and abundant phonon modes much more than those expected for a standard 122 structure (only two infrared active E_u modes are present for 122 structure). Chen *et al.* elaborated that both could be well explained by the blocked antiferromagnetism as determined by neutron scattering experiments. In this ordering pattern, each Fe site is ferromagnetically coupled with two neighboring Fe sites and antiferromagnetically coupled with the third neighboring Fe site. The Fe ion here is considered to have six 3d electrons and the Hund's rule coupling is assumed to be larger than the crystal-field splitting energy. Consequently, as shown in Fig. 5.13(f), five 3d electrons would fully occupy the five orbitals in a spin polarized direction, the rest one electron would occupy one of the orbital in the opposite spin direction, resulting in four empty orbitals. As the optical transitions do not involve spin-flip process, the possible d–d transitions (via hybridization with Se 4p orbitals) are indicated by the arrows, representing the transitions dominantly from the occupied states at one Fe site to the unoccupied states at nearest-neighboring Fe sites. It is easy to see that the α peak comes from the transitions between the neighboring ferromagnetic ordered sites and the β transition from the neighboring antiferromagnetic ordered sites. The double peaks represent the coexistence of the ferro- and antiferro-magnetic correlations between the neighboring sites. Due to the iron vacancy ordering, a $\sqrt{5} \times \sqrt{5} \times 1$ superlattice pattern forms, the crystallographic unit cell is actually enlarged as $K_2Fe_4Se_5$. Naturally one would expect to observe more phonon modes than that expected for 122 structure.

Clearly, the optical properties of different parent compounds of Fe-based superconductors are very much different. The significant variation of Fe moments and ordered patterns in different parent compounds suggests that the FS nesting is not the origin for the magnetic moment and magnetism. Nevertheless, the FSs are strongly affected by the magnetic or structure phase transition. When the magnetic wave vector matches with the wave vector connecting the hole and electron FSs, SDW gaps would form due to the band mixing. On the other hand, the FSs would not be altered if the magnetic wave vector does not match with the nesting wave vector. Apparently, some other interactions rather than the effect of FS nesting play important roles in the electronic properties of the Fe-based parent compounds. Most notably, the electron correlation effect plays a crucial role in the optical response as we shall discuss in Section 5.5.

5.4 Multi-Components *vs.* Extended Drude Model Analysis of Optical Conductivity

In the above section, we reviewed the evolution of the low-frequency optical spectra across the magnetic phase transition of Fe-pnictides and chalcogenides. Now we shall explain the general features of the optical conductivity in the paramagnetic phase. Like high-T_c cuprates, the optical conductivity of those materials do not show a simple Drude behavior. Even for the most metallic systems, for example, the 122 based compounds $Ba(Fe_{1-x}Ax)_2As_2$ (A = Co, Ni) and $Ba_{1-x}K_xFe_2As_2$, the in-plane optical conductivity decreases slowly with increasing frequency to a rather flat response in the mid-infrared region followed by a pronounced peak at about 5000 cm^{-1}.[70–73] For the 11 based compounds, the optical conductivity appears rather flat even in the low frequency regime.[53,62,74]

One technique to reproduce the optical conductivity is to decompose the spectrum into different components. Because these materials are multiband systems, the free carrier response can be described by separate contributions from the electron and hole pockets. Several groups[70–73] found that the low-energy conductivity of 122 based compounds could be described by two Drude components including a broad one with the width of several thousand wave numbers. The peak in the mid-infrared region could be well reproduced by adding a Lorentz component. Figure 5.14 shows the optical conductivity of several different iron pnictides in the normal state together with the components from the decomposition of Drude-Lorentz model for the whole 122 family.[70] The narrow Drude component represents the coherent quasiparticles, while the broad Drude component is an incoherent part, representing the overdamped quasiparticles. The coherent spectral weight is very small, only about 10% of the incoherent spectral weight.[71]

However, there is a different opinion on whether the broad band should be considered as a part of the Drude response. Tu *et al.*[72] examined the optical response of $BaFe_{1.85}Co_{0.15}As_2$. Although the conductivity above T_c at 27 K could be well reproduced by two Drude response, a "narrow" ($1/\tau \simeq 113$ cm^{-1}) and a "broad" one ($1/\tau \simeq 5700$ cm^{-1}), and a Lorentz oscillator at $\simeq 5200$ cm^{-1}, as shown in Fig. 5.15(a), they noticed the scattering rate for the broad Drude component is more than 50 times that of the narrow component. With the available Fermi velocity from the ARPES measurement, they estimated the mean free path $l_j = v_{F,j}\tau_j$ associated with the narrow and broad Drude components, respectively. The values they obtained are 58 Å for the narrow component, presumably from the electron pocket, and 0.4 Å for the broad component. The latter one is less than the shortest interatomic spacing, which places the value far below the Mott-Ioffe-Regel limit. They alerted that such a decomposition may not be physically meaningful.

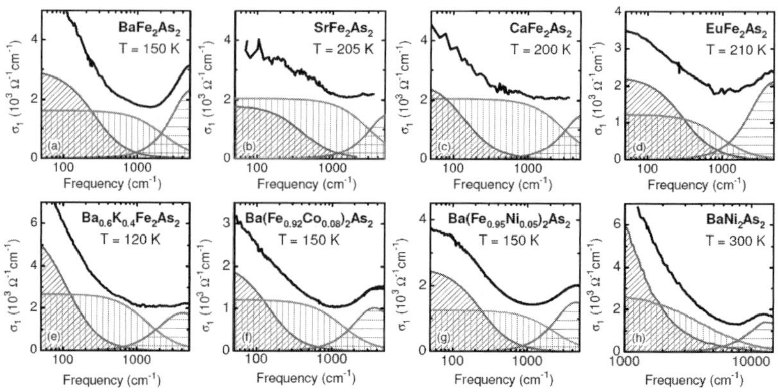

Figure 5.14 Comparison of the optical conductivity of different iron pnictides, in the normal state, indicates that the decomposition in two Drude terms (σ_N and σ_B shaded densely green and medium orange, respectively) plus oscillators in the midinfrared (e.g., one of them is indicated by sparsely hatched magenta) holds for the whole 122 family.[70]

Tu *et al.*[72] suggests that it is appropriate to include a low frequency bound excitation which may arise from localization effects or from interband transitions. They then considered the case of two Drude contributions as well as two low-frequency Lorentz oscillators. The results of the fit to the data at 27 K are shown in Fig. 5.15(b). The narrow Drude component at 27 K is only slightly broader than the same component in the previous approach, resulting in a slightly shorter mean free path for the electron pocket. Much of the spectral weight formerly associated with the broad Drude component has shifted to a Lorentzian centered at $\simeq$1100 cm^{-1}, resulting in a dramatic reduction in the scattering rate, $1/\tau$ $\simeq$600 cm^{-1} and a commensurate increase in the mean free path 3.5 Å. They noted that a fit by a single Drude component and two low-frequency Lorentz oscillators can also reproduce the results reasonably well. The width of the narrow Drude component is unchanged and the slight increase in strength reflects the fact that it has absorbed the spectral weight formerly associated with the broad Drude component. However, the plasma frequency for the single Drude component extracted from the fit is smaller, ω_p $\simeq$7840 cm^{-1}.[72] Nevertheless, we would like to remark that the exclusion

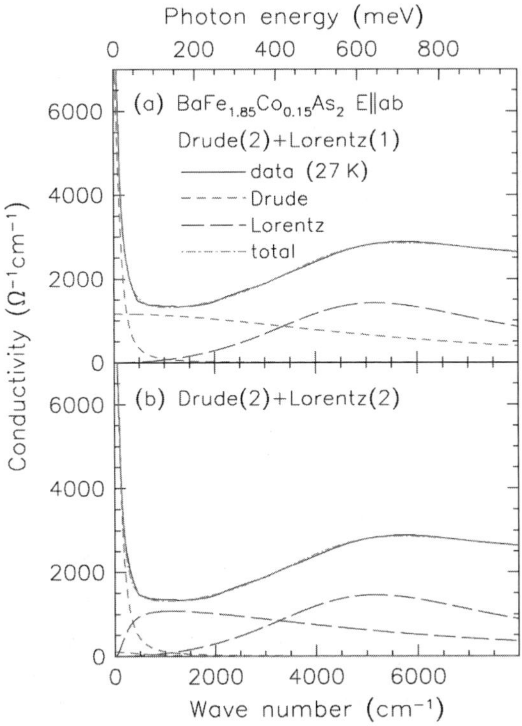

Figure 5.15 (a) The two-Drude and single Lorentz oscillator fit to the data at 27 K over a wide frequency range. The Drude responses can be characterized by narrow and broad bands with a single Lorentz oscillator centered at $\simeq$5200 cm^{-1}. (b) The fit to the data at 27 K using two Drude components and two Lorentz oscillators. A significant portion of the spectral weight of the broad Drude component has been transferred to the low-frequency Lorentz oscillator.[72]

of the broad tail contribution to the Drude part could cause an underestimation of the spectral weight of the intraband excitations, and thus the plasma frequency ω_p.

Like the case of cuprates, the low-frequency optical response of Fe-based systems could also be analyzed from a single component approach, i.e. the extended Drude model, where the slowly decaying optical conductivity is considered to be caused by a frequency-dependent scattering rate. From the measured optical conductivity, one could extract the scattering rate through equation 5.16.

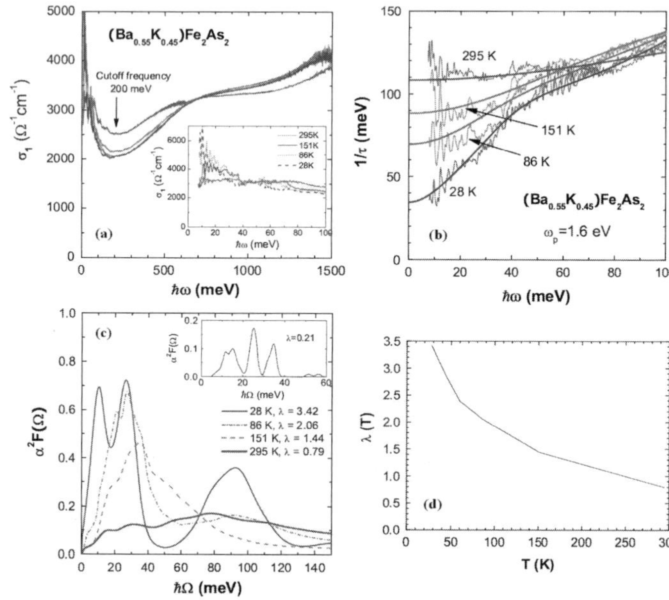

Figure 5.16 (a) The *ab*-plane optical spectra of $Ba_{0.55}K_{0.45}Fe_2As_2$ single crystals at different temperatures. (b) Scattering rate as a function of photon energy at different temperature. (c) Bosonic function $\alpha^2 F(\omega)$ extracted from the optical scattering rates using a maximum entropy inversion method. The inset is the calculated electron-phonon $\alpha^2 F(\omega)$ by Boeri *et al.* with $\lambda = 0.21$. (d) The temperature dependence of the mass renormalization parameter.[76]

Then, with some inversion schemes, one could further extract the Eliashberg spectra function $\alpha^2 F(\omega)$ from equation 5.20 or 5.21. The quantity reflects the coupling of charge carriers to certain bosonic modes.[24,75]

The optical spectra of $Ba_{0.55}K_{0.45}Fe_2As_2$ single crystals (T_c = 28 K) in the normal state were investigated by Yang *et al.*[76] Figure 5.16(a) and (b) show the conductivity and scattering rate spectra at several different temperatures. The extracted Eliashberg spectral functions $\alpha^2 F(\omega)$ are shown in Fig. 5.16(c). As the temperature increases, two peaks around 10 and 30 meV in $\alpha^2 F(\omega)$ spectrum shift in energy and rapidly merge into a single feature. The feature of $\alpha^2 F(\omega)$ evolves from a three peak structure at T = 28 K just above T_c to a singe broad peak structure at room temperature. The

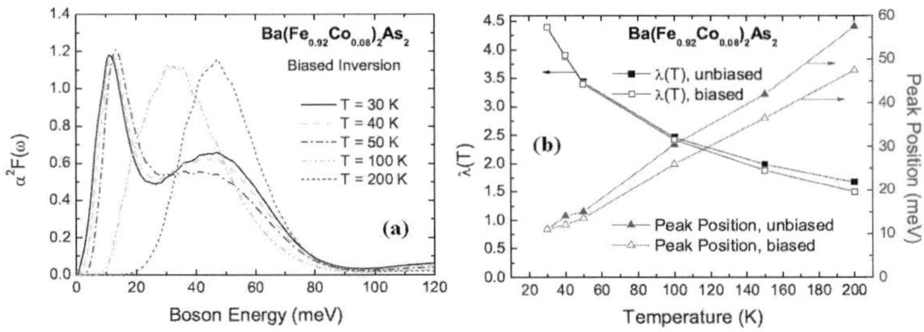

Figure 5.17 (a) The corresponding electron-boson spectral density $\alpha^2 F(\omega)$ as the result of a biased maximum entropy method inversion. (b) The mass enhancement factor $\lambda(T)$ for $Ba(Fe_{0.92}Co_{0.08})_2As_2$ as a function of temperature T (left-hand scale applies). The solid (black) squares indicate results of an unbiased maximum entropy method inversion and the open (black) squares refer to the biased maximum entropy method inversion. Furthermore, we show the shift of the position of the main peak with temperature from the two inversions (right-hand scale applies).[78]

frequency-dependent $\alpha^2 F(\omega)$ here disagrees with the behavior of electron-phonon coupling (shown in the inset of Fig. 5.16(c)).[77] The coupling constant

$$\lambda = \int \alpha^2 F(\Omega)/\Omega d\Omega, \tag{5.28}$$

is found to be enhanced monotonously to 3.42 as the temperature is lowered to $T = 28$ K, which is much larger than the electron-phonon coupling parameter $\lambda = 0.21$.[77] This study revealed that the charge carriers interact with a broad bosonic spectrum extending beyond 100 meV with a strongly temperature dependent coupling constant.

The Eliashberg analysis of the optical spectra of $Ba(Fe_{0.92}Co_{0.08})_2$ As_2 single crystals ($T_c = 25$ K) was reported by Wu *et al.*[78] The electron-boson spectral density $\alpha^2 F(\omega)$ is obtained from an inversion of the scattering rate by a maximum entropy method. As shown in Fig. 5.17(a), at 30 K, two peaks are present below 100 meV in $\alpha^2 F(\omega)$ spectrum. When the temperature increases, the peak initially at 10 meV shifts towards higher frequencies and the peak around 45 meV becomes weaker. Eventually, the double peak structure of $\alpha^2 F(\omega)$ evolves into a single peak feature at high temperature. In addition, the coupling constant $\lambda(T)$ increases monotonously with decreasing temperature and becomes much larger than the electron-phonon parameter $\lambda = 0.21$[77] essentially the same as the case in $Ba_{0.55}K_{0.45}Fe_2As_2$. In the figure, the two data sets are obtained from two different inversion procedures, so-called unbiased and biased maximum entropy method. The same trend and rather closed values indicate the robustness of the inversion technique by maximum entropy method. Wu *et al.*[78] alerted that those results bare some resemblance to but differ in other aspects from inelastic polarized neutron-scattering results for the dynamic spin susceptibility $\chi''(\mathbf{Q}_{AFM}, \omega)$ at the antiferromagnetic wave vector $\mathbf{Q}_{AFM} = (0.5, 0.5)$.

5.5 Electron Correlations in the Fe-Pnictides/Chalcogenides

5.5.1 *Kinetic Energy Reduction by Electron Correlations*

The electron correlation, impeding itineracy of electrons, would have the effect of reducing the the oscillator strength of the Drude component (proportional to the so-called optical kinetic energy) compared with band theory predictions. Then, the experimental kinetic energy of the electrons measured by optical spectroscopy is smaller than the kinetic energy obtained from first-principle band-structure calculations, as schematically shown in Fig. 5.18.[79] Such electron correlation would lead to an enhanced mass of low-energy emerging quasiparticle. The ratio between the experimental kinetic energy and the theoretic kinetic energy is close to unity in simple metals, but decreases to zero in strongly correlated Mott insulators. Thus, the ratio between the optical and band kinetic energies K_{exp}/K_{band} allows one to define the degree of electronic correlations.

The electron correlations also lead to a temperature dependent spectral weight transfer. Usually, the energy scale to which the spectral weight is transferred could be taken as another measure of the degree of electronic correlations.

Qazilbash *et al.*[80] examined the many-body effects experienced by the electron in LaFePO first. Figure 5.19 shows the real part of the *ab*-plane optical conductivity $\sigma_1(\omega)$ of LaFePO over a broad

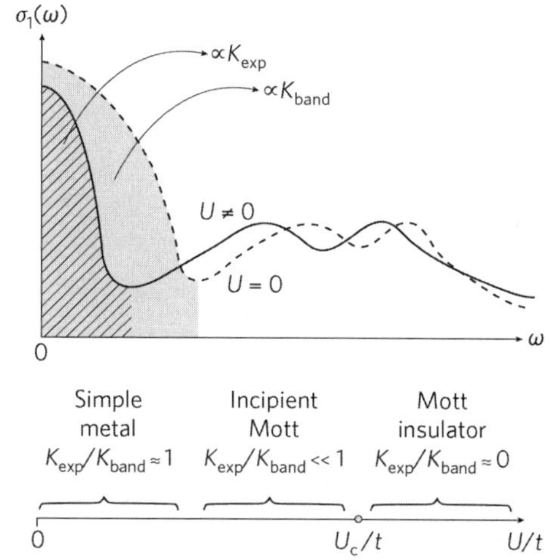

Figure 5.18 (a) Optical conductivity as a function of frequency. The area under the Drude peak is (for certain simplified energy dispersions) proportional to the kinetic energy of the coherent electrons near the Fermi energy, yielding K_{band} and K_{exp} for the non-interacting and interacting cases, respectively. The degree to which K_{exp}/K_{band} is smaller than one is a measure of the interaction induced reduction of the electrons' mobility. (b) Different regimes of correlation. U_c is the threshold interaction for a Mott localization transition in a system with an integer number of electrons partially occupying some electronic bands.[79]

frequency range. The calculated $\sigma_1(\omega)$ from band theory within the generalized gradient approximation using both experimental and relaxed phosphorous coordinates were also shown in the figure. The two choices do not significantly affect the Drude part but do affect the peak positions of the interband transitions. It can be immediately seen that the area under the Drude part of the measured $\sigma_1(\omega)$ is significantly less than that calculated within band theory. The ratio between the optical and band kinetic energies K_{exp}/K_{band} was then estimated to be $K_{exp}/K_{band} = \omega^2_{p,exp}/\omega^2_{p,band}$ $\approx$0.5. This value puts it in a regime close to an incoherent metal.

The reduction of the kinetic energy compared with band theory for FeAs-based compounds is even stronger than LaFePO. For example, the ratio of K_{exp}/K_{band} for LaFeAsO is about 0.30–0.38.[58] We remark that the band renormalization factors determined by

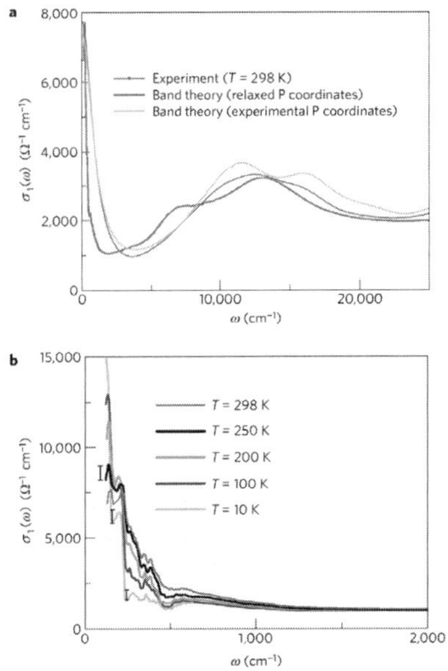

Figure 5.19 *ab*-plane optical conductivity of LaFePO. (a) Real part of the *ab*-plane optical conductivity $\sigma_1(\omega)$ of LaFePO as a function of frequency for $T = 298$ K. Also shown are $\sigma_1(\omega)$ plots calculated within band theory for two choices of the phosphorous (P) coordinate. (b) Systematic temperature dependence of $\sigma_1(\omega)$.[80]

optics is in good agreement with the values extracted from ARPES experiments.[39,82] A survey of the ratios for a number of different systems was done by Basov and Chubukov,[81] shown as Fig. 5.20.

Degiorgi and his colleague use a different approach to estimate the band renomalization effect.[83] They assume that the electron correlations drive the conductivity spectrum into a coherent Drude term and an incoherent component. The incoherent part, appearing as a mid-infrared (MIR) band, is still below the electronic interband transition energies. Then they suggested that the ratio of K_{exp}/K_{band} could be obtained by calculating the spectral weight ratio,

$$K_{exp}/K_{band} = \frac{\int_0^{\omega_{opt}} \sigma_1(\omega)d\omega}{\int_0^{\omega_{band}} \sigma_1(\omega)d\omega} \tag{5.29}$$

where ω_{opt} locates within the shallow minimum between the effective metallic part and the MIR band, an energy which separates

Table1| The ratio of the experimental kinetic energy K_{exp} extracted from optical measurements, as described in ref. 27, and K_{LDA} provided by band-structure calculations.

Superconductor	T_{cmax}	K_{exp}/K_{LDA} at T_{cmax}	Refs
CuSCs			
$Nd_{2-x}Ce_xCuO_4$	25	0.3	31
$Pr_{2-x}Ce_xCuO_4$	25	0.32	31
$La_{2-x}Sr_xCuO_4$	40	0.25	31
$YBa_2Cu_3O_{7-x}$	93.5	0.4	8
$Bi_2Sr_2CaCu_2O_{8+\delta}$	94	0.45	*
FeSCs			
LaFePO	7	0.5	27
$Ba(Fe_{1-x}Co_x)_2As_2$	23	0.35-0.5	27,**
$Ba_{1-x}K_xFe_2As_2$	39	0.3	43
Exotic SCs			
$CeCoIn_5$	2.3	0.17	44,45
Sr_2RuO_4	1.5	0.4	27
κ-$(BEDT\text{-}TTF)_2Cu(SCN)_2$	12	0.4	46
Electron-phonon SCs			
MgB_2	40	0.9	27
K_3C_{60}	20	0.96	47,48
Rb_3C_{60}	30	0.9	47,48

*D. van der Marel *et al.*, unpublished; **A. Schafgans *et al.*, unpublished.

Figure 5.20 A survey of the ratios of K_{exp}/K_{band} for several different systems. For more details, see reference 81 and references therein.

the upper limit of Drude response and the onset of the MIR absorption; and ω_{band} locates at the high-frequency end of the MIR feature, separating the upper limit of the incoherent part and the onset frequency of the electronic interband transitions. That is to say, the numerator contains only the coherent part, while the denominator contains the spectral weight from both coherent and incoherent parts. Equivalently, if the coherent and incoherent parts could be separated by Drude-Lorentz analysis, the ratio of K_{exp}/K_{band} could be calculated from following formula,

$$K_{exp}/K_{band} = \frac{\omega_p^2}{\omega_p^2 + (\omega_p^{MIR})^2} \tag{5.30}$$

where $\omega_p^2 = \sum_i \omega_{pi}^2$ are the plasma frequencies of the Drude terms, shaping the effective metallic part of $\sigma_1(\omega)$, and $(\omega_p^{MIR})^2 = \sum_j S_j^2$, S_j is the mode strength of the j-Lorentz component contributing to the MIR band in $\sigma_1(\omega)$. Nevertheless, we remark here that very often there is no unambiguous way to separate the contribution between an incoherent part of intraband excitations and true interband transitions.

5.5.2 *Effect of Hund's Coupling*

Soon after the identification of the magnetism in the parent compounds, there have been two schools of ideas in the study of the electronic structure and magnetic properties of Fe-based compounds: the itinerant electron and the local moment pictures. In the itinerant electron approach, the Fe-based compounds are considered as weakly correlated metallic systems that become magnetic through an SDW-type transition driven by the quasi-nested hole and electron Fermi surfaces. In the strongly coupling approaching, these materials are believed to be strongly or at least moderately correlated, the electrons are considerably localized with the local magnetic moments interacting via short-range superexchange. As was mentioned in section 5.3, a fully itinerant description is insufficient to account for the magnetism in Fe-pnictides and chalcogenides. The ordered moments in Fe-chalcogenides could be as high as $2\mu_B$ for FeTe and $3.3\mu_B$ for $K_xFe_{2-y}Se_2$, much larger than the values expected from an SDW origin. Very recently, it is found by Fe X-ray emission spectroscopy (XES)[84] that the local magnetic moments are present in the paramagnetic phase of all Fe pnictides and chalcogenides samples, indicating conclusively that the local moment does not have an itinerant origin. However, it is also found that the local moment detected by XES is about $1\mu_B$ for all Fe-pnictide compounds, which is still much smaller than the fluctuated local moment obtained from the band structure calculations ($2.4\mu_B$). This implied that there exist "missing" magnetic moments due to the itineracy of the electrons. It becomes very clear that a fully localized picture is also not correct.

Many argued that, in this multiple orbital systems, Hund's rule coupling plays a very important role in the magnetism and charge dynamics in Fe-pnictides/chalcogenides.[85–92] The reduction of the kinetic energy or the mass enhancement of the low-energy quasiparticle is not due to the Hubbard repulsion U, but caused by the Hund's coupling that tends to ferromagnetically align the spins of the electrons from different d orbitals of same Fe site. Indeed, recent optical studies[93,94] indicated that the correlation effect in those systems is different from other strongly correlated materials dominated by the on-site Hubbard U or Mott physics. The spectral feature could largely be linked with the effect of Hund's coupling.

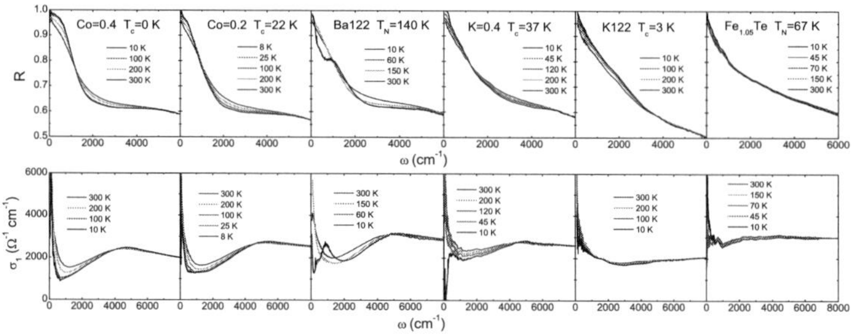

Figure 5.21 The evolution of the optical spectra of $BaFe_2As_2$ with Co- and K-doping. $Fe_{1.05}Te$ is added for comparison. Upper panel: R(ω) for the pure, Co- and K-doped $BaFe_2As_2$ and $Fe_{1.05}Te$ up to 6000 cm^{-1}. Lower panel: $\sigma_1(\omega)$ for for the pure, Co- and K-doped $BaFe_2As_2$ and $Fe_{1.05}Te$ up to 8000 cm^{-1}.[93] See also Color Insert.

Wang *et al.* examined the evolution of optical spectra with electron- and hole-doping in Ba122 system (Fig. 5.21).[93] Besides the free carrier response at low frequency, the optical conductivity spectra show a temperature-dependent gap-like suppression at rather high energy scale near 5000 cm^{-1} ($\sim$0.6 eV). The suppression is found to be doping-dependent. Compared with the parent $BaFe_2As_2$ compound, the high-energy gap-like suppression remains rather eminent in the Co-doped compounds. However, the feature becomes weaker in $K = 0.4$ doped case, and tends to disappear in the pure KFe_2As_2 sample with a T_c of only 3 K. The feature is also not visible in $Fe_{1.05}Te$. Although, at such a high energy scale, most of the spectral weight would come from the interband transitions, the temperature-dependent part must have a different origin.

Wang *et al.* explored several different possibilities for the emergence of a temperature-dependent peak at such high energy scale, and argued that it is unlikely to be caused by the correlation effect from Hubbard U interaction, because Hubbard U interaction would lead to enhancement of both coherent Drude peak and the incoherent component at lowered temperature. This was well examplified by V_2O_3.[95] The transfer of spectral weight from part of the Drude tail to higher frequency is substantially different from this expectation. It means that the effect from Hubbard U is small. The feature is also not likely to originate from the indirect interband transitions assisted by the spin-fluctuations. Instead, the feature

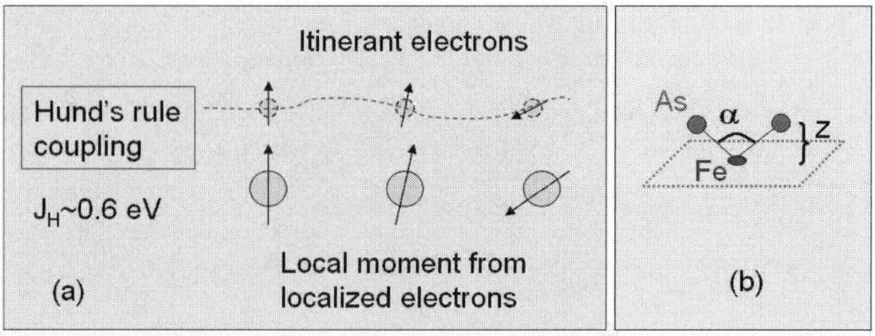

Figure 5.22 (a) A schematic picture about the Hund's coupling between itinerant electrons and local moments. The local moment tends to polarize the spin of the itinerant electrons in a ferromagnetic orientation. At low T, itinerant electrons tends to become more localized due to this coupling, leading to the spectral weigh transfer from low energy to higher energies. (b) the Hund's coupling strength changes with the As-Fe-As angle α, or the height of As atom relative to the Fe-layer **z**, in the structure.[93]

was suggested to be caused by the strong Hund's rule coupling effect between the itinerant electrons and localized moment arising from the multiple Fe 3d orbitals. According to band structural calculations, the crystal field splitting in those compounds is rather small.[96] Then, it was suggested that whether the electrons in the Fe-3d orbitals are itinerant or not depends on the relative value of electron kinetic energy to the Hund's rule coupling energy, i.e. if the kinetic energy of 3d electrons is larger (or smaller) than the Hund's coupling energy, those electrons will be itinerant (or localized). Of course, the itinerant electrons and local moments are not completely isolated, but are still coupled via the Hund's interaction, as illustrated in Fig. 5.22.

In fact, the energy levels of electrons in different orbitals are rather close from band structural calculations.[96] It is argued that the energy levels or bands that contribute to the local moments are rather close to the Fermi level where itinerant electrons dominate, so that the temperature change could affect their interaction. The energy scale of the suppression near 0.6 eV was identified as the local Hund's coupling energy scale. This value is indeed close to those estimated in a number of the theoretical works[85–91] and a resonant inelastic x-ray scattering (RIXS) experiment.[97] At lower temperature, the thermal excitations become weak, then a fraction of relatively itinerant electrons tend to become more localized by Hund's coupling to local moments. This explains the observed the spectral weight transfer from low energy to an energy scale higher than the Hund's coupling energy.

Note that the Fe atoms are not isolated, the coupling strength of itinerant electrons to the local moment formed by localized electrons would change with the environment of Fe atom. Experimentally, the high energy pseudogap feature in optical response changes with doping (Fig. 5.21), particularly in the K-doped case. According to the crystal structural characterizations, K-doping does not induce detectable change in the Fe-As bonding length, but leads to a continuous decrease of As-Fe-As angle α, or equivalently, an increase of the height of As atom (**z**) relative to the Fe-layer.[98] The Te height relative to the Fe-layer in FeTe is further increased compared to Fe-pnictides.[89] On the contrary, the Co-doping does not lead to an detectable change in the a-axis lattice parameter but only a slight decrease of c-axis parameter.[99] The increase of As height mainly make the Fe atom more isolated, leading to a narrowing of the Fe 3d bands. As a result, the effective Hund's coupling is enhanced.[89] Then the spectral weight suppression feature may not be seen within a temperature range change from 300 to 10 K. Higher temperature range would be required in order to see the spectral weight transfer structure.

Schafgans *et al.* investigated the optical conductivity of the parent $BaFe_2As_2$ and optimally doped $BaFe_{1.84}Co_{0.16}As_2$ compounds as a function of temperature over a broad spectral range.[94] By examining the energy scale and direction of spectral weight (SW) transfer, they also argued that Hund's coupling energy is the primary mechanism that gives rise to correlations. Figure 5.23 displays the ratio of the integrated SW at low temperature $K_{exp}(T, \Omega)$ to the room temperature value $K_{exp}(295K, \Omega)$, where the cut-off value Ω extends throughout the entire measured frequency range. This ratio emphasizes the relevant energy scale of SW transfer. If there is a transfer of SW from high to low energy, the SW ratio will exceed 1 at low energy and then smoothly approach 1 until the full energy scale of the Drude oscillator is reached. If there is a transfer of SW from low to high energy, the SW ratio will fall below 1 until the total energy scale of SW transfer is reached.

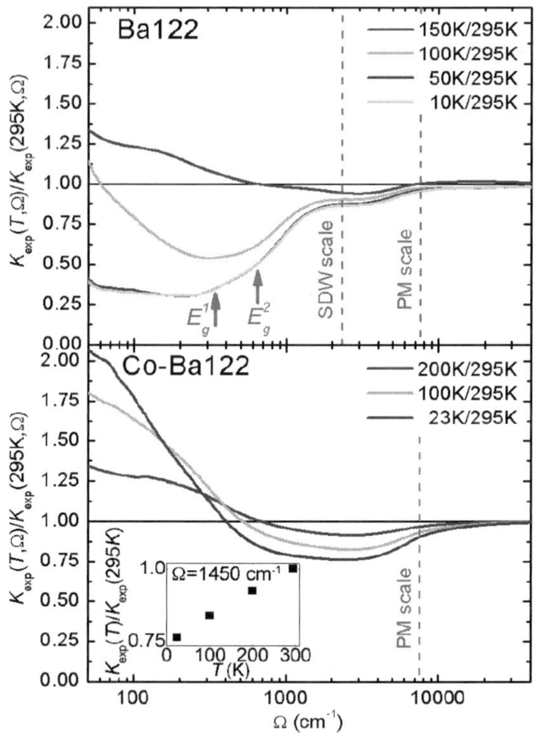

Figure 5.23 Ratio of the integrated conductivity $K_{exp}(T, \Omega)/K_{exp}$ (295 K,Ω) as a function of cutoff value Ω and temperature over the entire measured frequency range in Ba122 (top panel) and Co-Ba122 (bottom panel). The ratios are taken with respect to the $T = 295$ K data in order to display the temperature evolution of the PM and SDW SW transfer. The two vertical dashed bars show the energy scale below which all of the SW transfer takes place, for both the PM and SDW states. The energy scale of SW transfer in the PM state is $\approx$8000 cm^{-1} and in the SDW state is $\approx$2000 cm^{-1}. Inset: Temperature dependence of the kinetic energy of the Drude response, normalized to the room temperature value, at a cutoff $\Omega = 1450$ cm^{-1}.[94]

Schafgans *et al.*[94] first analyzed the SW ratios in the paramagnetic (PM) state. As temperature decreases, the Drude response narrows and some SW shifts to lower energy. This is in accordance with the expectation for a metal that causes the SW ratio to exceed 1 roughly below 600 cm^{-1}. Yet in the frequency range from 600 cm^{-1} to 8000 cm^{-1}, the ratios decrease to well below 1. There is a minimum in the PM state ratios near 3000 cm^{-1}, indicating a turning point in SW transfer: below this value, SW is depleted; while above this value, SW is amassed. At about 8000 cm^{-1}, the ratios return to $\sim$1 as all of the SW is recovered. Therefore, SW is transferred from the Drude response into the region between 3000 - 8000 cm^{-1} in both Ba122 and Co-Ba122. The energy scale of SW transfer is determined by observing the energy range over which most of the spectral weight is recovered, marked in Fig. 5.23 with vertical dashed lines. By this criterion, the energy scale for correlations in the PM state is about 8000 cm^{-1}. In the SDW state in Ba122, it is observed the superposition of two concomitant processes each with its own energy scale of SW transfer. The energy scale relevant for correlations in the PM state persists in the SDW state. In addition, there is SW transfer at lower energies, consistent with the formation of the SDW gaps. The SW is recovered roughly below 2000 cm^{-1} for the SDW gaps.

Schafgans *et al.*[94] pointed out that the energy scale of the SW transfer is not the energy scale of Huubard U. They also noted that the SW of the compounds does not significantly change with doping, while in a Mott insulator the effective number density of carriers increases linearly with doping. More importantly, they found an opposite direction of SW transfer. In strongly correlated metals derived from Mott insulators, the Drude oscillator strength grows with lowering temperatures due to SW transfer from high to low energy, over an energy scale of the Hubbard U/2, e.g. V_2O_3.[95] The situation in the 122 pnictides is quite different, where, as the temperature decreases, the SW is transferred from low to high energy. Schafgans *et al.* argued that the electron correlation mechanism in those materials was predominantly due to Hund's coupling: J$\approx$0.6–0.9 eV, which is approximately the observed energy scale of SW transfer.

5.6 Anisotropic Charge Dynamics

5.6.1 c-*Axis Optical Properties in Parent Compounds*

For quasi-two-dimensional layered materials, charge transport and dynamics within *ab*-plane would be different from those along

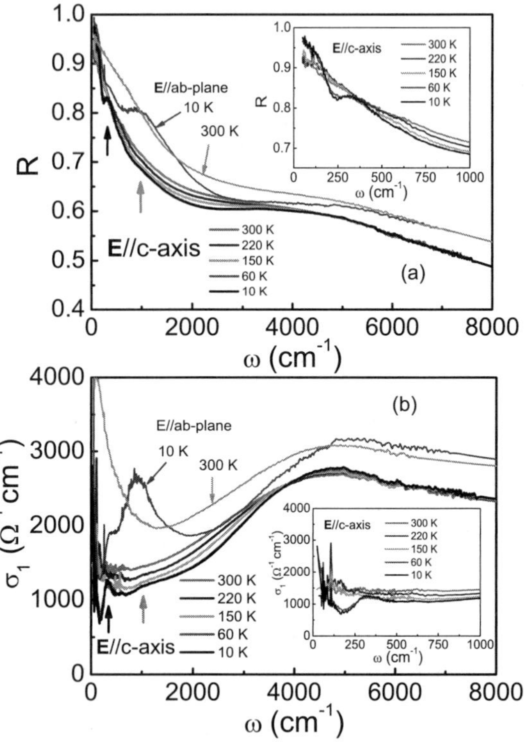

Figure 5.24 The c-axis $R(\omega)$ (a) and $\sigma_1(\omega)$ (b) at different temperatures below 7000 cm^{-1} for $BaFe_2As_2$. Inset: expanded plot of the low-ω $R(\omega)$ and $\sigma_1(\omega)$ spectra. The ab-plane $R(\omega)$ and $\sigma_1(\omega)$ at 300 K and 10 K are included for comparison. The two upward arrows indicate the anomalies associated with the SDW order. The very pronounced peak in $\sigma_1(\omega)$ for **E**|| ab-plane becomes extremely weak for **E**||**c**.[100]

c-axis. Optical studies of the anisotropic charge dynamics would be essential for a comprehensive understanding of these materials.

Measuring the c-axis optical response requires thick single crystal samples. Chen *et al.* studied the c-axis optical properties of AFe_2As_2 (A = Ba, Sr) and compared with the in-plane data.[100] Figure 5.24 displays the temperature dependences of $R(\omega)$ and $\sigma_1(\omega)$ below 7000 cm^{-1}. The in-plane data at 300 K and 10 K were also included for a comparison. The overall $R(\omega)$ along the c-axis is found to be quite similar to that in the ab-plane except for relatively lower values. This is dramatically different from the optical spectra of some high-T_c cuprates[101] and other layered compounds, for example, layered ruthenates,[102] where the c-axis R(ω) shows much lower values and quite different frequency-dependent behavior from the ab-plane. This observation suggests that the band structure of Fe-pnictide should be quite three-dimensional (3D), in contrast to the expectation based on its layered crystal structure. The difference in $\sigma_1(\omega)$ spectra between ab-plane and the c-axis at 300 K becomes larger at low frequencies. An extrapolation to zero frequency, or the dc conductivity, shows an anisotropy ratio of about 2.8. This value matches well with the anisotropy ratio determined directly from the dc resistivity measurement by a careful Montgomery technique (about 3 ± 1).[103]

The spectra at low temperatures appear more interesting. Different from the ab-plane optical response where two distinct energy gaps were observed in the SDW state, only the smaller energy gap could be seen clearly for **E**|| c-axis. The very pronounced energy gap structure seen at a higher energy scale for **E**|| ab-plane is almost invisible.

Understanding why two distinct gaps could be observed for **E**|| ab-plane but only a smaller one could be clearly seen for **E**|| c-axis would provide useful information about the band structure evolution across the transition. The data were explained by assuming the presence of two types of FSs in the system: the 2D cylinder like FSs and a large-size 3D ellipsoid like FS, as schematically presented in Fig. 5.25. Presence of the large-size 3D ellipsoid FS with dominant Fe-3d$_{3z^2-r^2}$ orbital is indicated by the recent *ab-initio* LDA+Gutzwiller calculations, where electron correlations are taken into account beyond LDA,[90] and the bulk Compton scattering measurements.[104] The 2D cylinder-like hole FSs around Γ and the 2D electron FSs around the corner of the Brillouin zone are well nested with a (π, π). With the development of magnetic order with (π, π) wave vector, those FSs are strongly affected with the formation of a large energy gap. On the other hand, the 3D ellipsoid FS does not show any nesting with other FSs. Nevertheless, with the formation of the SDW order the new magnetic Brillouin zone boundary would cut the large-size 3D ellipsoid FS and result in an energy gap at the intersecting points as shown in

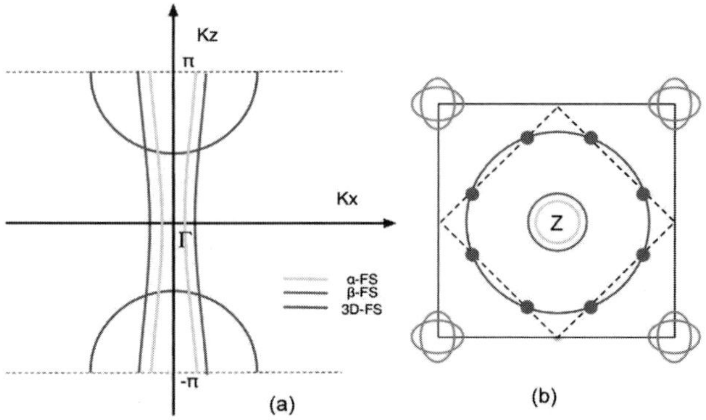

Figure 5.25 (e) The schematic plot of the Fermi surfaces along the k_z direction around Γ point. There are 2D cylinder-like Fermi surfaces (α and β bands) and 3D ellipsoid Fermi surface enclosing the Z point. (f) A top view of the Fermi surfaces. The new magnetic Brilliouin zone boundary (dashed lines) would cut the 3D FS and open gaps in the intersecting points (colored in red).[100]

Fig. 5.25. Considering that the Fermi velocities of the 2D cylinder-like FSs are mainly within the *ab*-plane, the Fermi velocity of the 3D FS has components both within the *ab*-plane and along the *c*-axis, Chen *et al.* ascribed the larger gap seen most clearly for **E**|| *ab*-plane to the nesting-driven gap opened on 2D FSs, the smaller one to the gap formed on 3D FS. Because the 2D FSs are more dramatically affected by the SDW instability than the 3D FS, a reduction of the anisotropy in the magnetic ordered state is expected.

5.6.2 *Anisotropic Optical Properties within* ab-*Plane*

It is known that the parent compounds of Fe arsenides also undergo a tetragonal to orthorhombic structural phase transition in accompany with the formation of long range magnetic order. With $BaFe_2As_2$ as an example, the *b*-axis is shorter than the *a*-axis in the orthorhombic phase below T_s. According to neutron measurements, the Fe spins align antiferromagnetically in the *a*-axis direction in the Fe-plane and ferromagnetically in the *b*-axis direction. Then, an in-plane anisotropy would develop below the magnetic/structural phase transition. However, cooling the sample

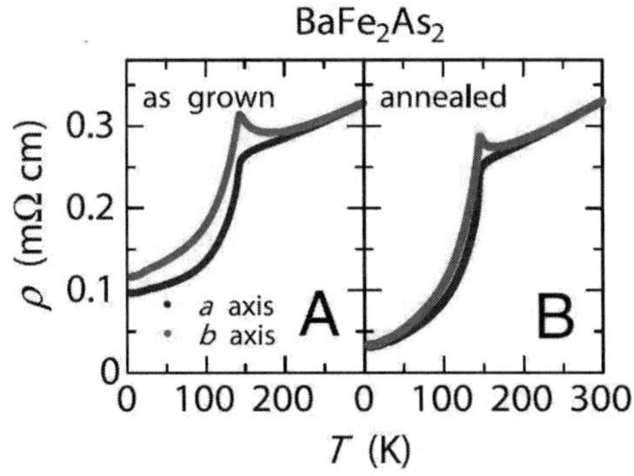

Figure 5.26 Temperature dependence of the in-plane resistivity of detwinned $BaFe_2As_2$, ρ_a (blue) and ρ_b (red). (A) For the as-grown crystal, anisotropy starts at approximately 220 K and persists down to the lowest measurement temperature. (B) Annealing decreases residual resistivity and increases magnetostructural transition temperature T_s slightly.[108]

at ambient pressure always leads to twinned domains in the magnetic phase. It is found that the application of a modest uniaxial pressure along the tetragonal [110] direction could remove the twinning in the cooling process.[105,106] Dc resistivity measurements revealed a smaller resistivity along the spin-AF direction with a longer a-axis than that along the spin-ferromagnetic direction with a shorter b-axis. The result is surprising because normally electrons would hop more easily between atoms along the shorter b-axis and FM spin alignment direction. Two groups reported polarized measurement on detwinned single crystals of $BaFe_2As_2$.[107,108] High quality data were achieved by Nakajima *et al.*[108]

Figure 5.26 displays the anisotropic resistivity for as-grown and vacuum-annealed samples.[108] The polarized optical measurement with $E \parallel a$ and $E \parallel b$ on the annealed crystal is shown in Fig. 5.27. The spectra in two polarizations are almost identical at temperatures above 200 K. The anisotropy is mainly seen below T_s where a rapid reduction in conductivity is observed below approximately 600 cm^{-1} in σ_a and below approximately 850 cm^{-1} in σ_b. It is found that the low-energy conductivity is more strongly suppressed in the b direction. This suppression, though affected by the rapid narrowing of Drude peak, is mainly caused by the opening

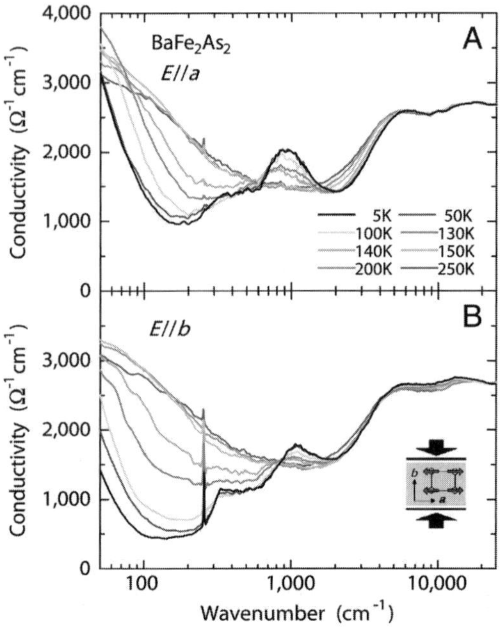

Figure 5.27 Temperature evolution of the optical conductivity spectrum of detwinned $BaFe_2As_2$ for polarization parallel to the *a* and *b* axes.[108]

of gaps. The opening of gaps is also evident from the appearance of a bump in σ_a (a cusp in σ_b) at approximately 340 cm^{-1}, as well as by the growth of a peak centered at 950 cm^{-1} in σ_a (1050 cm^{-1} in σ_b), which is formed by accumulating the low-energy conductivity that is reduced due to the gap. It is found that almost all the reduced conductivity is accumulated to form the peak in $\sigma_a(\omega)$. On the other hand, the missing SW in σ_b is transferred not only to the peak at 1050 cm^{-1} but also to a much higher-energy region. As a result, the anisotropy of conductivity persists up to 17000 cm^{-1} (approximately 2 eV), as shown in the spectra at $T = 5$ K in Fig. 5.28. Below 1350 cm^{-1}, $\sigma_b(\omega)$ is lower than $\sigma_a(\omega)$, whereas above 1350 cm^{-1} the anisotropy is reversed. Note that the gap feature appearing at approximately 340 cm^{-1} in both $\sigma_a(\omega)$ and $\sigma_b(\omega)$ plays a minor role in this SW transfer. Nakajima *et al.* found that the temperature-induced spetral weight transfer roughly extends to 9000 cm^{-1} for $\sigma_a(\omega)$ but to 17000 cm^{-1} for $\sigma_b(\omega)$. Apparently, the spectral weight

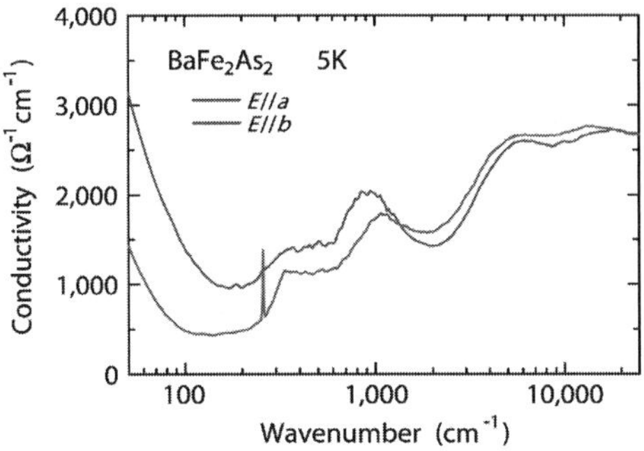

Figure 5.28 Comparison between the *a*- and *b*-axis optical conductivity spectra at 5 K above 50 cm^{-1}.[108]

transfer associated with the gap formation is not compatible with an ordinary SDW scenario based on the FS nesting.

Nakajima *et al.* tried to analyzed the data using a Drude-Lorentz model, as illustrated in Fig. 5.29. At high temperature (above T_s), the spectrum could be decomposed into a broad Drude component and a narrow Drude component, both contributing to the dc conductivity. In the orthorhombic stripe AF phase (T = 5 K), because of the partial gap opening, the width of the Drude component becomes extremely narrow. It is seen that the gap opens mainly in the broad Drude component. A small difference is the longer tail lasting above 100 cm^{-1} of the $\omega = 0$ peak in $\sigma_a(\omega)$, which requires an additional Drude component that probably originates from an ungapped portion of the Fermi surface after reconstruction. This component may explain the slightly lower resistivity in the *a* direction.

Nakajima *et al.* suggested that the difference between σ_b and σ_a in the magnetic ordered state comes mainly from the orbital effect after the band structural reconstruction below T_s. In particular, the Fe 3d$_{xz}$ and 3d$_{yz}$ orbitals are degenerate in energy in the tetragonal Fe sublattice. The degeneracy of Fe 3d$_{xz}$ and 3d$_{yz}$ orbitals are released in the low-T orthorhombic phase. The Fermi surface is composed of mainly Fe 3d$_{xz}$ orbital, whereas the energy bands composed of 3d$_{yz}$ orbital shift away from the Fermi level, thus forming gaps for electronic excitations. The distinction between the Fe 3d$_{xz}$ and 3d$_{yz}$ orbitals is suggested to be the major reason

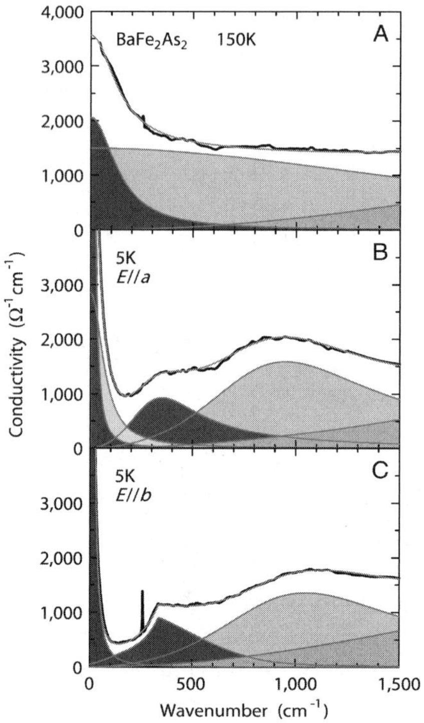

Figure 5.29 Decomposition of the low-energy conductivity spectrum below 1500 cm^{-1}. (A) The spectrum at 150 K in the tetragonal-paramagnetic phase is decomposed into a broad Drude component and a narrow Drude component, both contributing to the dc conductivity. (B and C) In the orthorhombicstripe AF phase ($T = 5$ K), the width of the narrow Drude component in A becomes extremely narrow, and the peak height reaches 30000 Ω^{-1} cm^{-1} in both spectra. A gap opens in the broad Drude component (in gray), but the gap in σ_b is deeper than that in σ_a.[108]

for the difference in the two polarizations along the a (x) and b (y) axes. Theoretical calculations by Yin *et al.*[88] using local density approximation combined with dynamical mean-field theory indicate that $\sigma_a(\omega)$ comes primarily from the Fe 3d_{xz} and 3d_{xy} orbitals, and $\sigma_b(\omega)$ is associated with the Fe 3d_{yz} and 3d_{xy} orbitals. Their results are qualitatively in agreement with the present experimental results.

Another important observation is that phonon mode near 257 cm^{-1} behaves very differently in two polarizations. The intensity of the mode in $\sigma_b(\omega)$ increases by a factor of 3~4 in comparison with that above T_s, whereas its intensity in $\sigma_a(\omega)$ decreases to an indiscernible level. The mechanism behind is not fully understood, although it was also suggested to be related to the orbital effect.

5.6.3 c-*Axis Optical Properties of Superconducting Compounds*

It is also very interesting to study the anisotropic charge dynamics for the superconducting samples and to compare the results with cuprates. Optical study on the high-T_c cuprates revealed rather peculiar phenomenon along the *c*-axis below T_c. Cuprates have 2D layered structures. The conducting CuO_2-planes are usually separated by the insulating blocking layers. In the normal state, those CuO_2-planes appear to be almost decoupled, leading to the non-metallic charge transport and dynamics along the *c*-axis. However, as soon as the systems enter into the superconducting state, the CuO_2-planes are coupled via the Josephson tunnelling effect, then a sharp reflectance edge arising from the condensed superconducting carriers immediately shows up in the *c*-axis polarization.[101,109–112] The plasma edge is referred to as the Josephson plasma edge, which is linked with the *c*-axis penetration depth. It offers a direct measure to the strength of the interlayer coupling. This behavior is also connected with the mechanism of high-temperature superconductivity, since the absence of this edge above T_c is considered as indicative of the confinement mechanism.

Fe-pnictide superconducting materials also crystalize in the layered structure with FeAs layers separated by alkaline earth metal ions (Ba^{2+}, Sr^{2+}) or other Re-O (Re=La, or rare-earth elements with trivalent Re^{3+} and negative divalent O^{2-}) layers which should be insulator-like. It is interesting to see whether or not the Fe-pnictides share similar anisotropic charge dynamical properties with cuprates. Cheng *et al.* performed *c*-axis polarized optical measurement on a $Ba_{0.67}K_{0.33}Fe_2As_2$ single crystal.[113] As shown in Fig. 5.30, the reflectance R(ω) spectra below 1200 cm^{-1} indicate a typical metallic temperature dependence: the low-ω R(ω) increases with decreasing temperature. Most noticeably, no sharp plasma edge

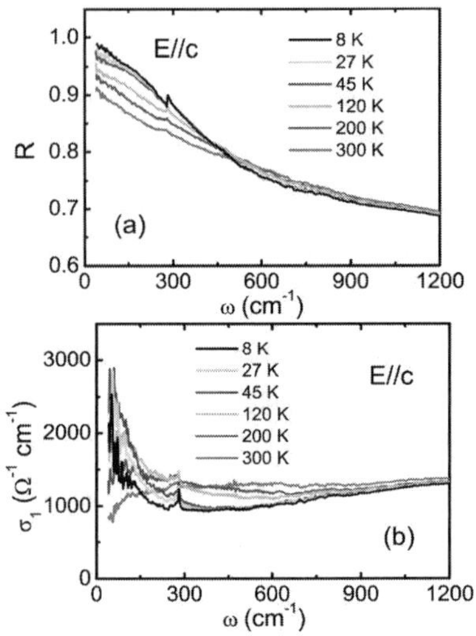

Figure 5.30 (a) optical reflectance R(ω) spectra at different temperatures below 1200 cm^{-1}. (b) the conductivity $\sigma_1(\omega)$ spectra of the sample at different temperatures.[113]

develops below T_c. This is in sharp contrast to the high-T_c cuprates where steep reflectance edge is seen in the superconducting state.

Interestingly, although the sample shows a Drude component centered at zero frequency at low temperature, a peak at finite frequency in $\sigma_1(\omega)$ is seen at high temperature. The peak locates near 300 cm^{-1} at 300 K and shifts gradually to lower frequency with decreasing temperature and eventually disappears below 45 K. It is worth noting that the presence of the conductivity peak at finite frequency does not necessarily mean the absence of the dispersive band. It is often seen in poor metallic materials where the quasi-particle peaks along the dispersive band are rather broad. It is a kind of dynamical localization effect.

A further interesting point is that the superconducting condensate probed with $E \parallel ab$-plane and $E \parallel c$-axis are very different. Figure 5.31 compares the c-axis R(ω) and $\sigma_1(\omega)$ at 45 K and 8 K with the ab-plane spectra. A full gap feature is seen for $E \parallel ab$-plane. On the other hand, rather samll difference between 45 K and 8 K is seen

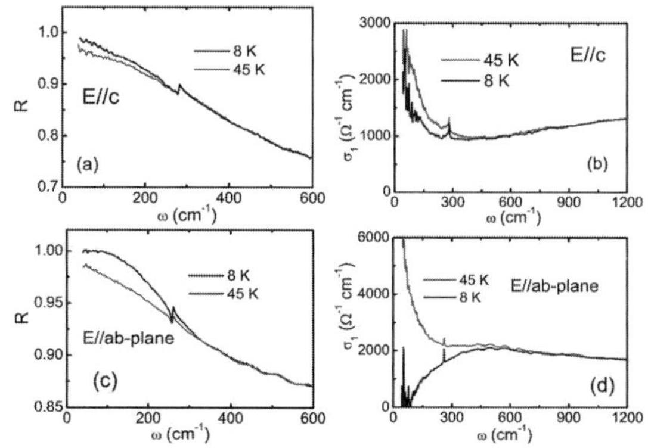

Figure 5.31 (a) and (b) the c-axis R(ω) and $\sigma_1(\omega)$ spectra at 8 K and 45 K at low frequencies. (c) and (d) the in-plane R(ω) and $\sigma_1(\omega)$ spectra at 8 K and 45 K. The in-plane data show an s-wave superconducting energy gap. By contrast, a significant residual Drude response is still observed for **E**|| c-axis.[113]

in c-axis polarized measurement. The data suggest the presence of a significant amount of quasiparticles which contribute specifically to the c-axis conductivity.

The data highly suggested the presence of nodes in FSs with quasiparticle excitations contributing dominantly to the c-axis transport. Since the c-axis polarized measurement mainly probes the three-dimensional FSs, the results strongly suggest the presence of horizontal nodes in the highly flared region as shown in Fig. 5.32, because vertical nodes on either 2D or 3D FSs would

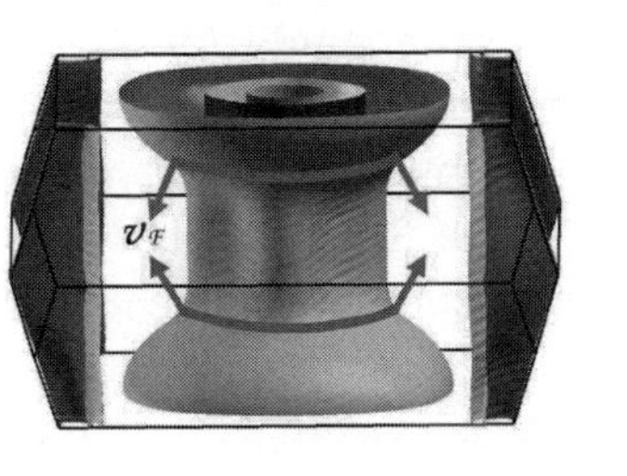

Figure 5.32 The Fermi surfaces of (Ba,K)Fe_2As_2 from LDA+Gutzwiller calculations.[90] Besides the 2D cylinder-like FSs, there is an additional 3D FS. Possible horizontal nodes may exist near the highly flared region of 3D FS which has maximum contribution to the c-axis conductivity.[113]

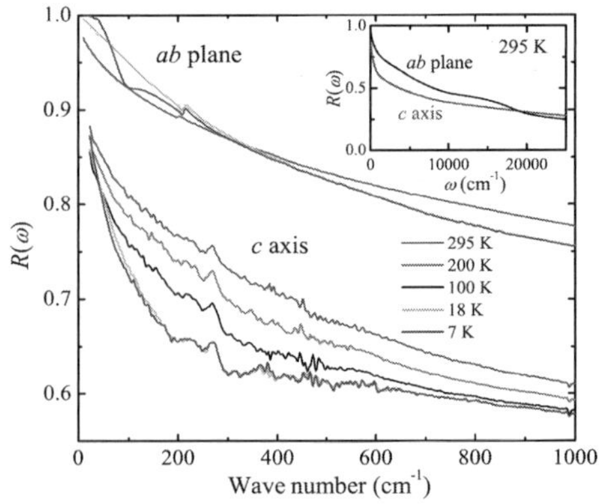

Figure 5.33 T-dependent c-axis reflectance $R(\omega)$ of $FeTe_{0.55}Se_{0.45}$ with $T_c \approx 14$ K. The ab-plane reflectance at 295, 18, and 6 K is also plotted for comparison. The inset shows the spectra over broad frequency up to 25 000 cm^{-1}.[119]

contribute dominantly to the ab-plane conductivity. The results are consistent with recent c-axis penetration depth and heat transport measurements,[114–117] which also suggested that the nodes contribute specifically to the c-axis transport. Neutron scattering experiment on the same batch of $Ba_{0.67}K_{0.33}Fe_2As_2$ crystal revealed that the low-energy spin excitations below T_c displays a sinusoidal modulation, indicating clear gap at momentum transfer $\mathbf{q}_z = 0$ but gapless behavior at $\mathbf{q}_z = \pi$,[118] thus being consistent with the optical results.

Moon *et al.* investigated the interplane c-axis electronic response of an iron chalcogenide superconductor $FeTe_{0.55}Se_{0.45}$ with $T_c \approx 14$ K.[119] Figure 5.33 displays c-axis reflectance $R(\omega)$ at different T. Although the spectral shape of c-axis $R(\omega)$ is similar to that of ab-plane, the value of the $R(\omega)$ is much lower in comparison of $Ba_{0.67}K_{0.33}Fe_2As_2$. Furthermore, the T dependence of c-axis $R(\omega)$ in the normal state is quite different from that of in-plane response. As T decreases, $R(\omega)$ decreases with decreasing T at all frequencies. This is different from $Ba_{0.67}K_{0.33}Fe_2As_2$, evidencing a non-metallic temperature dependence.

In accord with the reflectance spectrum, the conductivity spectra $\sigma_1(\omega)$ shows no Drude peak peak at any T, as shown in Fig. 5.34.

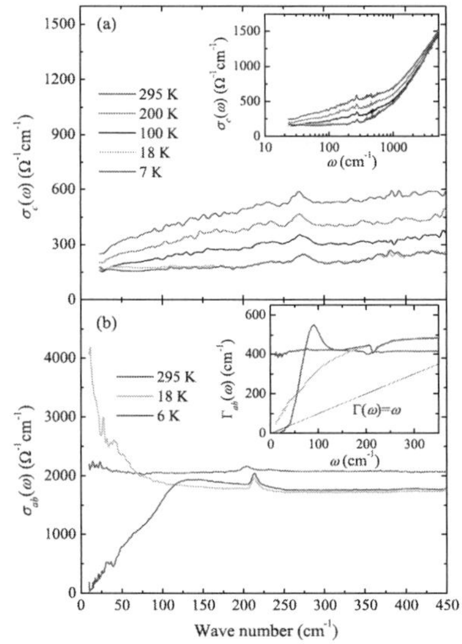

Figure 5.34 (a) T-dependent c-axis $\sigma_1(\omega)$. $\sigma_1(\omega)$ becomes clearly depressed with decreasing T, and the suppressed SW seems not to be recovered by 5000 cm^{-1}, as shown in the inset. (b) T-dependent ab-plane optical conductivity. The inset shows the frequency-dependent ab-plane scattering rate.[119]

Moreover, $\sigma_1(\omega)$ becomes suppressed as T is lowered and the spectral weight (SW) is transferred to higher energy. Just above the superconducting transition temperature, $\sigma_1(\omega)$ becomes almost frequency independent. The value of $\sigma_1(\omega)$ in the zero-frequency limit is about 160 Ω^{-1} cm^{-1} at 18 K, which is smaller than the ab-plane value by a factor of about 25. All these observations indicate that the normal-state c-axis electronic response of $FeTe_{0.55}Se_{0.45}$ is rather incoherent. It is more similar to underdoped cuprates.

Upon entering the superconducting state, R(ω) becomes larger (smaller) than R(ω) just above T_c below 37 cm^{-1} (between 37 and 200 cm^{-1}). Although R(ω) of $FeTe_{0.55}Se_{0.45}$ shows steep increase, it does not reach unity down to the lowest measurement frequency. Accordingly, while $\sigma_1(\omega)$ below T_c shows weak depression below 200 cm^{-1}, it does not exhibit a clear gaplike structure.

5.7 Optical Properties of Iron-Based Superconductors Below T_c

5.7.1 *Probing the Superconducting Energy Gaps*

Since optical spectroscopy is a powerful technique of probing energy gaps in the broken symmetry states, it is used to detect the pairing energy soon after the Fe-based superconductors were discovered. Optical signature of superconducting energy gap in Fe-pnictides was first seen in the polycrystalline samples of $LaFeAsO_{0.9}F_{0.1-\delta}$[120] and $NdFeAsO_{0.82}F_{0.18}$.[121] Much clearer and informative optical results about superconducting energy gap, possible pairing symmetry and condensed superfluid density were obtained when single crystal samples became available. Among the first obtained superconducting single crystal samples are K-doped $Ba_{0.6}K_{0.4}Fe_2As_2$.

The in-plane optical spectra of $Ba_{0.6}K_{0.4}Fe_2As_2$ single crystals ($T_c = 37$ K) were first reported by Li *et al.*[122] In the far-infrared region, a flat and close to unit reflectance and an almost zero conductivity (shown in Fig. 5.35(a) and (b) respectively) roughly below 150 cm^{-1} for $T \ll T_c$ provide evidence for the existence of an s-wave-like superconducting gap ($2\Delta_1^{opt} \approx 18.6$ meV). Moreover, a peak at 200 cm^{-1} is observed in the spectrum of the ratio of the $R_s(10\ K) / R_n(45\ K)$ (shown in Fig. 5.35(c)), indicating the existence of a second superconducting energy gap ($2\Delta_2^{opt} \approx 25$ meV) on the Fermi surface. The two energy gaps observed in the optical spectra match well with the ARPES results that a smaller gap ($\Delta_1^{ARPES} \simeq$ 6-8 meV) on the large hole-like Fermi surface and a larger gap ($\Delta_2^{ARPES} \simeq 12$ meV) on the two smaller hole- and electron-like FSs.[123]

The clear identification of a full superconducting energy gap below 150 cm^{-1} in $Ba_{0.6}K_{0.4}Fe_2As_2$ suggests an s-wave pairing symmetry for the Fe-based superconductors. From the missing area (Eq. 5.25), it is easy to get the London penetration depth $\lambda = 2080$ Å. The penetration depth could also be calculated from the imaginary part of optical conductivity in the low-frequency limit by Eq. 5.26. Then the value obtained is $\lambda \simeq 1950$ Å. The good agreement between the two methods suggests that the Ferrell-Glover-Tinkham (FGT)[32] is satisfied. In the inset of Fig. 5.35(b), it could be found that the

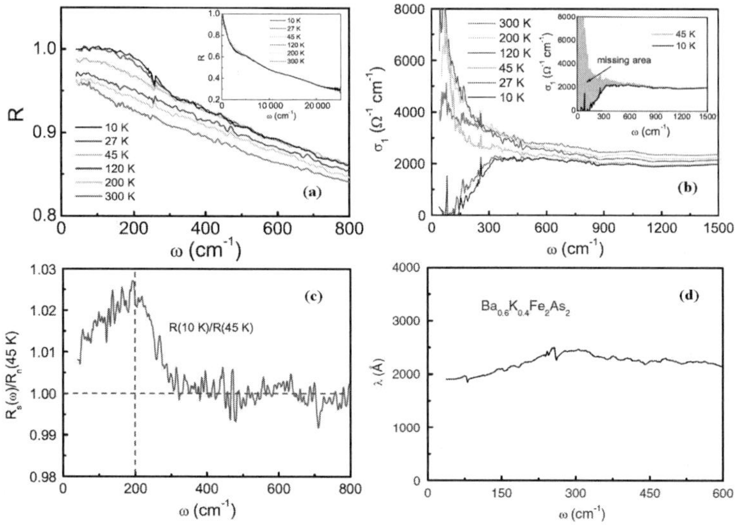

Figure 5.35 The *ab*-plane optical spectra of $Ba_{0.6}K_{0.4}Fe_2As_2$ single crystals: (a) Temperature dependence of reflectivity R(ω) in the far-infrared region. Inset: R(ω) over broad frequencies up to 250000 cm^{-1}. (b) Temperature dependence of the conductivity $\sigma_1(\omega)$. The shaded area in the inset represents the missing area due to the superconducting condensate. (c) The reflectance R(ω) at 10 K normalized to the values at 45 K in the normal state. A peak near 200 cm^{-1} is observed. (d) Frequency dependent London penetration depth for $Ba_{0.6}K_{0.4}Fe_2As_2$ single crystals at 10 K.[122] See also Color Insert.

missing spectral weight recovers around 600 cm^{-1}, indicating that for $Ba_{0.6}K_{0.4}Fe_2As_2$, superconducting condensate forms rapidly and the FGT sum rule is rapidly recovered. Thus, $Ba_{0.6}K_{0.4}Fe_2As_2$ is quite different from underdoped high-T_c cuprates that the recovery of the FGT sum rule goes to very high energy, or even violated.[33–37]

From the value of the superfluid density ($\rho_s = \omega_{ps}^2 = c^2/\lambda^2$) and T_c, Li *et al.* also found that $Ba_{0.6}K_{0.4}Fe_2As_2$ falls off the Uemura relation[124,125] which works well for the hole-doped cuprates in the underdoped region, but fits better to Homes' scaling relation for BCS dirty-limit superconductor.[37] The clear observation of the pairing energy gap and the good fit to the Homes' scaling would suggest that the *ab*-plane (FeAs-plane) of $Ba_{0.6}K_{0.4}Fe_2As_2$ is in the dirty limit. However, a different opinion exists. A recent work by Charnukha *et al.*[126] suggested that the $Ba_{0.6}K_{0.32}Fe_2As_2$ (T_c = 38.5 K) superconductor is not in the dirty but clean limit. They discussed the origin of superconductivity-induced infrared optical anomalies in the framework of a multiband Eliashberg theory with two distinct superconducting gap energies $2\Delta_A \approx 6k_BT_c$ and $2\Delta_B \approx 2.2k_BT_c$, and argued that an unusual suppression of the optical conductivity at energies up to 14 k_BT_c in the superconducting state could be ascribed to spin-fluctuation-assisted processes in the clean limit of the strong-coupling regime.

Following the work on the hole-doped $Ba_{1-x}K_xFe_2As_2$ superconductors, a large number of optical spectroscopic measurements have been done on the electron-doped $Ba(Fe_{1-x}Co_x)_2As_2$ superconducting single crystals.[70–73,127–130] Formation of superconducting energy gaps below T_c was observed by all research groups, however there were different opinions on the number of energy gaps manifested in the real part of conductivity spectra (ranged from one to three) as well as on the gap amplitudes. van Heumen *et al.* and Tu *et al.* found that the optical spectra of the optimally Co-doped superconductors at $T \ll T_c$ (Figs. 5.36 and 5.37) can be better simulated with a model of two isotropic s-wave gaps than a single gap model.[72,127] The two energy scalings of the gaps were obtained: $2\Delta_1 = 6.2\pm 0.8$ meV and $2\Delta_1 = 14\pm 2$ meV, which were consistent with those observed on the electron and hole Fermi surface, by ARPES.[131]

Besides the clear observation of the superconducting gaps in $BaFe_{1.85}Co_{0.15}As_2$, Tu *et al.* also estimated the superconducting plasma frequency ($\omega_{p,S} \sim 5200 \pm 400$ cm^{-1} and found that only 50% of the free carriers in the normal state collapsed into the superconducting condensate at the lowest temperature measured.[72] Based on the fact that for a clean superconductor the whole free carrier spectral weight would collapse into the superconducting condensate for $T \ll T_c$; while for the dirty-limit case ($2\Delta \ll 1/\tau$) only the low frequency part of the free carrier spectral weight contributes to the superconducting condensate, they also concluded that this material is not in the clean limit.

However, Kim *et al.* declared that the infrared spectra of $BaFe_{1.87}Co_{0.13}As_2$ at 5 K, shown in Fig. 5.38, could be also well simulated in terms of three isotropic s-wave superconducting gaps:

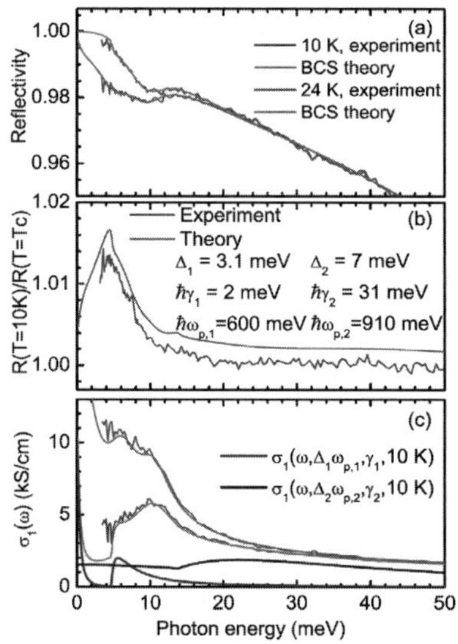

Figure 5.36 (a) Low energy experimental reflectivity (blue) and reflectivity obtained from the BCS simulation (violet) at 10 K. Also shown are the reflectivity (red) and Drude-Lorentzmodel result (green) near T_c for $BaFe_{2-x}Co_xAs_2$ with x = 0.14. (b) Ratio of 10 K reflectivity to the one at T_c (24 K) for the experimental data (blue) and simulation (for clarity, the simulation curve in red is multiplied by a factor of 1.002). (d) Comparison of the 10 K optical conductivity determined from experiment (blue) and simulation (violet) and the 24 K data (red) and Drude-Lorentz model (green).[127]

$2\Delta_1 = 53\ \text{cm}^{-1}, 2\Delta_2 = 80\ \text{cm}^{-1}, 2\Delta_3 = 157\ \text{cm}^{-1}$. The ratio of gap over T_c $2\Delta/k_B T_c = 3.1, 4.7$, and 9.2. They also extracted the value of the in-plane penetration depth of 270 nm.[128]

Gorshunov *et al.* measured the optical reflectivity and complex transmissivity of an epitaxially grown $Ba(Fe_{0.9}Co_{0.1})_2As_2$ thin film over a wide frequency range (4-35000 cm^{-1}).[129] As shown in Fig. 5.39(b), the real part of the optical conductivity for $T = 5$ K drops to almost zero at 30 cm^{-1}, indicating the opening of the superconducting gap on the Fermi surfaces. The shape of $\sigma_1(\omega)$ below T_c in the far-infrared region could be reproduced with a model of a single BSC-like gap. The size of the superconducting gap is estimated to be 3.7 meV, and thus $2\Delta/k_B T_c \approx 2.1$, which is lower

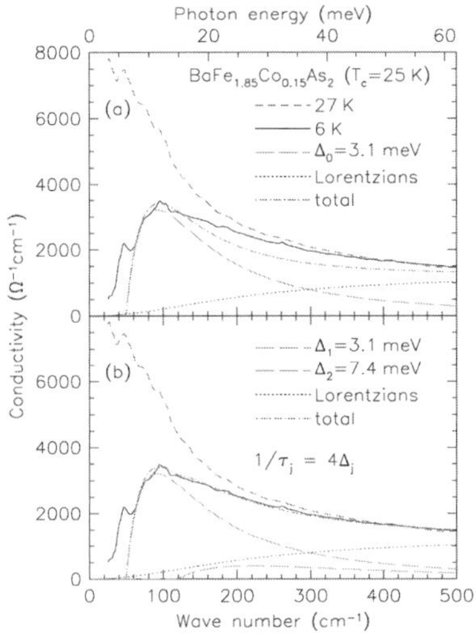

Figure 5.37 The in-plane optical conductivity of the iron-arsenic superconductor $BaFe_{1.85}Co_{0.15}As_2$ ($T_c = 25$ K) shown at 27 and 6 K (dashed and solid lines, respectively). (a) The optical conductivity with a single isotropic s-wave gap of $\Delta_0 = 3.1$ meV with a scattering rate $1/\tau = 4\Delta_0$ is calculated for $T \ll T_c$ (long-dashed line) and superimposed on the contribution from the bound excitations in the mid-infrared (dotted line); the linear combination of the two curves (dot-dash line) does not reproduce the data above ≈ 140 cm^{-1}. (b) The optical conductivity with isotropic s-wave gaps of $\Delta_1 = 3.1$ meV and $\Delta_2 = 7.4$ meV with the scattering rates $1/\tau_j = 4\Delta_j$ is calculated for $T \ll T_c$ and superimposed on the Lorentzian contribution; the linear combination of the three curves is in much better agreement with the measured data.[72]

than the value expected from BCS theory. Surprisingly, below 10 cm^{-1} the optical conductivity for $T = 5$ K is much larger than that expected from BCS theory, which was suggested to come from uncondensed quasiparticles.

Lobo *et al.* also claimed that besides the clear signature of superconducting gap in the optical spectra of single-crystalline $Ba(Fe_{0.92}Co_{0.08})_2As_2$ ($T_c = 22.5$ K), the very low frequency conductivity is observed to be higher than that from the thermally broken Cooper pairs.[130] Consequently, as shown in Fig. 5.40(b), a Drude component, a Lorentz term, and a Mattis-Bardeen term, should be involved to simulate the low-frequency optical conductivity at

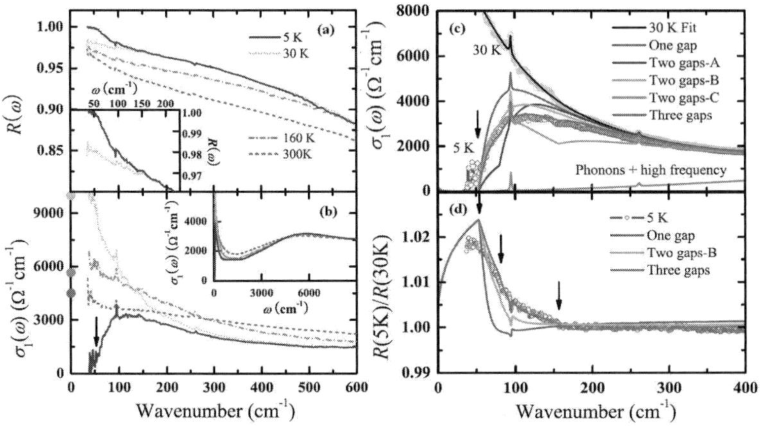

Figure 5.38 In-plane optical spectra of $BaFe_{1.87}Co_{0.13}As_2$ (a) Temperature dependence of far-infrared reflectivity and (b) the real part of the optical conductivity of $BaFe_{1.87}Co_{0.13}As_2$ single-crystals for light polarized in the a-b planes. Comparison of the experimental spectra of (c) conductivity and (d) reflectivity ratio R_T/R_{30K} with the spectra lines simulated with a BCS-type model of multiple isotropic gaps.[128]

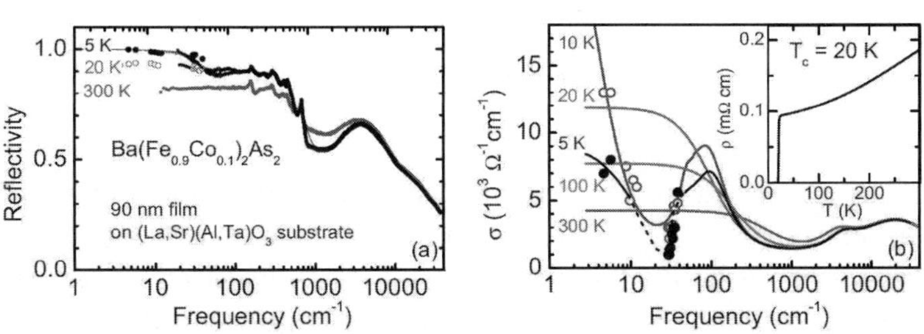

Figure 5.39 (a) Reflectivity of a 90 nm $Ba(Fe_{0.9}Co_{0.1})_2As_2$ film on a 1 mm (La,Sr)(Al,Ta)O_3 substrate measured in a wide frequency range at various temperatures. The dots between 4 and 40 cm^{-1} are calculated from the transmission measurements. (b) Optical conductivity of $Ba(Fe_{0.9}Co_{0.1})_2As_2$ obtained from the Drude-Lorentz analysis of the reflection and transmission spectra. The dots are directly calculated from transmission and phase measuring by the Mach-Zehnder interferometer.[129]

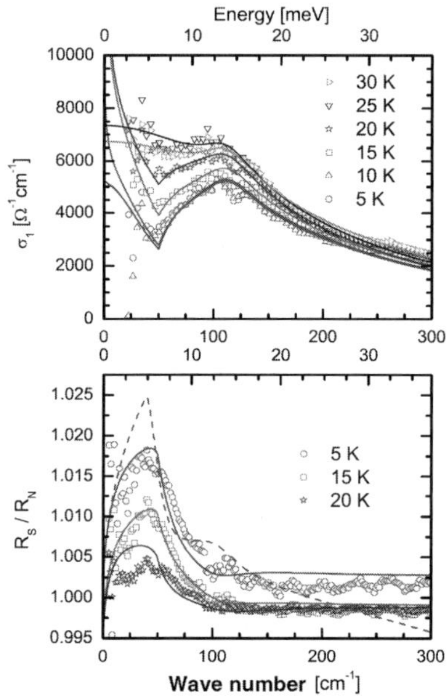

Figure 5.40 The top panel shows the measured σ_1 (symbols) at various temperatures and corresponding fits (solid lines) using Eq. 5.1. The symbols in the bottom panel are the measured superconducting-to-normal reflectivity ratios ($R_s/\ R_n$) and the simulations obtained by using the parameters that fit the optical conductivity. R_n is taken at 25 K. The blue dashed line is the best fit assuming that the superconducting state has no unpaired quasiparticles due to scattering pair breaking (no Drude term).[130]

15 K. The extra Drude term implies the presence of unpaired quasiparticles below T_c. They discussed possible origins of the extra absorption, including the gap anisotropy of the electron pocket, the impurity localized levels inside an isotropic gap, or pair-breaking due to interband impurity scattering in an $s\pm$ symmetric gap. Based on the remaining spectral weight in the superconducting state, they argued that it is more consistent with a highly anisotropic gap in the electron pocket, which could have nodes or a very small effective gap.

Wu *et al.* studied the in-plane infrared response of phosphorus-doped 122 superconducting single crystals. Many theoretical and experimental works have pointed out that line nodes might appear

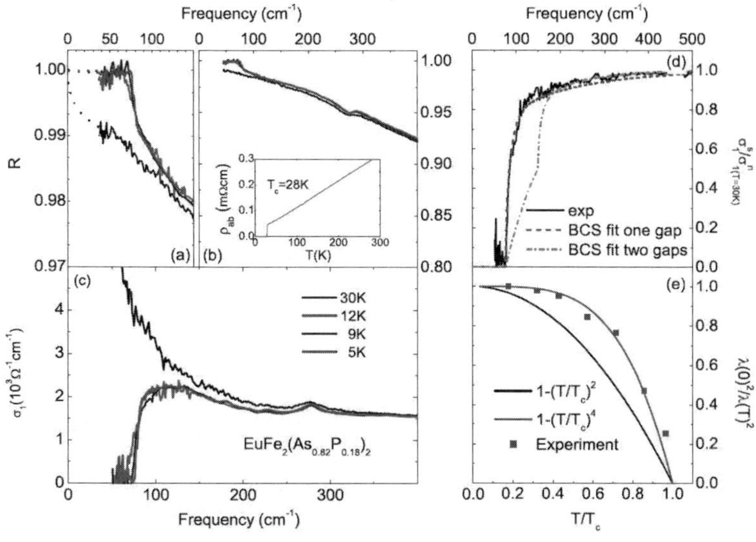

Figure 5.41 In-plane optical spectra of $EuFe_2(As_{0.82}P_{0.18})_2$ single crystals ($T_c = 28$ K) at different temperatures: (a) and (b) the reflectivity below T_c increases and then approaches unity at 74 cm^{-1}, which is a signature for a superconducting gap formation; (c) below T_c, the sudden drop of $\sigma_1(\omega)$ to zero indicates a nodeless s-wave superconducting gap. (d) The ratio of $\sigma_1(\omega)$ at $T = 30$ K and $\sigma_1(\omega)$ at $T = 5$ K. The dashed red line corresponds to the fit by the BCS model using a single gap of $2\Delta_0 = 74$ cm^{-1}, while the dashed gray line refers to the two-gap scenario with $2\Delta_1 = 74$ cm^{-1} and $2\Delta_2 = 150$ cm^{-1}. (g) Penetration depth as a function of temperature plotted in a renormalized fashion. The experimental data of $EuFe_2(As_{0.82}P_{0.18})_2$ can be described with 1-$(T/T_c)^4$.[132]

in the gap structure. Figure 5.41 shows the low-energy spectra of $EuFe_2(As_{0.82}P_{0.18})_2$ single crystals ($T_c = 28$ K) across the T_c.[132] Below T_c, the reflectivity in Fig. 5.41(a) and (b) increases abruptly around 100 cm^{-1} and then reaches unity at 74 cm^{-1}. Correspondingly, in Fig. 5.41(c) and (d), the optical conductivity below T_c drops steeply to zero at low frequency and can be fitted well by the model of a single BCS-like s-wave gap. The size of superconducting gap is estimated to be about $2\Delta = 74$ cm^{-1} and the ratio $2\Delta/k_B T_c = 3.7$, close to the value of BCS weak-coupling limit. The temperature dependence of the penetration depth obtained from the optical spectra agrees well with the relation 1-$(T/T_c)^4$,

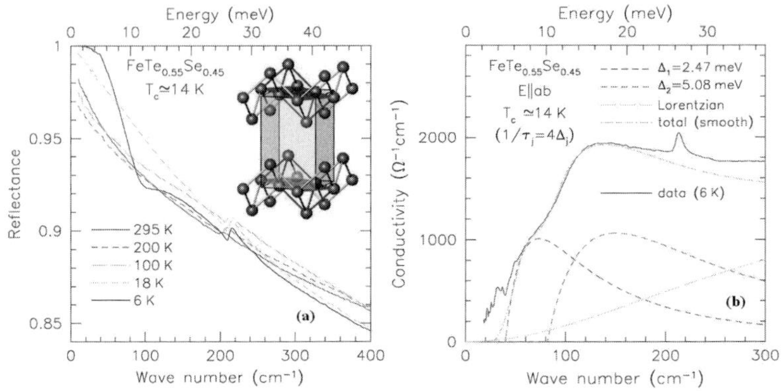

Figure 5.42 (a) In-plane reflectance and (b) conductivity spectra of $FeTe_{0.55}Se_{0.45}$ single crystals at different temperatures. The low-frequency optical conductivity at 6 K could be fitted well by the model of two superconducting gaps of $2\Delta_1 \simeq 5$ meV and $2\Delta_2 \simeq 10.2$ meV.[133]

i.e. the two-fluid model being commonly used for describing a BCS-like *s*-wave gap. The optical results would suggest that the superconducting gaps formed on different FSs have essentially the same energy scale and the gap structure has no nodal lines.

Superconducting energy gaps were also clearly observed in Fe-chalcogenide $FeTe_{0.55}Se_{0.45}$ single crystal samples.[133] The in-plane infrared spectra of $FeTe_{0.55}Se_{0.45}$ across T_c were shown in Fig. 5.42. The optical conductivity could be reproduced quite well using a Drude-Lorntz model (Eq. 5.15) at 295, 200 and 100 K, with fitted values of $\omega_{p,D} = 6400$ cm^{-1} and $1/\tau_D = 332$, 292 and 256 cm^{-1}, respectively. In order to fit the midinfrared component, Lorentzian oscillators at the somewhat arbitrary positions of 400, 850, and 3600 cm^{-1} were included, allowing the free-carrier component to be fit using a single Drude expression. The conductivity spectrum at 6 K could be reproduced by the model containing two BCS-like *s*-wave superconducting gaps: $2\Delta_1 \approx 5$ meV, $2\Delta_2 \approx 10.2$ meV. The superconducting plasma frequency determined is $\omega_{p,S} = 3000 \pm 200$ cm^{-1}, yielding the penetration depth of $\lambda_0 = 5300 \pm 300$ Å. Then, the ratio of $\omega_{p,S}^2/\omega_{p,D}^2 \leq 0.25$, indicating that less than one-quarter of the free-carriers in the normal state have condensed. The result suggests that this material is in the dirty limit. Indeed this material falls on the general scaling line predicted for a BCS dirty-limit superconductor.[133]

5.7.2 Josephson Coupling Plasmon in $K_xFe_{2-y}Se_2$

More recently, the superconductivity with T_c over 30 K has been discovered in a new Fe-chalcogenide system $A_xFe_{2-y}Se_2$ (A=alkaline metals or Tl, $x \leq 1$).[63] However, this system shows a number of surprising results. Optical measurements on superconducting samples appear very different from all other Fe-based superconducting systems.

Unlike other Fe-pnictides or chalcogenides where the superconductors develop from spin-density-wave (SDW) type metals with compensating hole and electron Fermi surfaces locating respectively at the Brillouin zone center and corners, the superconductivity in $A_xFe_{2-y}Se_2$ system was found to be in close proximity to an insulating phase with a so-called blocked checkerboard antiferromagnetic (AFM) structure.[19,64–66] The Fermi surface (FS) topologies of superconducting compounds are very different from previously known superconducting Fe-pnictides. Both band structure calculations[134,135] and ARPES studies[136,137] indicated that only the electron pockets are present in the superconducting compounds, while the hole bands sink below the Fermi level, indicating that the inter-pocket scattering between the hole and electron pockets is not an essential ingredient for superconductivity. More surprisingly, some experiments on $A_xFe_{2-y}Se_2$ (A = K, Rb, Cs, Tl), including muon-spin relaxation (μSR),[138,139] neutron diffraction,[19,65,66] Raman,[140]revealed a coexistence of superconductivity and very strong antiferromagnetism. On the other hand, some other measurements, e.g. magnetization,[141] TEM,[142] ARPES,[143] indicate that the superconductivity and magnetic order are phase separated in the superconducting samples.

Yuan *et al.* performed optical spectroscopy measurements on well-characterized superconducting $K_{0.75}Fe_{1.75}Se_2$ samples, as shown in Fig. 5.43.[144] Unexpectedly, they observed the development of a sharp upturn at very low frequency below T_c. Is this feature a superconducting energy gap? They argued against this opinion based on several reasons. First, although the sample is

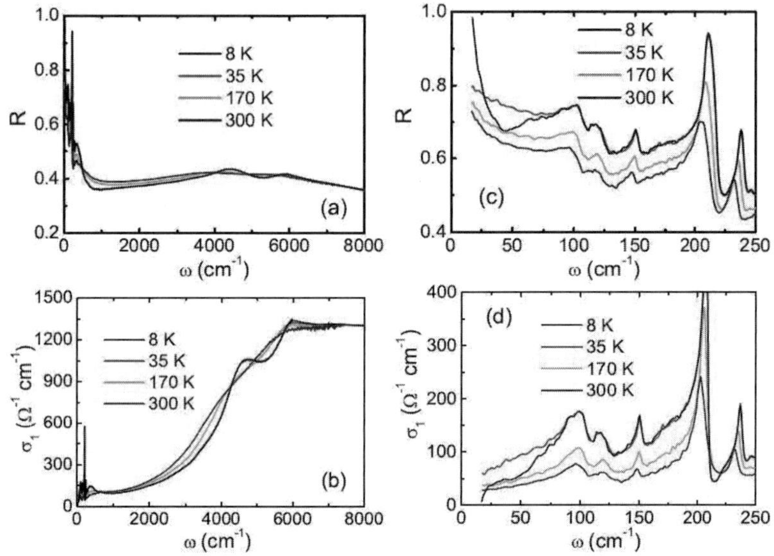

Figure 5.43 (a) and (b): optical reflectance $R(\omega)$ and conductivity $\sigma_1(\omega)$ spectra at different temperatures up to 8000 cm^{-1}. (c) and (d): An expanded plot of $R(\omega)$ and $\sigma_1(\omega)$ spectra below 250 cm^{-1}. A sharp plasma edge in $R(\omega)$ develops at low frequency in the superconducting state.[144]

superconducting, the reflectance values at the lowest measurement frequency limit in the normal state are far below the unit. Charge transport in the normal state is rather incoherent without showing a Drude-like response in $\sigma_1(\omega)$. This could be naturally explained if the sample contain parent insulating phase. Second, the reflectance spectral at 8 K shows a strong dip feature near 44 cm^{-1}. At this frequency, $R(\omega)$ value is much lower than that in the normal state (R(8 K)/R(35 K)$\approx$0.9). The dip is so pronounced that the feature is unlikely to be related to a superconducting gap. Third and most importantly, the superconducting gap (2Δ) amplitudes determined directly by the ARPES experiments on the electronic pockets are close to 18-20 meV,[136,137] which is much higher than the energy scale seen for the edge.

Yuan *et al.* argued that the relatively sharp feature is originated from the Josephson-coupling plasma edge. In fact, the reflectance edge corresponds to the zero-crossing of the real part of the

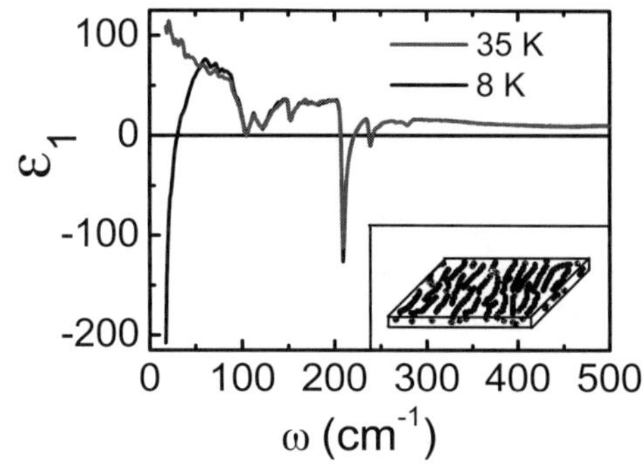

Figure 5.44 The real part of dielectric functions versus frequency at 35 and 8 K. $\epsilon_1(\omega)$ at 35 K is positive and increases further with decreasing frequency. However, the $\epsilon_1(\omega)$ crosses the zero at low frequency in the superconducting state. The inset shows a schematic picture of the nanoscale stripe-type phase separation between superconducting (white stripe) and insulating (black) regions.[144]

dielectric function $\epsilon_1(\omega)$, indicating that this edge is related to the collective plasma oscillation. They suggested that the sample has a nanoscale and possibly stripe-type phase separation between the superconducting and insulating phases. Then, Josephson-coupling plasmon would be expected to form below T_c.

The nanoscaled phase separation between superconductivity and magnetic insulator was further confirmed by the transmission electron microscopy (TEM) dark-field image technique in their study. Figures 5.45(a) shows an electron diffraction pattern taken along the [001] zone axis direction, in which the superstructure spots with both (1/5, 3/5, 0) and (1/2, 1/2, 0) wave vectors can be clearly seen. Figure 5.45(b) displays the dark-field TEM images by using one of the superstructure spot as indicated in Fig. 5.45(a). The well ordered regions with fine striped or speckled contrasts can be commonly observed. The length of a stripe could be in the range from several to several tens or even over 100 nanometers, while it width is usually less than 10 nanometers. This kind of complex contrast in Fig. 5.45(b) can be explained directly as the coexistence of Fe-vacancy ordered bright areas and Fe deficiency-free areas. Because the phase separation could not be uniform everywhere in the sample, the Josephson coupling strength would exhibit local

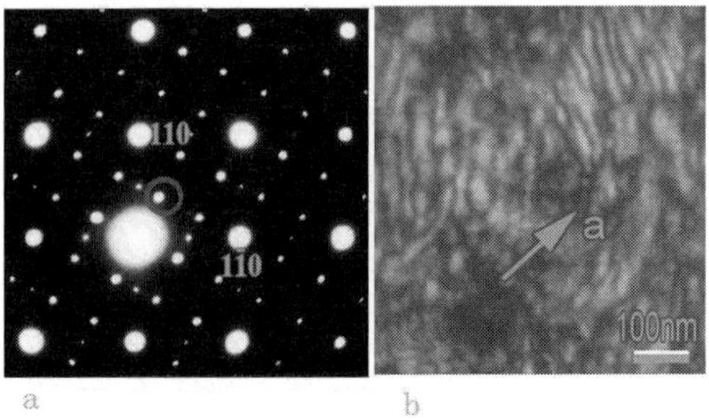

Figure 5.45 Phase separation in a $K_{0.75}Fe_{1.75}Se_2$ superconducting crystal. (a) Electron diffraction pattern showing the presence of superstructure spots along the [310] direction, the cycled spot is used for dark-field imaging. (b) Dark field image taken from a thin region of a $K_{0.75}Fe_{1.75}Se_2$ crystal. Stripe-type phase separation could be directly observed. The arrow **a** indicates the [100] direction.[144]

variations. As a result, the Josephson plasma edge could not be as sharp as that observed in the *c*-axis optical response in the cuprates.[101,109,110]

Since the Josephson coupling plasmon is a phenomenon related to the tunnelling of the condensed superconducting carriers, it should be easily influenced by the external magnetic field. Yuan *et al.*[144] also explored the effect of the magnetic field on the Josephson coupling plasmon edge, and found that the feature could be noticeably suppressed and shifted to lower frequency by a moderate magnetic field. Figure 5.46 shows far-infrared reflectance measurements for a similar sample under zero field and H = 8 T. The field is applied along the *c*-axis which is perpendicular to the electric field of the infrared radiation. From Fig. 5.46(a), it is seen that R(ω) at $T = 5$ K under the field of H = 8 T follows the normal-state R(ω) measured at 35 K down to much lower frequencies than that of H = 0. Then, a peak appears in the ratio plot of the zero-field reflectance at 35 K to the reflectance curves at 5 K under different fields. The peak shifts significantly to lower frequency side by a magnetic field of 8 T.

5.7.3 *Superconductivity-Induced Spectral Weight Transfer*

We have mentioned in section II that, for some unconventional superconductors due to the presence of strong electron correlations,

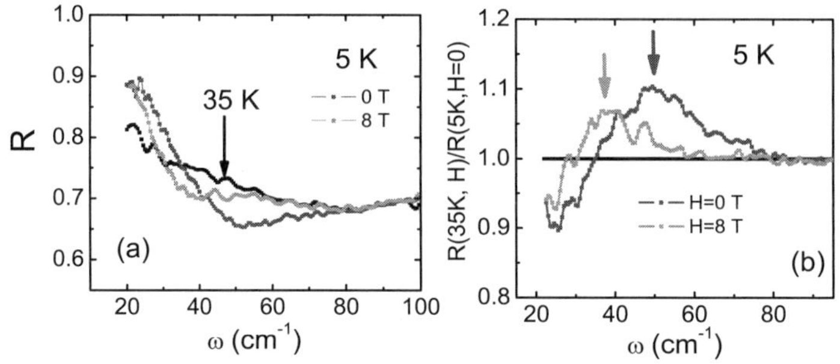

Figure 5.46 (a) The reflectance at 5 K under H = 0 and 8 T. The reflectance curve at 35 K under zero field is also shown as indicated by the black arrow. (b) The ratio of the zero-field reflectance at 35 K to the reflectance curves at 5 K under different fields. The arrows indicate the frequency shift of the peaks.[144]

the integral of conductivity up to a finite cutoff frequency (higher than the band width but still lower than the interband transitions) could be temperature dependent. In this situation, the simple low-frequency FGT sum rule would be violated, as explained in equation 5.27. The superconducting condensate comes from not only the low-frequency intraband excitations but also the interband transitions, possibly in near-infrared or visible regime (See Fig. 5.47).[29] The spectral weight transfer from high energy interband transitions to low energy (condensate) upon entering into superconducting state is due to the kinetic energy lowering. The kinetic energy of charge carriers depends on their effective mass. Carriers that interact strongly with other carriers or other degrees of freedom have a large effective mass and cannot respond well to low-frequency electric fields; hence, the low-frequency optical absorption is small, and optical absorption occurs mostly at high frequencies. On the condition that the effective mass decreases upon pairing, optical spectral weight would be transferred from high (optical) to low frequencies.[29] It is different from the ordinary BCS theory where the potential energy of the Cooper pairs is lowered through their interactions with the phonons.

There are a number of optical works on the high-T_c cuprates whether a fraction of the condensate comes from the high energy interband transitions, but the results are controversial.[33–37]

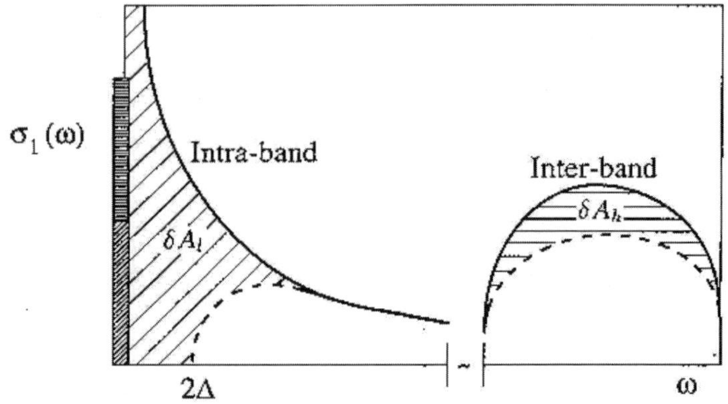

Figure 5.47 Real part of the conductivity in the normal (solid lines) and superconducting (dashed lines) states. The missed area from intraband (δA_1) and the interband (δA_h) transitions are shown as separate contributions to the δ-function at zero frequency.[29]

The observed superconductivity-induced anomalies in the optical response of highly conducting CuO_2 planes were argued to be dominated by the narrowing of the broad Drude peak caused by a superconductivity-induced modification of the scattering rate.[27,36]

Charnukha *et al.*[145] made careful study on the $Ba_{0.68}K_{0.32}Fe_2As_2$ single crystals ($T_c = 38.5$ K) and uncovered a superconductivity-induced suppression of an absorption band at 2.5 eV, two orders of magnitude above the superconducting gap energy $\Delta \approx 20$ meV. Figure 5.48b shows difference spectra between 40 and 10 K of the real parts of the optical conductivity and dielectric function. The suppressed band has a Lorentzian lineshape and appears abruptly across the superconducting transition, consistently in both $\Delta\sigma_1$ and $\Delta\varepsilon_1$, as shown in Fig. 5.48c for σ_1 (2.5 eV). The temperature dependence of the suppression (blue open circles) coincides with that of the far-infrared optical conductivity due to the opening of the superconducting gap (red filled circles). Thus, the onset of superconductivity not only modifies the low-energy quasiparticle response but also affects the overall electronic structure, including interband transitions in the visible range of the spectrum. As the SW is reduced, δA_H is not balanced in the vicinity of the absorption band (Fig. 5.48b), the data indicate a SW transfer over a wide energy range.

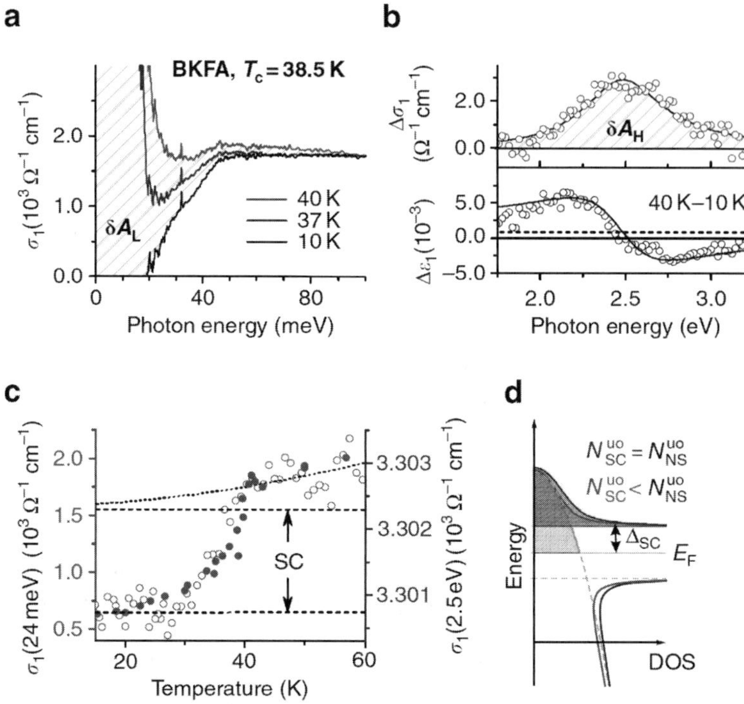

Figure 5.48 (a) Real part of optical conductivity of $Ba_{0.68}K_{0.32}Fe_2As_2$ and the missing area. (b) Difference spectra of the real part of the optical conductivity (top panel) and dielectric function (bottom panel) between 40 and 10 K, with a small background shift (horizontal dashed line) detected by temperature modulation measurements. Lorentzian fit to both spectra (black solid lines). (c) Temperature scan of σ_1 at 2.5 eV. Contribution of the normal-state dynamics (dotted line) was estimated to determine the magnitude of the SC-induced jump (dashed lines). (d) Density of states in the normal (NS; grey dashed line), conventional superconducting state (SC; blue line), and an unconventional state with a depletion of unoccupied states (UO; red line). Filled areas of respective colors represent total number of unoccupied states.[145]

The SW δA_H liberated from the absorption band upon cooling below T_c comprises only ~0.5% of the total SW δA_L (Fig. 5.48a) of the superconducting condensate, its contribution to the low-energy charge dynamics might be considered negligible. However, Charnukha *et al.*[145] argued that, assuming this additional high-

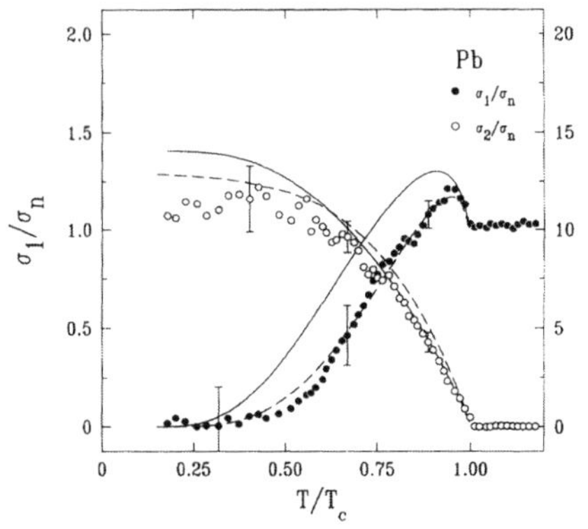

Figure 5.49 Temperature dependence of the complex conductivity $\sigma = \sigma_1 + \sigma_2$ of Pb as evaluated from the surface impedance measurement. The solid curves are the weak coupling Mattis-Bardeen prediction. The dashed curves are the strong-coupling Eliashberg prediction.[146]

energy SW contributes to the itinerant-carrier response below T_c, a simple estimate on the basis of the tight-binding nearest-neighbour approximation revealed that this would lead to a reduction in electronic kinetic energy of 0.60 meV per unit cell in the superconducting state. This is close to the condensation energy $\Delta F(0) = 0.36$ meV per unit cell obtained from specific heat measurements on the same sample. In comparison with the density function calculations, this band can be assigned to transitions from As-p to Fe-d orbitals crossing the Fermi level. The study revealed that As-p states deep below the Fermi level contribute to the formation of the superconductivity, which therefore pointed toward an electronic pairing mechanism.

5.7.4 *Coherent Peak Below T_c Probed by THz Spectroscopy*

According to the BCS theory, the frequency- and temperature-dependent conductivity in the superconducting state could be calculated from Eq. 5.24. Because of the case II coherence factors, the real part of conductivity shows a smooth rise for frequencies $\omega \geq 2\Delta$ at $T = 0$ K. Additionally, for frequencies far below 2Δ, the real part of conductivity has a "coherent peak" just below T_c. As an example, we show in Fig. 5.49 the temperature-dependent conductivity curve measured at 60 GHz for the strong coupling superconductor Pb.[146] Although surprising, the "coherent peak" is direct evidence for coherence effects in the superconductor, which is analogous to the Hebel-Slichter peak found in the nuclear spin lattice relaxation rate in NMR experiment.

Examining the "coherent peak" could be used to distinguish between different pairing symmetry. For d-wave superconductors like cuprates, the "coherent peak" just below T_c does not exist. It is partly due to the lack of a strong singularity in the d-wave density of states, and partly due to the vanishing of coherence factor for $\mathbf{q} = (\pi, \pi)$ since $\Delta_k \approx -\Delta_{k+\pi,\pi}$ for k near $(\pi, 0)$.[31]

Fischer *et al.*[147] measured the complex dynamical conductivity, $\sigma = \sigma_1 + \sigma_2$, of superconducting $Ba(Fe_{0.9}Co_{0.1})_2As_2$ ($T_c = 22$ K) at terahertz frequencies and temperatures T <30 K. In the frequency dependence of σ_1 below T_c, they observed clear signatures of the superconducting energy gap opening. The temperature dependence of σ_1 demonstrated a pronounced coherence peak at frequencies below 15 cm^{-1} (1.8 meV), shown in Fig. 5.50. Aguilar *et al.*[148] studied the high quality $BaFe_{1.84}Co_{0.16}As_2$ thin films using time-domain THz spectroscopy. Near T_c they also found evidence for a coherence peak and qualitative agreement with the weak-coupling Mattis-Bardeen form of the conductivity, as shown in

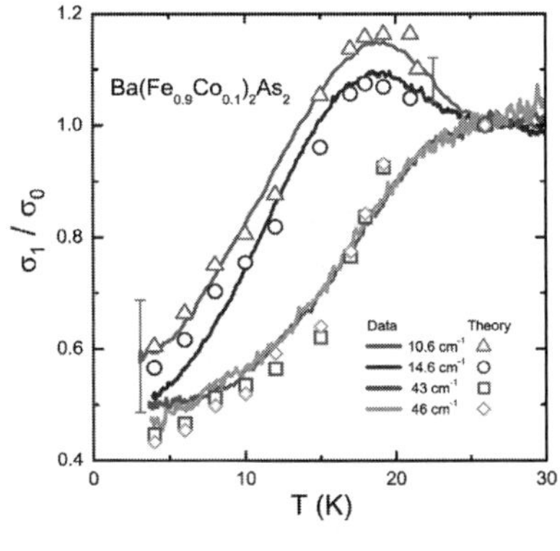

Figure 5.50 Temperature dependence of the real part of the complex conductivity in $Ba(Fe_{0.9}Co_{0.1})_2As_2$. The values are normalized to the respective normal-state conductivity at $T = 26$ K. Lines experiment and dots calculations with the two-gap model.[147]

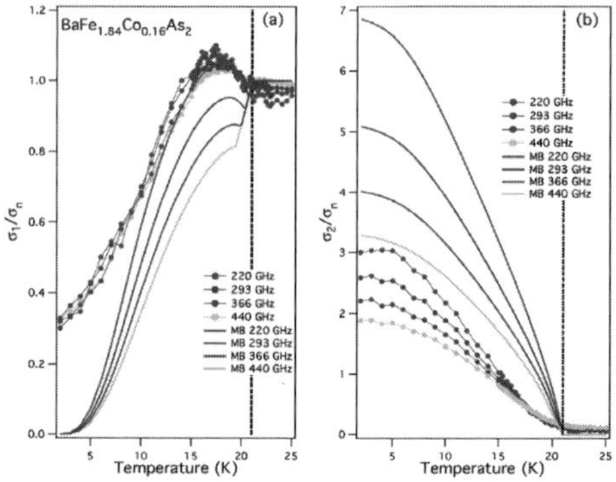

Figure 5.51 (a) Real and (b) imaginary parts of the THz conductivity as a function of temperature for different frequencies. Solid lines are calculations using the standard Mattis-Bardeen form.[148]

Fig. 5.51. Both works were believed to be consistent with multiband s± superconductivity. Nevertheless, it should be noted that in those compounds, the coherence effects manifested in terahertz spectroscopy and NMR are different. NMR measurements on similar Fe-pnictide superconductors revealed an absence of a Hebel-Slichter peak in the nuclear spin lattice relaxation rate.[149]

Acknowledgements

This work is supported by the National Science Foundation of China, the Knowledge Innovation Project of the Chinese Academy of Sciences, and the 973 Project of the Ministry of Science and Technology of China.

References

1. Y. Kamihara, H. Hiramatsu, M. Hirano, R. Kawamura, H. Yanagi, T. Kamiya, and H. Hosono, J. Am. Chem. Soc. **128**, 10012 (2006).
2. T. Watanabe, H. Yanagi, T. Kamiya, Y. Kamihara, H. Hiramatsu, M. Hirano, and H. Hosono, Inorg. Chem. **46**, 7719 (2007).

3. Y. Kamihara, T. Watanabe, M. Hirano, and H. Hosono, J. Am. Chem. Soc. **130**, 3296 (2008).
4. X. H. Chen, T. Wu, G. Wu, R. H. Liu, H. Chen, and D. F. Fang, Nature **453**, 761 (2008).
5. G. F. Chen, Z. Li, D. Wu, G. Li, W. Z. Hu, J. Dong, P. Zheng, J. L. Luo, and N. L. Wang, Phys. Rev. Lett. **100**, 247002 (2008).
6. Z. A. Ren, J. Yang, W. Lu, W. Yi, X. L. Shen, Z. C. Li, G. C. Che, X. L. Dong, L. L. Sun, F. Zhou, and Z. X. Zhao, Europhys. Lett. **82**, 57002 (2008).
7. C. de la Cruz, Q. Huang, J. W. Lynn, J. Li, W. R. Ii, J. L. Zarestky, H. A. Mook, G. F. Chen, J. L. Luo, N. L. Wang, and P. Dai, Nature **453**, 899 (2008).
8. M. Rotter, M. Tegel, D. Johrendt, Phys. Rev. Lett. **101**, 107006 (2008).
9. J. H. Chu, J. G. Analytis, C. Kucharczyk, and I. R. Fisher, Phys. Rev. B **79** 014506 (2009).
10. Jun Zhao, Q. Huang, Clarina de la Cruz, Shiliang Li, J. W. Lynn, Y. Chen, M. A. Green, G. F. Chen, G. Li, Z. Li, J. L. Luo, N. L. Wang, Pengcheng Dai, Nat. Mater. **7**, 953 (2008).
11. M. S. Torikachvili, S. L. Bud'ko, N. Ni, P. C. Canfield, Phys. Rev. Lett. **101**, 057006 (2008).
12. Y. Qiu, Wei Bao, Q. Huang, T. Yildirim, J. Simmons, J. W. Lynn, Y. C. Gasparovic, J. Li, M. Green, T. Wu, G. Wu, X. H. Chen, Phys. Rev. Lett. **101**, 257002 (2008).
13. Q. Huang, Y. Qiu, Wei Bao, J. W. Lynn, M. A. Green, Y. Chen, T. Wu, G. Wu, X. H. Chen, Phys. Rev. Lett. **101**, 257003 (2008).
14. J. Zhao, W. Ratcliff II, J. W. Lynn, G. F. Chen, J. L. Luo, N. L. Wang, Jiangping Hu, Pengcheng Dai , Phys. Rev. B **78**, 140504(R) (2008).
15. A. I. Goldman, D. N. Argyriou, B. Ouladdiaf, T. Chatterji, A. Kreyssig, S. Nandi, N. Ni, S. L. Bud'ko, P. C. Canfield, R. J. McQueeney, Phys. Rev. B **78**, 100506(R) (2008).
16. Shiliang Li, Clarina de la Cruz, Q. Huang, G. F. Chen, T.-L. Xia, J. L. Lou, N. L. Wang, Pengcheng Dai, Phys. Rev. B **80**, 020504(R) (2009).
17. Wei Bao, Y. Qiu, Q. Huang, M. A. Green, P. Zajdel, M. R. Fitzsimmons, M. Zhernenkov, M. Fang, B. Qian, E. K. Vehstedt, J. Yang, H. M. Pham, L. Spinu, Z. Q. Mao, Phys. Rev. Lett. **102**, 247001 (2009).
18. Shiliang Li, Clarina de la Cruz, Q. Huang, Y. Chen, J. W. Lynn, Jiangping Hu, Yi-Lin Huang, Fong-chi Hsu, Kuo-Wei Yeh, Maw-Kuen Wu, Pengcheng Dai, Phys. Rev. B **79**, 054503 (2009).
19. Wei Bao, Q. Huang, G. F. Chen, M. A. Green, D. M. Wang, J. B. He, X. Q. Wang, Y. Qiu, Chinese Phys. Lett. **28**, 086104 (2011).

20. W. Z. Hu, Q. M. Zhang, N. L. Wang Comments: brief review paper for the special issue of Physica C on iron-pnictide superconductors Journal-ref: Physica C **469**, 545–558 (2009).
21. D. C. Johnston, Adv. Phys. **59**, 803–1061 (2010).
22. F. Wooten, "Optical Properties of Solids", Academic Press, New York, (1972).
23. M. Dressel and G. Grüner, "Electrodynamics of Solids: Optical Properties of Electrons in Matter", Cambridge University Press, (2002).
24. D. N. Basov and T. Timusk, Rev. Mod. Phys. **77**, 721 (2005).
25. P. B. Allen, Phys. Rev. B **3**, 305 (1971).
26. S. V. Shulga, O. V. Dolgov, E. G. Maksimov, Physica C **178**, 266 (1991).
27. S. Maiti and A. V. Chubukov, Phys. Rev. B **81**, 245111 (2010) and also references in this paper.
28. R. Kubo, J. Phys. Soc. Jpn. **12**, 570 (1957).
29. J. E. Hirsch, Physica C **199**, 305 (1992); Science **295**, 2226 (2002).
30. M. Tinkham, Introduction to superconductivity, McGraw-Hill, Inc. (1975).
31. N. P. Armitage, arXiv:0908.1126.
32. A. Ferrell and R. E. Glover, Phys. Rev. 109, 1398 (1958); Tinkham and R. A. Ferrell, Phys. Rev. Lett. **2**, 331 (1959).
33. D. N. Basov, S. I. Woods, A. S. Katz, E. J. Singley, R. C. Dynes, M. Xu, D. G. Hinks, C. C. Homes, M. Strongin, Science **283**, 49 (2002).
34. H. J. A. Molegraaf, C. Presura, D. van der Marel, P. H. Kess, and M. Li, Science **295**, 2239 (2002).
35. A. F. Santander-Syro, R. P. S. M. Lobo, N. Bontemps, Z. Konstantinovic, Z. Z. Li, and H. Raffy, Europhys. Lett. **62**, 568 (2003).
36. A. V. Boris, N. N. Kovaleva, O. V. Dolgov, T. Holden, C. T. Lin, B. Keimer, and C. Bernhard, Science **304**, 708 (2004).
37. C. C. Homes, S. V. Dordevic, T. Valla, and M. Strongin, Phys. Rev. B **72**, 134517 (2005).
38. J. Dong, H. J. Zhang, G. Xu, Z. Li, G. Li, W. Z. Hu, D. Wu, G. F. Chen, X. Dai, J. L. Luo, Z. Fang, and N. L. Wang, Europhys. Lett. **83**, 27006 (2008)
39. Takafumi Sato, Pierre Richard, Kosuke Nakayama, Takashi Takahashi, and Hong Ding, "Angle-Resolved Photoemission Spectroscopy of Iron Pnictides", chapter in this book.
40. I. I. Mazin, D. J. Singh, M. D. Johannes, M. H. Du, Phys. Rev. Lett. **101**, 057003 (2008).

41. K. Kuroki, S. Onari, R. Arita, H. Usui, Y. Tanaka, H. Kontani, and H. Aoki, Phys. Rev. Lett. **101**, 087004 (2008).
42. A. V. Chubukov, D. Efremov, I. Eremin, Phys. Rev. B **78**, 134512 (2008).
43. Ying Ran, Fa Wang, Hui Zhai, Ashvin Vishwanath, Dung-Hai Lee, Phys. Rev. B **79**, 014505 (2009).
44. V. Cvetkovic and Z. Tesanovic, Europhys. Lett. **85**, 37002 (2009).
45. D. J. Singh, Physica C **469**, 418 (2009) and references therein.
46. T. Yildirim, Phys. Rev. Lett. **101**, 057010 (2008).
47. Q. Si and E. Abrahams, Phys. Rev. Lett. **101**, 076401 (2008).
48. F. J. Ma, Z. Y. Lu, and T. Xiang, Phys. Rev. B **78**, 224517 (2008).
49. C. Fang, H. Yao, W.-F. Tsai, J. P. Hu, S. A. Kivelson, Phys. Rev. B **77**, 224509 (2008); C. Xu, M. Mueller, S. Sachdev, Phys. Rev. B **78**, 020501(R) (2008).
50. J. Wu, P. Phillips, and A. H. Castro Neto, Phys. Rev. Lett. **101**, 126401 (2008).
51. W. Z. Hu, J. Dong, G. Li, Z. Li, P. Zheng, G. F. Chen, J. L. Luo, and N. L. Wang, Phys. Rev. Lett. **101**, 257005 (2008).
52. A. Akrap, J. J. Tu, L. J. Li, G. H. Cao, Z. A. Xu, and C. C. Homes, Phys. Rev. B. **80**, 180502(R) (2009).
53. J. N. Hancock, S. I. Mirzaei, J. Gillett, S. E. Sebastian, J. Teyssier, R. Viennois, E. Giannini, and D. van der Marel, Phys. Rev. B. **82**, 014523 (2010).
54. S. J. Moon, J. H. Shin, D. Parker, W. S. Choi, I. I. Mazin, Y. S. Lee, J. Y. Kim, N. H. Sung, B. K. Cho, S. H. Khim, J. S. Kim, K. H. Kim, and T. W. Noh, Phys. Rev. B. **81**, 205114 (2010).
55. G. F. Chen, W. Z. Hu, J. L. Luo, and N. L. Wang, Phys. Rev. Lett. **102**, 227004 (2009).
56. W. Z. Hu, G. Li, P. Zheng, G. F. Chen, J. L. Luo, and N. L. Wang, Phys. Rev. B. **80**, 100507(R) (2009).
57. J. Q. Yan, S. Nandi, J. L. Zarestky, W. Tian, A. Kreyssig, B. Jensen, A. Kracher, K. W. Dennis, R. J. McQueeney, A. I. Goldman, R. W. McCallum, and T. A. Lograsso, Appl. Phys. Lett. **95**, 222504 (2009).
58. Z. G. Chen, R. H. Yuan, T. Dong, and N. L. Wang, Phys. Rev. B. **81**, 100502(R) (2010).
59. T. Dong, Z. G. Chen, R. H. Yuan, B. F. Hu, B. Cheng, N. L. Wang, Phys. Rev. B. **82**, 054522 (2010).
60. Y. Xia, D. Qian, L. Wray, D. Hsieh, G. F. Chen, J. L. Luo, N. L. Wang, M. Z. Hasan, Phys. Rev. Lett. **103**, 037002 (2009).

61. Y. Zhang, F. Chen, C. He, L. X. Yang, B. P. Xie, Y. L. Xie, X. H. Chen, Minghu Fang, M. Arita, K. Shimada, H. Namatame, M. Taniguchi, J. P. Hu, and D. L. Feng, Phys. Rev. B. **82**, 165113 (2010).
62. G. F. Chen, Z. G. Chen, J. Dong, W. Z. Hu, G. Li, X. D. Zhang, P. Zheng, J. L. Luo, N. L. Wang, Phys. Rev. B. **79**, 140509(R) (2009).
63. Jiangang Guo, Shifeng Jin, Gang Wang, Shunchong Wang, Kaixing Zhu, Tingting Zhou, Meng He, and Xiaolong Chen, Phys. Rev. B. **82**, 180520(R) (2010).
64. M. H. Fang, H. Wang, C. Dong, Z. Li, C. Feng, J. Chen, and H. Q. Yuan, Europhys. Lett. **94**, 27009 (2011).
65. F. Ye, S. Chi, Wei Bao, X. F. Wang, J. J. Ying, X. H. Chen, H. D. Wang, C. H. Dong, and Minghu Fang, Phys. Rev. Lett. **107**, 137003 (2011).
66. Wei Bao, G. N. Li, Q. Huang, G. F. Chen, J. B. He, M. A. Green, Y. Qiu, D. M. Wang, and J. L. Luo, arXiv:1102.3674.
67. Xun-Wang Yan, Miao Gao, Zhong-Yi LU, Tao Xiang, Phys. Rev. Lett. **106**, 087005 (2011); Chao Cao, Jianhui Dai, Phys. Rev. B **83**, 193104 (2011).
68. Chao Cao, Jianhui Dai, Phys. Rev. Lett. **107**, 056401 (2011); Xun-Wang Yan, Miao Gao, Zhong-Yi LU, Tao Xiang, Phys. Rev. B **83**, 233205 (2011).
69. Z. G. Chen, R. H. Yuan, T. Dong, G. Xu, Y. G. Shi, P. Zheng, J. L. Luo, J. G. Guo, X. L. Chen, and N. L. Wang, Phys. Rev. B. **83**, 220507(R) (2011).
70. D. Wu, N. Barisic, P. Kallina, A. Faridian, B. Gorshunov, N. Drichko, L. J. Li, X. Lin, G. H. Cao, Z. A. Xu, N. L. Wang, and M. Dressel, Phys. Rev. B. **81**, 100512(R) (2010); N. Barisic, D. Wu, M. Dressel, L. J. Li, G. H. Cao, Z. A. Xu, Phys. Rev. B **82**, 054518 (2010).
71. M. Nakajima, S. Ishida, K. Kihou, Y. Tomioka, T. Ito, Y. Yoshida, C. H. Lee, H. Kito, A. Iyo, H. Eisaki, K. M. Kojima, S. Uchida, Phys. Rev. B **81**, 104528 (2010).
72. J. J. Tu, J. Li, W. Liu, A. Punnoose, Y. Gong, Y. H. Ren, L. J. Li, G. H. Cao, Z. A. Xu, and C. C. Homes, Phys. Rev. B. **82**, 174509 (2010).
73. A. Lucarelli, A. Dusza, F. Pfuner, P. Lerch, J. G. Analytis, J-H Chu, I. R. Fisher, and L. Degiorgi, New. J. Phys. **12**, 073036 (2010).
74. N. Stojilovic, A. Koncz, L. W. Kohlman, Rongwei Hu, C. Petrovic, and S. V. Dordevic, Phys. Rev. B. **81**, 174518 (2010).
75. J. Hwang, E. Schachinger, J. P. Carbotte, F. Gao, D. B. Tanner, and T. Timusk, Phys. Rev. Lett. **100**, 137005 (2008).
76. J. Yang, D. Huüvonen, U. Nagel, T. Rõõm, N. Ni, P. C. Canfield, S. L. Bud'ko, J. P. Carbotte, and T. Timusk, Phys. Rev. Lett. **102**, 187003 (2009).
77. L. Boeri, O. V. Dolgov, and A. A. Golubov, Phys. Rev. Lett. **101**, 026403 (2008).

78. D. Wu, N. Barišić, M. Dressel, G. H. Cao, Z-A. Xu, E. Schachinger, J. P. Carbotte, Phys. Rev. B. **82**, 144519 (2010).

79. Q. Si, Nat. Phys. **5**, 629 (2009).

80. M. M. Qazilbash, J. J. Hamlin, R. E. Baumbach, Lijun Zhang, D. J. Singh, M. B. Maple, D. N. Basov, Nat. Phys. **5**, 647 (2009).

81. D. N. Basov and Andrey V. Chubukov, Nat. Phys. **7**, 272 (2011).

82. M. Yi, D. H. Lu, J. G. Analytis, J.-H. Chu, S.-K. Mo, R.-H. He, X. J. Zhou, G. F. Chen, J. L. Luo, N. L. Wang, Z. Hussain, D. J. Singh, I. R. Fisher, Z.-X. Shen, Phys. Rev. B **80**, 024515 (2009).

83. L Degiorgi, New Journal of Physics **13**, 023011 (2011), and references cited.

84. H. Gretarsson, A. Lupascu, Jungho Kim, D. Casa, T. Gog, W. Wu, S. R. Julian, Z. J. Xu, J. S. Wen, G. D. Gu, R. H. Yuan, Z. G. Chen, N.-L. Wang, S. Khim, K. H. Kim, M. Ishikado, I. Jarrige, S. Shamoto, J.-H. Chu, I. R. Fisher, Young-June Kim, Phys. Rev. B **84**, 100509(R) (2011).

85. M. D. Johannes, Igor Mazin, Phys. Rev. B **79**, 220510(R) (2009).

86. K. Haule, J. H. Shim, and G. Kotliar, Phys. Rev. Lett. **100**, 226402 (2008).

87. K. Haule, G. Kotliar, New J. Phys. **11** 025021 (2009).

88. Z. P. Yin, K. Haule, G. Kotliar, Nat. Phys. **7**, 294–297 (2011).

89. Wei-Guo Yin, Chi-Cheng Lee, and Wei Ku, Phys. Rev. Lett. **105**, 107004 (2010).

90. G. T. Wang, Y. M. Qian, G. Xu, X. Dai, Z. Fang, Phys. Rev. Lett. **104**, 047002 (2010).

91. Chang-Youn Moon and Hyoung Joon Choi, Phys. Rev. Lett. **104**, 057003 (2010).

92. Z. P. Yin, K. Haule, G. Kotliar, Nat. Mater. **10**, 932 (2011).

93. N. L. Wang, W. Z. Hu, Z. G. Chen, R. H. Yuan, G. Li, G. F. Chen, and T. Xiang, J. Phys.: Condens. Matter **24**, 294202 (2012).

94. A. A. Schafgans, S. J. Moon, B. C. Pursley, A. D. LaForge, M. M. Qazilbash, A. S. Sefat, D. Mandrus, K. Haule, G. Kotliar, D. N. Basov, Phys. Rev. Lett. **108**, 147002 (2012).

95. M. J. Rozenberg, G. Kotliar, H. Kajueter, G. A. Thomas, D. H. Rapkine, J. M. Honig, and P. Metcalf, Phys. Rev. Lett. **75**, 105 (1995).

96. D. J. Singh, Physica C **469**, 418 (2009).

97. W. L. Yang, P. O. Velasco, J. D. Denlinger, A. P. Sorini, C-C. Chen, B. Moritz, W.-S. Lee, F. Vernay, B. Delley, J.-H. Chu, J. G. Analytis, I. R. Fisher, Z. A. Ren, J. Yang, W. Lu, Z. X. Zhao, J. van den Brink, Z. Hussain, Z.-X. Shen, T. P. Devereaux, Phys. Rev. B **80**, 014508 (2009).

98. M. Rotter, M. Pangerl, M. Tegel, and D. Johrendt, Angew. Chem. Int. Ed. **47**, 7949 (2008).
99. D. L. Sun, J. Z. Xiao, and C. T. Lin, Journal of Crystal Growth **321**, 55 (2011).
100. Z. G. Chen, T. Dong, R. H. Ruan, B. F. Hu, B. Cheng, W. Z. Hu, P. Zheng, Z. Fang, X. Dai, and N. L. Wang, Phys. Rev. Lett. **105**, 097003 (2010).
101. S. Uchida, K. Tamasaku, and S. Tajima, Phys. Rev. B. **53**, 14558 (1996).
102. T. Katsufuji, M. Kasai, and Y. Tokura, Phys. Rev. Lett. **76**, 126 (1996).
103. M. A. Tanatar, N. Ni, G. D. Samolyuk, S. L. Bud'ko, P. C. Canfield, and R. Prozorov, Phys. Rev. B. **79**, 134528 (2009).
104. C. Utfeld, J. Laverock, T. D. Haynes, S. B. Dugdale, J. A. Duffy, M. W. Butchers, J. W. Taylor, S. R. Giblin, J. G. Analytis, J. Chu, I. R. Fisher, M. Itou, Y. Sakurai, Phys. Rev. B. **81**, 064509 (2010).
105. M. A. Tanatar, E. C. Blomberg, A. Kreyssig, M. G. Kim, N. Ni, A. Thaler, S. L. Budko, P. C. Canfield, A. I. Goldman, I. I. Mazin, and R. Prozorov, Phys. Rev. B. **81**, 184508 (2010).
106. Jiun-Haw Chu, James G. Analytis, Kristiaan De Greve, Peter L. McMahon, Zahirul Islam, Yoshihisa Yamamoto, Ian R. Fisher, Science **329**, 824 (2010).
107. A. Dusza, A. Lucarelli, F. Pfuner, J.-H. Chu, I. R. Fisher, and L. Degiorgi, Europhys. Lett. **93**, 37002 (2011).
108. M. Nakajima, T. Liang, S. Ishida, Y. Tomioka, K. Kihou, C. H. Lee, A. Iyo, H. Eisaki, T. Kakeshita, T. Ito, S. Uchida, PNAS **108**, 12238 (2011).
109. K. Tamasaku, Y. Nakamura, and S. Uchinda, Phys. Rev. Lett. **69**, 1455 (1992).
110. D. N. Basov, T. Timusk, B. Dabrowski, and J. D. Jorgensen, Phys. Rev. B **50**, 3511 (1994).
111. H. Shibata and T. Yamada, Phys. Rev. B **56**, R14275 (1997).
112. T. Motohashi, J. Shimoyama, K. Kitazawa, K. Kishio, K. M. Kojima, S. Uchida, and S. Tajima, Phys. Rev. B **61**, R9269 (2000).
113. B. Cheng, Z. G. Chen, C. L. Zhang, R. H. Ruan, T. Dong, B. F. Hu, W. T. Guo, S. S. Miao, P. Zheng, J. L. Luo, G. Xu, Pengcheng Dai, N. L. Wang, Phys. Rev. B **83**, 144522 (2011).
114. C. Martin, H. Kim, R. T. Gordon, N. Ni, V. G. Kogan, S. L. Budko, P. C. Canfield, M. A. Tanatar, and R. Prozorov, Phys. Rev. B **81**, 060505 (2010).

115. J.-Ph. Reid, M. A. Tanatar, X. G. Luo, H. Shakeripour, N. Doiron-Leyraud, N. Ni, S. L. Budko, P. C. Canfield, R. Prozorov, and Louis Taillefer, Phys. Rev. B **82**, 064501 (2010).

116. J. G. Checkelsky, Lu Li, G. F. Chen, J. L. Luo, N. L. Wang, N. P. Ong, arXiv:0811.4668 (unpublished).

117. P. J. Hirschfeld and D. J. Scalapino, Physics **3**, 64 (2010).

118. C. L. Zhang, Meng Wang, Miaoyin Wang, Jun Zhao, Karol Marty, M. D. Lumsden, Songxue Chi, Sung Chang, J. W. Lynn, Huiqian Luo, Tao Xiang, Jiangping Hu, Pengcheng Dai, Scientifc Reports **1**, 115 (2011).

119. S. J. Moon, C. C. Homes, A. Akrap, Z. J. Xu, J. S. Wen, Z. W. Lin, Q. Li, G. D. Gu, and D. N. Basov, Phys. Rev. Lett. **106**, 217001 (2011).

120. G. F. Chen, Z. Li, G. Li, J. Zhou, D. Wu, J. Dong, W. Z. Hu, P. Zheng, Z. J. Chen, H. Q. Yuan, J. Singleton, J. L. Luo, and N. L. Wang, Phys. Rev. Lett. **101**, 057007 (2008).

121. A. Dubroka, K. W. Kim, M. Rössle, V. K. Malik, A. J. Drew, R. H. Liu, G. Wu, X. H. Chen, and C. Bernhard, Phys. Rev. Lett. **101**, 097011 (2008).

122. G. Li, W. Z. Hu, J. Dong, Z. Li, P. Zheng, G. F. Chen, J. L. Luo, and N. L. Wang, Phys. Rev. Lett. **101**, 107004 (2008).

123. H. Ding, P. Richard, K. Nakayama, T. Sugawara, T. Arakane, Y. Sekiba, A. Takayama, S. Souma, T. Sato, T. Takahashi, Z. Wang, X. Dai, Z. Fang, G. F. Chen, J. L. Luo, and N. L. Wang, Europhys. Lett. **83**, 47001 (2008).

124. Y. J. Uemura, G. M. Luke, B. J. Sternlieb, J. H. Brewer, J. F. Carolan, W. N. Hardy, R. Kadono, J. R. Kempton, R. F. Kiefl, S. R. Kreitzman, P. Mulhern, T. M. Riseman, D. Ll. Williams, B. X. Yang, S. Uchida, H. Takagi, J. Gopalakrishnan, A. W. Sleight, M. A. Subramanian, C. L. Chien, M. Z. Cieplak, Gang Xiao, V. Y. Lee, B. W. Statt, C. E. Stronach, W. J. Kossler, and X. H. Yu, Phys. Rev. Lett. **62**, 2317 (1989).

125. Y. J. Uemura, L. P. Le, G. M. Luke, B. J. Sternlieb, W. D. Wu, J. H. Brewer, T. M. Riseman, C. L. Seaman, M. B. Maple, M. Ishikawa, D. G. Hinks, J. D. Jorgensen, G. Saito, and H. Yamochi, Phys. Rev. Lett. **66**, 2665 (1991).

126. A. Charnukha, O. V. Dolgov, A. A. Golubov, Y. Matiks, D. L. Sun, C. T. Lin, B. Keimer, and A. V. Boris, Phys. Rev. B **84**, 174511 (2011).

127. E. van Heumen, Y. Huang, S. de Jong, A. B. Kuzmenko, M. S. Golden, and D. van der Marel, Europhys. Lett. **90**, 37005 (2010).

128. K. W. Kim, M. Rösle, A. Dubroka, V. K. Malik, T. Wolf, and C. Bernhard, Phys. Rev. B. **81**, 214508 (2010).

129. B. Gorshunov, D. Wu, A. A. Voronkov, P. Kallina, K. Iida, S. Haindl, F. Kurth, L. Schultz, B. Holzapfel, and M. Dressel, Phys. Rev. B. **81**, 060509(R) (2010).
130. R. P. S. M. Lobo, Y. M. Dai, U. Nage, T. Rõõm, J. P. Carbotte, T. Timusk, A. Forget, and D. Colson, Phys. Rev. B. **82**, 100506(R) (2010).
131. K. Terashima, Y. Sekiba, J. H. Bowen, K. Nakayama, T. Kawahara, T. Satob, P. Richarde, Y.-M. Xu, L. J. Li, G. H. Cao, Z.-A. Xu, H. Ding, and T. Takahashib, Proc. Natl Acad. Sci. USA **106**, 7330 (2009).
132. D. Wu, G. Chanda, H. S. Jeevan, P. Gegenwart, and M. Dressel, Phys. Rev. B. **83**, 100503(R) (2011).
133. C. C. Homes, A. Akrap, J. S. Wen, Z. J. Xu, Z. W. Lin, Q. Li, and G. D. Gu, Phys. Rev. B. **81**, 180508(R) (2010)
134. I. A. Nebrasov and M. V. Sadovskii, JETP Lett. **93**, 166 (2011).
135. X. W. Yan, M. Gao, Z.-Y. LU, T. Xiang, Phys. Rev. B. **84**, 054502 (2011).
136. Y. Zhang, L. X. Yang, M. Xu, Z. R. Ye, F. Chen, C. He, J. Jiang, B. P. Xie, J. J. Ying, X. F. Wang, X. H. Chen, J. P. Hu, and D. L. Feng, Nat. Mater. **10**, 273 (2011).
137. T. Qian, X.-P. Wang, W.-C. Jin, P. Zhang, P. Richard, G. Xu, X. Dai, Z. Fang, J.-G. Guo, X.-L. Chen, and H. Ding, Phys. Rev. Lett. **106**, 187001 (2011).
138. Z. Shermadini, A. Krzton-Maziopa, M. Bendele, R. Khasanov, H. Luetkens, K. Conder, E. Pomjakushina, S. Weyeneth, V. Pomjakushin, O. Bossen, and A. Amato, Phys. Rev. Lett. **106**, 117602 (2011).
139. V. Yu. Pomjakushin, D. V. Sheptyakov, E. V. Pomjakushina, A. Krzton-Maziopa, K. Conder, D. Chernyshov, V. Svitlyk, and Z. Shermadini, Phys. Rev. B **83**, 144410 (2011).
140. A. M. Zhang, J. H. Xiao, Y. S. Li, J. B. He, D. M. Wang, G. F. Chen, B. Normand, Q. M. Zhang, and T. Xiang, Phys. Rev. B **85**, 214508 (2012).
141. B. Shen, B. Zeng, G. F. Chen, J. B. He, D. M. Wang, H. Yang, H. H. Wen, Europhys. Lett. **96**, 37010 (2011).
142. Z. Wang, Y. J. Song, H. L. Shi, Z. W. Wang, Z. Chen, H. F. Tian, G. F. Chen, J. G. Guo, H. X. Yang, and J. Q. Li, Phys. Rev. B **83**, 140505(R) (2011).
143. F. Chen, M. Xu, Q. Q. Ge, Y. Zhang, Z. R. Ye, L. X. Yang, Juan Jiang, B. P. Xie, R. C. Che, M. Zhang, A. F. Wang, X. H. Chen, D. W. Shen, X. M. Xie, M. H. Jiang, J. P. Hu, and D. L. Feng, Phys. Rev. X **1**, 021020 (2011).
144. R. H. Yuan, T. Dong, Y. J. Song, P. Zheng, G. F. Chen, J. P. Hu, J. Q. Li, and N. L. Wang, Sci. Rep. **2**, 221 (2012).
145. A. Charnukha, P. Popovich, Y. Matiks, D. L. Sun, C. T. Lin, A. N. Yaresko, B. Keimer, and A. V. Boris, Nat. Commun. **2**, 219 (2011)

146. O. Klein, E. J. Nicolt, K. Holczer, and G. Griiner, Phys. Rev. B **50**, 6307 (1994).

147. T. Fischer, A. V. Pronin, J. Wosnitza, K. Iida, F. Kurth, S. Haindl, L. Schultz, B. Holzapfel, E. Schachinger, Phys. Rev. B **82**, 224507 (2010).

148. R. V. Aguilar, L. S. Bilbro, S. Lee, J. Jiang, J. D. Weiss, E. E. Hellstrom, D. C. Larbalestier, C. B. Eom, and N. P. Armitage, arXiv:1007.3677.

149. K. Ahilan, F. L. Ning, T. Imai, A. S. Sefat, R. Jin, M. A. McGuire, B. C. Sales, and D. Mandrus, Phys. Rev. B **78**, 100501(R) (2008).

Chapter 6

Antiferromagnetic Spin Fluctuations in the Fe-Based Superconductors

Shiliang Li[a] and Pengcheng Dai[a,b,*]

[a] *Beijing National Laboratory for Condensed Matter Physics, Institute of Physics, Beijing 100190, China*

[b] *Department of Physics and Astronomy, The University of Tennessee, Knoxville, Tennessee 37996-1200, USA*

*pdai.utk@gmail.com

Understanding the role of magnetism in unconventional superconductors is important for elucidating a microscopic mechanism of high-transition temperature (high-T_c) superconductivity. The Fe-based superconductors show rich phenomena related to the magnetism, including the long-range antiferromagnetic order and the associated spin waves in their parent nonsuperconducting materials. The optimal superconductivity is obtained through suppressing the antiferromagnetic order by doping charge carriers or other methods, while strong spin fluctuations persist in the superconducting samples. This chapter gives a brief review on the evolution of the antiferromagnetic spin fluctuations with doping studied by both elastic and inelastic neutron scattering techniques.

Iron-Based Superconductors: Materials, Properties, and Mechanisms
Edited by Nan-Lin Wang, Hideo Hosono, and Pengcheng Dai

ISBN 978-981-4303-22-4 (Hardcover), 978-981-4303-23-1 (eBook)
www.panstanford.com

6.1 Introduction

High-T_c superconductivity is undoubtedly one of the most important unresolved problems in modern condensed matter physics. The term itself used to be the exclusive title for copper oxide superconductors that still hold the record for the highest superconducting transition temperature. However, the discovery of the Fe-based superconductors with T_c up to 56 K[1–7] provided another opportunity to investigate the key ideas in copper oxides that may actually be fundamental to high-T_c superconductivity. A determination of the role of spin fluctuations for high-T_c superconductivity is one of those ideas, and thus immediately drew much attention after the discovery of a three-dimensional long-range antiferromagnetic (AF) order in LaFeAsO,[8] one of the parent compounds for Fe-based superconductors.[1]

In copper oxides, most of the important physics happens within the CuO_2 planes (Fig. 6.1(a)), where the parent compounds show long-range AF order due to the large Coulomb repulsion.[9] The superconductivity emerges with the destruction of the AF order by introducing sufficient holes or electrons onto the CuO_2 planes as shown in Fig. 6.1(c).[10] It is rather surprising that some of the Fe-based superconductors with layered structures composed of Fe*A* blocks (A = As, Se, P, ..., Fig. 6.1(b)) show similar phase diagrams as shown in Fig. 6.1(d),[45] while others have rather different electronic phase diagrams.[12–21] Since the parent compounds of Fe-based superconductors have static AF order regardless how superconductivity is achieved, it is of great interest to compare AF order and spin fluctuations between copper oxides and Fe-based materials as superconductivity is achieved via electron-hole doping.

From the perspective of charge, the copper and iron based systems show significant differences. For example, the parent compounds of cuprates are all Mott insulators[9] while most of the Fe-based superconductors are bad metals.[1] Moreover, most of the properties of the cuprates can be understood within a single band picture while the Fe-based superconductors have shown complicated phenomena related to multi-bands.[22] The pairing symmetry of the cuprates is d-wave while that of the Fe-based

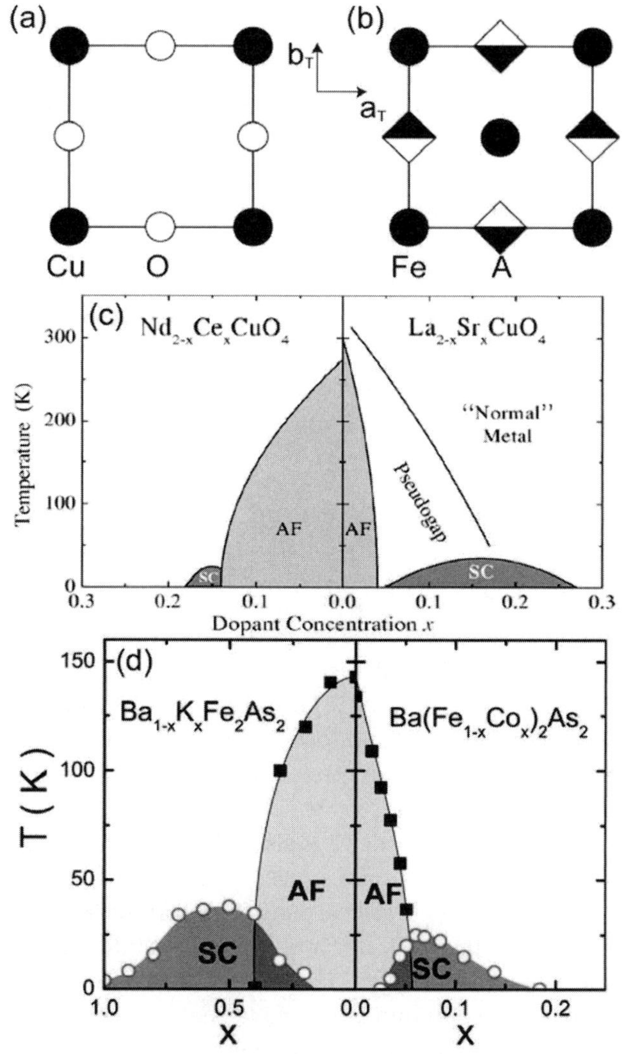

Figure 6.1 (a) The structure of the CuO_2 plane in cuprates. (b) The structure of FeA block in iron pnictides or chalcogenides. The diamond symbol labeled as "A" represents either a pnictogen (P, As) or a chalcogen (Se, Te). The half-white-half-black patterns within the diamond symbols indicate the "A" element is either above or below the Fe–Fe plane. (c) Phase diagram of the "214" family of cuprates.[10] (d) Phase diagram of the "122" family of iron pnictides. The data are taken from Ref. [16, 20].

superconductors is arguably *s*-wave.[23,24] Despite these significant differences, the spin systems of these two high-T_c classes show many common characteristics. It has been argued that phonons are most likely not responsible for pairing,[25,26] while the superconductivity may be mediated by the AF spin fluctuations.[27–33]

In this chapter, we give a brief overview on the current neutron scattering results of spin dynamics in the Fe-based superconductors and compared them with those in cuprates.[34,35] Section 6.2 introduces the long-range AF order and how it is suppressed by doping or other methods. Section 6.3 discusses the influence of superconductivity on the spin dynamics. Section 6.4 further reviews the spin excitations in the normal state of the superconducting materials.

6.2 Antiferromagnetism in Parent Compounds

6.2.1 *Long-Range Antiferromagnetic Order*

The magnetic structures of the long-range AF order in the parent compounds have been well discussed in Ref. [36, 37]; thus we will only give a brief summary here. It is worth noting that the definition of "parent" or "undoped" compounds is rather not as straightforward as that in cuprates. This is not only because some of the "undoped" compounds show superconductivity or properties other than long-range AF order but also from the fact that, arguably, that the Mott physics in cuprates gives a natural picture of doping holes or electrons into a classical Heisenberg antiferromagnetism[9] while the origin of the AF order and the effect of doping in the Fe-based superconductors are still under debate.[38,39] Here we broadly refer the "parent" compounds as those that show long-range AF order and exhibit phase diagrams similar to that in Fig. 6.1(c) upon partial substitution of one or more elements.

For all the LaFeAsO or "1111", $BaFe_2As_2$ or "122" and NaFeAs or "111" families, the magnetic orders of their parent compounds, if present, show a commensurate collinear AF structure as shown in Fig. 6.2(a). At temperatures above or at the formation of AF order, these materials exhibit a tetragonal-to-orthorhombic structure

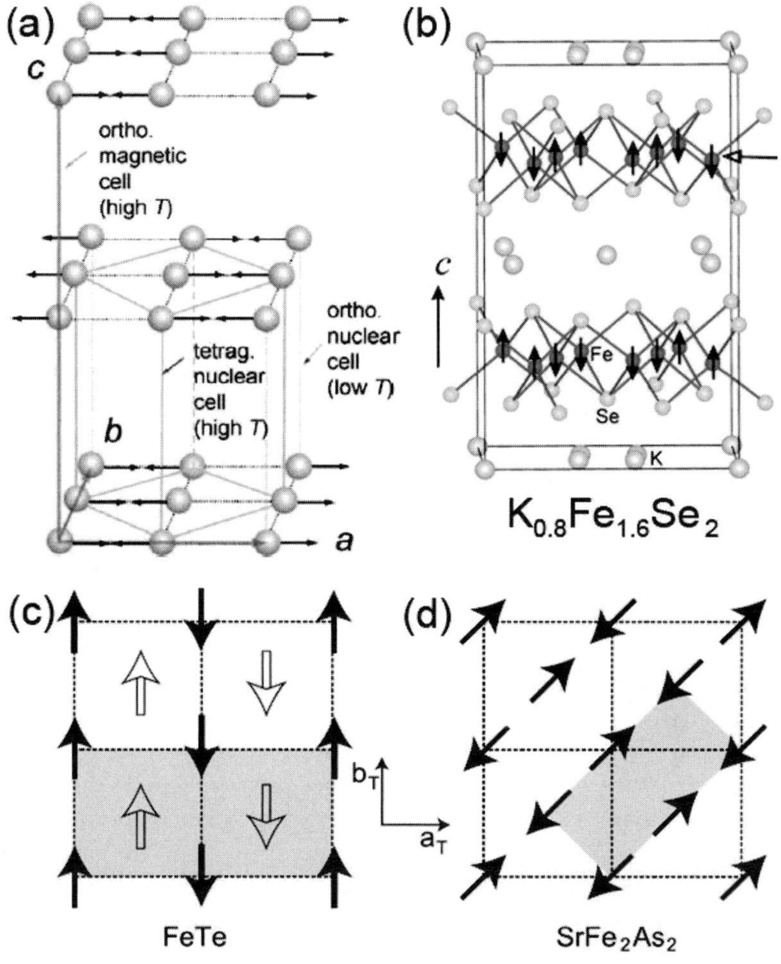

Figure 6.2 (a) Collinear magnetic structure of the parent compounds including "1111", "122" and "111" families[36] with a few exceptions as discussed in the text. (b) Magnetic structure of $K_{0.8}Fe_{1.6}Se_2$. (c) and (d) give a comparison of the in-plane magnetic structures between FeTe and $SrFe_2As_2$.[67] The blue areas represent the magnetic unit cells. The hollow arrows in (c) indicate that the relative directions between the two sublattices (hollow and solid arrows) cannot be conclusively determined.

transition that breaks the four-fold symmetry of the paramagnetic tetragonal phase.[40–42] In the notion of the orthorhombic structure, the moments are confined within the *a*–*b* plane and aligned along the *a*-axis. Along the *c*-axis, most spin configurations have antiparallel directions between two adjacent layers[8,43–48] as shown in Fig. 6.2(a) with the exception of Ce-1111[45] and Pr-1111[49] that have a parallel pattern. While the same AF order has been found in NaFeAs,[48] LiFeAs seems to be an exception since it shows no static magnetic order.[50] This may be caused by the poor nesting condition between hole and electron pockets[51] or very strong direct Fe–Fe interactions that may frustrate the AF order.[52] Although it was suggested that LiFeAs may have ferromagnetic spin fluctuations,[53] recent inelastic neutron scattering experiments on both the superconducting polycrystalline samples[54] and the non-superconducting single crystals[55] show that the magnetic fluctuations in LiFeAs are AF and similar as those in the other families of iron-based materials.

Besides the commonly known three families of iron arsenides mentioned above, some more complicated materials have also been fabricated, such as the so-called "42622" ($Sr_4A_2O_{6-x}Fe_2As_2$),[56–59] "32522" ($Sr_3Sc_2O_5Fe_2As_2$[60] and $Ba_3Sc_2O_5Fe_2As_2$[61]), $Sr_4(Sc,Ti)_3O_8Fe_2As_2$ and $Ba_4Sc_3O_{7.5}Fe_2As_2$,[61] the homologous series of $Ca_{n+1}M_nO_yFe_2As_2$[62] (n = 3,4,5, M = (Sc, Ti) and (Mg, Ti)) and $Ca_{n+2}(Al,Ti)_nO_yFe_2As_2$.[63] All these materials can be seen as the "122" structure with much larger distance between two FeAs blocks. Many of them show superconductivity or other non-magnetic ground states, but the magnetic structures seem to have the collinear arrangement discussed above.[64,65] The introduction of other elements such as V may lead to their own novel magnetic structures.[66] Due to the difficulty of growing single crystals, much of the physical properties in these materials have not been fully investigated yet.

While most of the parent compounds of the iron arsenides have simple commensurate AF structures if they exhibit the long-range AF order, those of iron chalcogenides show more complicated AF structures and crystal lattice distortions associated with the magnetic phases. Fig. 6.2(c) and 6.2(d) give a comparison of the in-plane magnetic structure between FeTe and $SrFe_2As_2$ in the notation of tetragonal unit cell, where both the collinear pattern and the directions of moments are rotated 45 degree away from those in the iron arsenides.[67] We note that such an in-plane structure is only a hypothesis in the stoichiometric limit. In reality, there are always extra irons ($Fe_{1+y}Te$) that drive the moments off the a–b plane[67,68] and result in an incommensurate structure for large y.[69,70] The incommensurate spin-density wave competes with the bicollinear commensurate ordering in the intermediate value of y.[70] The newly discovered "122" system of $RFe_2Se_2(R = \mathrm{K, Rb, Cs, Tl})$ also shows complicated nuclear structures due to large amount of Fe vacancies.[71,72] The only magnetic structure reported so far is for those that have a ratio of Fe:Se as 4:5 so that the Fe vacancies arrange orderly and the crystal has I4/m symmetry with a unit cell $\sqrt{5} \times \sqrt{5}$ that of the tetragonal unit cell of iron arsenides.[71,72] The magnetic structure can be seen as a regular AF structure with nearest "supermoments" pointing oppositely along c-axis. Each "supermoment" comprises four nearest-neighbor spins at iron sites arranged ferromagnetically. It is worth noting that AFe_4Se_5 holds the record for both the largest moment and the highest AF transition temperature T_N among all the Fe-based superconductors and it is insulating instead of a bad metal as in other Fe-based materials. Since the photoemission experiments have shown two electron-like Fermi surface sheets around M point that are distinct from that in all other Fe-based superconductors,[73–75] it is of great interest to study the spin dynamics in this system.

6.2.2 *Spin Waves*

While there had already been several attempts to study the spin waves of the Fe-based superconductors from the outset,[76–80] the whole spectra were first measured in $CaFe_2As_2$.[81] The data can be well fitted by a Heisenberg Hamiltonian consisting of effective in-plane nearest-neighbor (J_{1a} and J_{1b}), next-nearest-neighbor (J_2) and out-of-plane (J_c) exchange interactions (see Fig. 6.3(a)),

$$H = J_{1a}\sum_{i,j} S_i \cdot S_j + J_{1b}\sum_{i,j} S_i \cdot S_j + J_2\sum_{i,j} S_i \cdot S_j + J_c \times \sum_{i,j} S_i \cdot S_j - J_s\sum_{i} (S_i^z)^2 \tag{6.1}$$

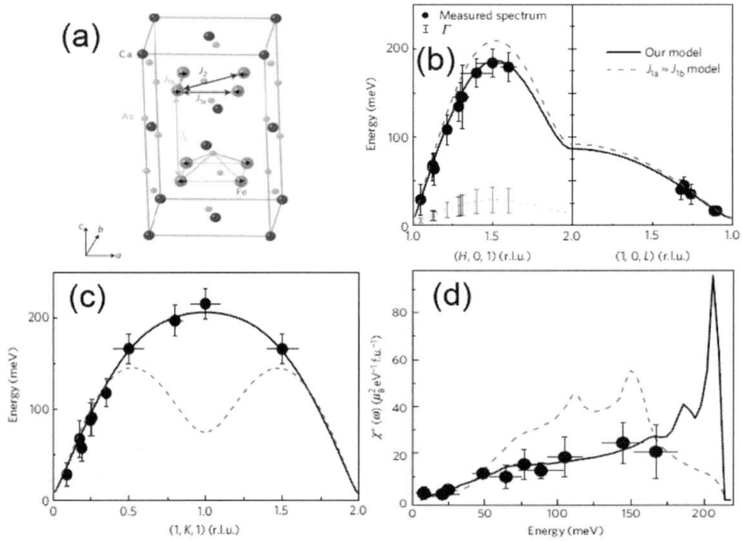

Figure 6.3 (a) Schematic diagram of the Fe spin ordering in $CaFe_2As_2$.[81] The definition of exchange couplings is also presented. (b), (c) Spin-wave dispersion relation along different high-symmetry directions.[81] (d) The energy dependence of the local susceptibility.[81] The black solid line and red broken line are the fitting results based on the model described in the text and $J_{la} = J_{lb}$ model, respectively.

where J_s is the single ion anisotropy. This model gives the dispersion relations as

$$E(q) = \sqrt{A_q^2 - B_q^2} \quad (6.2a)$$

$$A_q = 2S\left[J_{\text{lb}}\left(\cos(\pi K) - 1\right) + J_{\text{la}} + J_c + 2J_2 + J_s\right] \quad (6.2b)$$

$$B_q = 2S\left[J_{\text{la}}\cos(\pi H) + 2J_2\cos(\pi H)\cos(\pi K) + J_c\cos(\pi L)\right] \quad (6.2c)$$

where q is the reduced wave vector away from the AF zone center. The fitting results are shown in Fig. 6.3(b–d). The best fitting parameters are given as $SJ_{la} = 49.9 \pm 9.9$ meV, $SJ_{lb} = -5.7 \pm 4.5$ meV, $SJ_2 = 18.9 \pm 3.4$ meV, $SJ_c = 5.3 \pm 1.3$ meV. The anisotropy between the in-plane and out-of-plane exchange couplings of the Fe-based superconductors is much smaller than that of cuprates,[82,83] which results in a strong *c*-axis modulation of the spin dynamics at low energies. More importantly, the in-plane exchange couplings also show strong anisotropy, being positive and negative along a- and b-axes, respectively, which happens to be consistent with the in-plane arrangement of the moments as shown in Fig. 6.3(a). A similar model with $J_{1a} = J_{1b}$ can also lead to the collinear structure and explain the low-energy excitations,[80] but such model completely fails to fit data near zone boundaries. It also gives a wrong energy dependence of local susceptibility as shown in Fig. 6.3(d). The large in-plane anisotropy indicates that other degrees of freedom such as orbital should be involved.[84] The above results also suggest that there is no strong evidence for Stoner continuum since clear dispersions are observed up to zone boundaries as shown in Fig. 6.3(b) and (c). However, the authors find significant broadening of the spin wave near zone boundaries, which cannot be explained by a simple Heisenberg model and require the consideration of itinerant nature of electrons.

Measurements on $BaFe_2As_2$ confirm the results in $CaFe_2As_2$, giving similar exchange couplings with smaller SJ_2 and SJ_c but larger SJ_{1a} and SJ_{1b}.[85] In addition, the spin waves of $BaFe_2As_2$ are heavily damped along the orthorhombic a-axis direction, which could be an indication of Stoner continuum in this direction. Interestingly, the large in-plane anisotropy seems to persist above T_N since no significant softening near zone boundaries is observed in the paramagnetic state, although the low-energy spin excitations become weaker and broader as those in $CaFe_2As_2$.[86] It is suggested that this result is consistent with a nematic spin fluid that breaks the C_4 symmetry of the tetragonal phase.[87–89] However, the results may also be explained by considering orbital degrees of freedom or by including quadrupolar interactions not considered in a Heisenberg model.

As discussed above, the LiFeAs has been thought to be different from other iron arsenides. The inelastic neutron scattering experiment on nonsuperconducting $Li_{1-x}FeAs$ reveal that the low-energy spin excitations in this system also occur at (0.5,0.5).[55] However, the high-energy spin excitations near zone boundaries have different wave vectors from those of other compounds, which makes it impossible to fit the spectra with a simple Heisenberg Hamiltonian model. This suggests that both itinerancy and strong electron-electron correlations are essential to understand the spin excitations in this system.

The spin waves in $Fe_{1.05}Te$ can be fitted by a Heisenberg model,[90] which gives the exchange couplings differing from those of "122" materials. The large in-plane anisotropy of the nearest-neighbor couplings remains but with the value of $SJ_{la} = -17.5 \pm 5.7$ meV and $SJ_{lb} = -51.0 \pm 3.4$ meV. In addition, the next-next-nearest-neighbor exchange coupling SJ_3 must be considered with a fitted value of 6.8 ± 2.8 meV. These differences are expected to some extent since the AF structures between "122" and "11" systems are markedly different. However, the next-nearest-neighbor exchange couplings J_2 in both systems are almost the same, which seems to be insensitive to the lattice distortion and variation in the d-orbital components. Due to the extra irons in this system, it is shown that the spin gap in lower doping of interstitial iron disappears in $Fe_{1.141}Te$, whose spin structure has become incommensurate.[91] In addition, the spectra show a peak at 4 meV. It is interesting to see whether the high-energy spin excitations are also affected by the extra irons or not.

The spin waves in AFe_4Se_5 are strongly affected by the iron vacancies. It is shown that the whole spectra can be roughly divided into two parts: the low-energy acoustic mode arising mostly from AF interactions of the FM blocked spins, and the two high-energy optical modes associated with exchange interactions of iron spins within the FM blocks.[92] To fit the spectra with a Heisenberg model, the next-next-nearest-neighbor exchange coupling also need to be included as in the case of $Fe_{1.05}Te$. While the other exchange couplings show their materials-dependent characteristics, the next-nearest-neighbor exchange coupling J_2 is similar to those of other systems.

While the above results have shown many interesting features of the spin waves among various systems, there is still a lack of data from many other parent compounds that can give us a whole picture of the magnetic couplings in the Fe-based superconductors. As shown above, the current results suggest that the localized-moment picture[39,93] is able to explain the spin waves in the Fe-based superconductors by varying the strength of the exchange couplings. On the other hand, many studies based on Fermi nesting between electron and hole pockets are also able to reproduce the experimental results.[30–32,94–97] It is also proposed that the local and itinerant characteristics may present simultaneously.[98,99] In a sense, we still do not have a comprehensive understanding of the spin wave excitations in the parent compounds of Fe-based superconductors.

6.2.3 *Destruction of Antiferromagnetic Order*

It is of great interest to see how the AF order is suppressed upon charge carrier doping or the variation of other parameters like pressure since superconductivity can arise from such destruction of the long-range AF order. It has been found by neutron scattering experiments that in certain range of underdoped region, superconductivity may coexist with the AF order state, where the intensities of static AF Bragg peaks drop across T_c together with a redistribution of low-energy magnetic excitations.[100,101] The neutron itself cannot be used to determine whether such coexistence is due to spatially inhomogeneous samples with segregated magnetic and superconducting phases,[102] while measurements by some other techniques such as μSR seem to suggest that magnetism and superconductivity are able to coexist and compete on the nanometer length scale.[103]

In the 122 system, doping Co or Ni into the system will result in a separation of the structural and AF transitions with $T_s > T_N$.[104,105] However, Ru doping does not separate the structural and magnetic phase transitions.[106] Dynamically, it is found that the spin gaps at both $L = 1$ and $L = 0$ decrease with 4% Ni doping,[107] as shown in Fig. 6.4(a) and (b). In Heisenberg model, the anisotropy gaps at $L = 1$ and $L = 0$ can be written as $\Delta(L = 1) = 2S[(J_{la} + 2J_2 + J_c + J_s)^2 - (J_{la} + 2J_2 + J_c)^2]^{1/2}$ and $\Delta(L = 0) = 2S[(2J_{la} + 4J_2 + J_s)(2J_c + J_s)]^{1/2}$, respectively. The reduction of the spin gaps is thus mainly due to a rapid suppression of *c*-axis exchange coupling and magnetic single ion anisotropy, which suggests that electron doping results in a transition from 3D spin waves to quasi-2D spin excitations. It is interesting to note that the suppression of J_c also explains why the structural transition is separated from the magnetic transition upon doping in the 122 system if a simple model giving $T_N \sim J_2/\ln(J_2/J_c)$ is considered.[40]

Recently, it is found that the AF order becomes incommensurate along the transverse direction of the commensurate AF wavevector $Q_{AFM} = (0.5,0.5)$ within a narrow range $(0.056 \leq x \leq 0.059)$

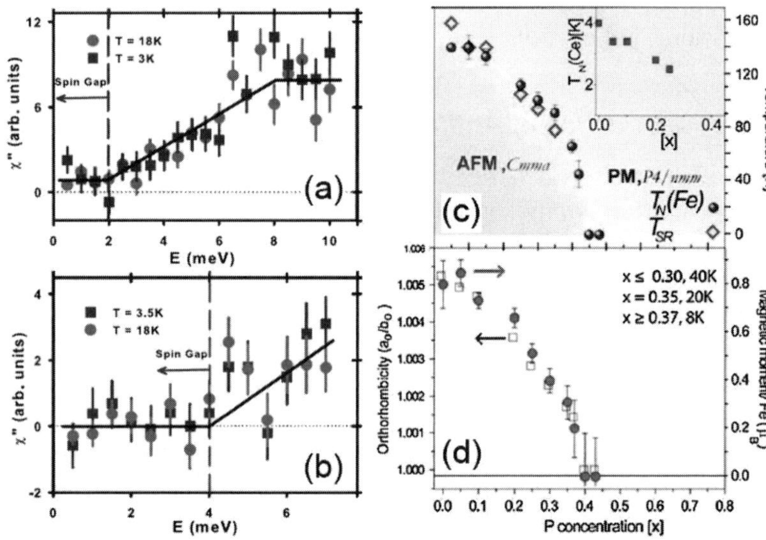

Figure 6.4 (a–b) Energy dependence of the local susceptibility χ'' at $L = 1$ and $L = 0$ in $BaFe_{1.96}Ni_{0.04}As_2$.[107] The values of spin gaps are defined as vertical broken lines. (c–d) Phosphor doping dependence of structural transition temperature T_{SR}, magnetic transition temperatures T_N (Fe) and T_N(Ce), Orthorhombicity and magnetic moment of Fe in $CeFeAs_{1-x}P_xO$.[117]

of Co doping in $Ba(Fe_{1-x}Co_x)_2As_2$.[108] The commensurate-to-incommensurate AF transition is believed to be first order in nature since both phases coexist in the $x = 0.056$ sample. The authors argue that the mismatch of hole and electron pocket areas favors the incommensurate AF order as in the case of Mn or Ru doped spin-density-wave compound Cr.[109,110]

In $Fe_{1+y}Se_xTe_{1-x}$, a spin-glass regime is found to exist for $0.1 \leq x \leq 0.4$ as shown in Fig. 6.5(a).[111] The excess irons within this system affect the properties of not just the parent compound $Fe_{1+y}Te$, but also the selenium-doped samples. Therefore, the regime of this glassy state may vary for different annealing processes that can result in different amount of extra Fe.[112] In any case, the neutron scattering measurements indicate that the spin-glass behavior is associated with the short-range spin density wave ordering coupled with a glassy structural distortion.[111] Upon increasing x,

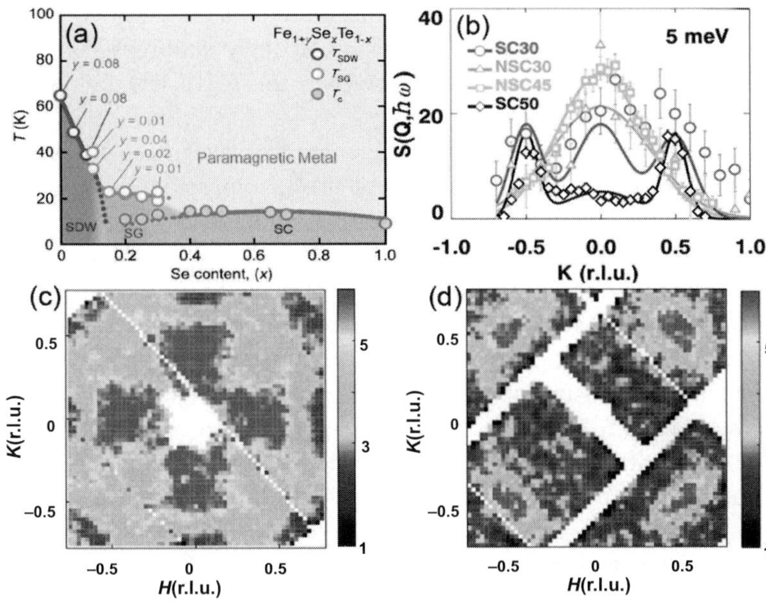

Figure 6.5 (a) Phase diagram of $Fe_{1+y}Se_xTe_{1-x}$.[111] (b) Magnetic excitations at 5 meV for various superconducting and nonsuperconducting samples.[113] (c–d) Mapping of the spin excitations in $x = 0.27$ and $x = 0.49$, respectively, at 6 ± 1 meV.[115] See also Color Insert.

the magnetic correlations at (0.5,0) are gradually suppressed, while those at (0.5,0.5) becomes stronger with the emergence of bulk superconductivity.[113–115] In fact, it is the superconductivity instead of Se doping that acts as the direct driving force to cause the redistribution of spin excitations as shown in Fig. 6.5(b), where the non-superconducting sample with 0.45 Se doping also shows a strong peak at (0.5,0) that is the same as those of low-doping samples.[113] The rearrangement of the in-plane low-energy spin excitations is nicely illustrated in Fig. 6.5(c) and (d). The excitations in the superconducting sample (Fig. 6.5(d)) show high anisotropy, which will be discussed later in Section 6.4.

Due to the lack of large single crystals, it is still unclear how the spin fluctuations evolve in many systems. Besides the introduction of charge carriers, there are at least two other controlling parameters that may induce superconductivity, the so-called "isovalent" doping and pressure. For example, early measurements on $CaFe_2As_2$ show that the AF spin fluctuations may be strongly suppressed in the collapsed tetragonal phase.[116] However, there are few follow up work in this direction.

The above discussion focuses on the crossover from the long-range AF order to superconductivity. It is also possible to suppress the AF order without inducing superconductivity. This is well illustrated in the case of $CeFeAs_{1-x}P_xO$,[117] where the AF order is destructed by the introduction of phosphor without charge carrier doping and the subsequent emergence of superconductivity. It is found that both the orthorhombic distortion and the AF order are simultaneously suppressed and disappear near $x \approx 0.4$ as shown in Fig. 6.4(c) and (d), suggesting the presence of a magnetic quantum critical point (QCP). The pnictogen height is found to be an important controlling parameter for their electronic and magnetic properties. These results are further confirmed by other methods and in other systems.[118,119]

In the 122 system, replacing Fe by either Cr or Mn does not lead to superconductivity but to other types of AFM. In $Ba(Fe_{1-x}Cr_x)_2As_2$, the stripe-like AF structure changes to a G-type AF order above $x \approx 0.3$.[120] For $Ba(Fe_{1-x}Mn_x)_2As_2$, the crossover happens at $x \approx 0.1$ and the AF structure at higher doping may be described by a two-Q ordering.[121] In both cases, the tetragonal-to-orthorhombic structural transition is firmly locked to the stripe-like AF transition and disappears when the AF structure changes.

6.3 Magnetic Excitations in the Superconducting State

The magnetic resonant spin excitations or neutron spin resonance is one of the key features that distinguish the unconventional superconductivity from the conventional superconductivity. Experimentally, the resonance can be broadly defined as an enhancement of the local susceptibility χ'' at a certain energy below T_c. Moreover, its magnetic scattering intensity decreases with increasing temperature below T_c, acting as an order parameter of superconductivity. It was first discovered in many classes of cuprates[34,35,122] and later on in heavy fermion superconductor $CeCoIn_5$.[123] Since the resonance

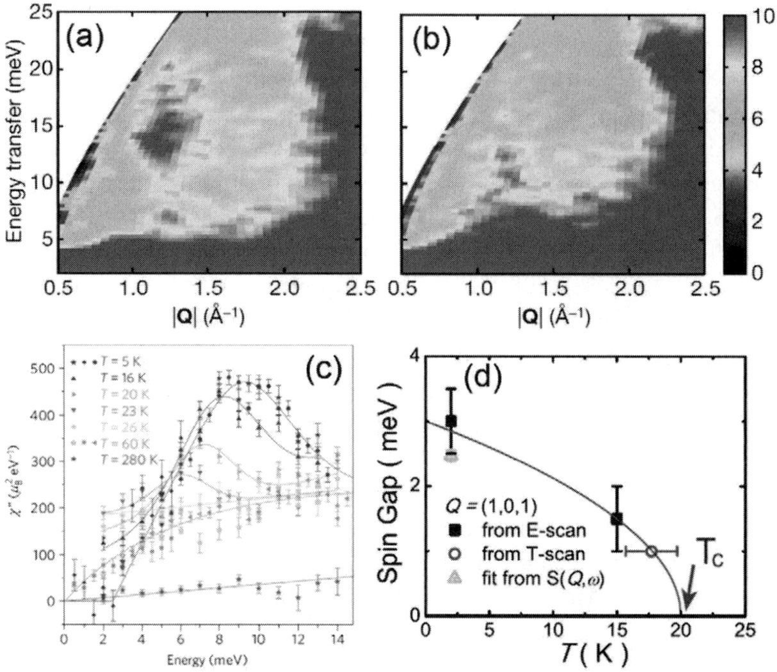

Figure 6.6 (a–b) The magnetic resonance in powder $Ba_{0.6}K_{0.4}Fe_2As_2$.[124] (c) Temperature dependence of the local susceptibility $\chi''(\omega)$ up to room temperature in $BaFe_{1.85}Co_{0.15}As_2$.[125] (d) Temperature dependence of spin gap in $BaFe_{1.9}Ni_{0.1}As_2$.[126] See also Color Insert.

is very important to understand the mechanism of many unconventional superconductors, the competition of searching for resonance immediately started after the discovery of antiferromagnetic order in the parents of the Fe-based superconductors.

6.3.1 *Magnetic Resonance*

The resonance is first found in powder $Ba_{0.6}K_{0.4}Fe_2As_2$ as shown in Fig. 6.6(a) and (b),[124] where the intensities of spin excitations at $|Q| \sim 1.15$ Å^{-1} and $E = 14$ meV show significant change across T_c. Such phenomenon is soon confirmed in the single crystals of Co and Ni doped Ba122,[125–131] where many of the detailed features not measured in powder experiments are revealed. Fig. 6.6(c) shows that there is no resonance peak above T_c (25 K) in $BaFe_{1.85}Co_{0.15}As_2$ and $\chi''(\omega)$ decreases smoothly with increasing temperature up to 280 K. Right below T_c, a peak of $\chi''(\omega)$ immediately shows up at about 5 meV. Both the peak intensity and the peak position, i.e., the resonance energy, increase quickly with decreasing temperature. Correspondingly, the spin gap also closes right at T_c, as shown in Fig. 6.6(d) measured in $BaFe_{1.9}Ni_{0.1}As_2$ ($T_c = 20$ K).

Since the spin fluctuations in parent compounds of the "122" system show some differences at odd and even L, it is interesting to see whether such L modulation exists in the spin excitations of superconducting samples. Measurements on $BaFe_{1.9}Ni_{0.1}As_2$ show that the resonance energies at $L = 0$ and $L = 1$ are indeed different,[127] being about 9 meV and 7 meV, respectively, as shown in Fig. 6.7(a) and (b). As to the Co-doped samples, it is initially suggested that the magnetic resonance is 2D-like in $BaFe_{1.84}Co_{0.16}As_2$.[129] However, the measurements on more under-doped samples reveal that the modulation of resonance energy along c-axis is also present but its amplitude decreases quickly with increasing Co doping, as shown in Fig. 6.7(c). Figure 6.7(d) sums up the results of both Ni and Co doped samples.[128] It is clear that the resonance energies at both $L = 0$ and $L = 1$ have a linear relationship with T_c, but the slopes are different. It is the resonance at $L = 1$ whose energy will become zero if T_c goes to zero. Moreover, it is the T_c instead of doping that determines the difference of the resonance energies between $L = 0$ and $L = 1$. Correspondingly, the spin gaps at $L = 0$ and $L = 1$ also show similar modulation.[126] However, in the hole-doped $Ba_{0.67}K_{0.33}Fe_2As_2$,[131] although the spin excitations still show a L modulation similar to those in electron-doped Ba122 samples, the spin gap at $L = 1$ becomes zero while that at $L = 0$ remains fairly large as shown in Fig. 6.7(e) and (f). It is not clear whether such different behavior of spin gaps comes from the intrinsic difference between hole and electron doping or not.

The magnetic resonance is also found in $FeSe_xTe_{1-x}$ system,[132,133] as shown in Fig. 6.8(a–d). As discussed in previous section, the in-plane magnetic excitations are centered at (0.5,0.5) instead of (0.5,0). Comparing to the "122" system, the resonance is clearly 2D-like as shown in Fig. 6.8(c). In addition, the spin excitations in "11" system are incommensurate within the a–b

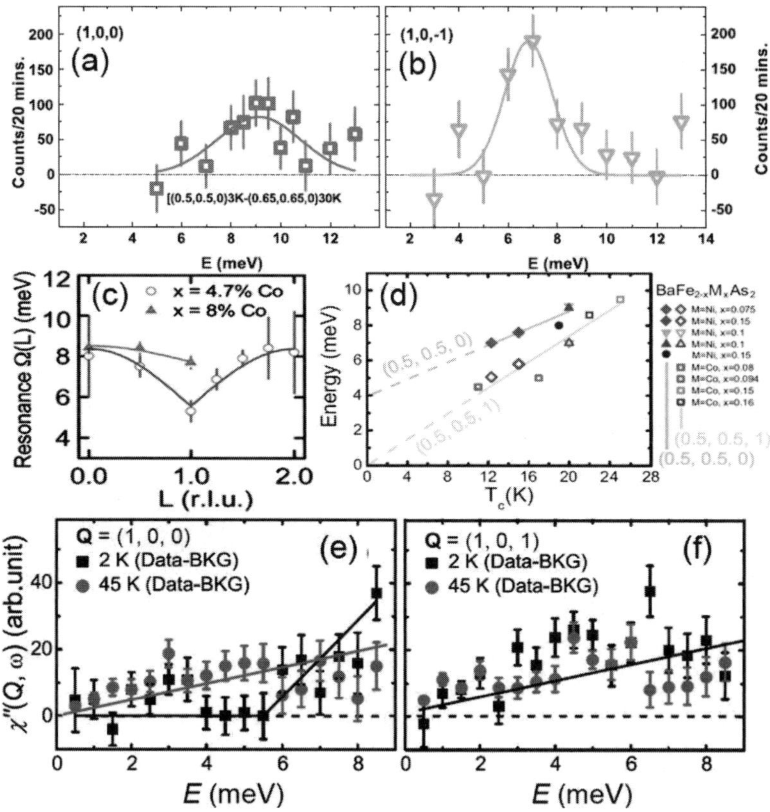

Figure 6.7 (a–b) Magnetic resonance in $BaFe_{1.9}Ni_{0.1}As_2$ at $L = 0$ and $L = 1$, respectively.[127] (c) Dispersions of the resonance energy along the L direction for $x = 4.7\%$ and 8.0% of $Ba(Fe_{1-x}Co_x)_2As_2$.[130] (d) T_c dependence of magnetic resonance in "122" system.[128] (e–f) Spin gaps at $L = 0$ and $L = 1$ in $Ba_{0.67}K_{0.33}Fe_2As_2$.[131]

plane,[115,134–137] which will be discussed in details in the next section. The magnetic resonance has also been reported in powder $LaFeAsO_{0.92}F_{0.08}$ at $|Q| \sim 1.15$ Å^{-1} and E $\sim$ 13 meV.[138] An experiment in powder LiFeAs also reveals a resonance-like enhancement of the $\chi''(\omega)$ at low energies below T_c.[54]

In cuprates, it was found that the resonance energy in most systems is proportional to T_c with a slope of 5.8 k_B. The resonance energies in the Fe-based superconductors and $CeCoIn_5$ also seem

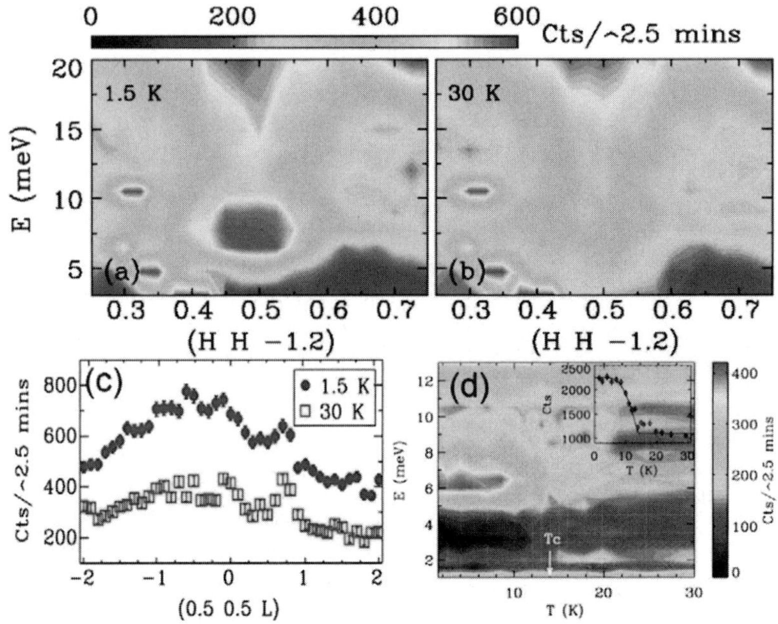

Figure 6.8 (a–b) Spin excitation spectrum of $Q = (H,H,-1.2)$ and energy at 1.5 K and 30 K, respectively, in $FeSe_{0.4}Te_{0.6}$. (c) Constant $E = 6.5$ meV scans along the $(0.5, 0.5, L)$ direction at 1.5 K and 30 K, respectively. (d) Temperature dependence of the magnetic resonance.[130] See also Color Insert.

to fall on this relationship, as shown in Fig. 6.9. However, detailed examination has questioned this simple scaling function and an alternate scaling has been proposed.[123,139] See more detailed discussion in Ref. [37].

Unlike the cuprates, where the theoretical understanding of the resonance is still under debate,[34,35] the resonance in the Fe-based superconductor can be well explained by the random-phase approximation (RPA) approach,[140,141] where a sign-reversed state s^{+-} is preferred. In fact, the discovery of the resonance has been treated as a experimental evidence for the s^{+-} pairing. However, it is argued that a s^{++} state without sign reversal is also possible to reproduce the experimentally observed broad peak.[142] More experiments are needed to distinguish these two pictures.[143,144]

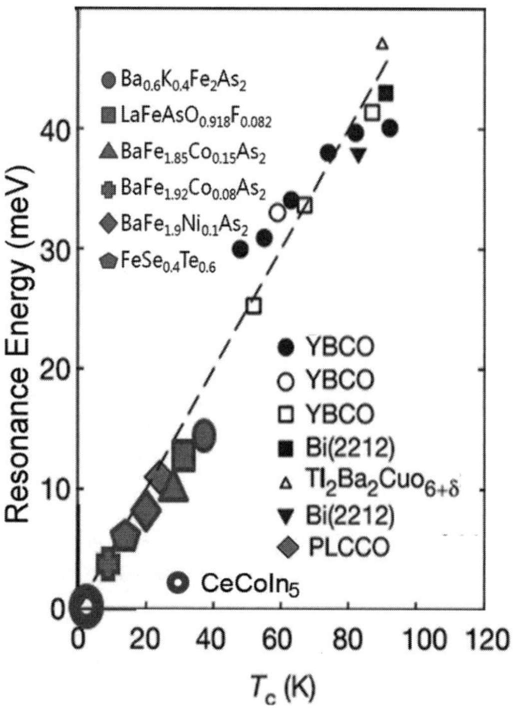

Figure 6.9 The relationship between T_c and resonance energy in unconventional superconductors. The data for cuprates are taken from Ref. [122] and the references therein. The resonance energy of $CeCoIn_5$ is from Ref. [123]. See text for the references reporting the resonance of the Fe-based superconductors.

6.3.2 *Field Effect on Magnetic Resonance*

Since the discovery of neutron spin resonance, many theories have been proposed to explain it and all of them treat the resonance as an excited pair of quasiparticles with spin $S = 1$.[135] Such a spin-1 triplet state will split into three different energy levels under a magnetic field as observed in the triplet state excited from a paramagnetic singlet state.[145] It has been proposed that the resonance in the Fe-based superconductors may be a doublet due to the lock-in of the moments within the $a - b$ plane.[146] In principle, the resonance peak will split into either three or two peaks under field measuring by unpolarized neutrons or one/two peaks

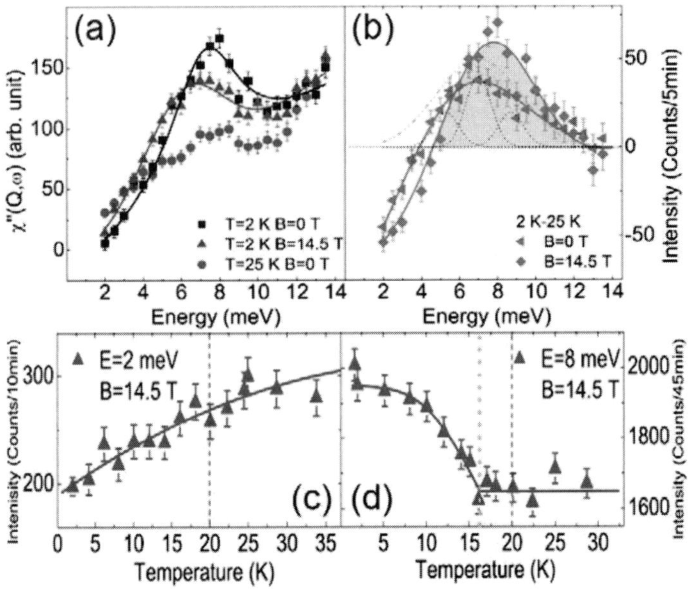

Figure 6.10 (a) Temperature and field dependence of $\chi''(Q,\omega)$ at Q = (0.5,0.5,0) in $BaFe_{1.9}Ni_{0.1}As_2$.[150] (b) Difference spectra of the neutron intensity between 2 K and 25 K (T_c+ 5 K) for 0 and 14.5 Tesla. The black dotted lines indicate the expected Zeeman splitting at 14.5 T field using the corresponding instrumental resolution. (c) Temperature dependence of scattering at 2 meV and 14.5 T. (d) Temperature dependence of the resonance peak intensity (8 meV). The red line indicates T_c at zero field.

in different polarized channels for polarized neutrons. However, no such splitting has ever been observed in cuprates.[147–149] The discovery of resonance in the Fe-based superconductors provided another opportunity to search for such effect.

The field effect of the resonance in $BaFe_{1.9}Ni_{0.1}As_2$ has been studied for magnetic field along *c*-axis.[150] Both the resonance energy and spin gap decrease with increasing field as shown in Fig. 6.10(a) and (b). In addition, the temperature dependence of the intensity of the resonance peak is consistent with the suppression of T_c under field as shown in Fig. 6.10(d). However, no obvious splitting is observed. Applying an in-plane field results in weaker suppression of the resonance,[151] which is consistent with the fact that the in-plane upper critical field is much higher than the out-of-plane one.

Contrary to the c-axis field, both the resonance energy and spin gap show no change upon in-plane field. Again, no splitting is observed. It is further shown that the scattering of the static AF order in $BaFe_{1.92}Ni_{0.08}As_2$ is enhanced when the intensity of the resonance is reduced under field.[152] These results suggest that the energy of the resonance is intimately connected to the electron paring energy.

Attempts on $FeSe_xTe_{1-x}$ system have been made for both in-plane[153] and out-of-plane[154] fields. A 7 Tesla in-plane field is found to be enough to reduce the intensity of the resonance. The temperature dependence of the resonance intensity also indicates a lowered T_c.[153] On the other hand, while the out-of-plane field also reduces the peak intensity of the resonance, it may also result in a splitting to three peaks.[154] Quantitative analysis shows that large in-plane anisotropy of 1.2(4) meV is required to better understand the data. However, a recent in-plane field experiment only shows a broadening of the resonance peak.[151] In addition, the suppression of the resonance intensity has a square root relationship with the magnetic field. At present, there is no conclusive evidence for field-induced Zeeman splitting of the resonance.

Here we summarize the peculiar properties of the resonance in the Fe-based superconductors that are different from those in cuprates. First, the enhancement of the spectra weight in the superconducting state is tightly associated with the opening of the spin gap just below the resonance peak. The spectral weight gain of the resonance appears to be entirely compensated by spectral weight loss of the sub-spin gap magnetic scattering. Second, no normal-state or residual peak of $\chi''(\omega)$ at the resonance energy has been observed in all the Fe-based superconductors, which is contrary to the case of underdoped cuprates.[155] It should be noted that the resonance in electron-doped cuprates actually shows aspects similar to the above two properties.[122,156] Third, the resonance energy and intensity are strongly affected by the c-axis correlations and have dispersion for different L-values. In cuprates, the resonance energy is L-independent. Finally, the resonance is either anisotropic[146] or quasi-isotropic,[157] while in the case of cuprates, it is isotropic.[158–160]

6.3.3 *Field-Induced Magnetization*

In the conventional superconductor, it has been shown that the field-induced magnetization due to the electron-spin polarization in the paramagnetic normal state is completely suppressed in the superconducting state,[161] which gives direct evidence that the Cooper pairs are in a singlet state. This method is also applied to the unconventional superconductors including the heavy-fermion superconductors,[162] cuprates[163] and $SrRuO_4$.[164] A similar experiment in $Ba(Fe_{0.935}Co_{0.065})_2As_2$[165] shows that the temperature dependence of the induced moment in the superconducting state is consistent with the singlet state of Cooper pairs. A large zero-field magnetization at low temperatures is suggested to come from the van Vleck susceptibility. The magnetization is found to be confined to the Fe atoms, which accords with the result in the paramagnetic state of the parent compound of $BaFe_2As_2$.[166]

6.4 Magnetic Excitations in the Normal State

The disappearance of the resonance in the normal state spin excitations spectra, while suggesting that iron-based superconductors are more Fermi-liquid-like, does not mean that electron/hole doping has no effect on the spin waves in the undoped nonsuperconducting parent compounds. In cuprates, the introduction of holes into the AF ordered CuO_2 planes can lead to modifications entirely different from the spin waves, such as incommensurate spin excitations and hourglass dispersion of the spin excitations in some materials.[34] Experiments on the Fe-based superconductors have also revealed that understanding the spin dynamics in the normal-state could give us further insights into the interplay among various channels including charge, orbital and spin. In this section, we will focus on the in-plane distribution of the low-energy spin excitations in the "122" and "11" systems.

6.4.1 *In-Plane Anisotropy in the "122" System*

As discussed in Section 6.2, the destruction of the long-range AF order is most likely due to the quick suppression of c-axis exchange coupling.[102] Therefore, the in-plane exchange couplings probably will not change much with increasing carrier doping so that the characteristic of the strongly anisotropic spin-wave velocity in the parent[81,85] compounds should persist in doped samples. This is indeed the case for nearly optimally doped $BaFe_{1.87}Co_{0.13}As_2$,[167] where the spin excitations are treated as a reminiscent of the parent compounds except for the part of the resonance. The results are confirmed in $BaFe_{1.852}Co_{0.148}As_2$[168] and $BaFe_{1.85}Co_{0.15}As_2$.[169] However, such anisotropy can be explained by either the local moment picture probably involving the nematic phase[168] or the iterant picture that can be calculated by the density functional theory.[169]

Surprisingly, the measurements on $Ba_{0.67}K_{0.33}Fe_2As_2$[131] show that the anisotropy rotates 90 degree compared to the Co-doped Ba122 samples. Fig. 6.11(a) and (b) show the comparison between the electron-doped Co-Ba122 and the hole-doped K-Ba122. In Fig. 6.11(b), the notion of reciprocal space is based on the orthorhombic unit cell in the parent compounds. Therefore, the spin excitations in electron-doped and hole-doped "122" materials elongate transversally ([H,1−H]) and longitudinally ([H,H]), respectively, in the notion of the tetragonal unit cell. It is proposed that a natural explanation may come from the different nesting conditions among two families due to the size and shape of the Fermi surfaces, as shown in Fig. 6.11(c). It should be noted that in KFe_2As_2,[170] the spin excitations become incommensurate along the [H,H] direction, which can also be explained by the multiorbital band structure.

6.4.2 *Incommensurate Magnetic Excitations in the "11" System*

The high-energy spin excitations of $Fe_{1+y}Se_xTe_{1-x}$[115] are mapped out as shown in Fig. 6.12(b). It is clear that the spin excitations can be treated as two incommensurate peaks centered at (0.5,0.5), whose incommensurability, i.e., the distance away from (0.5,0.5), decreases with decreasing energy. At low energies, such incommensurate character of the spin excitations remains,[135,136] as shown in Fig. 6.12(c) and (d). The direction of the incommensurability is the same as the anisotropy in electron-doped "122" materials, as

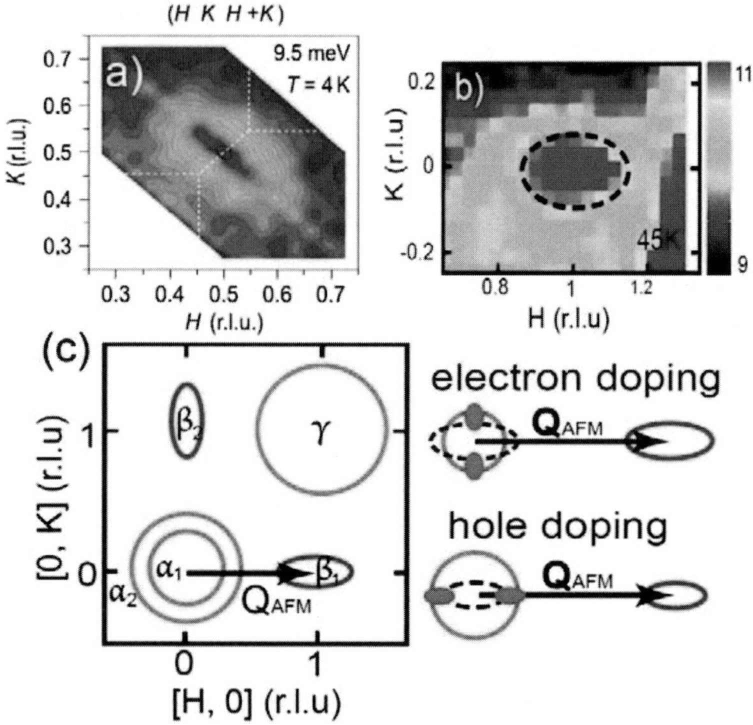

Figure 6.11 (a) The in-plane spin excitations in $BaFe_{1.85}Co_{0.15}As_2$ at 9.5 meV.[169] (b) Similar mapping in $Ba_{0.67}K_{0.33}Fe_2As_2$ integrated from 10 to 18 meV.[131] Note: the authors have used the notion of orthorhombic unit cell (see text). (c) Schematic diagram of Fermi surfaces and nesting in the "122" system.[131]

illustrated in Fig. 6.12(a). Figure 6.12(e) gives the dispersion by combining the high-energy and low-energy data. It is important to note that the incommensurability around the resonance energy shows significant difference below and above T_c.

While the above results illustrate a monotonic dispersion, further measurements at lower energies reveal that the low-energy spin excitations in $Fe_{1+y}Se_xTe_{1-x}$ system actually show an hourglass dispersion irrespective of whether the samples are superconducting or not.[137] The results are shown in Fig. 6.13(d) and (e), where the incommensurability decreases with decreasing energy but starts

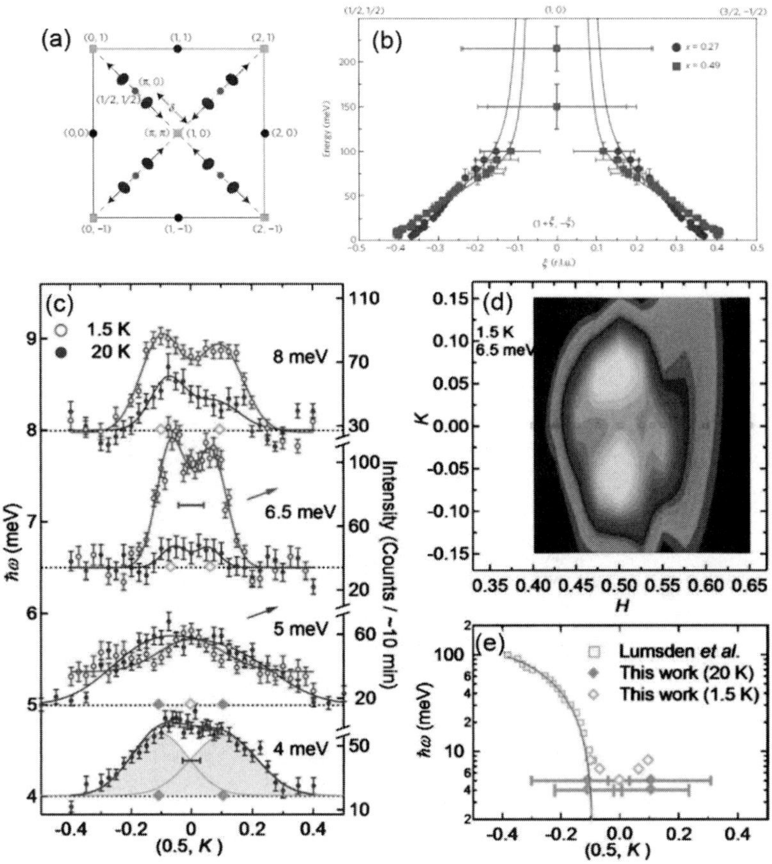

Figure 6.12 (a) Schematic diagram that shows the position of the magnetic excitations in $Fe_{1+y}Se_xTe_{1-x}$.[115] (b) High energy spin excitations for $x = 0.27$ and $x = 0.49$,[115] where $\xi = \delta/\sqrt{2}$. (c) Constant energy scans with several energies at 1.5 K (open red circles) and 20 K (filled blue circles) in $FeSe_{0.5}Te_{0.5}$.[135] The lines are fits to Gaussians. (d) Color contour map of the incommensurate peaks at 6.5 meV in $FeSe_{0.5}Te_{0.5}$. (e) Dispersion of the spin fluctuations in $FeSe_{0.5}Te_{0.5}$.

increasing below ~4.5 meV, an energy that is lower than that of the resonance in superconducting sample. Since such hourglass dispersion is observed in both the superconducting and non-superconducting samples, it should be associated with the normal-state properties. In cuprates, similar hourglass dispersion has been

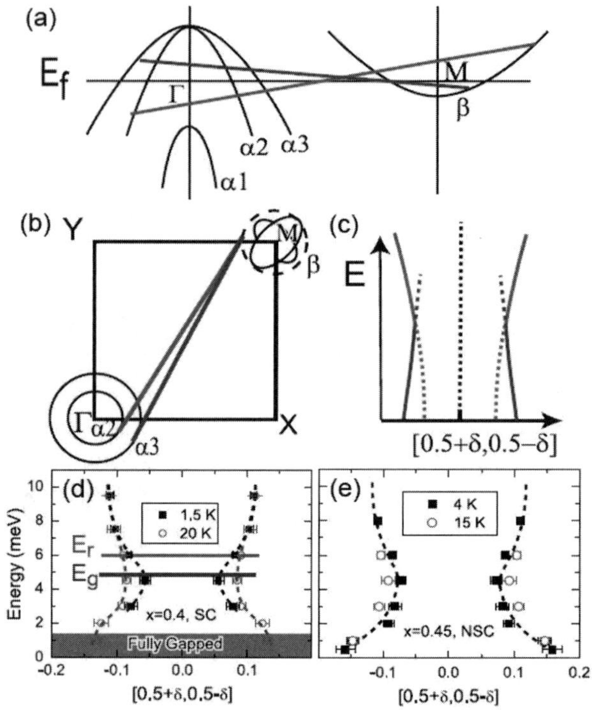

Figure 6.13 (a–c) Schematic diagram of the possible nesting conditions in $Fe_{1+y}Se_xTe_{1-x}$, where the blue and red lines represent two kinds of nesting that may give lower and upper branches of the spin dispersions in (c). (d–e) Hourglass dispersions in superconducting $FeSe_{0.4}Te_{0.6}$ and non-superconducting $FeSe_{0.45}Te_{0.65}$, respectively.[137]

interpreted as either a direct result of stripes phase or nesting upon the *d*-wave gap.[34] In the case of $Fe_{1+y}Se_xTe_{1-x}$, it is suggested that the nesting between two hole bands and one electron band can naturally explain the hourglass dispersion as shown in Fig. 6.13(a–c).[137]

6.5 Conclusion

The rich phenomena of spin fluctuations in the Fe-based superconductors undoubtedly suggest its intimate relationship with the superconductivity. Compared to cuprates, it seems that there are no other competing orders such as stripes[34] or orbital moments[171] that could complicate the spin spectra. From this point of view, the spin fluctuations in the Fe-based superconductors are somehow simpler than those in cuprates, although the multi-band nature may also bring us new phenomena. If the high-T_c superconductivity in both systems comes from the same mechanism, it is likely that similar behaviors should be found in both systems. While we do find many common phenomena, it is still not clear if spin excitations may have enough energy to drive the system into the superconducting state. Experimentally, the neutron scattering experiments need to cover more materials, especially those that show peculiar properties such as LiFeAs and AFe_2Se_2, and more quantitative results as those done in cuprates.[172,173] Only by having a comprehensive experimental results on different materials and entire phase diagram, one can hope to establish a solid base from which future high-T_c superconductivity theories can be derived.[174]

References

1. Y. Kamihara *et al.*, *J. Am. Chem. Soc.* **130**, 3296 (2008).
2. X. H. Chen *et al.*, Nature*(London)* **453**, 761 (2008).
3. G. F. Chen *et al.*, *Phys. Rev. Lett.* **100**, 247002 (2008).
4. Z.-A. Ren *et al.*, *Europhys. Lett.* **83**, 17002 (2008).
5. M. Rotter *et al.*, *Phys. Rev. Lett.* **101**, 107006 (2008).
6. H. H. Wen *et al.*, *Europhys. Lett.* **82**, 17009 (2008).
7. F.-C. Hsu *et al.*, *PNAS* **105**, 14262 (2008).
8. C. de la Cruz *et al.*, *Nature* **453**, 899 (2008).
9. P. A. Lee *et al.*, *Rev. Mod. Phys.* **78**, 17 (2006).
10. A. Damascelli, Z. Hussain and Z. X. Shen, *Rev. Mod. Phys.* **75**, 473 (2003).
11. J. Zhao *et al.*, *Nat. Mat.* **7**, 953 (2008).
12. Q. Huang *et al.*, *Phys. Rev. B* **78**, 054529 (2008).
13. A. J. Drew *et al.*, *Nat. Mat.* **8**, 310 (2009).
14. J. H. Chu *et al.*, *Phys. Rev. B* **79**, 014506 (2009).
15. S. Nandi *et al.*, *Phys. Rev. Lett.* **104**, 057006 (2010).
16. J. H. Chu *et al.*, *Phys. Rev. B* **79**, 014506 (2009).
17. S. Nandi *et al.*, *Phys. Rev. Lett.* **104**, 057006 (2010).

18. P. C. Canfield *et al.*, *Phys. Rev. B* **80**, 060501(R) (2009).
19. N. Ni *et al.*, *Phys. Rev. B* **82**, 024519 (2010).
20. H. Chen *et al.*, *EPL* **85**, 17006 (2009).
21. D. Johrendt *et al.*, *Physica C* **469**, 332 (2009).
22. D. J. Singh, *Physica C* **469**, 418 (2009).
23. H. Ding *et al.*, *Europhys. Lett.* **83**, 47001 (2008).
24. L. Zhao *et al.*, *Chin. Phys. Lett.* **25**, 4402 (2008).
25. L. Boeri *et al.*, *Phys. Rev. Lett.* **101**, 026403 (2008).
26. A. D. Christianson *et al.*, *Phys. Rev. Lett.* **101**, 157004 (2008).
27. I. I. Mazin *et al.*, *Phys. Rev. Lett.* **101**, 057003 (2008).
28. D. Parker *et al.*, *Phys. Rev. B* **78**, 134524 (2008).
29. P. A. Lee *et al.*, *Phys. Rev. B* **78**, 144517 (2008).
30. A. V. Chubukov *et al.*, *Phys. Rev. B* **78**, 134512 (2008).
31. V. Cvetkovic *et al.*, *Europhys. Lett.* **85**, 37002 (2009).
32. F. Wang *et al.*, *Phys. Rev. Lett.* **102**, 047005 (2009).
33. J. Wu *et al.*, *J. Phys.: Condens. Matter* **23**, 094203 (2011).
34. J. M. Tranquada *et al.*, *Handbook of High-Temperature Superconductivity* (Springer, New York, 2007), pp. 257–298.
35. M. Eschrig, *Adv. Phys.* **55**, 47 (2006).
36. J. W. Lynn and P. Dai, *Physica C* **469**, 418 (2009).
37. M. D. Lumsden and A. D. Christianson, *J. Phys: Condens. Matter* **22**, 203203 (2010).
38. D. J. Singh *et al.*, *Phys. Rev. Lett.* **100**, 237003 (2008).
39. Q. Si *et al.*, *Phys. Rev. Lett.* **101**, 076401 (2008).
40. C. Fang *et al.*, *Phys. Rev. B* **77**, 224509 (2008).
41. D. J. Singh, *Physica C* **469**, 418 (2009).
42. T. Yildirim, *Physica C* **469**, 425 (2009).
43. Y. Chen *et al.*, *Phys. Rev. B* **78**, 064515 (2008).
44. Q. Huang *et al.*, *Phys. Rev. Lett.* **101**, 257003 (2008).
45. J. Zhao *et al.*, *Phys. Rev. B* **78**, 140504 (2008).
46. K. Kaneko *et al.*, *Phys. Rev. B* **78**, 212502 (2008).
47. A. I. Goldman *et al.*, *Phys. Rev. B* **78**, 100506 (2008).
48. S. Li *et al.*, *Phys. Rev. B* **80**, 020504 (2009).
49. J. Zhao *et al.*, *Phys. Rev. B* **78**, 132504 (2008).
50. M. J. Pitcher *et al.*, *Chem. Comm.* 5918 (2008).

51. S. Borisenko *et al.*, *Phys. Rev. Lett.* **105**, 067002 (2010).
52. F. L. Pratt *et al.*, *Phys. Rev. B* **79**, 052508 (2009).
53. P. M. R. Brydon *et al.*, *Phys. Rev. B* **83**, 060501 (2011).
54. A. E. Taylor *et al.*, *Phys. Rev. B* **83**, 220514 (R) (2011).
55. M. Wang *et al.*, *Phys. Rev. B* **83**, 220515(R) (2011).
56. H. Ogino *et al.*, *Supercond. Sci. Technol.* **22**, 085001 (2009).
57. X. Zhu *et al.*, *Phys. Rev. B* **79**, 220512(R) (2009).
58. F. Han *et al.*, *Sci. Chi. Ser. G* **53**, 1202 (2010).
59. M. Tegel *et al.*, *Europhys. Lett.* **89**, 37006 (2010).
60. M. Tegel *et al.*, *B-Chem. Sci.* **64B**, 815 (2009).
61. N. Kawaguchi *et al.*, *App. Phys. Exp.* **3**, 063102 (2010).
62. H. Ogino *et al.*, *Appl. Phys. Lett.* **97**, 072506 (2010).
63. H. Ogino *et al.*, *Supercond. Sci. Technol.* **23**, 115005 (2010).
64. M. Tegel *et al.*, *Europhys. Lett.* **89**, 37006 (2010).
65. J. Munevar *et al.*, *Phys. Rev. B* **84**, 024527 (2011).
66. M. Tegel *et al.*, *Phys. Rev. B* **82**, 140507(R) (2010).
67. S. Li *et al.*, *Phys. Rev. B* **79**, 054503 (2009).
68. D. Fruchart *et al.*, *Mater. Res. Bull.* **10**, 169 (1975).
69. W. Bao *et al.*, *Phys. Rev. Lett.* **102**, 247001 (2009).
70. E. E. Rodriguez *et al.*, *Phys. Rev. B* **84**, 064403 (2011).
71. W Bao *et al.*, *Chin. Phys. Lett.* **28**, 086104 (2011).
72. F. Ye *et al.*, *Phys. Rev. Lett.* **107**, 137003 (2011).
73. D. Mou *et al.*, *Phys. Rev. Lett.* **106**, 107001 (2011).
74. L. Zhao *et al.*, *Phys. Rev. B* **83**, 140508(R) (2011).
75. X.-P. Wang *et al.*, *Europhys. Lett.* **93**, 57001 (2011).
76. J. Zhao *et al.*, *Phys. Rev. Lett.* **101**, 167203 (2008).
77. R. A. Ewings *et al.*, *Phys. Rev. B* **78**, 220501 (R) (2008).
78. R. J. McQueeney *et al.*, *Phys. Rev. Lett.* **101**, 227205 (2008).
79. K. Matan *et al.*, *Phys. Rev. B* **79**, 054526 (2009).
80. S. O. Diallo *et al.*, *Phys. Rev. Lett.* **102**, 187206 (2009).
81. J. Zhao *et al.*, *Nat. Phys.* **5**, 555 (2009).
82. B. Keimer *et al.*, *Z. Phys. B* **91**, 373 (1993).
83. J. M. Tranquada *et al.*, *Phys. Rev. B* **40**, 4503 (1989).
84. W. Lv *et al.*, *Phys. Rev. B* **80**, 224506 (2009).
85. L. W. Harriger *et al.*, *Phys. Rev. B* **84**, 054544 (2011).

86. S. O. Diallo *et al.*, *Phys. Rev. B* **81**, 214407 (2010).
87. C. Fang *et al.*, *Phys. Rev. B* **77**, 224509 (2008).
88. C. K. Xu *et al.*, *Phys. Rev. B* **78**, 020501(R) (2008).
89. J. H. Dai *et al.*, *PNAS* **106**, 4118 (2009).
90. O. J. Lipscombe *et al.*, *Phys. Rev. Lett.* **106**, 057004 (2011).
91. C. Stock *et al.*, *Phys. Rev. B* **84**, 045124 (2011).
92. M. Wang *et al.*, *Nat. Commun.* **2**, 580 (2011).
93. K. Seo *et al.*, *Phys. Rev. Lett.* **101**, 206404 (2008).
94. J. Dong *et al.*, *Europhys. Lett.* **83**, 27006 (2008).
95. I. Eremin *et al.*, *Phys. Rev. B* **81**, 024511 (2010).
96. J. Knolle *et al.*, *Phys. Rev. B* **81**, 140506(R) (2010).
97. J. Kang *et al.*, *Phys. Rev. B* **83**, 020505(R) (2011).
98. S. P. Kou *et al.*, *Europhys. Lett.* **88**, 17010 (2009).
99. M. D. Johannes *et al.*, *Phys. Rev. B* **79**, 220510(R) (2009).
100. D. K. Pratt *et al.*, *Phys. Rev. Lett.* **103**, 087001 (2009).
101. A. D. Christianson *et al.*, *Phys. Rev. Lett.* **103**, 087002 (2009).
102. Y. J. Uemura, *Nat. Mat.* **8**, 253 (2009).
103. P. Marsik *et al.*, *Phys. Rev. Lett.* **105**, 057001 (2010).
104. N. Ni *et al.*, *Phys. Rev. B* **78**, 214515 (2008).
105. L. J. Li *et al.*, *New J. Phys.* **11**, 025008 (2009).
106. M. G. Kim *et al.*, *Phys. Rev. B* **83**, 054514 (2011).
107. L. W. Harriger *et al.*, *Phys. Rev. Lett.* **103**, 087005 (2009).
108. D. K. Pratt *et al.*, *Phys. Rev. Lett.* **106**, 257001 (2011).
109. B. J. Sternlieb *et al.*, *Phys. Rev. B* **50**, 16438 (1994).
110. R. S. Eccleston *et al.*, *J. Phys.: Condens. Matter* **8**, 7837 (1996).
111. N. Katayama *et al.*, *J. Phys. Soc. Jpn.* **79**, 113702 (2010).
112. C. Dong *et al.*, arXiv:1012.5188.
113. Z. Xu *et al.*, *Phys. Rev. B* **82**, 104525 (2010).
114. T. J. Liu *et al.*, *Nat. Mat.* **9**, 718 (2010).
115. M. D. Lumsden *et al.*, *Nat. Phys.* **6**, 182 (2010).
116. D. K. Pratt *et al.*, *Phys. Rev. B* **79**, 060510(R) (2009).
117. C. de la Cruz *et al.*, *Phys. Rev. Lett.* **104**, 017204 (2010).
118. Y. Luo *et al.*, *Phys. Rev. B* **81**, 134422 (2010).
119. Y. Nakai *et al.*, *Phys. Rev. Lett.* **105**, 107003 (2010).
120. K. Marty *et al.*, *Phys. Rev. B* **83**, 060509(R) (2011).

121. M. G. Kim *et al.*, *Phys. Rev. B* **82**, 220503(R) (2010).
122. S. D. Wilson *et al.*, *Nature* **442**, 59 (2006).
123. C. Stock *et al.*, *Phys. Rev. Lett.* **100**, 087001 (2008).
124. A. D. Christianson *et al.*, *Nature* **456**, 930 (2008).
125. D. S. Inosov *et al.*, *Nat. Phys.* **6**, 178 (2010).
126. S. Li *et al.*, *Phys. Rev. B* **79**, 174527 (2009).
127. S. Chi *et al.*, *Phys. Rev. Lett.* **102**, 107006 (2009).
128. M. Wang *et al.*, *Phys. Rev. B* **81**, 174524 (2010).
129. M. D. Lumsden *et al.*, *Phys. Rev. Lett.* **102**, 107005 (2009).
130. D. K. Pratt *et al.*, *Phys. Rev. B* **81**, 140510(R) (2010).
131. C. Zhang *et al.*, *Sci. Rep.* **1**, 115 (2011).
132. Y. Qiu *et al.*, *Phys. Rev. Lett.* **103**, 067008 (2009).
133. H. A. Mook *et al.*, arXiv:0904.2178.
134. H. A. Mook *et al.*, *Phys. Rev. Lett.* **104**, 187002 (2010).
135. S. –H. Lee *et al.*, *Phys. Rev. B* **81**, 220502 (R) (2010).
136. D. N. Argyriou *et al.*, *Phys. Rev. B* **81**, 220503(R) (2010).
137. S. Li *et al.*, *Phys. Rev. Lett.* **105**, 157002 (2010).
138. S. Shamoto *et al.*, *Phys. Rev. B* **82**, 172508 (2010).
139. A. Mourachkine, *J. Low Temp. Phys.* **117**, 401 (1999).
140. T. A. Maier *et al.*, *Phys. Rev. B* **78**, 020514(R) (2008).
141. M. Korshunov *et al.*, *Phys. Rev. B* **78**, 140509(R) (2008).
142. S. Onari *et al.*, *Phys. Rev. B* **81**, 060504(R) (2010).
143. Y. Nagai *et al.*, *Phys. Rev. B* **83**, 220516(R) (2011).
144. T. A. Maier *et al.*, *Phys. Rev. B* **83**, 220505(R) (2011).
145. J. E. Lorenzo *et al.*, *Phys. Rev. B* **75**, 054418 (2007).
146. O. J. Lipscombe *et al.*, *Phys. Rev. B* **82**, 064515 (2010).
147. P. Dai *et al.*, *Nature* **406**, 965 (2000).
148. J. M. Tranquada *et al.*, *Phys. Rev. B* **69**, 174507 (2004).
149. S. D. Wilson *et al.*, *PNAS* **104**, 15259 (2007).
150. J. Zhao *et al.*, *Phys. Rev. B* **81**, 180505 (2010).
151. S. Li *et al.*, *Phys. Rev. B* **84**, 024518 (2011).
152. M. Wang *et al.*, *Phys. Rev. B* **83**, 094516 (2011).
153. J. Wen *et al.*, *Phys. Rev. B* **81**, 100513 (R) (2010).
154. W. Bao *et al.*, arXiv:1002.1617.
155. P. Dai *et al.*, *Science* **284**, 1344 (1999).

156. J. Zhao *et al., Phys. Rev. Lett.* **99**, 017001 (2007).
157. P. Babkevich *et al., Phys. Rev. B* **83**, 180506(R) (2011).
158. H. A. Mook *et al., Phys. Rev. Lett.* **70**, 3490 (1993).
159. H. F. Fong, *et al., Phys. Rev. B* **54**, 6708 (1996).
160. P. Dai *et al., Nature* **406**, 956 (2000).
161. C. G. Shull *et al., Phys. Rev. Lett.* **16**, 513 (1966).
162. C. Stassis *et al., Phys. Rev. B* **34**, 4382 (1986).
163. J. X. Boucherle *et al., Physica B* **192**, 25 (1993).
164. J. A. Duffy *et al., Phys. Rev. Lett.* **85**, 5412 (2000).
165. C. Lester *et al., Phys. Rev. B* **84**, 134514 (2011).
166. P. J. Brown *et al., Phys. Rev. B* **82**, 024421 (2010).
167. C. Lester *et al., Phys. Rev. B* **81**, 064505 (2010).
168. H.-F. Li *et al., Phys. Rev. B* **82**, 140503(R) (2010).
169. J. T. Park *et al., Phys. Rev. B* **82**, 134503 (2010).
170. C. H. Lee *et al., Phys. Rev. Lett.* **106**, 067003 (2011).
171. P. Bourges *et al.*, arXiv:1101.1786.
172. H. Woo *et al., Nat. Phys.* **2**, 600 (2006).
173. T. Dahm *et al., Nat. Phys.* **5**, 217 (2009).
174. A. Barth *et al., J. Supercond. Nov. Magn.*, **21**, 113 (2008).

Chapter 7

Review of NMR Studies on Iron-Based Superconductors

Kenji Ishida[a,b] and Yusuke Nakai[a,b,*]

[a] *Department of Physics, Graduate School of Science, Kyoto University, Kitashirakawa-Oiwakecho, Sakyo, Kyoto 606-8502, Japan*
[b] *TRIP (Transformative Research-Project on Iron Pnictides), JST (Japan Science and Technology Agency), Sanban-cho bldg., 5, Sanban-cho, Chiyoda, Tokyo 102-0075, Japan*
*nakai@scphys.kyoto-u.ac.jp, kishida@scphys.kyoto-u.ac.jp

This review presents summary of NMR experiments performed on various iron-based superconductors. The authors introduce NMR experiments which reveal magnetic properties in the normal and superconducting states. In the superconducting state, NMR measurements clarified that iron-based superconductors form spin-singlet pairing, although the orbital symmetry is controversial in some cases. From the comparison of the NMR results in various iron-based superconductors, the relationship between superconductivity and magnetic fluctuations probed with NMR is discussed.

7.1 Introduction

The discovery of iron oxypnictide superconductor LaFeAs($O_{1-x}F_x$) with T_c = 26 K[1] surprised physicists and chemists studying

Iron-Based Superconductors: Materials, Properties, and Mechanisms
Edited by Nan-Lin Wang, Hideo Hosono, and Pengcheng Dai

ISBN 978-981-4303-22-4 (Hardcover), 978-981-4303-23-1 (eBook)
www.panstanford.com

superconductivity all over the world. This is because Fe metal is a most familiar ferromagnet and the Fe site of LaFeAs($O_{1-x}F_x$) is thought to play an important role for such an unexpectedly high T_c. Shortly after this discovery, application of pressure increases T_c in LaFeAs($O_{1-x}F_x$) even further to 43 K,[2] and the replacement of the nonmagnetic La by magnetic rare earth elements enhances T_c up to 56.3 K, which is the current record of the iron-based superconductor.[3–6] This extraordinary high-T_c attained by iron-based superconductors has demonstrated that high-T_c superconductivity is more general than previously thought, and thus the investigation of physical properties on the iron-based superconductors is important in order to understand the mechanism of high-T_c superconductivity.

In this chapter, we review NMR results obtained on various iron-based superconductors, particularly focus on normal-state spin dynamics, and superconducting-gap structure suggested by NMR results. Before showing detailed NMR results on iron-based superconductors, we briefly introduce how NMR gives valuable information about static and dynamic properties of electronic spins in compounds.

7.2 NMR Basics

7.2.1 *NMR Hamiltonian*[7–10]

Nuclear spins in condensed matter are weakly coupled with an external bath, usually electronic spins in metallic compounds. This weak coupling gives rise to small perturbations of the nuclear-spin Hamiltonian, and induces the nuclear-spin relaxation as well as the decoherence of the nuclear-spin system. In other words, thanks to this coupling, nuclear-spin can provide valuable information about the electronic degree of freedom in compounds. The nuclear-spin Hamiltonian in correlated systems is expressed as the sum of four interactions:

$$\mathscr{H} = \mathscr{H}_{\text{Zeeman}} + \mathscr{H}_{\text{eQ}} + \mathscr{H}_{\text{dip}} + \mathscr{H}_{\text{e-n}}. \tag{7.1}$$

The first term describes the Zeeman interaction, $\mathscr{H}_{\text{Zeeman}} = -\gamma_n \hbar \boldsymbol{I} \cdot \boldsymbol{H_0}$, here γ_n, $\boldsymbol{I}$ and $\boldsymbol{H_0}$ are the gyromagnetic ratio, the nuclear spin and the local magnetic field, respectively. This term is usually dominant for nuclei in a strong external field. The second term is the electric quadrupole interaction, which can be significant for nuclei with $I > 1/2$ when those are located in the non-cubic symmetry, but vanishes for $I = 1/2$. This interaction is given by

$$\mathscr{H}_{\text{eQ}} = \frac{eQV_{zz}}{4I(2I-1)}\left[(3I_z^2 - \boldsymbol{I}^2) + \eta(I_x^2 - I_y^2)\right]. \tag{7.2}$$

where e is the electron charge, Q is the quadrupole moment, $V_{\alpha\beta}$ is the component of the electric field gradient (EFG) tensor, and $\eta = (V_{xx} - V_{yy})/V_{zz}$ is the asymmetry parameter of the EFG. When $I = 3/2$, corresponding to As nucleus, the NQR frequency in zero field is defined as $\nu_Q = \nu_z\sqrt{1+\eta^2/3}$ with $\nu_z = eQV_{zz}/2h$. For an axial symmetric site, $\eta = 0$, and there are $I - 1/2$ quadrupolar resonances at frequencies $n\nu_z$ with $n = 1, \ldots I - 1/2$. If $\eta > 0$, the separation of the resonance frequencies is no longer equal. The EFG interactions are characterized by ν_z, η and the direction of the principal axis of the EFG interaction. In practice, this means that the degeneracy of the nuclear spin manifold is lifted even in zero field, and the spin resonance can be observed even in zero field. This is the nuclear quadrupole resonance (NQR), which is in contrast for NMR observed in magnetic field. NQR measurements are particularly important for superconductors, since the nuclear spin-lattice relaxation can be measured in zero field, without the influence of vortices.

The third interaction is the dipolar interaction between nuclear spins in a lattice, and is given by

$$\mathscr{H}_{\text{dip}} = -\frac{1}{2}\sum_{j,k=1}^{N}\left[\frac{\boldsymbol{\mu_j}\cdot\boldsymbol{\mu_i}}{r_{j,k}^3} - \frac{3(\boldsymbol{\mu_j}\cdot\mathbf{r}_{j,k})(\boldsymbol{\mu_k}\cdot\mathbf{r}_{j,k})}{r_{i,j}^5}\right]. \tag{7.3}$$

Here, $\mathbf{r}_{j,k}$ is the distance between the nuclei moments at the j and k sites with moments $\mu = \gamma\hbar\boldsymbol{I}$. In most solids the dipolar interaction gives rise to a static broadening of the lineshape, and in most cases the broadening is temperature independent and is typically estimated as a few G depending on the structural details of the particular compound.

The fourth term, consisting of following three terms, is the magnetic part of the nuclear-electron interaction and is given by

$$\mathscr{H}_{\text{e-n}} = \gamma\hbar\boldsymbol{I}\sum_i(2\mu_B)\left[\frac{\boldsymbol{l}_i}{r_i^3} - \left(\frac{\boldsymbol{S}_i}{r_i^3} - 3\frac{\mathbf{r}_i(\boldsymbol{S}_i\cdot\mathbf{r}_i)}{r_i^5}\right) + \frac{8}{3}\pi\boldsymbol{S}_i\delta(\mathbf{r}_i)\right] \tag{7.4}$$

where $\mathbf{S}_i$ and $\mathbf{l}_i$ are the spin and angular momentum with respect to the nucleus of the ith electron. The first term is the orbital interaction, which gives rise to the orbital (chemical) shift in materials where orbital angular momentum is unquenched. The second and third terms are the dipolar and contact hyperfine interaction, which are usually the dominant interaction for resonance shift and relaxation phenomena in metals and superconductors. All terms produce additional fields at the nuclear site. In strongly correlated electron superconductors, hyperfine-field interaction is the primary interaction between nuclei and quasiparticles, from which superconducting gap properties can be known. In addition, from analyses of the NMR spectrum in the superconducting state, we can know the penetration depth λ and vortex-lattice structure since the magnetization produced by Meissner current in superconductors creates a specific resonance-peak pattern depending on the vortex lattice.

In most cases, the hyperfine interaction can be expressed by,

$$\mathscr{H}_{\rm hf} = \boldsymbol{I} \cdot \mathbf{A} \cdot \boldsymbol{S}(\mathbf{r} = 0) + \sum_i \mathbf{B}_i \boldsymbol{I} \cdot \boldsymbol{S}(\mathbf{r}_i), \tag{7.5}$$

where $\mathbf{A}$ is an on-site hyperfine coupling constant coupling typically dominated by direct coupling and core-polarization effects by s-orbits, and $\mathbf{B}_i$ is a transferred hyperfine coupling from the neighboring spin $\mathbf{S}(\mathbf{r}_i)$. In the NMR studies on cuprate and iron-based superconductors, the latter coupling is very important, as described later for iron-based superconductors.

7.2.2 *Knight Shift and Nuclear Spin-Lattice Relaxation Rate in Metals*

In a simple metallic compound, the dominant hyperfine interaction is the on-site direct coupling between the nuclear spins and the quasiparticle spins, which is assumed to be isotropic and is given with a constant value A, as is shown in Fig. 7.1. The onsite hyperfine coupling interaction is given by,

$$\mathscr{H}_{\rm hf} = A\boldsymbol{I} \cdot \boldsymbol{S} = AI_zS_z + \frac{1}{2}A\left(I_+S_- + I_-S_+\right) \tag{7.6}$$

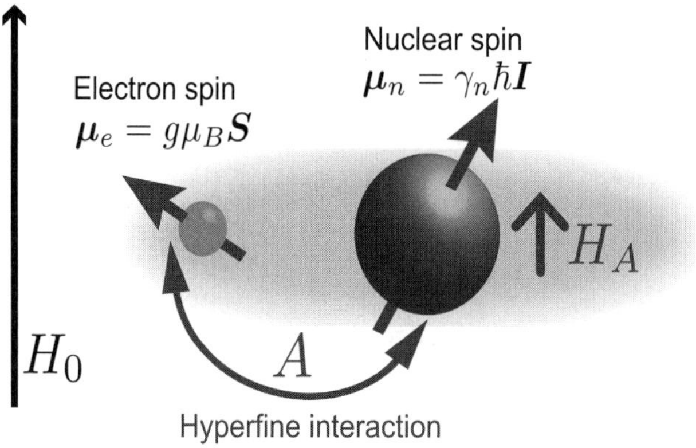

Figure 7.1 Hyperfine coupling between nuclear spin $\boldsymbol{I}$ and electron spin $\boldsymbol{S}$ under magnetic field $\boldsymbol{H}_0$. The hyperfine coupling constant is A. The nuclear spin experiences the additional field produced by electron spins.

The diagonal term AI_zS_z gives rise to a shift of the resonance frequency, which is known as Knight shift (K). K is defined as

$$K \equiv \left(\frac{\omega_{\text{res}} - \omega_0}{\omega_0}\right)_{H_0=const.} = \left(\frac{H_0 - H_{\text{res}}}{H_{\text{res}}}\right)_{\omega_0=const.} \tag{7.7}$$

$$= \frac{A\langle S\rangle}{\gamma_n \hbar\, H_{\text{res}}} = \frac{A}{\gamma_n \hbar \mu_B N_A}\chi \equiv \frac{A_{\text{hf}}}{\mu_B N_A}\chi, \tag{7.8}$$

where ω_0 is the resonance frequency of an isolated nucleus ($\omega_0 = \gamma_n H_0$), and ω_{res} and H_{res} are the resonant frequency and magnetic field of a metallic compound we measured at H_0 and ω_0, respectively. Thermal average of the spin component $\langle S\rangle$ induced by magnetic field is related to the magnetic susceptibility χ with the form of $\chi \equiv M/H = N_A\mu_B\langle S\rangle/H$. Here, N_A is the Avogadro's number. For the unified unit, we usually use the hyperfine coupling constant A_{hf}, which is defined as $A_{\text{hf}} = A/\gamma_n\hbar$. In a conventional metal, χ is given by the Pauli susceptibility, $\chi_{\text{Pauli}} = \mu_B^2 N(E_F)N_A$ with density of states at Fermi energy $N(E_F)$, and thus $K = A_{\text{hf}}\mu_B N(E_F)$ is temperature independent.

The off diagonal terms of the hyperfine coupling can induce a spin-flip exchange between the electronic spin and the nuclear spin. This process does not shift the resonance frequency, but is related to the dynamics of the electronic spin. In general, electron spins scatter from one nucleus to another, but maintain the same energy, since the nuclear Zeeman energy is orders of magnitude of lower than the Fermi energy, and the electron spins within $k_B T$ of E_F can flip by absorbing the nuclear Zeeman energy. By using time-dependent perturbation theory, the spin-lattice relaxation rate $1/T_1$ can be written as:

$$\frac{1}{T_1} = \gamma_n^2 A_{\mathrm{hf}}^2 \int_0^{\infty} \langle S_+(t) S_-(0) \rangle \, e^{i\omega_0 t} dt, \tag{7.9}$$

where $\langle S_+(t)S_-(0)\rangle$ indicates a thermal averaged correlation function of the spin dynamics. In a conventional metal, we consider the transition from an initial electron state |k ↑> to a final state |k' ↓>, where both the states are on the Fermi surface. $1/T_1$ in this case can be written as

$$\frac{1}{T_1} = \frac{\gamma_n^2 A_{\mathrm{hf}}^2}{2} \int_0^{\infty} N(E_i) N(E_f) f(E_i)(1 - f(E_f)) dE_i, \tag{7.10}$$

where $E_f - E_i = \hbar\omega_0$, and $f(E)$ is the Fermi distribution function. Since $f(E)(1 - f(E)) \approx k_B T \delta(E - E_F)$, we find that $1/T_1 \sim A_{\mathrm{hf}}^2 N(E_F)^2 k_B T$, and thus the measurement of T_1 give information about $N(E_F)$. From the relationship between K and $1/T_1$, it is obvious that $T_1 T K^2$ is a constant, which is known as the Korringa relation. The Korringa constant is expressed as, $(T_1 T K^2)_0 = \mu_B^2/\pi k_B \hbar \gamma_n^2 \equiv S_0$, which is valid for non-interacting systems. In a realistic compound, the ratio between S_0 and observed $T_1 T K^2$, $R \equiv S_0/T_1 T K^2$ provides important information about magnetic fluctuations as discussed later.

Toru Moriya showed that the expression (7.9) can be rewritten with the dynamical susceptibility $\chi(\boldsymbol{q}, \omega)$.[11] By employing the fluctuation-dissipation theorem, $1/T_1$ can be described as,

$$\frac{1}{T_1} = \frac{\gamma_n^2 k_B T}{\mu_B^2} \lim_{\omega \to 0} \sum_{\boldsymbol{q}} A(\boldsymbol{q})^2 \frac{\chi''(\boldsymbol{q}, \omega)}{\omega}, \tag{7.11}$$

where $A(\boldsymbol{q})$ is the Fourier transform of the hyperfine interaction, and the sum of $\boldsymbol{q}$ is taken over the first Brillouin zone. $A(\boldsymbol{q})$ is $\boldsymbol{q}$ independent when on-site contact type interactions are dominant, but $A(\boldsymbol{q})$ is $\boldsymbol{q}$-dependent and shows a peak or vanishes at a certain $\boldsymbol{q}$ when the transferred hyperfine coupling between neighboring sites are present.

The self-consistent renormalization (SCR) theory developed by Moriya and coworkers has been employed to analyze NMR results

Table 7.1 Temperature dependence $1/T_1$ and electrical resistivity ρ anticipated from the SCR theory

	FM ($Q = 0$)		AFM ($Q \neq 0$)	
	3D	2D	3D	2D
$1/T_1 \propto$	$T/(T - T_C)$	$T/(T - T_C)^{3/2}$	$T/(T - T_N)^{1/2}$	$T/(T - T_N)$
$\rho \propto$	$T^{5/3}$	$T^{4/3}$	$T^{3/2}$	T

obtained in a metallic compound with magnetic instability.[12] In the theory, the dynamical susceptibility is expanded in the $\boldsymbol{q}$ space around the magnetic-ordering wave vector $\boldsymbol{Q}$, and is assumed to be expressed as

$$\frac{1}{\chi(\boldsymbol{Q} + \boldsymbol{q}, \omega)} = \frac{1}{\chi(\boldsymbol{Q})} + A\boldsymbol{q}^2 + iC\frac{\omega}{q^{\theta}}, \tag{7.12}$$

where $\theta = 1(0)$ when the magnetic ordering is ferromagnetism $\boldsymbol{Q} = 0$ (antiferromagnetism $\boldsymbol{Q} \neq 0$). The parameters of A and C are determined self-consistently to minimize the free energy. The SCR theory predicts various physical quantities consistently when the electronic system is close to a magnetic instability. The temperature dependence of various quantities above the Curie temperature T_C and Néel temperature T_N is described using the renormalized dynamical susceptibility, and depends on the dimensionality and θ. The temperature dependence of $1/T_1$ and resistivity ρ anticipated from the SCR theory is listed in Table 7.1.

7.2.3 *Knight Shift and Nuclear Spin-Lattice Relaxation Rate in the Superconducting State*[13]

In the superconducting state, the Knight-shift behavior determined by the spin susceptibility critically depends on the spin symmetry of the Cooper pairs, which can be either even parity (spin singlet) pairing with net spin $S = 0$ or odd parity (spin triplet) pairing with spin $S = 1$. For spin-singlet pairing, since there is no coupling between nuclei and the Cooper pairs, Knight shift vanishes at $T = 0$ and thus is determined by the thermally excited quasiparticles. The spin susceptibility of the superconducting state χ_{SC} of a spin-singlet superconductor is expressed as:

$$\chi_{SC}(T) = -4\mu_B^2 \int_0^\infty N_S(E) \frac{df(E)}{dE} dE, \qquad (7.13)$$

where $N_S(E)$ is the density of states of quasiparticles in the superconducting state, which is described with the density of states in the normal state $N(E_F)$ and the superconducting gap $\Delta(\theta, \phi)$ as

$$N_S(E) = N(E_F) \frac{E}{\sqrt{E^2 - \Delta(\theta, \phi)^2}}. \qquad (7.14)$$

The Knight shift vanishes as $T \to 0$, irrespective of the applied field direction, and its temperature dependence follows a Yosida function.[14]

For spin-triplet pairing, the $S = 1$ pairs do couple to the nuclear spins, and the Knight-shift behavior below T_c depends strongly on the orientation of the appled magnetic field with respect to the spin of the Cooper pairs. In spin-triplet superconductivity, the direction of the Cooper-pair spin is defined with $\boldsymbol{d}$ vector perpendicular to a plane in which the spin expectation value remains finite. When the $\boldsymbol{d}$ vector is tightly pinned to the crystalline axis due to the spin-orbit coupling, the Knight shift perpendicular to the $\boldsymbol{d}$ vector remain unchanged and the Knight shift parallel to the $\boldsymbol{d}$ vector vanishes following the Yosida function in principle. Indeed unchanged Knight shift behavior was reported in a few superconductors, such as UPt_3[15] and Sr_2RuO_4,[16,17] which are considered as spin-triplet superconductors.

$1/T_1$ in the superconducting state is also determined by thermally excited quasiparticles, since there are no fluctuations of the hyperfine fields that couple to the nuclei in a fully gapped superconductor at $T = 0$. The hyperfine-field fluctuations experienced by the nuclei in the superconducting state are different from those in the normal state, since the excitation of a BCS state cannot be interpreted by the scattering between up-spin and down-spin electrons, but the scattering between unpaired quasiparticles through the exchange between Cooper pairs. This scattering process has profound consequence for $1/T_1$ below T_c, and gives rise to the Hebel-Slichter coherence peak just below T_c in a conventional s-wave superconductor.[18]

Here the superconducting coherence effect, which is important for understanding $1/T_1$ behavior just below T_c in iron-based superconductors, is introduced. As mentioned above, the contribution of unpaired qasiparticles is not enough in the spin scattering process in the superconducting state, but the contribution of the Cooper pairs should be included. An initial state $|i\rangle$ is assumed to consist of an unpaired up-spin quasiparticle with wave number k and one Cooper pair with k'. After spin-scattering between the quasipartice and nuclear spins, the wave numbers are exchanged, and the final state $|f\rangle$ is described as a down-spin quasiparticle with $-k'$ and a Cooper pair with k by preserving energy. The initial and final states are expressed with the creation and annihilation operators c_k^* and c_k, and the probability of the presence of a Cooper pair with k h_k as

$$|i\rangle = \left[\sqrt{1-h_{k'}} + \sqrt{h_{k'}}c_{k'\uparrow}^* c_{-k'\downarrow}^*\right] c_{k\uparrow}^* |\psi_0\rangle \tag{7.15}$$

$$|f\rangle = \left[\sqrt{1-h_k} + \sqrt{h_k}c_{k\uparrow}^* c_{-k\downarrow}^*\right] c_{-k'\downarrow}^* |\psi_0\rangle, \tag{7.16}$$

where ψ_0 represents the vacuum state, and h_k is determined as $h_k = (1-\varepsilon_k/E_k)/2$ with $E_k = \sqrt{\varepsilon_k^2 + \Delta_k^2}$. The spin-scattering Hamiltonian is described with creation and annihilation operators as

$$\mathscr{H}_{\rm hf} = A\sum_{k,k'}\frac{1}{2}\left(I_+ c_{k'\downarrow}^* c_{k\uparrow} + I_- c_{k'\uparrow}^* c_{k\downarrow}\right). \tag{7.17}$$

The transition probability from $|i\rangle$ to $|f\rangle$ induced by $\mathscr{H}_{\rm hf}$ is derived as,

$$|\langle f|\mathscr{H}_{\rm hf}|i\rangle|^2 \propto \frac{1}{2}\left(\sqrt{1-h_k}\sqrt{1-h_{k'}} + \sqrt{h_k}\sqrt{h_{k'}}\right)^2 \tag{7.18}$$

$$\propto \frac{1}{2}\left(1 + \frac{\varepsilon_k\varepsilon_k'}{E_k E_{k'}} + \frac{\Delta_k\Delta_{k'}}{E_k E_{k'}}\right) \tag{7.19}$$

The second term of the above equation is cancelled out when this term is integrated over the Fermi surface, because ε_k is the energy measured from the Fermi energy. Since the reverse scattering process (scattering from the down-spin quasiparticle to up-spin quasiparticle) is also present, the transition probability becomes twice. By taking the superconducting coherence factor into account, $1/T_1$ in the superconducting state is described as

$$\left(\frac{1}{T_1}\right)_{\rm S} \propto A^2\int\left(1+\frac{\Delta^2}{EE'}\right)N_{\rm S}(E)f(E)N_{\rm S}(E')\left(1-f(E')\right)dE\,dE' \tag{7.20}$$

$$\propto A^2 \int \left(N_S(E)^2 + M_S(E)^2\right) f(E)\,(1 - f(E))\,dE. \quad (7.21)$$

Here, M_S is called as the "anomalous" density of state related with the superconducting coherence effect, which is given by

$$M_S = \int_{\Omega=F.S.} N \frac{\Delta}{\sqrt{E^2 - \Delta^2}} d\Omega, \quad (7.22)$$

where the integration is taken over the Fermi surfaces. It should be noted that the (Hebel-Slichter) coherence peak of $1/T_1$ just below T_c, which is observed in a conventional s-wave superconductor, originates from not only the singularity of $N_S(E)$ at $E = \Delta$, but also from the presence of M_S due to the coherence factor. In contrast, M_S vanishes in unconventional superconductors since the integration of Δ over the Fermi surfaces becomes zero due to a sign-change superconducting gap. This is also the case in the $s_{\pm}$-wave symmetry, where the sign of the superconducting gap is opposite between different Fermi surfaces. In addition, since the singularity of $N_S(E)$ in unconventional superconductors at $E = \Delta$ is much weaker than that in s-wave superconductors, the coherence peak observed in $1/T_1$ just below T_c is much smaller or absent in a non s-wave superconductor.

Temperature dependence of Knight shift and $1/T_1$ far below T_c, typically $T \leq T_c/3$ is determined by the superconducting-gap structure, since the gap shows temperature dependence down to $T_c/2$. Most of unconventional superconductors possesses the angle-dependent superconducting gap $\Delta(\theta, \phi)$ with respect to their Fermi surface, and N_S is calculated by the integration over the Fermi surface. The angle dependence of $\Delta(\theta, \phi)$ in typical unconventional superconductors together with an s-wave superconductor with an isotropic gap is listed, and $N_S(E)$ in a corresponding superconducting state is represented in Fig. 7.2.

$\Delta(\theta, \phi) = \Delta_0$	for s-wave,
$\Delta(\theta, \phi) = \Delta_0 \sin\theta e^{-i\phi}$	for p-wave axial state with point nodes,
$\Delta(\theta, \phi) = \Delta_0 \cos\theta$	for p-wave polar state with a line node,
$\Delta(\theta, \phi) = \Delta_0 \cos 2\phi$	for two-dimensional d-wave with line nodes.

In unconventional superconductors, gapless quasiparticle excitations are possible at gap nodes, and thus quasiparticle states

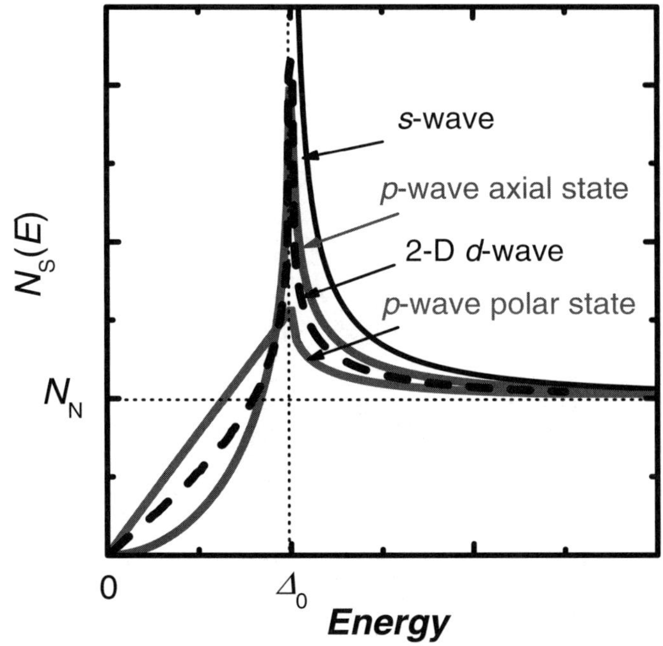

Figure 7.2 Quasiparticle density of state in various superconducting states. Finite $N_S(E)$ exists at energies lower than the maximum gap size, when superconducting gap has nodes. The energy dependence of $N_S(E)$ at low energy determines the exponents for the power-law temperature dependence of physical quantities.

are present within Δ_0, which is in contrast with an s-wave superconductor. The energy dependence of $N_S(E)$ near E_F (e.g. $N_S \propto E(E^2)$ when a superconducting gap has a line (point) node) determines the low-temperature dependence of $1/T_1$, which is calculated as

$$\begin{aligned} 1/T_1 &\propto \int_0^\infty E^n f(E)[1 - f(E)]dE \\ &\propto T^{n+1} \int_0^\infty (E/T)^n f(E/T)[1 - f(E/T)]d(E/T) \propto T^{n+1} \end{aligned}$$

and is listed below:

$$1/T_1 \propto \begin{cases} \exp(-\Delta/T) & \text{for a finite gap} \\ T^3 & \text{for a line-node gap} \\ T^5 & \text{for a point-node gap} \end{cases} .$$

The temperature dependence of $1/T_1$ calculated with above $N_S(E)$,

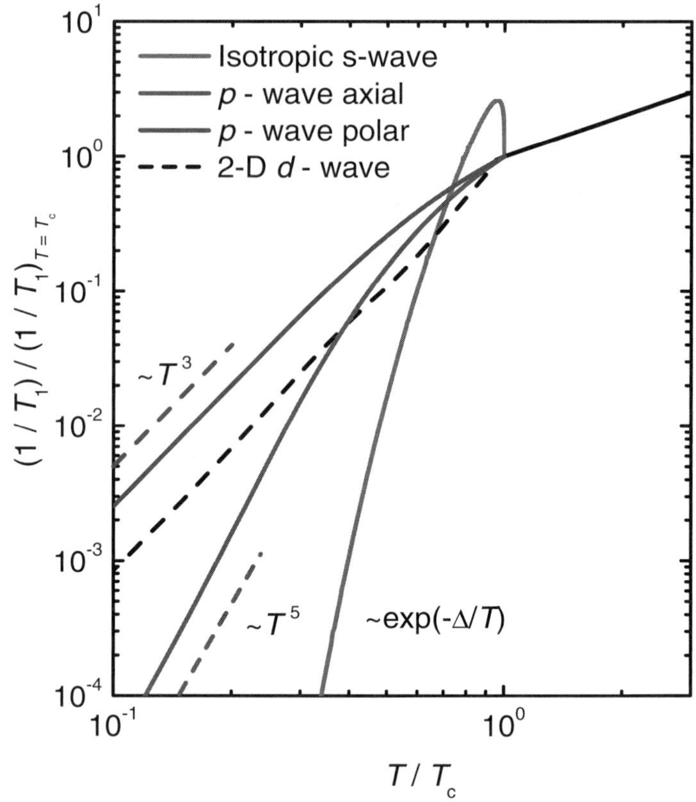

Figure 7.3 Temperature dependence of $1/T_1$ in the superconducting state, which is calculated with $N_S(E)$ shown in Fig. 7.2. In the calculation, the magnitude of the superconducting gap is taken with $\Delta_0 = 2k_B T_c$, and the temperature dependence of the superconducting gap with the BCS model is adoped.

which is renormalized by T_c and by $1/T_1$ at T_c, is shown in Fig. 7.3. $1/T_1$ measurement at low temperatures can help to distinguish between conventional *s*-wave paring and unconventional pairing states. However, it should be kept in mind that, for $1/T_1$ measurements by NMR, vortex motions and additional relaxation processes at the normal vortex cores generally make the interpretation of $1/T_1$ a highly complicated affair well below T_c. In addition, since it was reported that external fields suppress the Hebel-Slichter peaks in some cases,[19,20] external-field depenence of $1/T_1$ and/or $1/T_1$ measurements by NQR are important to know intrinsic $1/T_1$ behavior at low temperatures.

In unconventional superconductors, nodal structure is quite sensitive to the impurity scattering which mixes k vectors. This scattering is induced by nonmagnetic impurities and/or crystal imperfection and dislocation observed in less quality samples. The scattering modifies the energy dependence of $N_S(E)$ near E_F, and resultantly induces the residual density of state at E_F.[21] This changes the temperature dependence of various physical properties at low temperatures,[22] which has been actually observed in various unconventional superconductors.[23] Therefore, nonmagnetic impurity effect against superconductors is one of the most crucial tests to classify unconventional superconductors.

7.3 NMR Experimental Results on Iron-Based Superconductors

It is well-known that there are four main crystal structures in iron-based superconductors, which are shown in Fig. 7.4. Recently, Ogino and coworkers succeeded in growing new iron-based superconductors with thick blocking layers formed by perovskite-type oxides.[25] The crystal structures of this new compounds are expressed as $(M_2Pn_2)(AE_{n+2}M'_nO_{3n})$ or $(M_2Pn_2)(AE_{n+1}M'_nO_{3n-1})$ where M = Fe or Ni, Pn = P or As, AE = Ca, Sr, Ba, and M' = Sr, V, Cr, (Mg, Ti) *etc.* with $n = 2, 3, \cdots$. Most of these compounds with M = Fe exhibit superconductivity with T_c's of 35 $\sim$ 47K. In this paper, we focus on the main four crystal structures.

7.3.1 *$LaFeAs(O_{1-x}F_x)$ and $LaFeAsO_{1-\delta}$ with "1111" Structure*

Superconductivity in $LaFeAs(O_{1-x}F_x)$ emerges at around $x = 0.04$,[1] and oxygen-deficiency induces superconductivity in $LaFeAsO_{1-\delta}$,[6] where maximum T_c is reported at $x \sim 0.1$ and $\delta \sim 0.35$, respectively. The F-doping and oxygen deficiency, both corresponding to electron doping, suppresses the nesting between the hole and electron

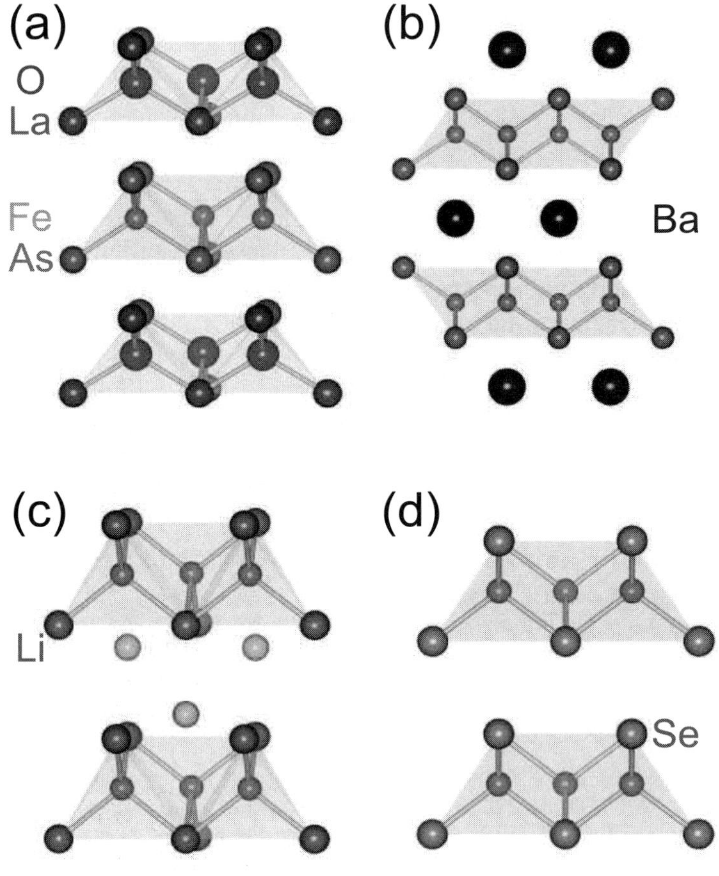

Figure 7.4 Crystal structures of (a) LaFeAsO (refered to as "1111" in the article), (b) BaFe2As2 ("122"), (c) LiFeAs ("111"), and (d) FeSe ("11").[24] Figure reprinted from K. Ishida *et al.*: *J. Phys. Soc. Jpn.* **78**, 062001 (2009). Copyright 2009 by the Physical Society of Japan. See also Color Insert.

pocket FSs and would alter low-lying excitations originating from the nesting-related magnetic fluctuations.[26]

7.3.1.1 LaFeAsO: parent compound

First, NMR results obtained in parent LaFeAsO are introduced below. LaFeAsO shows a structural transition at $T_S \sim 160$ K

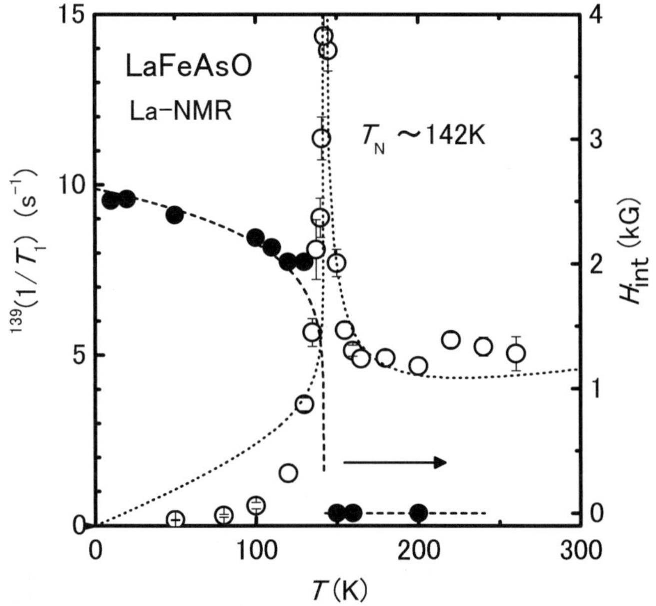

Figure 7.5 T-dependence of $1/T_1$ (○) and the internal magnetic field (•) of undoped LaFeAsO ($x = 0$).[29] The dotted line is a fit to the SCR theory for weak itinerant antiferromagnets (see text). Figure reprinted from Y. Nakai *et al.*: *New J. Phys.* **11**, 045004 (2009). Copyright 2009 by the IOP Publishing Ltd.

from tetragonal ($P4/nmm$) to orthorombic ($Cmma$) symmetry and antiferromagnetic (AFM) transition at $T_N \sim 140$ K.[27] Nakai *et al.* performed ^{139}La-NMR on poly-crystalline powdered LaFeAsO, and reported temperature dependences of $1/T_1$ and internal field at the La-site determined from the splitting of NMR spectra below T_N.[28,29] These temperature dependences are shown in Fig. 7.5. $1/T_1$ exhibits a clear divergence at $T_N = 142$ K, and the overall temperature dependence of $1/T_1$ in Fig. 7.5 can be fitted to the SCR theory for a 3-D itinerant antiferromagnet:[12]

$$\frac{1}{T_1} = \begin{cases} aT + bT/\sqrt{T - T_N} & T > T_N \\ cT/M(T) & T < T_N, \end{cases} \tag{7.23}$$

where $a = 0.005$ (sK)$^{-1}$, $b = 0.13$ s^{-1}K$^{-1/2}$ and $c/M_0 = 0.02$ (sK)$^{-1}$ are fitting parameters, and $M(T) = M_0(1 - T/T_N)^{0.15}$ is an AFM order parameter as mentioned above. The first term in Eq. (7.23) comes from usual Korringa relaxation expected in a metal. It is noteworthy that $1/T_1$ related to the q-averaged dynamical susceptibility shows a divergence at 142 K whereas a clear anomaly is not observed in the static susceptibility. These are the characteristic features of an itinerant antiferromagnet, in which the ordered $\boldsymbol{q}$-vector is far from $\boldsymbol{q} = \boldsymbol{0}$.[30]

Below T_N, internal field appears at the La-site and abruptly develops with decreasing temperature. Since the internal field arises from the ordered Fe moment M, temperature dependence of the internal field gives information about the growth of the ordered moment against temperature. The dashed line in Fig. 7.5 is a fitting to the following expression, $M(T)/M_0 = [1 - (T/T_N)]^{0.15}$. The growth of $M(T)/M_0$ below T_N is much steeper than the 3-D mean field value (0.5), and the transition to the AFM state is almost first-order like. The divergence of $1/T_1$ is affected by a structural phase transition occurring at 165 K upon cooling, which was observed by synchrotron X-ray diffraction in the same batch sample.[31] $1/T_1$ starts to increase just below 165 K, where the resistivity shows a kink due to the structural transition. The temperature derivative of the resistivity shows a sharp peak at T_N determined by the peak of $1/T_1$. The $1/T_1$ result indicates that a characteristic frequency of the magnetic fluctuations start to decrease at the structural transition T_S, and becomes static below T_N. The relationship between the structural and magnetic transition strongly suggests the importance of the spin-lattice coupling, and the direction of the ordered moments would be determined due to the crystal distortion.

In the magnetically ordered state, Mukuda *et al.* succeeded in observing ^{75}As-NMR signals of LaFeAsO in a zero field,[32] which is affected by the internal field at the As site arising from the ordered moments. The spectrum obtained at 2.2.K is shown in Fig. 7.6. From the analyses of these signals, the NQR frequency $\nu_Q = 8.8$ MHz and $\mu_B H_{\text{int}} = 1.60$ T pointing to the c-axis are derived. The direction of the internal field is consistent with the presence of the off-diagonal hyperfine coupling field induced by the stripe-type AFM ordered moments in the ab-plane as observed in $BaFe_2As_2$. The internal field $\mu_B H_{\text{int}} = 1.60$ T for LaFeAsO is nearly equal to $\mu_B H_{\text{int}} = 1.5$ T for $BaFe_2As_2$.

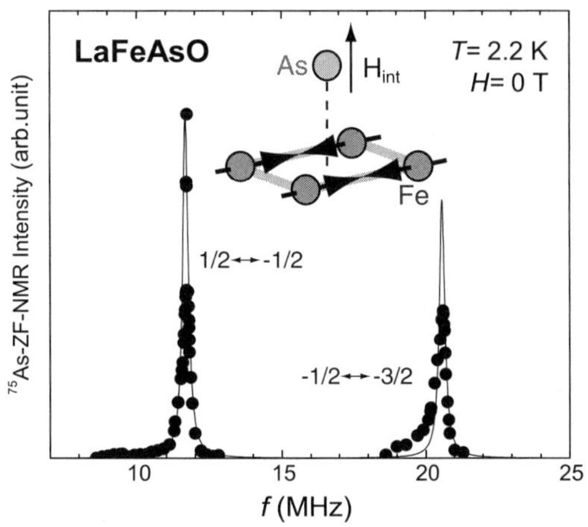

Figure 7.6 Zero-field ^{75}As-NMR spectrum for undoped LaFeAsO at 2.2 K well below T_N.[32] Two narrow spectra at 11.7 MHz (central peak) and 20.6 MHz (satellite peak) can be simulated using the NQR frequency ν_Q = 8.8 MHz and $H_{\rm int}$ = 1.60 T at ^{75}As site along the *c*-axis, as shown by the solid curve. Note that $H_{\rm int}$ at the As site is the off-diagonal pseudodipolar field induced by the AFM moment at the Fe site, whose direction is in the *ab*-plane, which forms the stripe-type AFM structure. Figure reprinted from H. Mukuda *et al.*: *J. Phys. Soc. Jpn.* **78**, 084717 (2009). Copyright 2009 by the Physical Society of Japan.

7.3.1.2 Normal state of LaFeAs($O_{1-x}F_x$) and LaFeAs$O_{1-\delta}$

We move to ^{75}As ($I = 3/2$) NMR results obtained in superconducting LaFeAs($O_{1-x}F_x$)[28,29] and LaFeAs$O_{1-\delta}$. Nakai *et al.* reported the T-dependence of $1/T_1T$ at the As site in LaFeAs($O_{1-x}F_x$) with various x, as shown in Fig. 7.7.[28,29] Similar results were also reported by Lang *et al.*,[33] Nakano *et al.*,[34] and Kobayashi *et al.*,[35] and oxygen-deficiency dependence of $1/T_1T$ observed in LaFeAs$O_{1-\delta}$ is reported by Mukuda *et al.*[32] The $1/T_1T$ results indicate that spin dynamics in the normal state systematically evolve with the F and oxygen-deficiency content. As discussed above, a clear critical slowing down due to the AFM ordering with $\boldsymbol{Q}_{\rm stripe}$ is observed in undoped LaFeAsO. For $x = 0.04$, $1/T_1T$, which is the sum of low-lying dynamical susceptibility $\chi(\boldsymbol{q})$ over the first Brillouin zone, follows a Curie-Weiss temperature dependence down to 30 K. In contrast, the magnetic susceptibility $\chi(\boldsymbol{q} = \boldsymbol{0})$ of LaFeAs($O_{1-x}F_x$) decreases

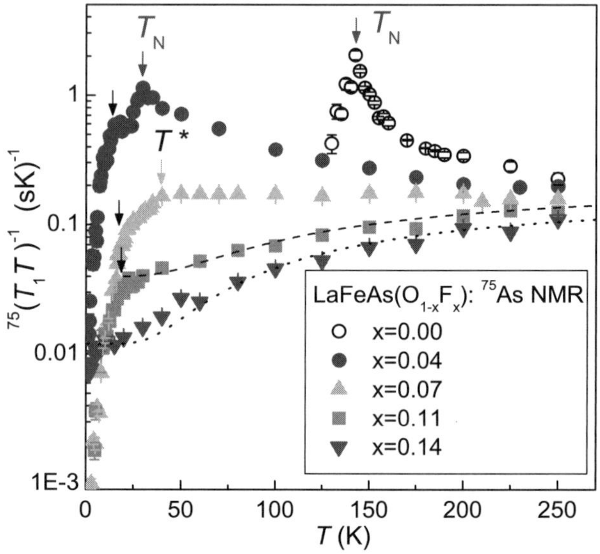

Figure 7.7 Doping dependence of ^{75}As $1/T_1T$.[29] The black (green) arrows denote T_c (T^*) (see text). The broken (dotted) line is a fitting to the data for $x = 0.11$ (0.14). The red arrows indicate T_N.

with decreasing temperature.[36] Their contrasting behavior is a clear indication of the development of AFM fluctuations away from $\boldsymbol{q} = \boldsymbol{0}$ at $x = 0.04$. At $x = 0.07$, $1/T_1T$ remains nearly constant down to $T^* \simeq 40$ K, then decreases rapidly. T^* can be ascribed neither to a magnetic anomaly nor to a SC transition; invariant ^{75}As NMR spectra through T^* rules out the former, and the absence of Meissner signal excludes the latter. The reduction of $1/T_1T$ below T^*, approximately 20 K higher than T_c, is reminiscent of the pseudogap behavior observed in the cuprates.[37] The pseudogap-like behavior is more pronounced for $x = 0.11$ and 0.14, where $1/T_1T$ decreases on cooling, approaching a nearly constant value, as shown in Fig. 7.8. By fitting the data to $1/T_1T = a + b\exp(-\Delta_{PG}/T)$, we obtained $a = 0.04$ ($s^{-1}K^{-1}$), $b = 0.19$ ($s^{-1}K^{-1}$) and $\Delta_{PG} = 172 \pm 12$ K for $x = 0.11$, and $a = 0.012$ ($s^{-1}K^{-1}$), $b = 0.18$ ($s^{-1}K^{-1}$) and $\Delta_{PG} = 165 \pm 15$ K for $x = 0.14$, yielding almost the same pseudogap energies Δ_{PG} for $x = 0.11$ and 0.14 within experimental uncertainty. Based on these results along with those reported previously, the

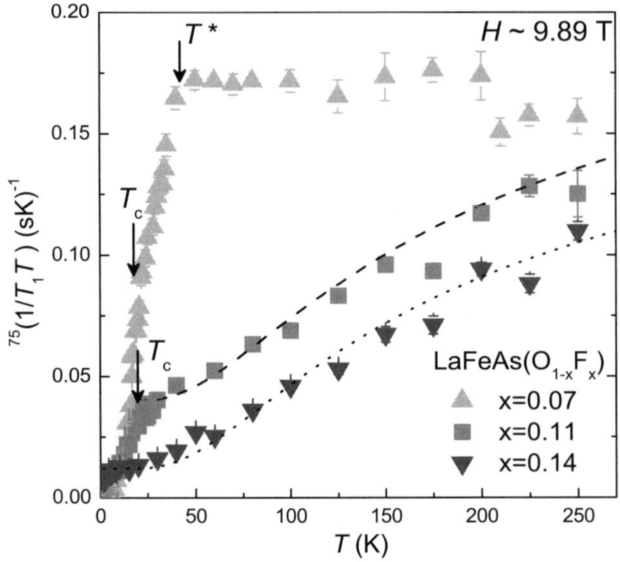

Figure 7.8 Temperature-dependence of $^{75}(T_1T)^{-1}$ at x =0.07, 0.11, and 0.14.[29] The broken (dotted) line is a fitting to $^{75}(T_1T)^{-1}$ for $x = 0.11$ ($x = 0.14$) (see text). Figure reprinted from Y. Nakai *et al.*: *New J. Phys.* **11**, 045004 (2009). Copyright 2009 by the IOP Publishing Ltd.

phase diagram for LaFeAs($O_{1-x}F_x$) is shown in Fig. 7.9. The doping dependence of $1/T_1T$ indicates that the nature of the pseudogap in LaFeAs($O_{1-x}F_x$) differs from that in the cuprates in the following respects: (1) In the cuprates, $1/T_1T$ decreases from temperatures well above T_c and no clear anomaly is observed at T_c. In contrast, a clear anomaly in $1/T_1T$ is found at T_c for $x = 0.11$, and Korringa behavior (T_1T = const.), suggestive of a Fermi-liquid state, is observed at low temperatures, which may be related to the T^2 behavior of resistivity. Considering the multi-band nature of LaFeAs($O_{1-x}F_x$) and electron doping, some parts of holelike FSs are related with the pseudogap behavior while the others contribute to the Fermi-liquid state, since the hole bands would be filled by electron doping. (2) The pseudogap behavior in LaFeAs($O_{1-x}F_x$) becomes more pronounced with F-doping, opposite to the behavior observed in the cuprates. The pseudogap behavior in the cuprates is most pronounced near the AFM phase boundary, and it is most

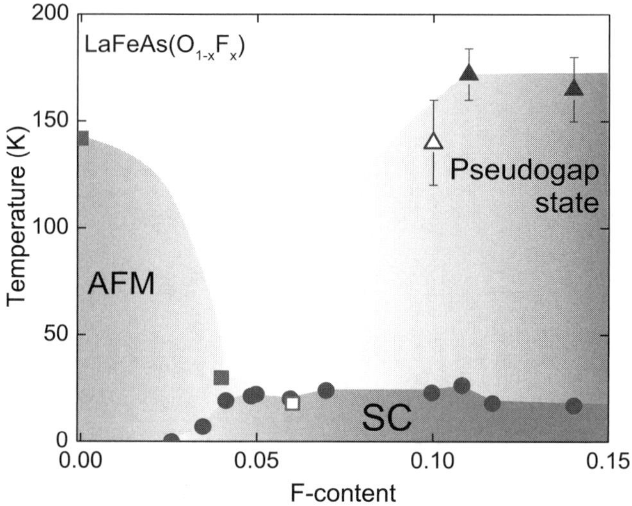

Figure 7.9 Phase diagram of $LaFeAs(O_{1-x}F_x)$.[29] The closed (open) triangle designates the pseudogap energy determined from $1/T_1$ of ^{75}As (Knight shift of ^{19}F cited from Ref. 38). The blue triangle indicates T^* below which the pseudogap behavior was observed. The closed square indicates the magnetic ordering temperature T_N determined from ^{139}La and ^{75}As NMR. The open square indicates the onset temperature of spin-glass like magnetism determined from μSR.[39] T_c (closed circle) is determined from the temperature where ρ becomes half as that at the onset temperature.[1] Figure reprinted from Y. Nakai *et al.*: *New J. Phys.* **11**, 045004 (2009). Copyright 2009 by the IOP Publishing Ltd.

likely that AFM correlations may be responsible for the pseudogap behavior. In $LaFeAs(O_{1-x}F_x)$, however, low-energy AFM correlations are unlikely to yield the pseudogap behavior since no apparent AFM fluctuations are observed via ^{75}As NMR for $x = 0.11$ and 0.14. Furthermore, almost the same Δ_{PG} is observed through ^{57}Fe NMR measurements, suggesting that the pseudogap does not have $\boldsymbol{q}$-dependence.[40] From a theoretical point of view, Ikeda suggested that the pseudogap behavior in $1/T_1T$ may originate from band structure effects near the Fermi energy.[41] The existence of a high density of states (DOS) just below the Fermi level, which is assigned to a hole FS around Γ' consisting of $d_{x^2-y^2}$ orbitals in the *unfolded* Brillouin zone, gives rise to a T-dependent DOS, and the calculated T-dependence of $1/T_1T$ is consistent with the experimental results.

Thus, the pseudogap behavior may originate from its characteristic band structure in $LaFeAs(O_{1-x}F_x)$, but other possibility and/or additional effects are also considered.

As for spin correlations in the normal state, the inelastic neutron scattering measurement revealed that two-dimensional magnetic fluctuations possess the stripe correlations ($\boldsymbol{Q}_{\text{stripe}} = (\pi, 0)$).[42] The presence of such a stripe correlation gives a clear explanation to the anisotropy of $1/T_1$ at the As site ($R \equiv {}^{75}(1/T_1T)_{H\perp c}/{}^{75}(1/T_1T)_{H\parallel c}$) in the normal state, where ${}^{75}(1/T_1T)_{H\parallel c}$ ($(1/T_1T)_{H\perp c}$) is $(1/T_1T)$ of ^{75}As measured in magnetic fields parallel (perpendicular) to the c-axis.[43,44] Kitagawa *et al.* reported the temperature dependence of ${}^{75}(1/T_1T)_{H\perp c}$ and ${}^{75}(1/T_1T)_{H\parallel c}$ for $LaFeAs(O_{1-x}F_x)$ ($x = 0.07, 0.11$ and 0.14) as shown in Fig. 7.10, and pointed out that R in the normal state is almost the same in $x = 0.07$ and 0.11 as shown in Fig. 7.10, while temperature dependence of $1/T_1T$ in the two samples is different with each other.[45] We introduce the analyses of the $1/T_1T$ anisotropy done by Kitagawa *et al.*, and show that the anisotropy of $1/T_1T$ gives information about the $\boldsymbol{q}$ space in magnetic-fluctuation spectrum.[43–45]

From the ^{75}As NMR spectrum in the antiferromagnetically ordered state, it is assumed that hyperfine fields at the As site $\boldsymbol{H}_{\text{hf}}^{\text{As}}$ are determined by the sum of the fields from the four nearest-neighbor Fe electron spins $\boldsymbol{S}$

$$\boldsymbol{H}_{\text{hf}}^{\text{As}} = \sum_{i=1}^{4} \boldsymbol{B}_i \cdot \boldsymbol{S}_i = \tilde{A}\boldsymbol{S}, \tag{7.24}$$

where $\boldsymbol{S}_i$ is the Fe electron spin at the ith Fe site, $\boldsymbol{B}_i$ is the hyperfine coupling tensor between the As nucleus and ith Fe site, and $\tilde{A}$ is the hyperfine coupling tensor ascribed to the four nearest-neighbor Fe electron spins. Following the discussion,[43] $\tilde{A}$ can be described as follows in the orthorhombic notation:

$$\tilde{A} = \begin{pmatrix} A_a & C & B_1 \\ C & A_b & B_2 \\ B_1 & B_2 & A_c \end{pmatrix}. \tag{7.25}$$

A_i is a diagonal term for the i direction ($i = a, b$, and c), which is related to the Knight-shift components. $B_{1\{2\}}$ components are related with the stripe $(\pi, 0)$ $\{(0, \pi)\}$ AF correlations and C is related with the checkerboard (π, π) AF correlations.

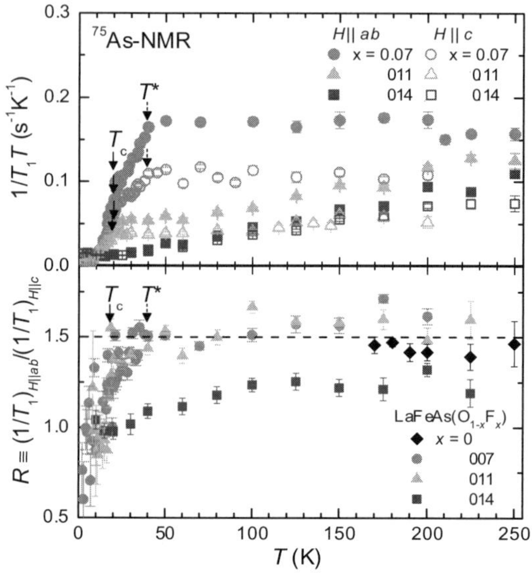

Figure 7.10 (Upper panel) T dependence of $1/T_1T$ for $H \parallel ab$ and $H \parallel c$ in $\mathrm{LaFeAs(O_{1-x}F_x)}$.[45] The solid (dashed) arrows represent T_c (T^*). At $x = 0.07$ and 0.11, normal state $1/T_1T$ was anisotropic, while clear anisotropy in $1/T_1T$ was not observed at $x = 0.14$. (Lower panel) T dependence of the anisotropy of $1/T_1$ [$R \equiv (1/T_1)_{H\parallel ab}/(1/T_1)_{H\parallel c}$]. R in the normal state is $\simeq 1.5$ at $x = 0.07$ and 0.11 as well as in LaFeAsO, whereas R is $\simeq 1$ at $x = 0.14$. Figure reprinted from S. Kitagawa *et al.*: *Phys. Rev. B* **81**, 212502 (2010). Copyright 2010 by the American Physical Society.

In general, $1/T_1$ can be described in terms of fluctuating hyperfine fields perpendicular to the applied magnetic field parallel to the z-axis

$$\begin{aligned}\left(\frac{1}{T_1}\right)_z &= \frac{(\mu_0\gamma_N)^2}{2}\int_{-\infty}^{\infty} dt e^{i\omega_{\mathrm{res}}t}(\langle H_{\mathrm{hf,x}}(t), H_{\mathrm{hf,x}}(0)\rangle \\ &\quad +\langle H_{\mathrm{hf,y}}(t), H_{\mathrm{hf,y}}(0)\rangle) \\ &= (\mu_0\gamma_N)^2(|H_{\mathrm{hf,x}}(\omega_{\mathrm{res}})|^2 + |H_{\mathrm{hf,y}}(\omega_{\mathrm{res}})|^2), \qquad (7.26)\end{aligned}$$

where $|X(\omega)|^2$ denotes the power spectral density of a time-dependent random variable $X(t)$. Since $(\pi, 0)$ and $(0, \pi)$ AFM correlations cannot be distinguished in the tetragonal phase, the following three types of spin correlations are considered uncorrelated (UC) fluctuations centered at $q \simeq (0, 0)$, stripe $(\pi, 0)$, and checkerboard (π,π) correlations. From Eqs. (7.24)–(7.26), $1/T_1$ at the As site can be described as follows:

$$\begin{pmatrix} (1/T_1)_{H\|a} \\ (1/T_1)_{H\|b} \\ (1/T_1)_{H\|c} \end{pmatrix} \propto \begin{pmatrix} |A_b S_b(\omega_{\rm res})|^2 + |A_c S_c(\omega_{\rm res})|^2 \\ |A_c S_c(\omega_{\rm res})|^2 + |A_a S_a(\omega_{\rm res})|^2 \\ |A_a S_a(\omega_{\rm res})|^2 + |A_b S_b(\omega_{\rm res})|^2 \end{pmatrix}$$

for UC fluctuations, and

$$\propto \begin{pmatrix} |B_1 S_a(\omega_{\rm res})|^2 \\ |B_1 S_a(\omega_{\rm res})|^2 + |B_1 S_c(\omega_{\rm res})|^2 \\ |B_1 S_c(\omega_{\rm res})|^2 \end{pmatrix}$$

for stripe $(\pi,0)$, and

$$\propto \begin{pmatrix} |C\ S_a(\omega_{\rm res})|^2 \\ |C\ S_b(\omega_{\rm res})|^2 \\ |C\ S_a(\omega_{\rm res})|^2 + |C\ S_b(\omega_{\rm res})|^2 \end{pmatrix}$$

for checkerboard (π,π).

Assuming that fluctuating Fe spin components are isotropic in the spin space, $|S_a(\omega_{\rm res})| = |S_b(\omega_{\rm res})| = |S_c(\omega_{\rm res})|$, the $R \equiv (1/T_1)_{H\|ab}/(1/T_1)_{H\|c}$ is calculated as follows:

$$R = \begin{cases} \frac{(A_b^2+A_c^2+A_c^2+A_a^2)/2}{A_a^2+A_b^2} = \frac{A_{ab}^2+A_c^2}{2A_{ab}^2} = 0.5 + 0.5\left(\frac{A_c}{A_{ab}}\right)^2 & \text{for UC,} \\ \frac{(B_1^2+2B_1^2)/2}{B_1^2} = 1.5 & \text{for } (\pi,0), \\ \frac{(C^2+C^2)/2}{2C^2} = 0.5 & \text{for } (\pi,\ \pi), \end{cases} \tag{7.27}$$

where $(1/T_1)_{H\|ab} = \frac{(1/T_1)_{H\|a}+(1/T_1)_{H\|b}}{2}$ and $A_{ab} \equiv A_a = A_b$ in the tetragonal phase. The anisotropy of the diagonal A_i terms in the hyperfine coupling tensor, which can be obtained from the plot of the Knight shift against the bulk susceptibility is estimated as $A_c/A_{ab} \simeq 0.7$ in LaFeAsO. It was found that the value of A_c/A_{ab} does not change largely in iron pnictides, although the diagonal hyperfine-coupling terms of LaFeAsO are approximately two times larger than those of $A\mathrm{Fe_2As_2}$ (A = Ba, Sr).[43,44] The value of B_1, which is related to the stripe $(\pi,0)$ correlations, can be known from the internal field at the As site $H_{\rm int}$ in the AFM state, since the ordered Fe moments pointing to the a-axis with $(\pi,0)$ correlations give rise to the internal field along the c-axis at the As site. In LaFeAsO, it was reported that $H_{\rm int}$ = 1.60 T[32] and $\mu = 0.36\mu_B$.[27] These values lead to B_1 =

$4.4\text{T}/\mu_\text{B}$ in LaFeAsO, which is also nearly two times larger than $B_1 = 1.72\text{T}/\mu_\text{B}$ in $BaFe_2As_2$[43] and $B_1 = 2.08\text{T}/\mu_\text{B}$ in $SrFe_2As_2$.[44]

The observed A_c/A_{ab} value ($\simeq 0.7$) in LaFeAsO, which is smaller than 1, cannot make R to be larger than 1 as expected from Eq. (7.27). Therefore, the observed $R \simeq 1.5$ in LaFeAsO above the structural-transition temperature T_S is consistently interpreted as the presence of the local $(\pi, 0)$ AF correlations, which were actually detected by the neutron-scattering measurement.[42] Similarly, $R \simeq 1.5$ observed at $x = 0.07$ and 0.11 also suggests the presence of the local stripe correlations, although any pronounced development of spin fluctuations was not detected from the temperature dependence of $1/T_1T$.

The anisotropy of $1/T_1$ becomes approximately 1.0 at $x = 0.14$, indicating that the stripe $(\pi, 0)$ correlations are weak or absent. The local stripe correlations can be induced by the nesting between the hole and electron FSs. Therefore, the weakness or absence of the stripe correlations at $x = 0.14$ is ascribable to a worse nesting condition, since the hole FSs become smaller with F doping and would disappear at a critical concentration. The concentration of $x \simeq 0.14$ could correspond to a critical one, since R changes abruptly from $x = 0.11$. Kitagawa *et al.* suggested that the presence of the local stripe correlations originating from the nesting between the hole and electron Fermi surfaces might be linked to high-T_c superconductivity in iron-based superconductors.[45]

Hydrostatic pressure effects on $1/T_1T$ were investigated by Nakano *et al.*[34] They reported the relationship between AFM fluctuations and superconductivity from the pressure dependence of $1/T_1$ in $LaFeAs(O_{1-x}F_x)$. Temperature dependence of $1/T_1T$ for $x = 0$, 0.05, 0.08, 0.10 and 0.14, measured at ambient pressure and 3.0 GPa, is shown in Fig. 7.11.[34] From the pressure dependence of $1/T_1T$, the critical doping level at which low-frequency AFM fluctuation vanishes is around the optimally doped region, and the fluctuation is enhanced by applying pressure in the underdoped regime. They pointed out that the AFM fluctuation is not important for the high T_c mechanism since the increase of T_c is small in the underdoped region, but that the density of states at the electron pocket is important since T_c remarkably increases in the overdoped region where $1/T_1T$ just above T_c is increased by applying pressure.[34]

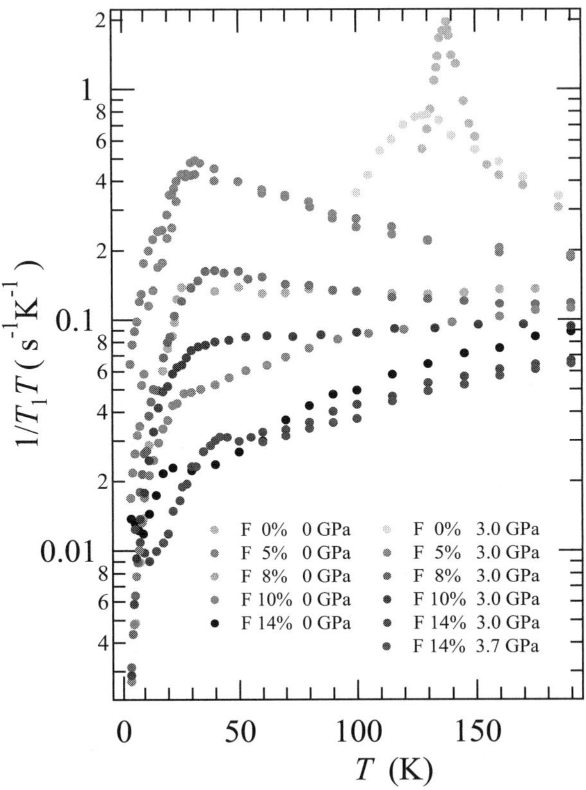

Figure 7.11 Temperature dependence of $1/T_1T$ for $x = 0$ measured at ambient pressure and 3.0 GPa.[34] For the $x = 0.14$ samples $1/T_1T$ measured at 3.7 GPa is also plotted. Figure reprinted from T. Nakano *et al.*: Phys. Rev. B **82**, 172502 (2010). Copyright 2010 by the American Physical Society.

7.3.1.3 Superconducting state of LaFeAs($O_{1-x}F_x$) and LaFeAs$O_{1-\delta}$

Spin susceptibility in the superconducting state of the La-"1111" compounds was investigated from the Knight shift measurements at the As and Fe sites, which are reported by Kawabata *et al.* and Terasaki *et al.*, respectively.[40,46] Figure 7.12(a) shows the temperature dependence of the ^{75}As Knight shift below T_c measured by using a powder sample of LaFeAs($O_{0.89}F_{0.11}$) ($T_c \sim 26$ K).[46] A similar temperature dependence of the Knight shift in the SC state was

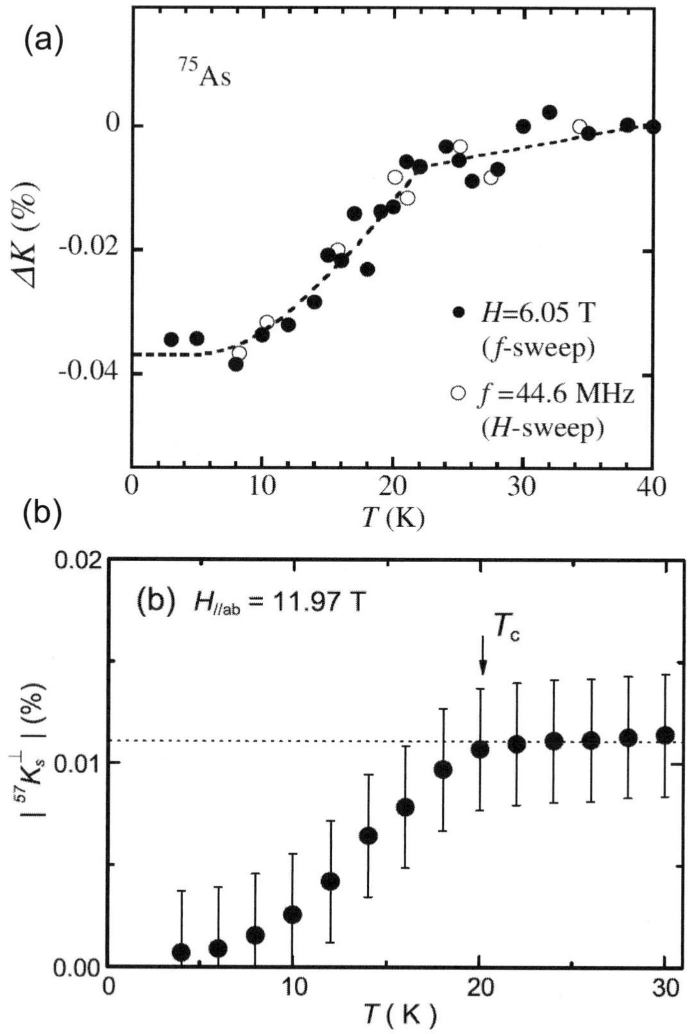

Figure 7.12 (a) Temperature dependence of Knight shift at the As-site. The Knight-shift value is normalized at 40 K.[46] Figure reprinted from A. Kawabata *et al.*: *J. Phys. Soc. Jpn.* **77** (2008) 103704. Copyright 2009 by the Japanese Physical Society (b) Temperature dependence of the spin-part of the Knight shift $|^{57}K_s^{\perp}|$ at the Fe site. The spin part of the Knight shift is evaluated from temperature dependences of the Knight shift at various sites.[40] Figure reprinted from N. Terasaki *et al.*: *J. Phys. Soc. Jpn.* **78** (2009) 013701. Copyright 2009 by the Japanese Physical Society.

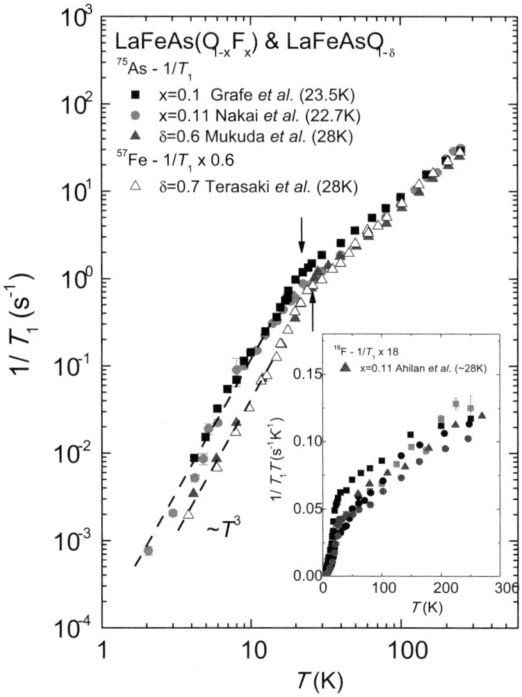

Figure 7.13 Temperature dependence of $1/T_1$ of ^{75}As in LaFeAs($O_{0.9}F_{0.1}$), LaFeAs($O_{0.89}F_{0.11}$), and LaFeAs$O_{1-\delta}$ with similar doping level.[28,38,47,48] Inset: $1/T_1T$ in these samples.

observed at the Fe site in LaFeAs$O_{0.7}$ with $T_c = 28$ K, as shown in Fig. 7.12(b).[40] Terasaki *et al.* estimated the spin part of the Knight shift by comparing the Knight shifts at various different sites, and showed that the spin-part of the Knight shift decreases to almost zero at $T = 0$.[40] These results indicate that the superconducting pair in the La-"1111" compounds is in the spin-singlet state. Figure 7.13 shows temperature dependence of $1/T_1$ measured in LaFeAs($O_{1-x}F_x$) and LaFeAs$O_{1-\delta}$ with similar doping levels.[28,38,47,48] $1/T_1$ in the SC state decreases suddenly without showing a (Hebel-Slichter) coherence peak just below T_c, and power-law temperature dependence of $1/T_1$ ($1/T_1 \sim T^3$) was observed for all the samples. Although absence of a coherence peak and that T^3 dependence observed in $1/T_1$ strongly suggest the presence of linenodes as discussed in

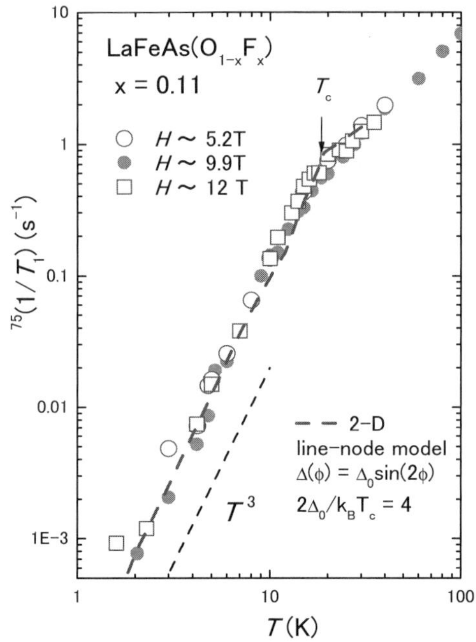

Figure 7.14 T-dependence of $1/T_1$ of ^{75}As in LaFeAs($O_{0.89}F_{0.11}$) under 5.2, 9.9 and 12 T.[29] The dashed curve is a calculation with assuming a line node gap $\Delta(\phi) = \Delta_0 \sin(2\phi)$ with $\Delta_0 = 2k_B T_c$. Figure reprinted from Y. Nakai *et al.*: *New J. Phys.* **11**, 045004 (2009). Copyright 2009 by the IOP Publishing Ltd.

former chapter, the following NMR results are inconsistent with the presence of linenodes. It is known in anisotropic superconductors with nodes that applied magnetic field induces the extra relaxation originating from the Volvik effect.[49] It was pointed out that the quasiparticle state are extended outside vortex cores with ungapped spectrum in a *d*-wave superconductor, which gives rise to a field-induced relaxation rate. However such field-induced relaxation was not observed in LaFeAs($O_{1-x}F_x$) as shown in Fig. 7.14. It was found that the T^3 behavior holds down to 4 K ($\sim 0.2T_c$) in the field range of $5.2 - 12$ T within the experimental error (typically $\Delta T_1 \sim 10\%$), indicative of the absence of the field-induced $1/T_1$. It was also pointed out that a residual density of states (DOS) suggested from the low-temperature Korringa behavior was not observed in these samples. The residual DOS is easily induced by tiny amount of impurities and crystal imperfections in non s-wave superconductors with linenodes, and was observed in most of unconventional superconductors.[50,51] Therefore, it seems that the absence a residual DOS is contrary to non s-wave models, since LaFeAs($O_{1-x}F_x$) is considered to possess the randomness related to F substitution. In addition, experimental results such as penetration-depth measurement in the "1111" superconductors suggest the presence of a finite superconducting gap, which is seemingly inconsistent with the NMR results.[52] To reconcile with these experimental results, $s_\pm$ state with anisotropic gaps is considered as a most promising state in LaFeAs($O_{1-x}F_x$) at present. In this state, the superconducting gap is sign reverse between the hole and electron FSs across the nesting vector.[26,53] In a case of the anisotropic s-wave gap ($\Delta(\phi) = |\Delta_0 \sin(2\phi)|$)[54] and s_{++}-wave gap, a coherence peak remains since the anomalous DOS M_S does not vanish when superconducting gap function is integrated over the FSs. Since the resistivity follows T^2 dependence in the wide temperature range between T_c and 200 K in $x = 0.11$, indicative of a Fermi liquid, it is hard to consider that the quasi-particle damping, which suppresses a coherence peak just below T_c, is working in $x = 0.11$. Alternatively, the absence of a coherence peak is consistently understood by the sign-changed superconducting gaps realized in the $s_\pm$ state.[55–58] Furthermore, it was reported that the T^3 dependence in the superconducting state can be reproduced with the $s_\pm$ state with anisotropic superconducting gaps at the electronic FSs together with some impurity effect.[59]

As for the superconducting gap structure, Kawasaki *et al.* reported ^{75}As NQR studies on superconducting LaFeAs($O_{0.92}F_{0.08}$) with $T_c = 23$ K. They showed that $1/T_1$ in the superconducting state possesses a kink at $T \sim T_c/2$, below which temperature dependence is changed.[60] The temperature dependence of $1/T_1$ shown in Fig. 7.15 cannot be fit by simple power law nor exponential, but can be understood in terms of two superconducting gaps. They analyzed the temperature dependence by assuming two gaps, $N_{S,i}(E) = N_{S,i}\frac{E}{\sqrt{E^2-\Delta_i(\theta,\phi)^2}}$ ($i = 1$, and 2) with the parameter of the two fraction $\alpha = N_{S,1}/(N_{S,1} + N_{S,2})$, and estimated the magnitude of

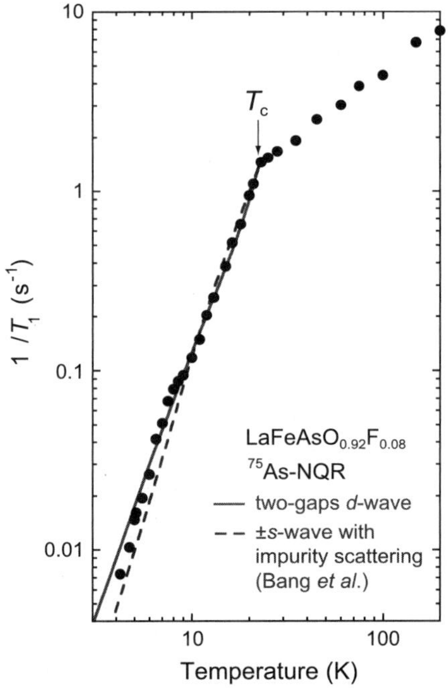

Figure 7.15 The temperature dependence of $^{75}(1/T_1)$ in $LaFeAsO_{0.92}F_{0.08}$ measured at zero magnetic field.[60] Experimental error is within the size of the symbols. The solid curve is a two-gap fit assuming a d-wave symmetry, and the dotted curve is a simulation assuming two s-wave gaps that change signs with impurity scattering. The parameters in the two models are shown in Table 7.2. The solid arrow indicates T_c. Figure reprinted from S Kawasaki *et al.*: *Phys. Rev. B* **78** (2008) 220506(R). Copyright 2008 by the American Physical Society.

two gaps as shown in Table 7.2.[60] They suggested that the multiple-gap feature is universal in the oxypnictide superconductors, which is probably associated with the multiple electronic band structures in this class of materials, and pointed out that low-temperature measurements of $1/T_1$ using high-quality samples are important, since an exponential temperature dependence of $1/T_1$ should be observed in a impurity-free sample with the $s_{\pm}$ pairing.

Mukuda *et al.* reported $1/T_1$ in heavily electron-doped $LaFeAs(O_{1-x}F_x)$ with $T_c = 5$ K, and showed that a coherence peak of $1/T_1$ partially recovers just below T_c, which is incontrast

Table 7.2 The sizes of two superconducting gaps, fraction of the larger gap, and the superconducting gap function, which are adopted for fitting the experimental results shown in Fig. 7.15

Larger gap $\Delta_1(0)$	Smaller gap $\Delta_2(0)$	α	Gap function
4.2 $k_B T_c$	1.6 $k_B T_c$	0.6	$\Delta_0 \cos(2\phi)$ (2-D d wave)
3.75 $k_B T_c$	1.5 $k_B T_c$	0.38	isotropic $s_\pm$ wave

to previous results reported for optimally doped $LaFeAs(O_{1-x}F_x)$ with $T_c = 28$ K.[61] Temperature dependence of $1/T_1T$ in heavily and optimally doped $LaFeAs(O_{1-x}F_x)$, which is normalized by the values of T_c and $1/T_1T$ at T_c is shown in Fig. 7.16. They suggested that, from the temperature dependence of $1/T_1T$ in the normal state, the interband scattering between the hole and electron FSs is strongly suppressed in the heavily doped $LaFeAs(O_{1-x}F_x)$, where the hole Fermi surfaces almost vanish. They point out that the coherence effect on the Cooper pair partially recovers in the heavily electron-doped $LaFeAs(O_{1-x}F_x)$ since a sign-nonconserving interband scattering process becomes weaker and comparable to sign conserving intraband scattering. From these contrasting temperature dependencies of $1/T_1$ between the optimal and heavily overdoped $LaFeAs(O_{1-x}F_x)$, they suggested that a sign reversal of gap functions on the different FSs ($s_\pm$-wave) is realized in $LaFeAs(O_{1-x}F_x)$, and that interband scattering on well-nested FSs is essential for stabilizing the $s_\pm$-wave state and enhancing T_c up to 28 K in LaFeAsO-based superconductors.

7.3.2 *NMR Study in "122" System*

Superconductivity in AFe_2As_2 (A = Ba, Sr and Ca) with the "122" structure appears where antiferromagnetism is suppressed via chemical substitution or pressure. The well-known systems are $(Ba_{1-x}K_x)Fe_2As_2$[62,63] and $Ba(Fe_{1-x}Co_x)_2As_2$,[64,65] in which superconductivity is induced by hole and electron doping, respectively. Phase diagram of the two systems is shown in Fig. 7.17. It is noteworthy that superconductivity is observed in the end compound KFe_2As_2, where 0.5 holes are introduced per Fe. Comparison beween hole and electron dopings suggests that superconductivity

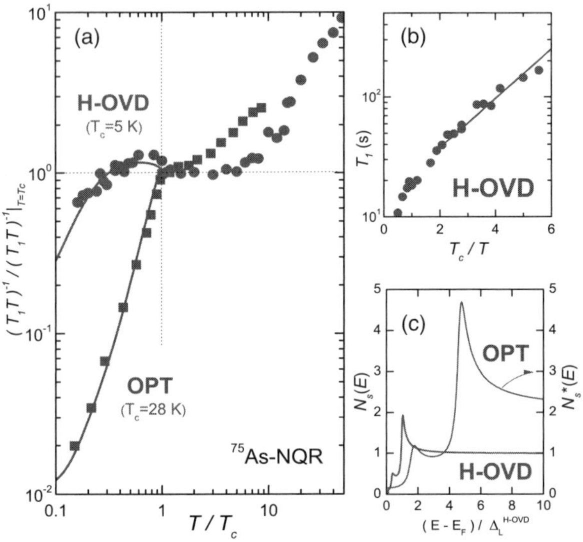

Figure 7.16 (a) $1/T_1T$ normalized by the value at $T = T_c$ for heavily overdoped (H-OVD) and optimally doped (OPT) $LaFeAs(O_{1-x}F_x)$ as a function of the normalized temperature T/T_c. (b) Plot of T_1 vs T_c/T enables us to evaluate $2\Delta_0/k_BT_c \sim 0.95$ roughly from the slope. Solid curves in (a) for OPT and OVD are simulations based on the two-fully-gapped $s_\pm$-wave model with parameters $2\Delta_L/k_BT_c$. (c) DOS used in the simulation. Note that $N_S(E)$ of OPT is displayed as $N_S(E)/0.47$ using the ratio of the normal-state DOS of OPT and H-OVD just above T_c. Figure reprinted from H. Mukuda *et al.*: *J. Phys. Soc. Jpn.* **79**, 113701 (2010). Copyright 2010 by the Physical Society of Japan.

is robust (weak) against hole (electron) doping, and that the superconducting phase diagram is electron-hole asymmetric. In addition, isovalent substitution of P for As brings superconductivity in $BaFe_2(As_{1-x}P_x)_2$, the phase diagram of which is shown in Fig. 7.25.[66,67] In these systems, the existence of a quantum critical point of AFM magnetism has been suggested, and much attention has been paid to the relationship between quantum critical magnetic fluctuations and superconductivity as discussed in heavy-fermion superconductors.[68–71] In this section, we introduce NMR results on the parent compound $BaFe_2As_2$[43] and subsequently review NMR studies on isovalent substituted $BaFe_2(As_{1-x}P_x)_2$,[72,73] electron-doped $Ba(Fe_{1-x}Co_x)_2As_2$,[69] and hole-doped $(Ba_{1-x}K_x)Fe_2As_2$.[74,75]

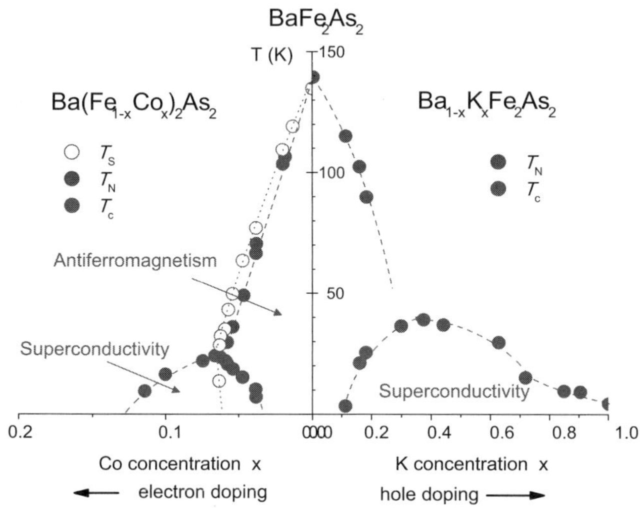

Figure 7.17 Phase diagram of hole doped $(Ba_{1-x}K_x)Fe_2As_2$[63] and electron doped $Ba(Fe_{1-x}Co_x)_2As_2$.[65]

7.3.2.1 $BaFe_2As_2$[43]

Careful ^{75}As-NMR measurements on a single-crystal of $BaFe_2As_2$ were performed by Kitagawa *et al.*[43] In the following, their NMR results are introduced.

Figure 7.18 shows ^{75}As-NMR spectra at the fixed frequency of 48.41 MHz obtained by sweeping magnetic field parallel and perpendicular to the *c*-axis at several different temperatures. The spectra at $T = 141$ K are representative of the paramagnetic phase, while the spectra at the other temperatures belong to the AFM ordered phase. Since ^{75}As nuclei have spin 3/2, the NMR frequencies are generally expressed by a perturbation series when the magnetic Zeeman interaction dominates over the quadrupole interaction,

$$\nu_{m\leftrightarrow m-1} = \mu_0\,{}^{75}\gamma\, H_{\text{eff}} + \frac{1}{2}\nu^c\left(m - \frac{1}{2}\right) \times \left(3\cos^2\theta - 1 + \frac{\nu^a - \nu^b}{\nu^c}\sin^2\theta\cos 2\phi\right) + (\text{2nd order correction}). \tag{7.28}$$

The first term is the Zeeman frequency, where $^{75}\gamma = 7.29019$ MHz/T is the nuclear gyromagnetic ratio for ^{75}As and H_{eff} is the

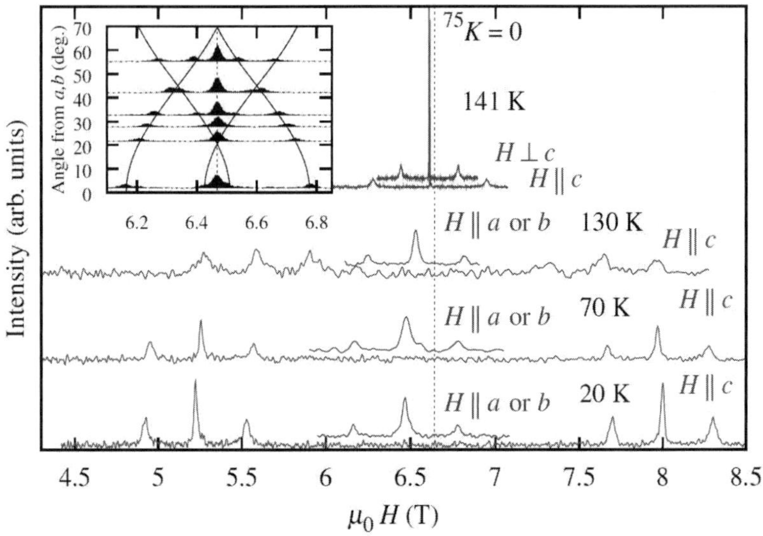

Figure 7.18 ^{75}As-NMR spectra obtained by sweeping the magnetic field at the fixed frequency of 48.41 MHz.[43] In the paramagnetic state at 141 K, the spectrum consists of a sharp central line and two broader satellite lines. Below $T_N = 135$ K, the internal fields along the c-axis due to commensurate AFM order splits the NMR spectra for $H \parallel c$ (red lines) and shifts the spectra to lower fields for $H \parallel a$- or b-axis (blue lines). The inset shows angular variation of the NMR spectrum at 20K for the field rotated in the ab-plane. The two sets of satellite lines originate from the twinned structural domains in the orthorhombic phase. The satellite positions are fitted to Eq. (2) with the quadrupole parameters; $\nu^c = 2.21$ MHz, $|\nu^a - \nu^b|/\nu^c = 1.18$ (solid lines). Figure reprinted from K. Kitagawa *et al.*: *J. Phys. Soc. Jpn.* **77**, 114709 (2008). Copyright 2008 by the Physical Society of Japan.

effective field at the As nuclei. In the paramagnetic state, the effective field is expressed as $H_{\text{eff}} = (1 + K^i)H$, where H is the external field and K^i is the Knight shift along the i-axis. The second term represents the first order quadrupolar shift for the three nuclear transitions $I_z = m \leftrightarrow m - 1$ ($m = 3/2$, $1/2$, or $1/2$). The explicit expression for the second order correction can be found in a standard textbook.[76] The quadrupole splitting parameters ν^i ($i = a, b, c$) are related to the electric field gradient (EFG) tensor $V_{\alpha\beta} = \partial^2 V/\partial r_\alpha\, \partial r_\beta$ at the As nuclei as $\nu^i = eV_{ii}Q/2\hbar$, where Q is the nuclear quadrupole moment of ^{75}As. Here the crystal axes are defined with respect to the orthorhombic ($Fmmm$) unit cell. The $mm2$ symmetry of the As sites in the orthorhombic phase guarantees that the crystalline a, b, and c axes are the principal axes of EFG tensor. The polar angle (θ, ϕ) specifies the direction of the magnetic field relative to the crystalline axes. In the tetragonal phase, the NMR spectrum is independent of ϕ since $\nu_a = \nu_b = -\nu_c/2$. As shown in Fig. 7.18, the NMR spectra at $T = 141$ K show a sharp central line ($m = 1/2$) with the full width at half maximum (FWHM) of 3.5 kHz and two broader satellite lines ($m = 3/2$ and $1/2$) with the FWHM of 100 kHz. The quadrupole splitting ν_c is determined from the separation between the two satellite lines for $H \perp c(\theta = \pi/2)$, which is a half of the separation for $H \parallel c(\theta = 0)$ as expected. The value of ν_c is plotted against temperature in Fig. 7.19.

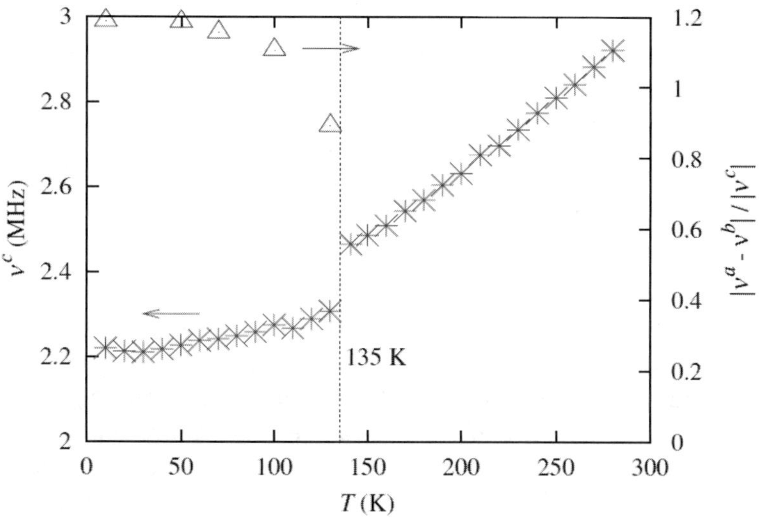

Figure 7.19 Nuclear quadrupole splitting frequency along the c-axis, ν_c, and the asymmetry parameter $|\nu_a - \nu_b|/|\nu_c|$ are plotted as a function of temperature.[43] The asymmetry parameter is zero in the tetragonal phase. Its emergence below 135 K and a jump in ν_c indicate the first-order structural transition into the orthorhombic phase. Figure reprinted from K. Kitagawa *et al.*: *J. Phys. Soc. Jpn.* **77**, 114709 (2008). Copyright 2008 by the Physical Society of Japan.

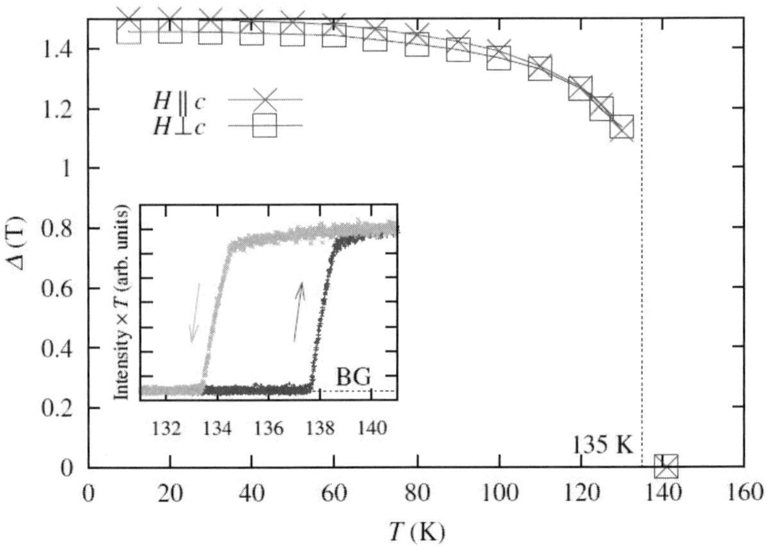

Figure 7.20 The temperature dependence of the internal field at the As nuclei associated with the AFM order.[43] The inset shows the intensity of the central line at the resonance position of the paramagnetic phase for $H \perp c$. The abrupt change with hysteresis is the evidence for the first-order magnetic transition. Figure reprinted from K. Kitagawa *et al.*: *J. Phys. Soc. Jpn.* **77**, 114709 (2008). Copyright 2008 by the Physical Society of Japan.

When the temperature is lowered below $T_N = 135$ K, the NMR spectrum develops a two-fold splitting for $H \parallel c$. This is the direct evidence for a two-sublattice AFM order. The AFM moments generate a spontaneous internal field with alternating sign $\pm\Delta$, therefore $H_{\text{eff}} = H \pm \Delta$ in Eq. (7.28). The temperature dependence of Δ determined from the separation between the split center lines is shown in Fig. 7.20 by crosses. No splitting was observed when the field is applied along the *a* or *b*-axis. Instead the whole spectrum is shifted to lower fields as seen in Fig. 7.18. This indicates that the internal field is parallel to the *c*-axis. When $H \perp c$, H_{eff} is given by the magnitude of the vector sum of the mutually orthogonal external and internal fields, $H_{\text{eff}} = \sqrt{H^2 + \Delta^2}$, giving positively shifted unsplit resonance lines. In Fig. 7.20, Δ determined also from the shift of the central line for $H \parallel a$ or b is plotted by squares. The values of Δ for the two field orientations show good agreement, indicating that the magnitude and the direction of the AFM moment are independent of the field direction. The behavior in the vicinity of the phase transition was investigated by recording the peak intensity of the Fourier transformed spectrum at the resonance position of the paramagnetic phase with changing temperature as shown in the inset of Fig. 7.20. The magnetic transition was found to be very sharp with the transition width less than 1 K, indicating good homogeneity of the crystal. It also shows a clear hysteresis in temperature about 4 K wide, providing conclusive evidence for the first-order transition. This is also shown from the discontinuous development of the internal field as well as a discontinuous structural transition suggested by a pronounced change of symmetry of EFG. As shown in the inset of Fig. 7.18, the quadrupole splitting at 20 K varies with the field direction (ϕ) in the ab-plane, indicating that $\nu_a \neq \nu_b$. This is consistent with the orthorhombic symmetry. The angular variation of the NMR spectra displayed in the inset of Fig. 7.18 shows two branches of satellite lines, which are shifted by 90°. This comes from twinned orthorhombic domains. The ϕ dependence of the satellite lines can be fit to Eq. (7.28). The extrema of the angular variation correspond to the field directions parallel to either a- or b-axis. The asymmetry parameter of EFG defined by $|\nu_a - \nu_b|/\nu_c$ is plotted in Fig. 7.19. The asymmetry parameter, which is zero in the tetragonal phase, is observed to develop abruptly below the transition temperature, and a jump in ν_c is also observed across the transition. These results establish the simultaneous structural and magnetic transition. The large value of the asymmetry parameter, exceeding one in the low temperature phase, implies that the principal axis for the largest EFG changes by 90° across the transition. This suggests that the band structure or the nature of the Fe-As bonds becomes highly anisotropic in the ab-plane.

From the ^{75}As-NMR spectra, they give valuable information about the spin structure in the AFM ordered state. There are two major experimental constraints: (i) the internal field at the As sites is parallel to the c-axis and (ii) the AFM state has a simple two-sublattice structure. There are two possible spin structures compatible with the NMR experiments as shown in Fig. 7.21,

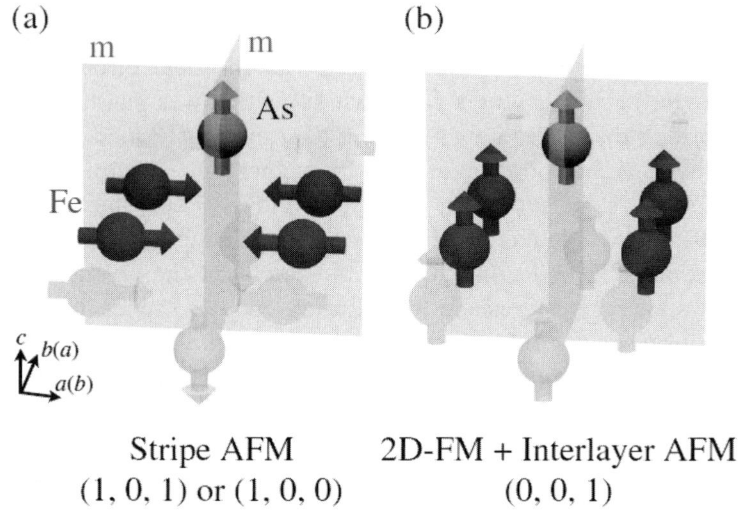

Figure 7.21 Possible spin configurations in the AFM ordered state in $BaFe_2As_2$, which is compatible with the ^{75}As-NMR results.[43] Only one Fe-As layer is drawn for simplicity. The internal field at the As sites are shown by grey arrows. There are two candidates: (a) the stripe AFM order and (b) AFM stacking of the ferromagnetic layers. The former case (a) is realized in $BaFe_2As_2$ (see text). Figure reprinted from K. Kitagawa *et al.*: *J. Phys. Soc. Jpn.* **77**, 114709 (2008). Copyright 2008 by the Physical Society of Japan. See also Color Insert.

which generate a finite *c*-component of the internal field. In the first configuration (a), the AFM order is described by the wave vector (101) or (100), forming stripes with the AFM moments directed perpendicular to the stripe. (Note that the orthorhombic *a*- and *b*-axes cannot be distinguishable by NMR spectra.) In the second configuration (b), ferromagnetic layers in the *ab*-plane stack antiferromagnetically along the *c*-axis. The second structure, however, is highly unlikely since the temperature dependence of the susceptibility rules out ferromagnetic correlation.

They measured the spin-lattice relaxation rate divided by temperature $1/T_1T$, which is shown in Fig. 7.22. In the paramagnetic phase, $1/T_1T$ shows anisotropic temperature dependence. Substantial enhancement is seen for $H \parallel a$ or b with decreasing temperature, while $1/T_1T$ is nearly constant for $H \parallel c$. Across the first-order

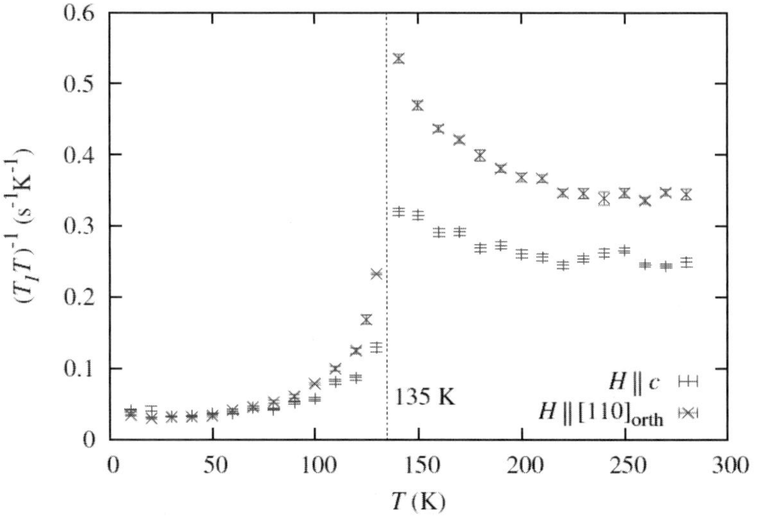

Figure 7.22 Nuclear spin-lattice relaxation rate divided by temperature, $1/T_1T$, is plotted as a function of temperature for two field-orientations.[43] The behavior, $T_1T = const.$, at low temperatures indicates the itinerant (Fermi liquid) character of the material. We should remark that the upturn of $1/T_1T$ in the paramagnetic phase is observed only for $H \perp c$. This indicates anisotropic development of the stripe-type AFM spin fluctuations perpendicular to both the *c*-axis and the stripe. Figure reprinted from K. Kitagawa *et al.*: *J. Phys. Soc. Jpn.* **77**, 114709 (2008). Copyright 2008 by the Physical Society of Japan.

structural/magnetic transition, $1/T_1T$ drops discontinuously by a factor of three. At low temperatures, $1/T_1T$ approaches a constant value for the two field orientation expected for a Fermi liquid. This is consistent with the metallic behavior of the resistivity. The low temperature phase is thus characterized as an itinerant AFM phase. Although the AFM order may be driven by FS nesting, the NMR results demonstrate that the order is commensurate and a sizable density of states at the Fermi level remains down to $T = 0$.

7.3.2.2 NMR in the normal state of $BaFe_2(As_{1-x}P_x)_2$[73]

Nakai *et al.* have studied systematically isovalent-substituted $BaFe_2(As_{1-x}P_x)_2$ with the "122" structure.[73] They advocated that

$BaFe_2(As_{1-x}P_x)_2$ is an ideal system to investigate relationship between magnetic fluctuations and superconductivity. since isovalent P-substitution is not expected to add carriers,[67] and thus $BaFe_2(As_{1-x}P_x)_2$ maintains the compensation condition, i.e. the volume of the hole FSs is equal to that of the electron FSs. Nakai *et al.* performed ^{31}P NMR measurements on $BaFe_2(As_{1-x}P_x)_2$ for $0.2 \leq x \leq 0.64$ in order to investigate the normal-state spin dynamics.[73]

^{31}P-NMR spectra ($^{31}\gamma/2\pi = 17.237$ MHz/T) were obtained by sweeping frequency in a fixed magnetic field of 4.12 T. Samples consisting of mosaic of single crystals were used for their measurements.[67]

Figure 7.23 displays ^{31}P NMR spectra of $BaFe_2(As_{1-x}P_x)_2$ crystals.[73] Each ^{31}P NMR spectrum consists of a single line, ruling out microscopic inhomogeneity caused by P substitution.

Knight shift K, which measures the effective field produced by electron at a concerned nucleus, is described as $K = K_{\text{spin}} + K_{\text{orb}}$; K_{spin} is the spin part of K and is related to the uniform spin susceptibility $\chi(\boldsymbol{q} = \boldsymbol{0})$, which is proportional to the density of states at the Fermi energy $N(E_F)$. K_{orb} is the orbital shift, which is unrelated to $\chi(\boldsymbol{q} = \boldsymbol{0})$, and is estimated to be $\sim 0.018\%$ from the plot of $\sqrt{T_1 T}$ against K at 270 K for different x shown in the inset of Fig. 7.23. This indicates that K_{spin} accounts for 86% of the observed Knight shift for $x = 0.33$. By assuming $K_{\text{spin}} \propto \chi(\boldsymbol{q} = \boldsymbol{0}) = \mu_B^2 N(E_F)$, the P-substitution dependence of K_{spin} at 270 K suggests that the change in $N(E_F)$ would be at most 10% for $x \leq 0.64$, which is quantitatively consistent with the result of their band calculation.[73] The calculation indicates that $N(E_F)$ barely changes for $x < 0.5$, and then decreases for $x > 0.5$. K is almost temperature independent for $x \leq 0.56$ as seen in Fig. 7.23 (b), and the absolute value of K in the normal state decreases only slightly upon P substitution. These data indicate that P substitution does not produce significant changes in $\chi(\boldsymbol{q} = \boldsymbol{0})$ and $N(E_F)$.

Significant low-energy AFM fluctuations are probed at the optimal concentration $x \sim 0.3$ via $1/T_1$ measurements. The Korringa law $T_1 T K^2 =$ const. is inapplicable to $BaFe_2(As_{1-x}P_x)_2$ near the AFM phase since Curie-Weiss (CW) behavior is observed in $1/T_1 T$

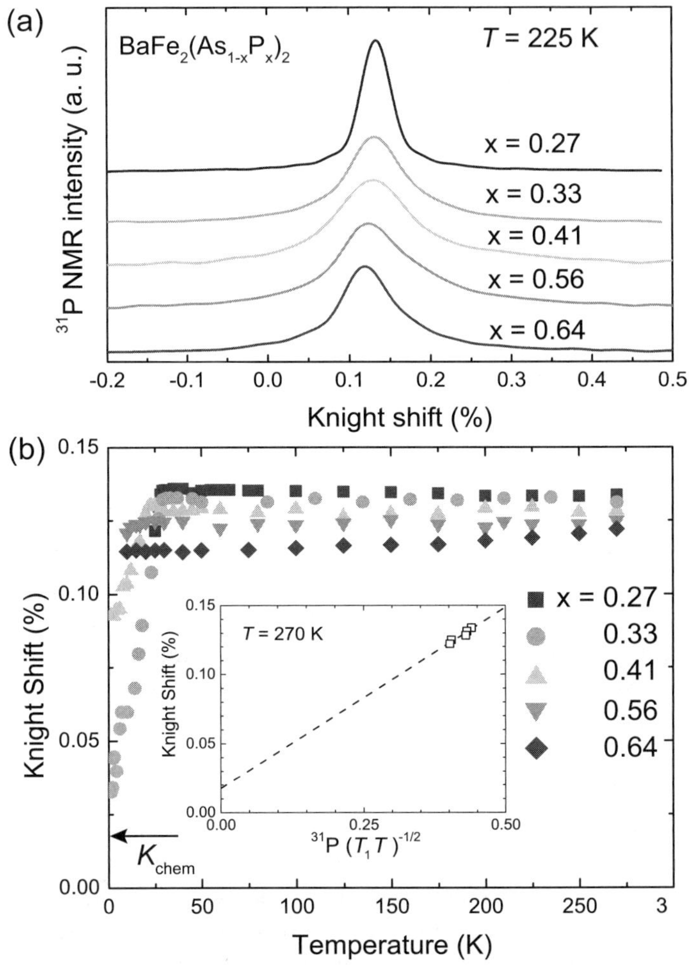

Figure 7.23 P substitution evolution of (a) ^{31}P NMR spectra and (b) ^{31}P Knight shift (^{31}K) determined at the spectral peak, obtained in a mosaic of the single crystals at 4.12 T.[73] (b) The arrows indicate the orbital shift K_{orb} (see text). The abrupt decrease in ^{31}K at low temperatures is due to the onset of superconductivity. Inset: ^{31}K vs $(T_1T)^{-1/2}$ at 270 K for different P concentrations x. Figure reprinted from Y. Nakai *et al.*: *Phys. Rev. Lett.* **105**, 107003 (2010). Copyright 2010 by the American Physical Society.

(see Fig. 7.24).[73] For $x = 0.33$, $1/T_1T$ increases significantly down to T_c whereas the Knight shift is constant. These $1/T_1T$ and K data demonstrate convincingly that AFM fluctuations with finite $\boldsymbol{q}$ continue to grow down to T_c at the optimal concentration. The AFM fluctuations observed in $BaFe_2(As_{1-x}P_x)_2$ are enhanced significantly as P concentration is reduced towards the maximum T_c ($x \simeq 0.33$), as evidenced by the rapid increase in $1/T_1T$ from conventional Fermi-liquid (FL) behavior at $x = 0.64$ (where $1/T_1T$ and K are almost constant). The crossover from the FL to CW behavior in $1/T_1T$ correlates perfectly with the change in the resistivity results;[67] As the system evolves from a FL ($x = 0.71$) towards the maximum T_c ($x = 0.33$) near the AFM phase, the temperature dependence of the resistivity changes from T^2 to T-linear, one hallmark of non FL behavior. The exponent of the temperature dependence of the resistivity is shown as a contour plot in Fig. 7.25(a).[73] Specifically, the CW behavior of $1/T_1T$ and the T-linear resistivity at $x = 0.33$ can be explained by the existence of two-dimensional (2D) AFM spin fluctuations in the theory of nearly AFM metals.[77] Indeed, the evolution of the AFM spin excitations measured by $1/T_1T$ upon P substitution can be fit consistently by the equation expected from the same theory,[77] $1/T_1T = a + b/(T + \theta)$ (see solid lines in Fig. 7.24). Such 2D AFM fluctuations were experimentally observed in the parent $BaFe_2As_2$ via neutron scattering experiments.[78] According to band calculations,[79] substantial 2D AFM fluctuations can be generated by the interband spin excitations between the multiple FSs predominantly derived from Fe d electrons. These experimental results observed in $BaFe_2(As_{1-x}P_x)_2$ indicate that the 2D AFM fluctuations of a quantum-critical nature have a clear correlation with the enhancement of quasiparticle effective mass and T_c as summarized in Fig. 7.25,[73] and that the Weiss temperature θ obtained from the fitting increases with P substitution and becomes almost zero near $x = 0.33$ where the maximum T_c is achieved. $\theta = 0$ K implies that the dynamical susceptibility probed by $1/T_1T$ measurement diverges at absolute zero, or that the magnetic correlation length continues to increase down to $T = 0$ K. Their NMR result strongly suggests the presence of an AFM

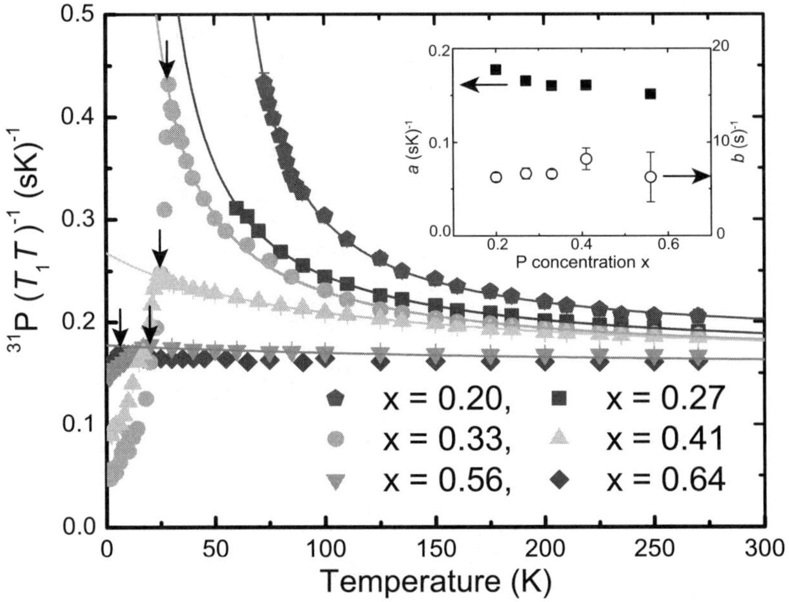

Figure 7.24 The ^{31}P nuclear spin-lattice relaxation rate divided by temperature $(T_1T)^{-1}$ for $BaFe_2(As_{1-x}P_x)_2$ at 4.12 T.[73] Solid lines represent fits to $(T_1T)^{-1} = a + b(T + \theta)^{-1}$ (see text). As AFM fluctuations are suppressed as inferred from the suppression of $(T_1T)^{-1}$, T_c (denoted by arrows) also decreases. Inset: Fitting parameters of $(T_1T)^{-1}$. The fitting parameters a and b are plotted against P concentration x. The a and b weakly depend on x, but θ shows a strong x dependence (see Fig. 7.25). The small value of θ at $x = 0.33$ ($\theta \sim 0$) are insensitive to the fitting parameters of a and b. Figure reprinted from Y. Nakai *et al.*: *Phys. Rev. Lett.* **105**, 107003 (2010). Copyright 2010 by the American Physical Society.

quantum critical point near the maximum T_c in proximity to the AFM phase boundary. As P concentration is varied towards the optimal concentration ($x \sim 0.33$) from $x = 0.64$ where the FL state is observed, the magnetic fluctuations become dramatically enhanced as θ decreases. Importantly, the quasiparticle mass and T_c increase as θ approaches 0 K. This strongly suggests that the AFM quantum-critical fluctuations lead to strong mass renormalization and unconventional superconductivity.

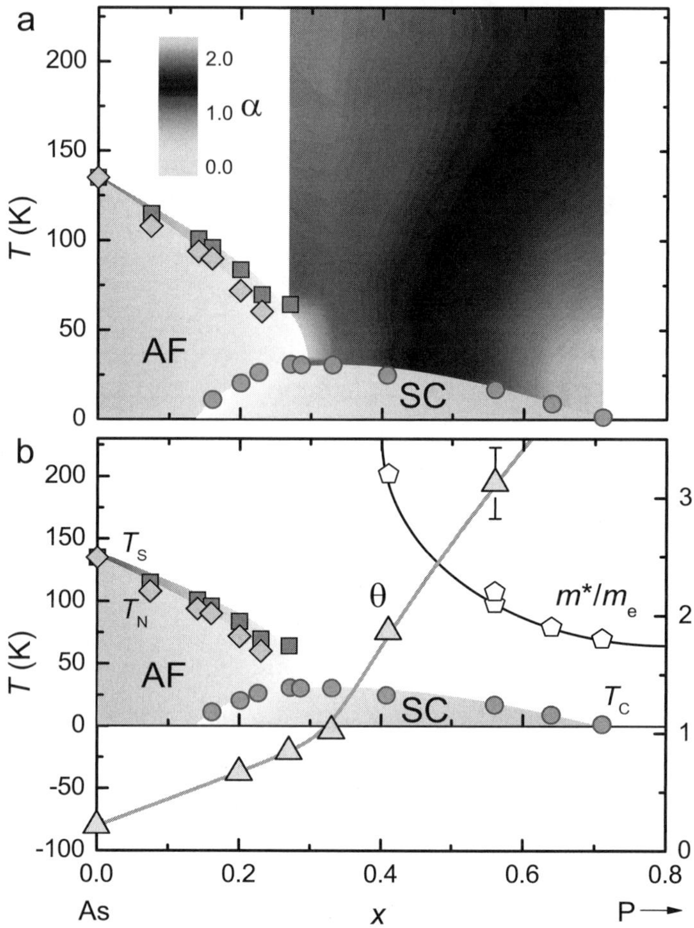

Figure 7.25 Temperature - P concentration phase diagram of $BaFe_2(As_{1-x}P_x)_2$.[73] T_c, T_s, and T_N denote the SC, orthorhombic-tetragonal, and AFM transition temperatures, respectively. (a) Colors represent the exponent, $\alpha = d(\ln \Delta\rho_{ab})/d(\ln T)$, at zero magnetic field for the resistivity curves in Ref. 67, where $\Delta\rho_{ab} = \rho_{ab} - \rho_{ab}(T \rightarrow 0\ \text{K})$. (b) The triangles represent θ obtained by fitting $(T_1T)^{-1}$ data to the equation, $(T_1T)^{-1} = a + b(T+\theta)^{-1}$ (error bars are within the symbol size). T-linear resistivity, suggesting quantum-critical behavior, is observed in proximity to a magnetic instability signaled by $\theta \simeq 0$ at $x = 0.33$. These quantum-critical fluctuations are found to correlate with the enhancement of the effective mass as denoted by the pentagons (taken from Ref. 80). Figure reprinted from Y. Nakai *et al.*: *Phys. Rev. Lett.* **105**, 107003 (2010). Copyright 2010 by the American Physical Society. See also Color Insert.

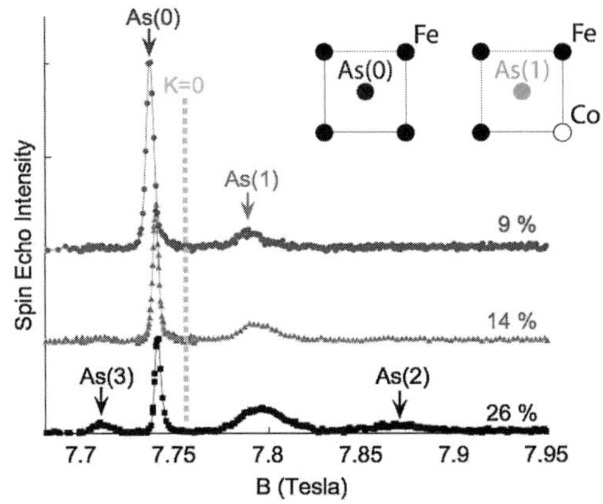

Figure 7.26 Field-swept ^{75}As NMR line shapes of overdoped Ba($Fe_{1-x}Co_x)_2As_2$ with $x = 0.09$ at 20 K ($> T_c$), $x = 0.14$ at 4.2 K ($> T_c$), and nonsuperconducting $x = 0.26$ at 4.2 K with external magnetic field $B \parallel c$-axis. The NMR frequency is fixed at $f = 56.555$ MHz. The gray dashed line marks the expected resonance position for $K = 0$. The systematic increase in the relative intensity of the As(N) peaks (N = 1–3) for larger x. Inset: schematic representations of the Fe and Co coordinations of the As(0) and As(1) sites. Figure reprinted from F. L. Ning *et al.*: *Phys. Rev. Lett.* **104**, 037001 (2010). Copyright 2010 by the American Physical Society.

7.3.2.3 NMR in the normal state of $Ba(Fe_{1-x}Co_x)_2As_2$[69]

Ning *et al.* have studied electron-doped $Ba(Fe_{1-x}Co_x)_2As_2$, and reported systematic ^{75}As-NMR results.[69] Figure 7.26 shows the representative field-swept ^{75}As NMR line shapes corresponding to the central transition for single crystalline samples of $x =$ 0.09, 0.14, and 0.26.[69] The sharp main peak in the NMR line shape, As(0), arises from As sites with all four nearest neighbor (NN) sites occupied by Fe^{2+} ions. Additional broad peaks are observed for all concentrations. From systematic measurements of the NMR line shapes at different magnetic fields, the cause of the line splitting is second order nuclear quadrupole effects, and the Knight shifts of different peaks are comparable. Three additional peaks in Fig. 7.26 are assigned as As(1), As(2), and As(3) sites with 1, 2, and 3 of the NN Fe sites occupied by Co from the intensity ratio of these peaks.[69] They confirmed that spin dynamics measured at the As(1)

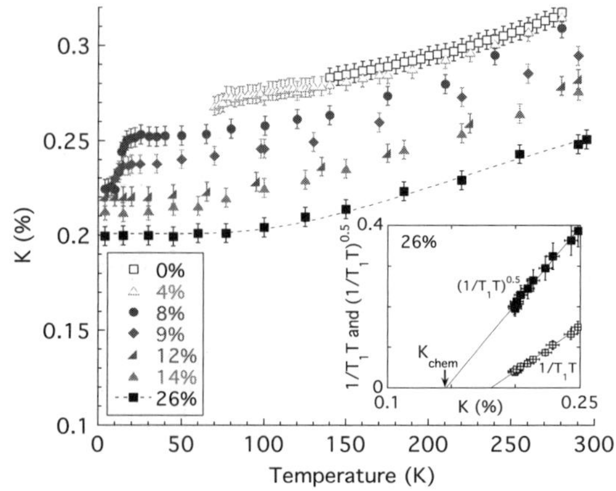

Figure 7.27 The $\boldsymbol{q} = \boldsymbol{0}$ uniform susceptibility with measured by ^{75}As NMR Knight shift K for the As(0) sites in representative compositions. The dashed curve is a fit of $x = 0.26$ data to an activation form, $K = 0.20 + 0.23 \times \exp(-\Delta/T)$, with $\Delta/k_B = 450$ K. Inset: $(1/T_1T)^{0.5}$ and $1/T_1T$ for $x = 0.26$ plotted as a function of K with temperature as the implicit parameter. Figure reprinted from F. L. Ning *et al.*: *Phys. Rev. Lett.* **104**, 037001 (2010). Copyright 2010 by the American Physical Society.

sites show qualitatively the same temperature and concentration dependencies as at the As(0) sites.

Figure 7.27 shows the temperature dependence of the Knight shift K thus determined for $x = 0$, 0.04, 0.08, 0.09, 0.12, 0.14 and 0.26.[69] In the metallic state above T_{SDW} and T_c, all compositions exhibit qualitatively the same behavior, K decreases monotonically with decreasing temperature, then levels off below 50 K. NMR Knight shift is related to the local electron spin susceptibility χ_{spin} by $K = K_{\text{spin}} + K_{\text{orb}}$; $K_{\text{spin}} = A_{\text{hf}}\chi_{\text{spin}}/N_A\mu_B$ is the spin contribution to the Knight shift, where $A_{\text{hf}} = 1.9$ T/μ_B is the hyperfine coupling constant between ^{75}As nuclear spins and surrounding electrons. The temperature independent orbital shift is $K_{\text{orb}} = 0.22\%$ for $x = 0.08$, but has a small concentration dependence. Figure 7.27 indicates that χ_{spin} shows qualitatively the same behavior for all compositions regardless of the nature of the ground state.

Figure 7.28 shows temperature dependence of $1/T_1T$ measured at the As(0) sites.[69] As seen in Fig. 7.28, $1/T_1T$ in the nonsuperconducting $x = 0.26$ is in sharp contrast to $1/T_1T$ in the optimally

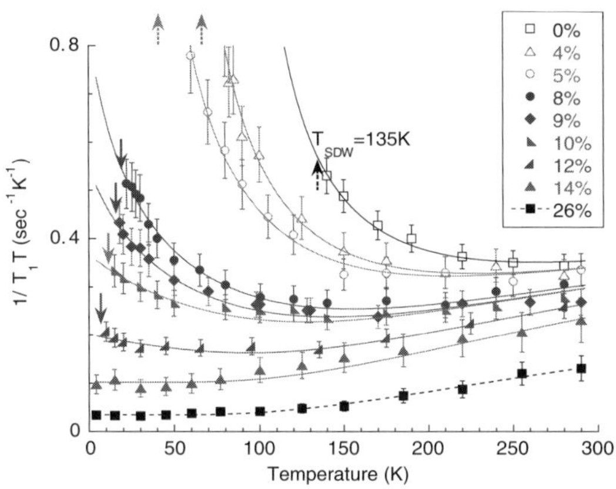

Figure 7.28 $1/T_1T$ measured at the As(0) sites for various concentrations x with magnetic field B applied along the ab-plane. Solid and dashed arrows mark T_c and T_{SDW}, respectively. Solid and dashed curves are the best fits with (below $x = 0.14$) and without (for $x = 0.26$) a Curie-Weiss term arising from AFM spin fluctuations. (See text.) Figure reprinted from F. L. Ning *et al.*: *Phys. Rev. Lett.* **104**, 037001 (2010). Copyright 2010 by the American Physical Society.

doped superconductor $x = 0.08$; $1/T_1T$ of the former levels off to a very small constant value below 50 K. From the Korringa relation of $(1/T_1T)^{0.5} = (\text{const}) \times K_{\text{spin}}$ observed in the whole temperature range between 4.2 and 290 K (inset to Fig. 7.27), it is considered that the nature of spin excitations in $x = 0.26$ is consistent with a FL picture where only intraband electron excitations exist. This is in good agreement with the fact that in-plane resistivity varies as $\rho_{ab} \sim T^2$ in $x = 0.26$. From the extrapolation of the linear fit to the $(1/T_1T)^{0.5}$ vs K ($(1/T_1T)$ vs K) plot in the inset of Fig. 7.27, K_{orb} is estimated as 0.15% (0.18%) and the net spin contribution to the Knight shift below 50 K can then be determined as $K_{\text{spin}} = K - K_{\text{orb}} = 0.2 - 0.15(0.18) = 0.05\ (0.02)\%$, and $\chi_{\text{spin}} \sim 1.5(0.6) \times 10^{-4}$ (emu/mol Fe) in $x = 0.26$. $1/T_1T$ exhibits an upturn for $x = 0.12$ due to the growth of AFM spin fluctuation as seen in Fig. 7.28. Further reduction of the doping level results in divergence of $1/T_1T$ toward T_{SDW} due to the growth of AFM spin fluctuations with $\mathbf{q} = \mathbf{Q}_{\text{stripe}}$.

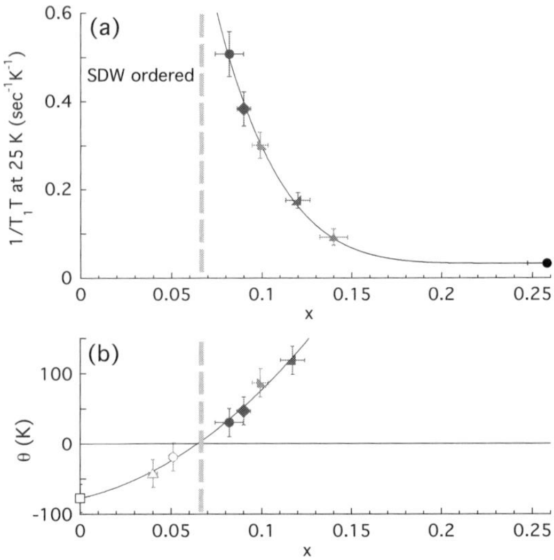

Figure 7.29 The concentration dependence of (a) the strength of paramagnetic spin fluctuations as measured by $1/T_1T$ observed at 25 K, and (b) Weiss temperature θ obtained from the fit in Fig. 7.28. Solid curves are guides for the eyes. Figure reprinted from F. L. Ning *et al.*: *Phys. Rev. Lett.* **104**, 037001 (2010). Copyright 2010 by the American Physical Society.

The systematic evolution of spin dynamics was clearly observed by plotting the concentration x dependence of $1/T_1T$ observed at 25 K, as shown in Fig. 7.29(a).[69] The strength of spin fluctuations at 25 K, as reflected by the magnitude of $1/T_1T$, shows only a mild concentration dependence from $x = 0.26$ down to $x \sim 0.15$, but grows dramatically below $x \sim 0.15$. They found that the growth of spin fluctuations with decreasing x correlates with that of T_c in Fig. 7.17. The $1/T_1T$ data clearly establish that (i) robust AFM spin fluctuations remain even in the optimum ($x = 0.08$) and slightly overdoped ($x = 0.09 - 0.10$) superconducting samples, and (ii) it is unlikely that AFM spin fluctuations and the superconducting mechanism compete with each other. In order to gain additional insight into the relation between AFM spin fluctuations and superconductivity, the $1/T_1T$ data is fitted with a simple phenomenological two-component model, $1/T_1T = (1/T_1T)_{\text{inter}} + (1/T_1T)_{\text{intra}}$, where the contributions of the interband AFM spin fluctuations is expressed with a Curie-Weiss term, $(1/T_1T)_{\text{inter}} = C/(T+\theta)$, and the intraband contributions may be represented by the same phenomenological activation form, $(1/T_1T)_{\text{intra}} = \alpha + \beta\exp(-\Delta/k_BT)$ by taking the same $\Delta/k_B = 450$ K for all compositions as determined from the fit of K in $x = 0.26$, since the Knight shift data show nearly identical temperature dependence except for constant offsets. While $\alpha = 0.11$ and $\beta = 0.63$ are fixed, the essential aspects of the temperature and concentration dependences of the $1/T_1T$ data are captured as shown in Fig. 7.28. The obtained values of the Weiss temperature θ are summarized in Fig. 7.29(b). The negative value of θ for $x \leq 0.05$ implies these samples are in the magnetic ground state. On the other hand, the relatively large positive value of $\theta \sim 119$ K for $x = 0.12$ reflects the fact that the overdoped sample is far from magnetic instabilities, indicating that the growth of AFM spin fluctuations is only modest. The small positive value of $\theta = 31$ K for $x = 0.08$ is evidence for the close proximity of the optimally doped superconducting phase with a magnetic instability; i.e., high T_c superconductivity is realized near a quantum critical point, where one expects $\theta = 0$. As for the Curie term in $(1/T_1T)_{\text{inter}}$, $C = 24 \pm 4$ (sec^{-1}) is independent of x from $x = 0$ to 0.1, then decreases to $C = 12$ (sec^{-1}) for $x = 0.12$ and $C = 0$ (sec^{-1}) for $x = 0.14$. Therefore, the contribution of the Curie-Weiss term associated with interband transitions becomes negligibly small for $x = 0.14$.

The NMR results in $Ba(Fe_{1-x}Co_x)_2As_2$ are quite similar to those in $BaFe_2(As_{1-x}P_x)_2$ except for the temperature dependencies of K and $(1/T_1T)_{\text{intra}}$. Such different T-dependence would be attributable to the T-dependent DOS at E_F resulting from the electron doping, as discussed above.

7.3.2.4 NMR in the normal state of $(Ba_{1-x}K_x)Fe_2As_2$[74,75]

NMR strudies on the hole-doped $(Ba_{1-x}K_x)Fe_2As_2$ have been reported by several authors.[74,75] Yashima *et al.* reported temperature dependence of $1/T_1T$ at the As and Fe sites in poly-crystalline $(Ba_{0.6}K_{0.4})Fe_2As_2$ with the maximum $T_c = 38$ K in this system, which is shown in Fig. 7.30.[74] The $1/T_1T$ at both sites are enhanced with decreasing temperature down to T_c, although their Knight shift

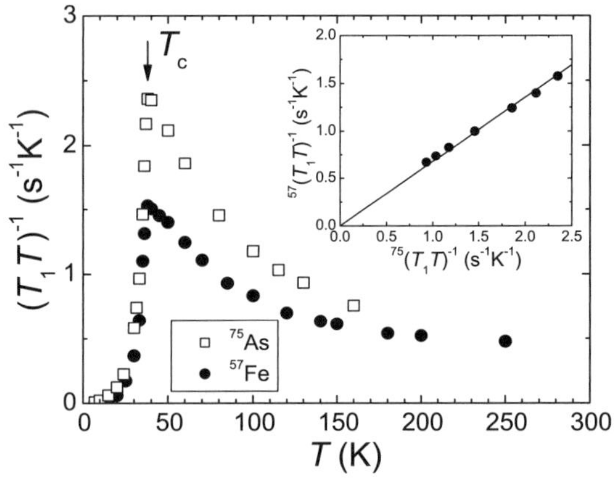

Figure 7.30 Temperature dependence of ^{57}Fe-$(T_1T)^{-1}$ in $Ba_{0.6}K_{0.4}Fe_2As_2$, along with that in ^{75}As-$(T_1T)^{-1}$. The inset shows the plot of ^{57}Fe-$(T_1T)^{-1}$ against ^{75}As-$(T_1T)^{-1}$ with temperature as the implicit parameter; $^{57}(T_1T)^{-1} =^{75} (T_1T)^{-1} \sim 0.68$. Figure reprinted from M. Yashima *et al.*: *J. Phys. Soc. Jpn.* **78**, 103702 (2009). Copyright 2009 by the Physical Society of Japan.

is temperature independent. These point to the development of AFM spin fluctuations, except for the $\boldsymbol{q} = \boldsymbol{0}$ mode, upon cooling. The ratio of $(T_1T)^{-1}$ at the Fe and As sites ^{57}Fe-$(T_1T)^{-1}/^{75}$As-$(T_1T)^{-1}$ is ~ 0.68 and is almost constant in the temperature range from T_c (= 38 K) to 100 K, as shown in the inset of Fig. 7.30.[74] The ratio exclude a possibility that ferromagnetic spin fluctuations with the $\boldsymbol{q} = \boldsymbol{0}$ mode are predominant, since the ratio of $^{57}(T_1T)^{-1}/^{75}(T_1T)^{-1} \simeq 0.038$ is expected using the ratio of the hyperfine coupling constant $(^{57}A_{\rm hf})/(^{75}A_{\rm hf}) \sim 1$ when the spin fluctuations are ferromagnetic, and this is obviously not the case. The same temperature dependence between $^{57}(T_1T)^{-1}$ and $^{75}(T_1T)^{-1}$ can be consistently interpreted in terms of the predominance of the off-diagonal terms of the hyperfine coupling tensor and the stripe correlation with $Q = (\pi, 0)$ and $(0, \pi)$ expected from the interband nesting between the hole and electron FSs. Fukazawa *et al.* reported temperature dependence of $1/T_1$ in heavily hole-doped KFe_2As_2. $1/T_1$ in the normal state follows $T^{0.8}$ between 5 and 160 K, suggesting that weak AFM fluctuations still remain although AFM fluctuations are much suppressed by the

K doping.[75] It should be noted that superconductivity and AFM fluctuations are observed even in the end compound KFe_2As_2 in which 0.5 holes are introduced per Fe. Superconductivity and AFM spin fluctuations are robust against the hole doping, which is in contrast with the electron doping where superconductivity and AFM fluctuations disappear around 0.14 electrons/Fe.

As seen above, strong spin fluctuations with the stripe AFM correlations are observed in all three "122"-type superconductors with maximum T_c concentration, which is in contrast with $LaFeAs(O_{1-x}F_x)$ and $LaFeAsO_{1-\delta}$.

7.3.2.5 NMR results on the superconducting state of "122" compounds

Here, we review NMR results in the superconducting state of "122" superconductors. First we introduce the results in $(Ba_{1-x}K_x)Fe_2As_2$. Figure 7.31(a) shows the temperature dependence of the ^{57}Fe-NMR spectra in $(Ba_{0.6}K_{0.4})Fe_2As_2$ obtained at $H = 12$ T parallel to the ab-plane.[74] The Knight shift $^{57}K_{ab}$ stays almost constant above T_c, followed by a steep increase upon cooling below T_c, as shown in Fig. 7.31(b). The Knight shift comprises a spin component and an orbital component, denoted as $K_{\rm spin}$ and $K_{\rm orb}$, respectively. Here, $K_{\rm spin} = A_{\rm hf}/(\mu_B N_A)\chi_s(T)$ depends on T, but $K_{\rm orb}$ does not. Although the onset of the SC diamagnetism could decrease the ^{57}Fe-NMR spectral shift, the fact that $^{57}K_{ab}(={}^{57}K_{{\rm spin},ab}+{}^{57}K_{{\rm orb},ab})$ increases markedly below T_c indicates that $\chi_{{\rm spin},ab}(T)$ is reduced because of the negative value of $^{57}A_{hf}$ along the ab. This result demonstrates that $(Ba_{0.6}K_{0.4})Fe_2As_2$ is a spin-singlet superconductor. Furthermore, since the ^{57}Fe-NMR spectral width is almost invariant below T_c, as shown in Fig. 7.31(a), the superconducting diamagnetic shift at $H \sim 12$ T might be very small. If $^{57}K_{ab,orb} = 1.38\%$ is assumed in this compound, $^{57}K_{ab,s}$ goes down to zero at $T = 0$, as displayed in Fig. 7.31(c).

$1/T_1$ in the superconducting state of $(Ba_{0.6}K_{0.4})Fe_2As_2$ was measured at the Fe and As sites. Figure 7.32 shows temperature dependence of $1/T_1$ of ^{57}Fe and ^{75}As. In the superconducting state, $(1/T_1)$ at both the sites steeply decreases upon cooling, without a coherence peak just below T_c, as seen in the La-based "1111"

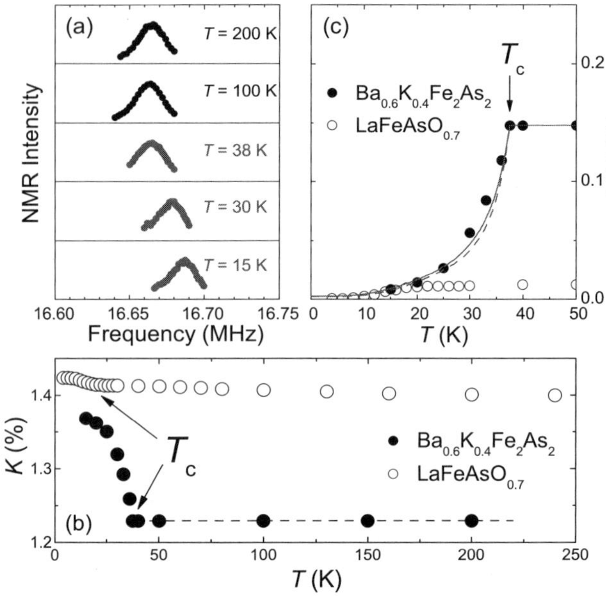

Figure 7.31 Temperature dependence of (a) ^{57}Fe-NMR spectrum and (b) Knight shift (^{57}K) at $H = 11.966$ T parallel to the *ab*-plane in $Ba_{0.6}K_{0.4}Fe_2As_2$. (c) The spin component $^{57}K_{\rm spin}$ in ^{57}K deduced by subtracting $K_{\rm orb} \sim 1.38\%$, revealing a spin-singlet SC state similar to that in $LaFeAsO_{0.7}$. The solid and dashed lines indicate the calculated results assuming Model A and Model B with the same fitting parameters as those in $1/T_1$, respectively. See in the text. Figure reprinted from M. Yashima *et al.*: *J. Phys. Soc. Jpn.* **78**, 103702 (2009). Copyright 2009 by the Physical Society of Japan.

superconductors. The slightly different temperature dependence was observed at low temperatures; temperature dependence of $^{75}(1/T_1)$ becomes weaker than that of $^{57}(1/T_1)$ below 20 K. The origin of the difference is not clear at present, but it might be related with the fluctuations of tiny amount of magnetic-impurity spin, since this effect is expected to be larger for a ^{75}As nuclear spin with the larger gyromagnetic ratio than that for a ^{57}Fe nuclear spin.

Yashima *et al.* analyzed the temperature dependence of $^{57}(1/T_1)$ below T_c in $(Ba_{0.6}K_{0.4})Fe_2As_2$ and $LaFeAsO_{0.7}$ by applying two-gap models,[74] following previous reports.[60,81] In these models, the two Fermi surfaces (FS1 and FS2) are assumed, and possess different

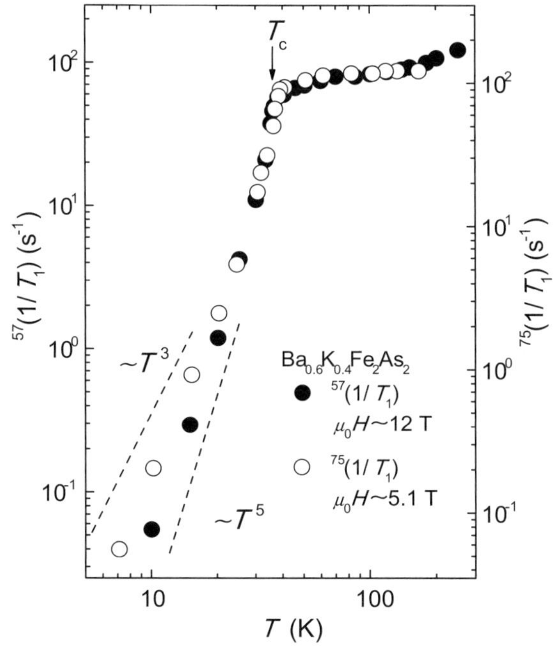

Figure 7.32 Temperature dependence of $1/T_1$ of ^{57}Fe and ^{75}As in polycrystalline $Ba_{0.6}K_{0.4}Fe_2As_2$.[74]

superconducting gaps. One of their model is an anisotropic two-full-gap $s\pm$-wave model (Model A), in which one of the gaps is isotropic ($\Delta^{FS,1} = \Delta_0$) and the other anisotropic ($\Delta^{FS,2}(\phi) = (\Delta_0 + \Delta_{\min})/2 + (\Delta_0 - \Delta_{\min}) \cos 2\phi/2$). Their second model is an isotropic two-full-gap $s\pm$-wave model (Model B), in which two isotropic full gaps open on FS1 and FS2, respectively. They calculated temperature dependence of $1/T_1$ using these two models to explain the experimental results of $(Ba_{0.6}K_{0.4})Fe_2As_2$ and $LaFeAO_{0.7}$. In the calculations, they assumed that the temperature dependence of the superconducting gap is the same as the conventional BCS theory and that the coherence factor is absent due to the sign-changed superconducting gaps. The calculations based on both the models can reproduce the experimental results, if appropriate parameters are adopted. The parameters are listed in Table 7.3, where η is a smearing factor. The analysis using similar two-gap models was

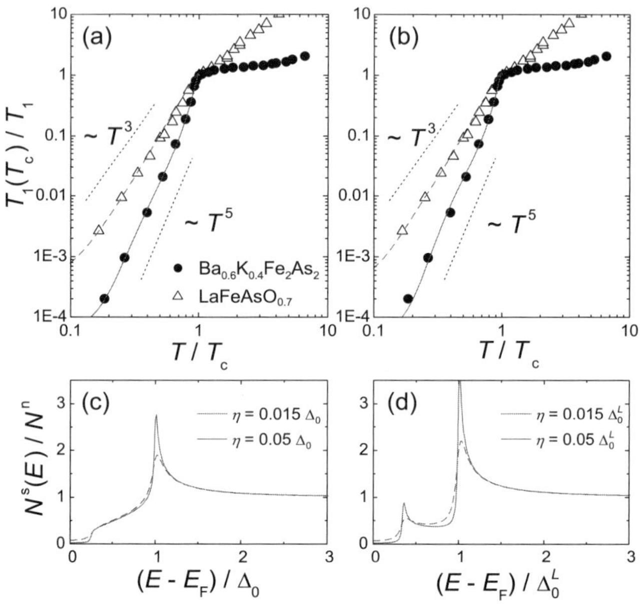

Figure 7.33 Simulations based on (a) Model A and (b) Model B. The experimental $^{57}(1/T_1)$ data for $Ba_{0.6}K_{0.4}Fe_2As_2$ and $LaFeAsO_{0.7}$ are normalized by the value at T_c. Either model can qualitatively reproduce the experimental results, assuming reasonable fitting parameters (see Table I). The energy dependences of the density of states used in these simulations are based on (c) Model A and (d) Model B. Figure reprinted from M. Yashima *et al.*: *J. Phys. Soc. Jpn.* **78**, 103702 (2009). Copyright 2009 by the Physical Society of Japan.

done by Matano *et al.* to fit the $1/T_1$ result measured in the single-crystal $Ba_{0.72}K_{0.28}Fe_2As_2$, and the similar conclusion was derived from their analyses.[82]

Quite different temperature dependence of $1/T_1$ was observed in the superconducting state of $BaFe_2(As_{0.67}P_{0.33})_2$, which possesses the maximum T_c (= 31 K) in $BaFe_2(As_{1-x}P_x)_2$.[72] Figure 7.34(c) shows the temperature dependence of $1/T_1$ of ^{31}P in $BaFe_2(As_{0.67}P_{0.33})_2$.[72] $^{31}(1/T_1)$ decreases sharply just below T_c and there is no evidence for a coherence peak. It was found that $1/T_1$ follows T below ~4 K down to 100 mK, showing the existence of a residual DOS at zero energy. Below about 10 K (~ $0.3T_c$), $1/T_1$ depends on the magnetic field [see Fig 7.34(c)]. By extrapolating

Table 7.3 The sizes of the superconducting gaps, $N_{FS1}/(N_{FS1}+N_{FS2})$, and smearing factor η/Δ_0 estimated by assuming Model A and Model B for $Ba_{0.6}K_{0.4}Fe_2As_2$ and $LaFeAsO_{0.7}$

Model	Fitting parameters	$Ba_{0.6}K_{0.4}Fe_2As_2$ $T_c = 38$ K	$LaFeAsO_{0.7}$ $T_c^{H=0} = 28$ K
A	$2\Delta_0/k_B T_c$	9.2	4.6
	Δ_{min}/Δ_0	0.25	0.25
	$\frac{N_{FS1}}{N_{FS1}+N_{FS2}}$	0.3	0.3
	η/Δ_0	0.015	0.05
B	$2\Delta_0^L/k_B T_c$	9.4	4.4
	Δ_0^S/Δ_0^L	0.35	0.35
	$\frac{N_{FS1}}{N_{FS1}+N_{FS2}}$	0.7	0.7
	η/Δ_0	0.015	0.05

to zero field, $T_1T = const.$ behavior is still observed below ~4 K, excluding the possibility that the $T_1T = const.$ behavior is attributable to applied fields such as vortex dynamics. The existence of a residual DOS in $H \to 0$ inferred from NMR is consistent with thermal-conductivity results on $BaFe_2(As_{0.67}P_{0.33})_2$ that reveal a finite thermal conductivity divided by temperature κ_0/T at $T \to 0$ under zero field.[79] The presence of the residual density of states may also be related to the recent specific-heat results on $Ba(Fe_{1-x}Co_x)_2As_2$.[83]

There are two possible contributions to the field dependence of $1/T_1$ arising from quasiparticle excitations in the mixed state: localized quasiparticles inside vortex cores and delocalized quasiparticles originating from nodes in the SC gap. Since the applied fields are much smaller than $H_{c2} \approx 52$ T, localized quasiparticles inside vortex cores are unlikely to contribute to the relaxation at these fields ($H/H_{c2} \leq 0.22$). In fact, the thermal-conductivity measurements also show a sizable field dependence of $\kappa_0(H)/T$ at low fields, attaining nearly 70% of the normal-state value even at $0.2H_{c2}$, indicative of the predominance of delocalized quasiparticles. These results suggest that the field dependence of $1/T_1$ is attributable to the contribution from delocalized quasiparticles originating from nodes in the gap. As shown in Fig. 7.35, the existence of a residual DOS in the SC state of $BaFe_2(As_{0.67}P_{0.33})_2$ differs strikingly

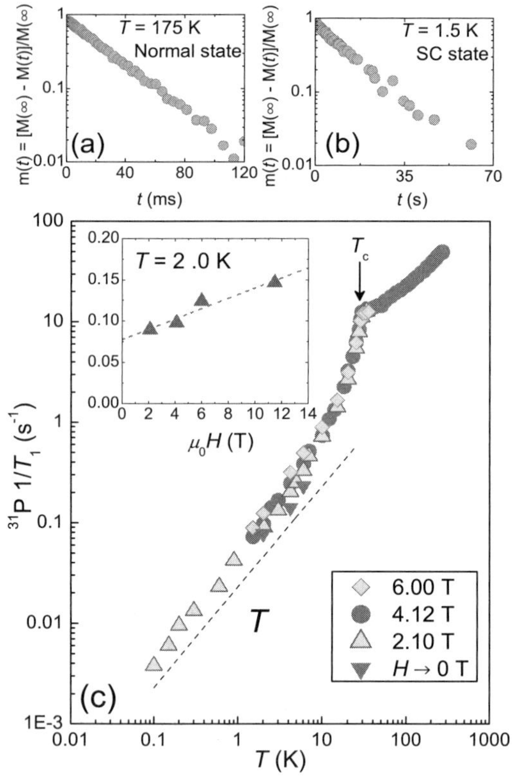

Figure 7.34 (a) and (b) show ^{31}P nuclear-spin relaxation curves for 175 K and 1.5 K in 4.12 T, exhibiting clear single exponential behavior.[72] (c) ^{31}P $1/T_1$ under 2.1, 4.12, and 6.0 T. Inset in panel (c) shows ^{31}P $1/T_1$ vs magnetic field at 2 K; a linear fit is used to extrapolate $1/T_1$ for $H \rightarrow 0$ T. Figure reprinted from Y. Nakai *et al.*: *Phys. Rev. B* **81**, 020503 (R) (2010). Copyright 2010 by the America Physical Society.

from other iron arsenides, including "1111"[28,29,40,47,48,60,81,84] and "122" superconductors[74,75,82,85] shown above, instead resembling the cuprates and heavy fermion superconductors, which have nodes in their SC gap.

As discussed above, $1/T_1$ in the superconducting state of $(Ba_{0.6}K_{0.4})Fe_2As_2$ can be understood by a fully gapped state, and was analyzed in terms of $s_{\pm}$-wave superconductivity with two gaps.[74] In the $s_{\pm}$ models with unitary scatterings, in-gap states at zero energy

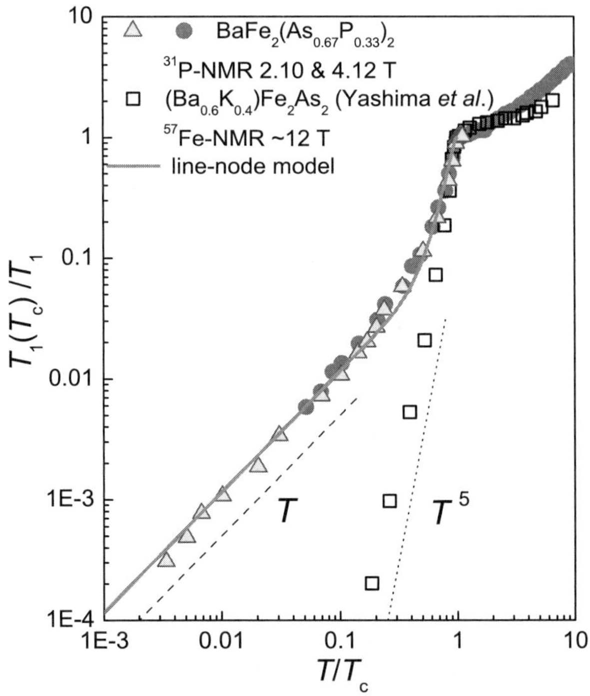

Figure 7.35 Normalized ^{31}P $1/T_1$ for $BaFe_2(As_{0.67}P_{0.33})_2$ and $(Ba_{0.6}K_{0.4})Fe_2As_2$.[74] The solid curve shows a calculation for a line-node gap function $\Delta(\phi) = \Delta_0 \sin(2\phi)$ with $2\Delta_0 = 6k_B T_c$, $N_{res}/N_0 = 0.34$.[72] Figure reprinted from Y. Nakai *et al.*: *Phys. Rev. B* **81**, 020503 (R) (2010). Copyright 2010 by the America Physical Society.

can be induced without severely reducing T_c and can in principle account for the $1/T_1 \propto T$ behavior. However, the penetration depth $\Delta\lambda(T) \propto T$ at low temperatures observed in $BaFe_2(As_{0.67}P_{0.33})_2$ is incompatible with these models, but is consistent with line nodes in the gap. In a symmetry-imposed nodal superconductor such as described by a linenode model $\Delta = \Delta_0 \sin(2\phi)$, $1/T_1 \propto T^3$ in the clean limit, but impurity scattering induced near gap nodes gives rise to an enhanced $1/T_1$ that is proportional to temperature at low temperatures. In fact, the $1/T_1$ data of $BaFe_2(As_{0.67}P_{0.33})_2$ are well fit by the line-node model with $2\Delta_0/k_B T_c = 6$ and $N_{res}/N_0 = 0.34$, where N_{res}/N_0 is the ratio of a residual DOS in the SC state to that in the normal state (see Fig. 7.35).[72] Therefore, the NMR, penetration-depth, and thermal conductivity results on $BaFe_2(As_{0.67}P_{0.33})_2$ are consistent with the existence of line nodes in the superconducting gap, although the estimation of the residual DOS depends on the experimental techniques.

The strikingly different behavior of $1/T_1$ in the SC state between $BaFe_2(As_{0.67}P_{0.33})_2$ and $(Ba_{0.6}K_{0.4})Fe_2As_2$ is surprising because they have comparable T_c values and similar Fermi surfaces. This implies that high-T_c superconductivities with and without nodes are nearly degenerate and that subtle differences can significantly modify the nature of superconductivity.[53,86–89] Changes in the pnictogen height may lead to nodal superconductivity, but its T_c should be much lower than that of fully gapped $s_\pm$ states,[86] and this is not the case in $BaFe_2(As_{0.67}P_{0.33})_2$. One notable difference inferred from band calculations[67,79] is that $BaFe_2(As_{0.67}P_{0.33})_2$ may be more three dimensional than $(Ba_{0.6}K_{0.4})Fe_2As_2$, suggesting that a three-dimensional model is required. To clarify this problem, identifying the location of the node on Fermi surfaces will be crucial, and the systematic measurements of $1/T_1$ on other samples with various P concentration are needed in order to know how the superconducting gap evolves with the P doping.

7.3.3 *NMR Study in "111" System, LiFeAs and NaFeAs*

NMR studies on superconducting LiFeAs with the "111" structure have been reported.[90–92] Figure 7.36 shows the temperature dependencies of the Knight shift and $1/T_1T$ measured by ^{75}As-NMR and NQR in LiFeAs with $T_c \sim 17$ K, which are reported by Li *et al.*[90] The Knight shift decreases with decreasing temperature, but almost constant below $T = 50$ K. The temperature dependence of $1/T_1T$ shows similar behavior as the Knight shift; it also decreases with decreasing temperature and becomes constant below 50 K. The similar behaviors of Knight shift and $1/T_1T$ were also reported by Jeglic *et al.*,[91] and the appreciable development of AFM fluctuations was not observed in superconducting LiFeAs. However, Ma *et al.* reported that the temperature dependence of $1/T_1T$ shows sample dependence.[92] $1/T_1T$ of a sample with better quality of superconducting LiFeAs shows a small upturn from 50 K to T_c and

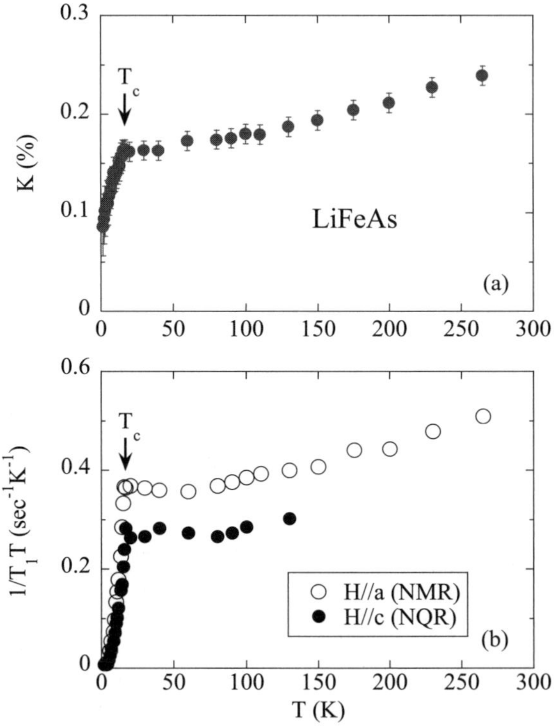

Figure 7.36 (a) The Temperature dependence of the Knight shift. (b) The temperature dependence of the $^{75}(1/T_1T)$ measured with NMR and NQR, respectively. Experimental error is within the size of the symbols. Figure reprinted from Z. Li *et al.*: *J. Phys. Soc. Jpn.* **79**, 083702 (2010). Copyright 2010 by the Physical Society of Japan.

a sharp drop due to the onset of superconductivity, whereas $1/T_1T$ in a less quality sample, with a smaller superconducting volume and a lower T_c, increases dramatically with decreasing temperature, which is close to a diverging behavior. This behavior is a clear indication that the system is close to a magnetic ordering. Although the enhancement of $1/T_1T$ was not observed in Fig. 7.36, it should be noted that the anisotropy of $1/T_1$, $T_{1,c}/T_{1,ab}$ is 1.33, which can be understood as a signature of the stripe AFM correlations as discussed above. Therefore, the stripe correlations might be related with the superconductivity in LaFeAs as well.

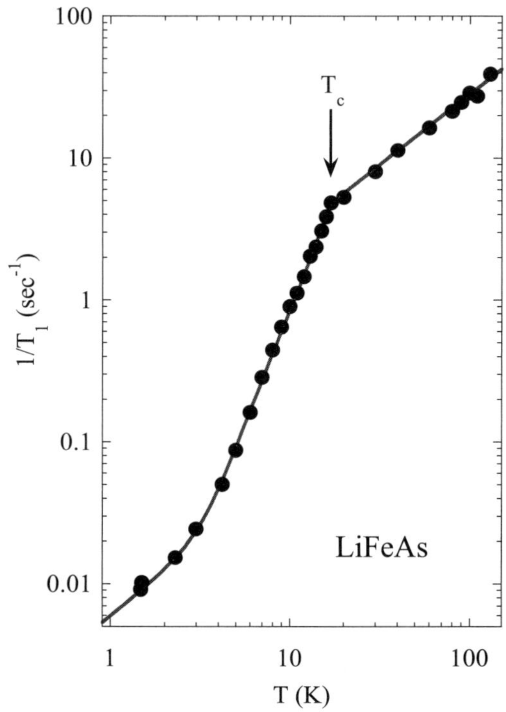

Figure 7.37 The temperature dependence of $^{75}(1/T_1)$ measured with ^{75}As-NQR for LiFeAs. Experimental error is within the size of the symbols. The curve below T_c is fit to the $s_{\pm}$-wave model with $\Delta_1^+ = 3.0k_BT_c$, $\Delta_2^- = 1.3k_BT_c$, and the impurity scattering rate $\eta = 0.26k_BT_c$. The straight line above T_c indicates the relation $T_1T = const.$ Figure reprinted from Z. Li *et al.*: *J. Phys. Soc. Jpn.* **79**, 083702 (2010). Copyright 2010 by the Physical Society of Japan

Figure 7.37 shows the temperature dependence of $1/T_1$ measured with ^{75}As-NQR at zero magnetic field, which can be determined with a single component relaxation after a saturating pulse.[90] $1/T_1$ shows no coherence peak just below T_c, which is similar in La-"1111" and "122" iron-based superconductors shown above. The absence of a coherence peak has also been reported by other groups.[91,92] Below $T \sim T_c/4$, $1/T_1$ becomes proportional to temperature, suggestive of the existence of a residual DOS. Although the behavior of $1/T_1$ can be qualitatively accounted for by a d-wave model with impurity scattering, Li *et al.* showed that a $s_{\pm}$-wave model with two different-size gaps, in which impurity scattering is introduced, can fit the experimental data nicely as shown by the solid curve in Fig. 7.37. Here, fitting parameters of the two gaps are $\Delta_1^+ = 3.0k_B T_c$ and $\Delta_2^- = 1.3k_B T_c$, and the impurity-scattering parameter is $\eta = 0.26k_B T_c$, respectively. The larger gap in LiFeAs is smaller and the η is slightly larger than those in $LaFeAsO_{0.92}F_{0.08}$. It should be noted that the $1/T_1$ behavior in the superconducting state of LiFeAs can be understood by the same two-SC-gap model with different fitting parameters.

NMR studies on the non-superconductive NaFeAs was reported by Kitagawa *et al.*[93] and Ma *et al.*[94]. NaFeAs exhibits successive phase transitions: tetragonal to orthorhombic structural transition at $T_O \sim 50$ K and AFM transition at $T_{AF} \sim 40$ K, and only filamentary superconductivity at $T_{SC} = 23$ K are observed. The structural and magnetic transitions are significantly lower than those in LaFeAsO and $BaFe_2As_2$. Details of these transitions were carefully investigated by ^{23}Na NMR, since a signal-to-noise ratio of ^{23}Na-NMR is higher than that of ^{75}As-NMR. Figures 7.38(a), (b) and (c) are the experimental results obtained from the ^{23}Na-NMR spectrum.[93] Figure 7.38(b) is the angular dependence of the satellite resonance frequencies in the *ab*-plane at 50 K, which is well expressed by the first-order perturbation of the quadrupole interaction with the asymmetric parameter η of the electric field gradient [see in Eq. (7.28)]. The onset temperature for the splitting of the satellite lines, which corresponds to the appearance of η, agrees with T_O determined with resistivity measurements. The AFM transition is characterized by the symmetric splitting into two lines for $H \parallel c$, as seen in Fig. 7.38(a), but no line splitting occurs for $H \perp c$. This anisotropic splitting can be understood by the staggered hyperfine fields at the Na/As sites along the c direction, $\boldsymbol{H}_{\text{hf}} = (0, 0, \Delta)$, which is generated by stripe AFM order, as observed in $BaFe_2As_2$. If the off-diagonal terms of the hyperfine coupling tensor related with the AFM moments have the same value as $BaFe_2As_2$ and $SrFe_2As_2$, the antiferromagnet moment in NaFeAs is estimated to be 0.28–0.34 μ_B at 20 K, which is consistent with the report by Ma *at al.*,[94] but is significantly larger than $0.09 \pm 0.04\mu_B$ obtained by neutron experiments.[95] It should be noted that the spectral shape exhibits

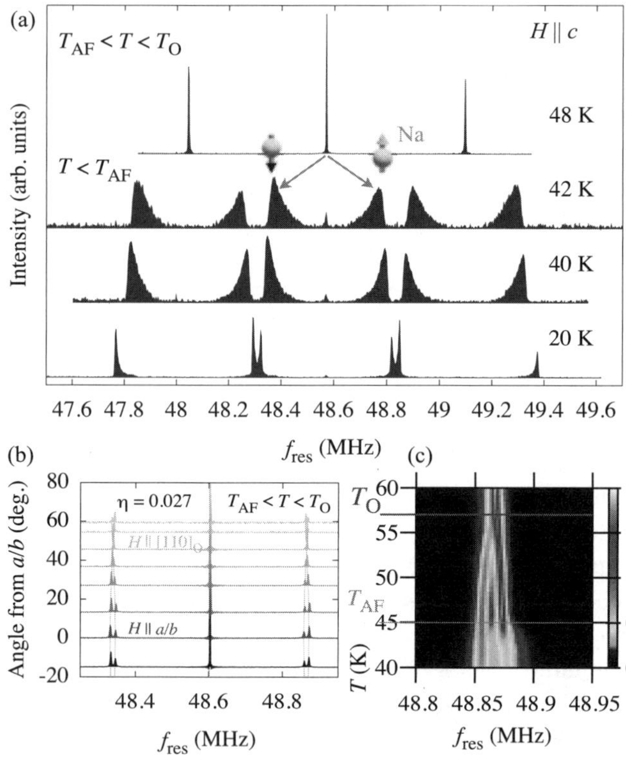

Figure 7.38 Frequency-swept ^{23}Na-NMR spectra at 4.25 T (a) for $H \parallel c$ at different temperatures and (b) for $H \perp c$ along various directions in the ab-plane at $T = 50$ K. The spectra for $H \parallel c$ show incommensurate modulation immediately below T_{AF}. The satellite lines split for $H \perp c$ due to twinning in the orthorhombic structure. Their angular dependence in the ab-plane is well reproduced by Eq. (7.2) with $\eta = 0.027$ (dashed lines). (c) A color plot of the satellite intensity as a function of temperature for $H \parallel a$ or b, showing the tetragonal-to-orthorhombic structural transition at $T_O = 57$ K. Figure reprinted from K. Kitagawa *et al.*: *J. Phys. Soc. Jpn.* **80**, 033705 (2011). Copyright 2011 by the Physical Society of Japan.

an unusual temperature dependence near T_{AF}. The splitting spectra just below T_{AF} show an asymmetric distribution with a sharp edge at the maximum Δ as seen in Fig. 7.38(a). This characteristic spectral broadening indicates an incommensurate modulation of the hyperfine field, but the spectral shape changes gradually to normal discrete peaks at low temperatures. This spectral evolution suggests the crossover from incommensurate to commensurate spin modulation along the SDW ordered moments.

Ma *et al.* recently performed high-pressure NMR studies on NaFeAs and found that the T_{AF} increases by $\sim$ 7 K and the magnetic moments increases by $\sim$ 30% under 2.5 GPa pressure.[94] It seems that the hydrostatic pressure effect on T_{AF} and the moments in NaFeAs is opposite to that in the "1111" and "122" compounds.

7.3.4 *NMR Study in "11" System, FeSe*

Electronic properties of undoped FeSe, which is not magnetically ordered and exhibits superconductivity at $T_c \sim 9$ K[96] was investigated by Imai *et al.* using ^{77}Se NMR.[97] They prepared β-$Fe_{1+\delta}Se$ with a nearly defect-free composition of $\delta = 0.01 \pm 0.02$ and non-superconductive $\delta = 0.03$. From the transport, specific-heat, and SQUID measurements, $T_c \sim 9$ K for $\delta = 0.01$ was identified, but superconductivity was not observed down to $T \sim 0.4$ K in the $\delta = 0.03$ sample. They also investigated the pressure dependence of the superconductivity in the $\delta = 0.01$ sample, and found that T_c rises roughly linearly from $\sim$ 9 K at ambient pressure to $\sim$ 22K under the application of 2.2 GPa.

Figure 7.39 shows the temperature dependencies of ^{77}Se NMR Knight shift $K = K_{spin} + K_{orb}$ in the normal state above T_c.[97] K_{spin} is proportional to the uniform spin susceptibility χ_{spin} and K_{orb} is constant, hence the results reflect the temperature dependence of χ_{spin}. Figure 7.39 show that χ_{spin} changes little between $P = 0$ and 2.2 GPa, and that the nonsuperconducting $\delta = 0.03$ sample also exhibits nearly identical χ_{spin}.[97] These findings conclude that the physical parameters that control χ_{spin} (e.g., the density of electronic states, Fe spin-spin exchange interaction J, etc.) may not have a direct link with the superconducting mechanism.

Figure 7.40 shows the temperature dependence of $1/T_1T$ in the superconducting $\delta = 0.01$ under pressure, and nonsuperconducting $\delta = 0.03$.[97] $1/T_1T$ observed for the superconducting $\delta = 0.01$ is strikingly similar to that of the optimally electron-doped $Ba(Fe_{1.92}Co_{0.08})_2As_2$; $1/T_1T$ decreases with T down to 100 K, then begins to increase toward T_c. Since K is nearly temperature indepen-

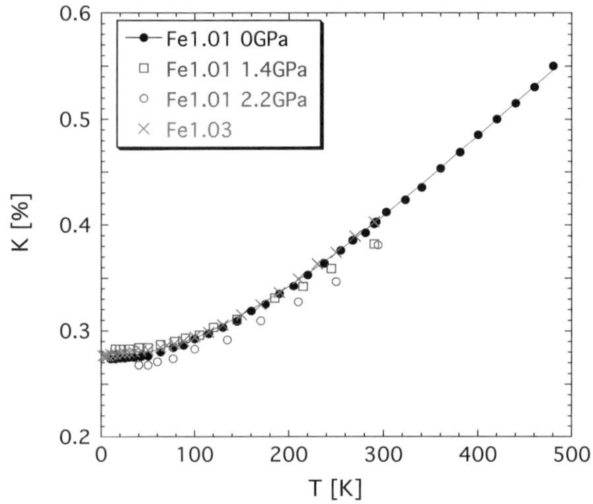

Figure 7.39 The temperature, pressure, and concentration dependencies of ^{77}Se NMR Knight shift $K = K_{\text{spin}} + K_{\text{orb}}$ in the normal state above T_c. K_{spin} is proportional to the uniform spin susceptibility χ_{spin}, and K_{orb} is temperature independent, hence the temperature dependence of K reflects that of χ_{spin}. Figure reprinted from T. Imai *et al.*: *Phys. Rev. Lett.* **102**, 177005 (2009). Copyright 2009 by the America Physical Society.

dent below 100 K, the latter implies that some $\boldsymbol{q} \neq \boldsymbol{0}$ AFM modes of spin fluctuations are strongly enhanced toward T_c. The similarities of K and $1/T_1T$ between $Fe_{1.01}Se$ and $Ba(Fe_{0.92}Co_{0.08})_2As_2$ conclude that superconducting FeSe is also in close proximity to a magnetic instability. It should be noted that $1/T_1T$ measured in 0 and 0.7 GPa shows a sharp peak exactly at T_c as determined by *ac* susceptibility measurements. This means that superconductivity sets in at T_c after AFM spin fluctuations are enhanced, and the opening of the superconducting energy gap suddenly suppresses low frequency spin fluctuations. Unlike typical isotropic BCS s-wave superconductors with a full gap, $1/T_1T$ measured in 0 and 0.7 GPa does not exhibit a Hebel-Slichter coherence peak just below T_c. Instead, as shown in the inset to Fig. 7.40, $1/T_1$ dives below T_c, exhibiting a power-law-like behavior. The $1/T_1$ behavior below T_c is in good agreement with the previous report of FeSe by Kotegawa *et al.*, which showed that $1/T_1$ follows T^3 dependence below T_c.[98] This finding is consistent with NMR reports on various

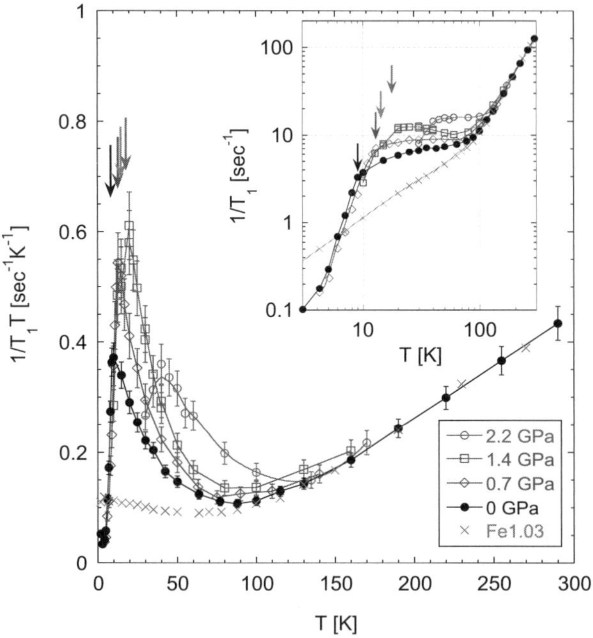

Figure 7.40 $1/T_1T$ for superconducting β-$Fe_{1.01}Se$ under various pressures, and for nonsuperconducting β-$Fe_{1.03}Se$ in P = 0 GPa. $1/T_1T$ reflects the spin fluctuation susceptibility averaged over various wave-vector modes $\boldsymbol{q}$. Inset: A log-log plot of $1/T_1$. Vertical arrows mark (from left to right) T_c for 0, 0.7, 1.4, and 2.2 GPa in the applied magnetic field. Figure reprinted from T. Imai *et al.*: *Phys. Rev. Lett.* **102**, 177005 (2009). Copyright 2009 by the America Physical Society

other iron-based superconductors. Imai *et al.* suggested that AFM spin fluctuations are positively linked with the superconducting mechanism from the following two experimental facts; (1) $1/T_1T$ in the nonsuperconductive $Fe_{1.03}Se$ increases very little below 100 K, indicating that the enhancement of AFM spin fluctuations below 100 K is much weaker than in the superconducting $Fe_{1.01}Se$, and (2) both T_c and spin fluctuations grow under pressure. They pointed out that $1/T_1T$ would be suppressed if spin fluctuations with $\boldsymbol{q} \neq \boldsymbol{0}$ are genuinely competing against the superconducting mechanism.

By applying further pressure above 1 GPa, paramagnetic NMR signals becomes weaker below a peak of $1/T_1T$, suggesting that magnetic phase transition or spin freezing occurs at that temperature. This is also consistent with the fact that superconducting FeSe occurs in close proximity to a magnetic instability.

Quite similar pressure effect was reported on NMR results in superconducting $FeSe_{0.5}Te_{0.5}$ at ambient pressure ($T_c = 14$ K) and 2GPa ($T_c = 22$ K).[99] Shimizu *et al.* performed ^{125}Te NMR at $P = 0$ and 2 GPa, and found that AFM fluctuations probed with $1/T_1T$ are enhanced at 2 GPa upon lowering T towards T_c, although $1/T_1T$ at ambient pressure monotonically decreases with T. It seems that the pressure-induced evolution of AFM fluctuations is common to the iron-chalcogenide superconductors. Similar enhancement of AFM fluctuations towards T_c was also reported in superconducting $FeSe_{0.33}Te_{0.67}$ at ambient pressure ($T_c \sim 14$ K).[100] Although AFM fluctuations in $FeSe_{0.33}Te_{0.67}$ above T_c differ from those in $FeSe_{0.5}Te_{0.5}$ at $P = 0$, T_c is the same in two samples. The relationship between AFM fluctuations and superconductivity seems to be different in the pressure dependence and chemical substitution system. The relevance between AFM fluctuations and superconductivity is also an important issue in $FeSe_{1-x}Te_x$. It is added that $1/T_1$ in $FeSe_{0.5}Te_{0.5}$ and $FeSe_{0.33}Te_{0.67}$ shows a absence of a coherence peak just below T_c and T^n dependence in the superconducting state, consistent with other iron-based superconductors.

7.3.5 *NMR Study in $K_xFe_{2-y}Se_2$*

Finally, we pick up NMR studies on recently discovered $K_xFe_{2-y}Se_2$ with $T_c = 33$ K.[101] This superconductor is particularly intriguing because T_c is a relatively high transition temperature and very recently angle resolved photoemission spectroscopy (ARPES) measurements on $K_xFe_{2-y}Se_2$ revealed that unlike iron-arsenide high-T_c systems, all the hole bands near the center of the first Brillouin zone are filled by electrons donated by intercalated K.[102,103] In many other iron-arsenide systems, the presence of the FSs nesting between the hole bands and electron bands near the zone edge have been considered to gives rise to AFM spin fluctuations, which is invoked to be an important ingredient of the high-T_c mechanisms. Therefore,

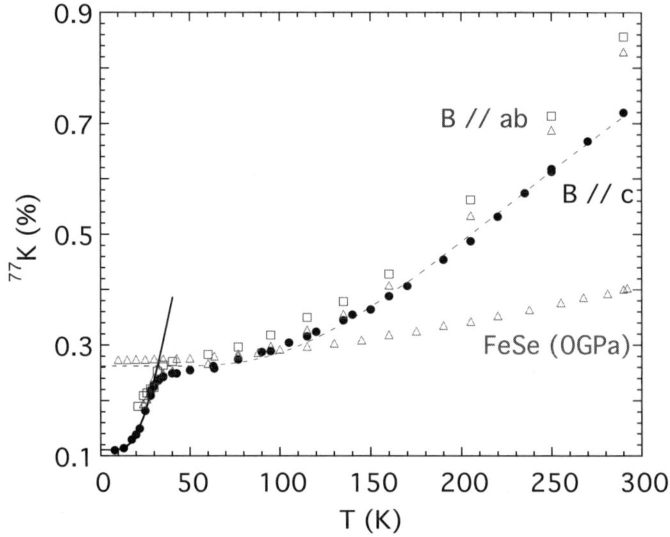

Figure 7.41 The temperature dependence of the uniform $\boldsymbol{q} = \boldsymbol{0}$ component of the static magnetic spin susceptibility χ_{spin} in $K_xFe_{2-y}Se_2$ (T_c = 33 K) as measured by ^{77}Se NMR Knight shift ^{77}K. Filled and open symbols represent the results with magnetic field 8.3 T applied along the crystal c-axis ($^{77}K_c$) and ab-plane ($^{77}K_{ab}$), respectively. In the figure, the temperature independent residual shift, which is estimated to be 0.1%, is subtracted from observed ^{77}K. A sharp drop below T_c for both orientations indicates a formation of the spin-singlet pairing. The solid curve represents a fit with an activation form $^{77}K_{\text{spin}} \sim \exp(-\Delta_{sc}/k_BT)$ with the superconducting energy gap $\Delta_{\text{sc}}/k_BT_c = 2.78$, which is larger than the BCS value of 1.76. The dashed curve represents a phenomenological fit with a pseudogap $^{77}K_{\text{spin}} \sim \exp(-\Delta/k_BT)$ with $\Delta/k_B = 435$ K. For comparison, the normal state results above $T_c = 9$ K of β-$Fe_{1.01}Se$ are also plotted. Figure reprinted from D. A. Torchettiet *al.*: *Phys. Rev. B* **83**, 104508 (2011). Copyright 2011 by the America Physical Society

$K_xFe_{2-y}Se_2$ provides a counter example to the significance of the presence of the hole bands.

^{77}Se NMR measurements in $K_xFe_{2-y}Se_2$ have been performed by Yu *et al.*,[104] Kotegawa *et al.*,[105] and Torchetti *et al.*[106] independently. Figure 7.41 shows the temperature dependence of ^{77}Se NMR Knight shift ^{77}K measured with a magnetic field applied along the c-axis ($^{77}K_c$) and ab-plane ($^{77}K_{ab}$).[106] In the figure, the temperature independent residual shift, which is estimated to be 0.1%, is subtracted from the observed ^{77}K. In the normal state above T_c, $\chi_{\rm spin}$ reflected by the ^{77}K measurements decreases monotonically with temperature. Similar behavior was previously reported for many iron-based superconductors, $LaFeAsO_{1-x}F_x$, $Ba(Fe_{1-x}Co_x)_2As_2$, LiFeAs and FeSe, as shown above. The temperature dependence of ^{77}K can be fitted by a phenomenological pseudogap model with an activation form $^{77}K_{\rm spin} = ^{77}K - ^{77}K_0 \sim \exp(-\Delta/k_B T)$, as shown by a solid curve in Fig. 7.41, where K_0 is a constant term in K. The best fit of the data from T_c to 290K yields $\Delta/k_B = 435$ K, and this value is comparable to the previous estimate of $\Delta/k_B = 450 - 711$ K in $Ba(Fe_{1-x}Co_x)_2As_2$ and much greater than $\Delta/k_B = 140 - 172$ K in $LaFeAsO_{1-x}F_x$.[29,38,69] The mechanism of the pseudogap behavior in $\chi_{\rm spin}$ above T_c has been controversial, but as already discussed, one of the plausible scenarios is changes in the effective DOS of low-energy spin excitations. In canonical Fermi liquid systems, if the chemical potential μ is located on the lower energy side of a sharp peak of the density of states, $\chi_{\rm spin}$ could decrease at lower temperatures when the DOS becomes smaller. In this scenario, the ^{77}K results shown in Fig. 7.41 imply that there is a large, sharp peak of the DOS somewhat above the Fermi energy E_F. However, other possibilities such as short-range AFM correlations and frustration effects between nearest and next-nearest neighbor Fe-Fe exchange interactions or more exotic scenarios are also possible. It is noteworthy that the pseudogap behavior in $K_xFe_{2-y}Se_2$ is most prominent among the iron-based superconductors, possibly giving a hint for the mystery of the pseudogap behavior in the future.

Below T_c, ^{77}K suddenly begins to decrease from $^{77}K_c \sim 0.25\%$, then levels off at $K_{\rm orb} = 0.11 \pm 0.01\%$, and $^{77}K_{ab}$ also decreases immediately below T_c. These observations in the superconducting state below T_c are consistent with the suppression of both $\chi_{ab,\rm spin}$ and $\chi_{c,\rm spin}$ expected for the singlet pairing of Cooper pairs.

Figure 7.42 presents the temperature dependence of $^{77}(1/T_1T)$ measured in the field of 8.3 T applied along the c-axis (preliminary measurements in the field applied along the ab-plane showed nearly identical behavior at least up to 160 K).[106] $^{77}(1/T_1T)$ decreases with temperature from 290 K down to T_c. The observed temperature dependence is qualitatively similar to that of ^{77}K.

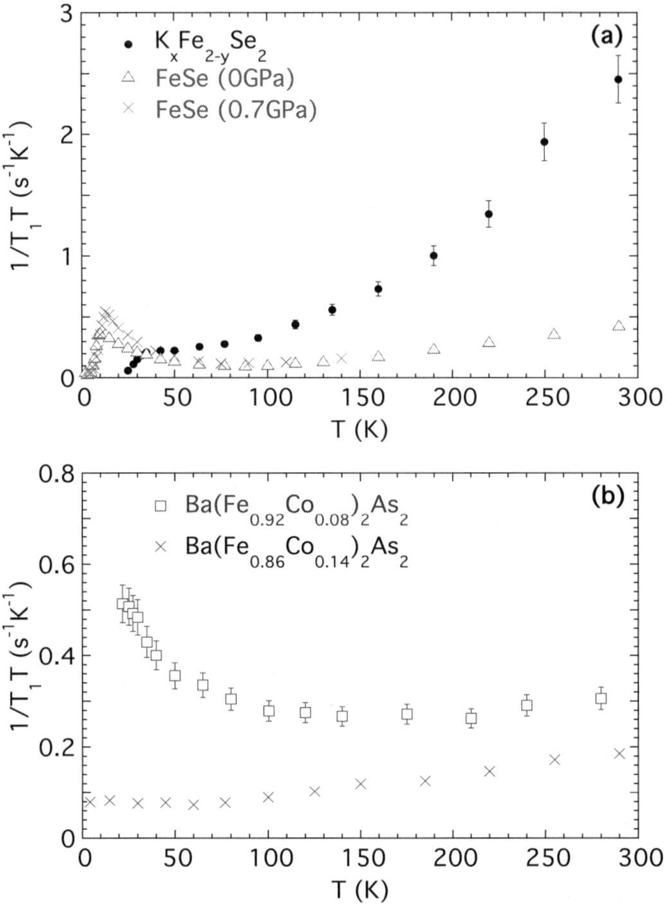

Figure 7.42 (a) The temperature dependence of $^{77}(1/T_1T)$ observed for $K_xFe_{2-y}Se_2$ is compared with the results for FeSe in 0 GPa ($T_c = 9$ K) and 0.7 GPa ($T_c = 14$ K). (b) Representative results of $^{75}(1/T_1T)$ observed at ^{75}As sites of $Ba(Fe_{1-x}Co_x)_2As_2$ with $x = 0.08$ ($T_c = 25$ K) and $x = 0.14$ ($T_c \sim 0$ K) with an external magnetic field applied along the *ab*-plane. Notice the qualitative similarity between $K_xFe_{2-y}Se_2$ and the overdoped $Ba(Fe_{0.86}Co_{0.14})_2As_2$; both of these materials lack the hole bands near the zone center. Figure reprinted from D. A. Torchetti*et al.*: *Phys. Rev. B* **83**, 104508 (2011). Copyright 2011 by the America Physical Society.

Since $^{77}(1/T_1T)$ measures all wave-vector $\boldsymbol{q}$ modes of spin fluctuations while ^{77}K picks up the uniform $\boldsymbol{q} = 0$ mode exclusively, there is no a priori reason to expect such analogous behaviors. In some iron-based high T_c superconductors $1/T_1T$ and K exhibit qualitatively different temperature dependencies near T_c. For comparison, $^{77}(1/T_1T)$ observed for FeSe are plotted in the same Fig. 7.42.[106] $1/T_1T$ is enhanced toward T_c in both FeSe (P = 0 and 0.7 GPa), providing an evidence for enhancement of the low-frequency components of the AFM spin fluctuations toward T_c at some finite wave vectors in FeSe. In contrast, $^{77}(1/T_1T)$ in $K_xFe_{2-y}Se_2$ indicates that this is not the case, and are qualitatively similar to overdoped, nonsuperconducting $Ba(Fe_{0.86}Co_{0.14})_2As_2$ [as shown in Fig. 7.42(b)]. The absence of AFM fluctuations in the latter was interpreted as the consequence of the lack of the nested Fermi surfaces between the hole and electron bands, because overdoped electrons fill up the hole bands at the zone center. Since ARPES data suggest that $K_xFe_{2-y}Se_2$ does not possess a hole Fermi surface near the Brillouin-zone center,[103] the absence of low-frequency AFM fluctuations related with the nesting of the Fermi surfaces may be consistent with such an interpretation.

Figure 7.43 shows the temperature dependence of $^{77}(1/T_1)$ reported by Kotegawa *et al.*[105] The steep decrease in $^{77}(1/T_1)$ was observed below T_c without any sign of a coherence peak, and its temperature dependence is T^6-like just below T_c, approaches T^3-like below $T_c/2$, and then starts to deviate from T^3 below $T_c/5$. Such power-law behavior in $1/T_1$ has been observed in most Fe-based superconductors. The absence of a coherence peak and non exponential temperature dependence of $1/T_1$ clearly excludes the possibility of a single isotropic gap pointed out from the ARPES measurement.[102] The dashed curve is a calculation using a single isotropic gap of $\Delta_0 = 4k_BT_c$ ignoring the coherence effect. This curve disagrees with the data for the low-temperature region. Non exponential behavior suggests that multigaps with different magnitudes of gap size or anisotropic superconductivity with a finite DOS in the SC gap is realized in $K_xFe_{2-y}Se_2$. The red solid curve is the results of a calculation based on two isotropic gaps, which are expected on the $s_\pm$ state adopted above in $LaFeAs(O_{1-x}F_x)$ and $(Ba_{1-x}K_x)Fe_2As_2$. However it should be noted that the sign

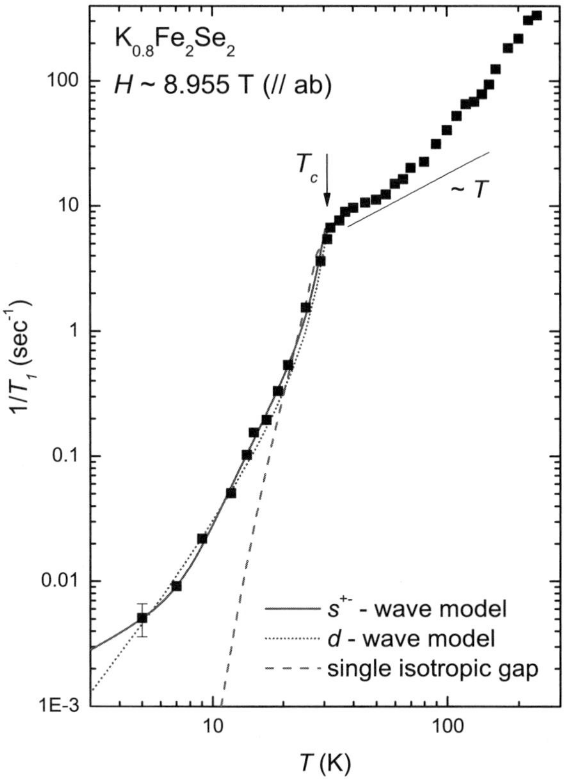

Figure 7.43 Temperature dependence of $1/T_1$ for $K_0.8Fe_{2-y}Se_2$. $1/T_1$ does not show a coherence peak just below T_c. Non exponential behavior of $1/T_1$ excludes a single isotropic gap. The red solid curve is the best-fitting result obtained using the $s_{\pm}$-wave model (See in the text). The blue dotted curve (the green dashed curve) was obtained using the *d*-wave model with a single superconducting gap. Figure reprinted from H. Kotegawa *et al.*: *J. Phys. Soc. Jpn.* **80**, 043708 (2011). Copyright 2011 by the Physical Society of Japan.

reversal in the SC gap of $K_xFe_{2-y}Se_2$ might be different from that in the above two compounds, since the recent ARPES measurements suggest that holelike FSs at the zone center are completely filled with electron doping and only electron-like FSs exist at the zone edge. In such a case, the sign reversal of SC gaps might occur between the adjacent electron-like FSs at the zone edges, which is classified as a nodeless d-wave state. The $1/T_1$ results are reproduced well using two isotropic gaps with $\Delta_1 = 4.2k_B T_c$, $\Delta_2 = 1.7k_B T_c$, $N_1/N_2 = 0.7/0.3$, and smearing factor $\eta = 0.048\Delta_1$. $\Delta_1 = 4.2k_B T_c$ indicates the superconductivity in the strong-coupling regime, and is comparable to that of $(Ba_{0.6}K_{0.4})Fe_2As_2$. However, the temperature dependence of $1/T_1$ can be also reproduced with a two-dimensional d-wave model $[\Delta(\phi) = \Delta_0 \cos(2\phi)]$ with nodes along the k_z direction together with a residual DOS. The calculation is shown by the blue dotted curve, in which $\Delta_0 = 5k_B T_c$ is adopted, and the $1/T_1$ at the lowest temperature gives the upper limit of the residual DOS of 7%. However, the small residual DOS of $\sim 3 - 7\%$ seems to be incompatible with the nodal d-wave model, since a deficiency of irons would give rise to a large residual DOS in a d-wave superconductor, in general. Although the temperature dependence of $1/T_1$ in the superconducting state excludes a single isotropic gap and suggests that the two isotropic-gap model is more plausible than the d-wave symmetry, but it is impossible to identify the symmetry realized in $K_xFe_{2-y}Se_2$ solely from the present $1/T_1$ result.

7.4 Summary

We have shown NMR results obtained in various iron-based superconductors. On the basis of the NMR results reported so far, following results are derived.

(1) In undoped compounds, except for LiFeSe and FeSe, the structual and AFM transitions are observed. LaFeAsO and NaFeAs exhibit AFM ordering at T_N, $\sim$ 15 K below the structural transition T_S. Since AFM fluctuations are enhanced below T_S, the magnetic ordering is intimately coupled with the structural transition. On the other hand, the two transitions occur simultaneously with a first-order character in $BaFe_2As_2$ and other "122" componds. The antiferromagnetic $\boldsymbol{q}$ vectors, which are $(\pi, 0)$ and $(0, \pi)$ in the "*unfolded*" Brillouin zone, corresponding to a stripe-type magnetic structure, are suggested also from the analyses of the ^{75}As-NMR spectrum.

(2) From the F-concentration and temperature dependence of $1/T_1T$, the low-energy magnetic fluctuations are strongly suppressed by electron doping, and the pseudogap behavior is observed in $1/T_1T$ for $LaFeAs(O_{1-x}F_x)$ and $LaFeAsO_{1-\delta}$ with the maximum T_c. The pseudogap behavior becomes pronounced with F (electron)-doping. The doping dependence of the psuedogap behavior is opposite to that observed in the cuprate superconductors, in which the pseudogap behavior is significant in the underdoped region. It seems that the pseudogap behavior is ascribed to the character of the band structure near E_F, rather than short-range AFM correlations in the iron-based superconductors. The similar pseudogap behaviors have been observed in heavily electron-doped $Ba(Fe_{1-x}Co_x)_2As_2$, superconducting LiFeAs and $K_xFe_{2-y}Se_2$. However, other possibilities such as frustration effects between nearest and next nearest Fe spins are also suggested.

(3) $1/T_1T$ in $LaFeAs(O_{1-x}F_x)$ depends significantly on F-concentration x, whereas T_c is insensitive to x within a range from $x = 0.04$ to 0.14. Almost a constant value of T_c against the F-concentration, irrelevant to the drastic change of $1/T_1T$ in the normal state, suggests that the low-energy spin fluctuations probed through $1/T_1$ measurements do not play a vital role for the superconductivity in $LaFeAs(O_{1-x}F_x)$.

(4) In contrast with $LaFeAs(O_{1-x}F_x)$, the NMR measurements on superconducting Ba-"122" with various chemical substitution (P, Co, and K doping) reveal that two-dimensional AFM fluctuations are notably enhanced on approaching the AFM phase from the SC dome, and that the magnetically ordered temperature θ deduced from $1/T_1T$ vanishes at optimal doping for the P and Co doped systems. These results strongly indicate that quantum critical points are present near optimal doping, and strongly suggest that the low-energy AFM fluctuations correlate with T_c in high-T_c superconductivity in Ba-"122". Similar tendency has been observed in superconducting FeSe. Therefore, we suggest that superconducting "122" and "11" compounds share the commonality with heavy-fermion superconductors which are also in the proximity of magnetic ground states.

(5) The relationship between superconductivity and magnetic fluctuations probed with NMR measurements is quite different between $LaFeAs(O_{1-x}F_x)$ and Ba-"122". The difference of the relationship is also suggested from their diagrams of the two systems [see in Figs. 7.9 and 7.17]. The AFM phase disappears abruptly, and the SC phase with a nearly constant T_c appears in $LaFeAs(O_{1-x}F_x)$. In contrast, T_N is gradually suppressed by doping, and the SC phase appears as if it covers a magnetic quantum critical point in $BaFe_2(As_{1-x}P_x)_2$ and $Ba(Fe_{1-x}Co_x)_2As_2$. However, since $1/T_1$ can detect only low-energy magnetic fluctuations (typically $\sim$ mK order), and could be insensitive to detect important magnetic fluctuations for Cooper pairing if their characteristic energy exceeds the limitation of NMR experiments, inelastic neutron experiments over wide $\boldsymbol{q}$ and energy space are needed for uncovering magnetic fluctuations thoroughly, particularly on "1111" compounds.

(6) $1/T_1$ in the SC state in all iron-based superconductors shows the absence of a coherence peak just below T_c and nonexponential temperature dependence at low temperatures, excluding conventional superconductivity with a finite gap. The power-law T dependence of $1/T_1$ well below T_c observed except for $BaFe_2(As_{0.67}P_{0.33})_2$ can be understood with an anisotropic superconducting gap and/or multiple superconducting gaps with an impurity effect whereas the absence of field-induced extra relaxation and of a residual DOS at low temperatures seem to be incompatible with d-wave superconductivity. The absence of a coherence peak in $1/T_1$ just below T_c can be understood with the $s_{\pm}$-wave model, except for $K^xFe_{2-y}Se_2$.

(7) $1/T_1$ result in the SC state on $BaFe_2(As_{0.67}P_{0.33})_2$ shows the existence of a residual DOS at zero energy despite its high purity, which is consistent with the presence of nodes in the SC gap inferred from the thermal conductivity and penetration-depth results. This suggests that strikingly different SC gaps appear in $BaFe_2(As_{1-x}P_x)_2$ despite comparable high-T_c values, analogous phase diagrams, and similar FSs with $(Ba_{1-x}K_x)Fe_2As_2$. Since the FSs in $BaFe_2(As_{1-x}P_x)_2$ become three dimensional with substituting P for As, the dimensionality of the FSs might be related with the SC gap structure.

(8) Quite recently, $K_xFe_{2-y}Se_2$ with a relatively high-T_c of 32 K was discovered, and has been clarified to possess quite different FS character: ARPES measurements suggest that the holelike FSs at the zone center are completely filled with electron doping and only the electron-like FSs exist at the zone edge. Knight-shift and $1/T_1T$ behaviors in the normal state show the most prominent pseudogap behavior among the iron-based superconductors, and show the absence of AFM fluctuations in the normal state. These results can be interpreted as the consequence of the lack of the holelike FSs, which is consistent with the ARPES results. The absence of a coherence peak in $1/T_1$ just below T_c may need a different interpretation from the $s_\pm$-wave model, since the holelike FSs are absent. The $s_\pm$ pairing driven by the interband nesting between holelike and electronlike FSs might not work in the superconductivity.

Acknowledgment

The authors thank to Shunsaku Kitagawa, and Tetsuya Iye for supporting this manuscript. The authors were supported by a Grant-in-Aid for "Transformative Research-project on Iron Pnictides (TRIP)" from Japan Science and Technology Agency (JST), a Grant-in-Aid for Scientific Research on Innovative Areas "Heavy Electrons" (No. 20102006) from The Ministry of Education, Culture, Sports, Science, and Technology (MEXT) of Japan, a Grant-in-Aid for the Global COE Program "The Next Generation of Physics, Spun from Universality and Emergence" from MEXT of Japan, and the Grants-in-Aid for Scientific Research from Japan Society for Promotion of Science (JSPS).

References

1. Y. Kamihara, T. Watanabe, M. Hirano, and H. Hosono, *J. Am. Chem. Soc.* **130**, 3296, (2008).
2. H. Takahashi, K. Igawa, K. Arii, Y. Kamihara, M. Hirano, and H. Hosono, *Nature*, **453**, 376, (2008).

3. C. Wang, L. Li, S. Chi, Z. Zhu, Z. Ren, Y. Li, Y. Wang, X. Lin, Y. Luo, S. Jiang, X. Xu, G. Cao, and Z. Xu. **83**, 67006, (2008).
4. Z.-A. Ren, W. Lu, J. Yang, W. Yi, X.-L. Shen, Z.-C. Li, G.-C. Che, X.-L. Dong, L.-L. Sun, F. Zhou, and Z.-X. Zhao, *Chin. Phys. Lett.* **25**, 2215, (2008).
5. H. Kito, H. Eisaki, and A. Iyo, *J. Phys. Soc. Jpn.* **77**, 063707, (2008).
6. Z.-A. Ren, G.-C. Che, X.-L. Dong, J. Yang, W. Lu, W. Yi, X.-L. Shen, Z.-C. Li, L.-L. Sun, F. Zhou, and Z.-X. Zhao, *EPL.* **83**, 17002, (2008).
7. C. P. Slichter, *Principles of Magnetic Resonance (3rd Enlarged and Update Edition).* (Springer, New York, Berlin, Heidelberg, 1996).
8. K. Asayama, *Nuclear Magnetic Resonance in Itinerant Electron System -Magnetism and Superconductivity.* (Shokabo, Japan, 2002).
9. R. E. Walstedt, *The NMR Probe of High-T_c Materials.* (Springer, Berlin, Heidelberg, 2008).
10. N. J. Curro, *Rep. Prog. Phys.* **72**, 026502, (2009).
11. T. Moriya, *Prog. Theor. Phys.* **16**, 641, (1956).
12. T. Moriya, *Spin Fluctuations in Itinerant Electron Magnetism.* Springer-Verlag, (1985).
13. D. E. MacLaughlin, *Magnetic Resonance in the Superconducting state.* ACADEMIC PRESS (Solid State Physics, Vol. 31), New York, San Francisco, London, (1976).
14. K. Yosida, *Phys. Rev.* **110**, 769, (1958).
15. H. Tou, Y. Kitaoka, K. Asayama, N. Kimura, Y. Ōnuki, E. Yamamoto, and K. Maezawa, *Phys. Rev. Lett.* **77**, 1374, (1996).
16. K. Ishida, H. Mukuda, Y. Kitaoka, K. Asayama, Z. Q. Mao, Y. Mori, and Y. Maeno, *Nature,* **396**, 658, (1998).
17. K. Ishida, H. Mukuda, Y. Kitaoka, Z. Q. Mao, H. Fukazawa, and Y. Maeno, *Phys. Rev. B.* **63**, 060507 (R), (2001).
18. L. C. Hebel and C. P. Slichter, *Phys. Rev.* **113**, 1504, (1959).
19. N. Okubo and Y. Masuda, *Phys. Rev. Lett.* **20**, 1475–1477, (1968).
20. V. A. Stenger, C. H. Pennington, D. R. Buffinger, and R. P. Ziebarth, *Phys. Rev. Lett.* **74**, 1649–1652, (1995).
21. S. Schmitt-Rink, K. Miyake, and C. M. Varma, *Phys. Rev. Lett.* **57**, 2575, (1986).
22. M. Sigrist and K. Ueda, *Rev. Mod. Phys.* **63**, 239, (1991).
23. K. Ishida, Y. Kitaoka, N. Ogata, T. Kamino, K. Asayama, J. R. Cooper, and N. Athanassopoulou, *J. Phys. Soc. Jpn.* **62**, 2803–2818.
24. K. Ishida, Y. Nakai, and H. Hosono, *J. Phys. Soc. Jpn.* **78**, 062001, (2009).

25. H. Ogino, K. Machida, A. Yamamoto, K. Kishio, J. ichi Shimoyama, T. Tohei, and Y. Ikuhara, *Supercond. Sci. and Technol.* **23**, 115005, (2010).
26. I. I. Mazin, D. J. Singh, M. D. Johannes, and M. H. Du, *Phys. Rev. Lett.* **101**, 057003, (2008).
27. C. de la Cruz, Q. Huang, J. W. Lynn, J. Li, W. R. II, J. L. Zarestky, H. A. Mook, G. F. Chen, J. L. Luo, N. L. Wang, and P. Dai, *Nature*, **453**, 899, (2008).
28. Y. Nakai, K. Ishida, Y. Kamihara, M. Hirano, and H. Hosono, *J. Phys. Soc. Jpn.* **77**, 073701, (2008).
29. Y. Nakai, S. Kitagawa, K. Ishida, Y. Kamihara, M. Hirano, and H. Hosono, *New J. Phys.*, **11**, 045004, (2009).
30. Y. Kitaoka and Yasuoka, *J. Phys. Soc. Jpn.* **48**, 1460, (1980).
31. T. Nomura, S. W. Kim, Y. Kamihara, M. Hirano, P. V. Sushko, K. Kato, M. Takata, A. L. Shluger, and H. Hosono, *Supercond. Sci. Technol.* **21**, 125028, (2008).
32. H. Mukuda, N. Terasaki, N. Tamura, H. Kinouchi, M. Yashima, Y. Kitaoka, K. Miyazawa, P. Shirage, S. Suzuki, S. Miyasaka, S. Tajima, H. Kito, H. Eisaki, and A. Iyo, *J. Phys. Soc. Jpn.* **78**, 084717, (2009).
33. G. Lang, H.-J. Grafe, D. Paar, F. Hammerath, K. Manthey, G. Behr, J. Werner, and B. Buüchner, *Phys. Rev. Lett.* **104**, 097001, (2010).
34. T. Nakano, N. Fujiwara, Y. Kamihara, M. Hirano, H. Hosono, H. Okada, and H. Takahashi, *Phys. Rev. B.* **82**, 172502, (2010).
35. Y. Kobayashi, E. Satomi, S. C. Lee, and M. Sato, *J. Phys. Soc. Jpn.* **79**, 093709, (2010).
36. R. Klingeler, N. Leps, I. Hellmann, A. Popa, C. Hess, A. Kondra, J. Hamann-Borrero, G. Behr, V. Kataev, and B. Buechner, *Phys. Rev. B.* **81**, 024506, (2010).
37. M. Takigawa, A. P. Reyes, P. C. Hammel, J. D. Thompson, R. Heffner, Z. Fisk, and K.C.Ott, *Phys. Rev. B.* **43**, 247, (1991).
38. T. Imai, K. Ahilan, F. Ning, M. A. McGuire, A. S. Sefat, R. Jin, B. C. Sales, and D. Mandrus, *J. Phys. Soc. Jpn.* **77 Suppl. C**, 47, (2008).
39. S. Takeshita, R. Kadono, M. Hiraishi, M. Miyazaki, A. Koda, Y. Kamihara, and H. Hosono, *J. Phys. Soc. Jpn.* **77**, 103703, (2008).
40. N. Terasaki, H. Mukuda, M. Yashima, Y. Kitaoka, K. Miyazawa, P. M. Shirage, H. Kito, H. Eisaki, and A. Iyo, *J. Phys. Soc. Jpn.* **78**, 013701, (2009).
41. H. Ikeda, *J. Phys. Soc. Jpn.* **77**, 123707, (2008).

42. M. Ishikado, R. Kajimoto, S. Shamoto, M. Arai, A. Iyo, K. Miyazawa, P. M. Shirage, H. Kito, H. Eisaki, S. Kim, H. Hosono, T. Guidi, R. Bewley, and S. M. Bennington, *J. Phys. Soc. Jpn.* **78**, 043705, (2009).
43. K. Kitagawa, N. Katayama, K. Ohgushi, M. Yoshida, and M. Takigawa, *J. Phys. Soc. Jpn.* **77**, 114709, (2008).
44. K. Kitagawa, N. Katayama, K. Ohgushi, and M. Takigawa, *J. Phys. Soc. Jpn.* **78**, 063706, (2009).
45. S. Kitagawa, Y. Nakai, T. Iye, K. Ishida, Y. Kamihara, M. Hirano, and H. Hosono, *Phys. Rev. B* **81**, 212502, (2010).
46. A. Kawabata, S. C. Lee, T. Moyoshi, Y. Kobayashi, and M. Sato, *J. Phys. Soc. Jpn.* **77**, 103704, (2008).
47. H.-J. Grafe, D. Paar, G. Lang, N. J. Curro, G. Behr, J. Werner, J. Hamann-Borrero, C. Hess, N. Leps, R. Klingeler, and B. Büchner, *Phys. Rev. Lett.* **101**, 047003, (2008).
48. H. Mukuda, N. Terasaki, H. Kinouchi, M. Yashima, Y. Kitaoka, S. Suzuki, S. Miyasaka, S. Tajima, K. Miyazawa, P. Shirage, H. Kito, H. Eisaki, and A. Iyo, *J. Phys. Soc. Jpn.* **77**, 093704, (2008).
49. G.-q. Zheng, H. Ozaki, Y. Kitaoka, P. Kuhns, A. P. Reyes, and W. G. Moulton, *Phys. Rev. Lett.* **88**, 077003, (2002).
50. K. Ishida, H. Mukuda, Y. Kitaoka, Z. Q. Mao, Y. Mori, and Y. Maeno, *Phys. Rev. Lett.* **84**, 5387, (2000).
51. N. J. Curro, T. Calwell, E. D. Bauer, L. A. Morales, M. J. Graf, Y. Bang, A. V. Balatsky, J. D. Thompson, and J. L. Sarrao, *Nature* **434**, 622, (2005).
52. K. Hashimoto, T. Shibauchi, T. Kato, K. Ikada, R. Okazaki, H. Shishido, M. Ishikado, H. Kito, A. Iyo, H. Eisaki, S. Shamoto, and Y. Matsuda, *Phys. Rev. Lett.* **102**, 017002, (2009).
53. K. Kuroki, S. Onari, R. Arita, H. Usui, Y. Tanaka, H. Kontani, and H. Aoki, *Phys. Rev. Lett.* **101**, 087004, (2008).
54. L. S. Borkowski and P. J. Hirschfeld, *Phys. Rev. B* **49**, 15404, (1994).
55. D. Parker, O. V. Dolgov, M. M. Korshunov, A. A. Golubov, and I. I. Mazin, *Phys. Rev. B* **78**, 134524, (2008).
56. A. V. Chubukov, D. V. Efremov, and I. Eremin, *Phys. Rev. B* **78**, 134512, (2008).
57. M. M. Parish, J. Hu, and B. A. Bernevig, *Phys. Rev. B* **78**, 144514, (2008).
58. Y. Bang and H.-Y. Choi, *Phys. Rev. B* **79**, (2009).
59. Y. Nagai, N. Hayashi, N. Nakai, H. Nakamura, M. Okumura, and M. Machida, *New J. Phys.* **10**, 103026, (2008).

60. S. Kawasaki, K. Shimada, G. F. Chen, J. L. Luo, N. L. Wang, and G. qing Zheng, *Phys. Rev. B* **78**, 220506 (R), (2008).

61. H. Mukuda, M. Nitta, M. Yashima, Y. Kitaoka, P. M. Shirage, H. Eisaki, and A. Iyo, *J. Phys. Soc. Jpn.* **79**, 113701, (2010).

62. M. Rotter, M. Tegel, and D. Johrendt, *Phys. Rev. Lett.* **101**, 107006, (2008).

63. R. R. Urbano, E. L. Green, W. G. Moulton, A. P. Reyes, P. L. Kuhns, E. M. Bittar, C. Adriano, T. M. Garitezi, L. Bufaiçal, and P. G. Pagliuso, *Phys. Rev. Lett.* **105**, 107001, (2010).

64. A. S. Sefat, R. Jin, M. A. McGuire, B. C. Sales, D. J. Singh, and D. Mandrus, *Phys. Rev. Lett.* **101**, 117004, (2008).

65. S. Nandi, M. G. Kim, A. Kreyssig, R. M. Fernandes, D. K. Pratt, A. Thaler, N. Ni, S. L. Bud'ko, P. C. Canfield, J. Schmalian, R. J. McQueeney, and A. I. Goldman, *Phys. Rev. Lett.* **104**, 057006, (2010).

66. S. Jiang, H. Xing, G. Xuan, C. Wang, Z. Ren, C. Feng, J. Dai, Z. Xu, and G. Cao, *J. Phys. Condens. Matter* **21**, 382203, (2009).

67. S. Kasahara, T. Shibauchi, K. Hashimoto, K. Ikada, S. Tonegawa, R. Okazaki, H. Shishido, H. Ikeda, H. Takeya, K. Hirata, T. Terashima, and Y. Matsuda, *Phys. Rev. B* **81**, 184519, (2010).

68. J. H. Chu, J. G. Analytis, C. Kucharczyk, and I. R. Fisher, *Phys. Rev. B* **79**, 014506, (2008).

69. F. L. Ning, K. Ahilan, T. Imai, A. S. Sefat, M. A. McGuire, B. C. Sales, D. Mandrus, P. Cheng, B. Shen, and H. H. Wen, *Phys. Rev. Lett.* **104**, 017204, (2010).

70. C. de la Cruz, W. Z. Hu, S. Li, Q. Huang, J. W. Lynn, M. A. Green, G. F. Chen, N. L. Wang, H. A. Mook, Q. Si, and P. Dai, *Phys. Rev. Lett.* **104**, 017204, (2010).

71. J. Dai, Q. Si, J. X. Zhu, and E. Abrahams, *Proc. Natl. Acad. Sci. U.S.A.* **106**, 4118, (2009).

72. Y. Nakai, T. Iye, S. Kitagawa, K. Ishida, Y. M. S. Kasahara, T. Shibauchi, and T. Terashima, *Phys. Rev. B* **81**, 020503 (R), (2010).

73. Y. Nakai, T. Iye, S. Kitagawa, K. Ishida, H. Ikeda, S. Kasahara, H. Shishido, T. Shibauchi, Y. Matsuda, and T. Terashima, *Phys. Rev. Lett.* **105**, 107003, (2010).

74. M. Yashima, H. Nishimura, H. Mukuda, Y. Kitaoka, K. Miyazawa, P. M. Shirage, K. Kihou, H. Kito, H. Eisaki, and A. Iyo, *J. Phys. Soc. Jpn.* **78**, 103702, (2009).

75. H. Fukazawa, T. Yamazaki, K. Kondo, Y. Kohori, N. Takeshita, P. M. Shirage, K. Kihou, K. Miyazawa, H. Kito, H. Eisaki, and A. Iyo, *J. Phys. Soc. Jpn.* **78**, 033704, (2009).
76. G. C. Carter, L. H. Bennett, and D. J. Kahan, *Metallic Shifts in NMR.* **(Progress in Materials Science vol 20)**, 1977, (Oxford: Pergamon).
77. T. Moriya, Y. Takahashi, and K. Ueda, *J. Phys. Soc. Jpn.* **59**, 2905, (1990).
78. K. Matan, R. Morinaga, K. Iida, and T. J. Sato, *Phys. Rev. B* **79**, 054526, (2009).
79. K. Hashimoto, M. Yamashita, S. Kasahara, Y. Senshu, N. Nakata, S. Tonegawa, K. Ikada, A. Serafin, A. Carrington, T. Terashima, H. Ikeda, T. Shibauchi, and Y. Matsuda, *Phys. Rev. B* **81**, 220501, (2010).
80. H. Shishido, A. F. Bangura, A. I. Coldea, S. Tonegawa, K. Hashimoto, S. Kasahara, P. M. C. Rourke, H. Ikeda, T. Terashima, R. Settai, Y. Ōnuki, D. Vignolles, C. Proust, B. Vignolle, A. McCollam, Y. Matsuda, and T. S. A. Carrington, *Phys. Rev. Lett.* **104**, 057008, (2010).
81. K. Matano, Z. A. Ren, X. L. Dong, L. L. Sun, Z. X. Zhao, and G. qing Zheng, *Euro. Phys. Lett.* **83**, 57001, (2008).
82. K. Matano, G. L. Sun, D. L. Sun, C. T. Lin, M. Ichioka, and G. qing Zheng, *Euro. Phys. Lett.* **87**, 27012, (2009).
83. G. Mu, B. Zeng, P. Cheng, Z. Wang, L. Fang, B. Shen, L. Shan, C. Ren, and H. Wen, *Chin. Phys. Lett.* **27**.
84. Y. Kobayashi, A. Kawabata, S. C. Lee, T. Moyoshi, and M. Sato, *J. Phys. Soc. Jpn.* **78**, 073704, (2009).
85. H. Fukazawa, Y. Yamada, K. Kondo, T. Saito, Y. Kohori, K. Kuga, Y. Matsumoto, S. Nakatsuji, H. Kito, P. M. Shirage, K. Kihou, N. Takeshita, C. H. Lee, A. Iyo, and H. Eisaki, *J. Phys. Soc. Jpn.* **78**, 083712, (2009).
86. K. Kuroki, H. Usui, S. Onari, R. Arita, and H. Aoki, *Phys. Rev. B* **79**, 224511, (2009).
87. V. Mishra, G. Boyd, S. Graser, T. Maier, P. J. Hirschfeld, and D. J. Scalapino, *Phys. Rev. B* **79**, 094512, (2009).
88. K. Seo, B. A. Bernevig, and J. Hu, *Phys. Rev. Lett.* **101** (20), 206404 (Nov, 2008).
89. A. V. Chubukov, M. G. Vavilov, and A. B. Vorontsov, *Phys. Rev. B* **80**, 140515, (2009).
90. Z. Li, Q.-Q. L. C.-Q. J. M. I. Y. Ooe, X-C. Wang, and G. q. Zheng, *J. Phys. Soc. Jpn.* **79**, 083702, (2010).
91. P. Jeglič, A. Potočnik, M. Klanjšek, M. Bobnar, M. Jagodič, K. Koch, H. Rosner, S. Margadonna, B. Lv, A. M. Guloy, and D. Arčon, *Phys. Rev. B* **81**, 140511, (2010).

92. L. Ma, J. Zhang, G. F. Chen, and W. Yu, *Phys. Rev. B* **82**, 180501, (2010).
93. K. Kitagawa, Y. Mezaki, K. Matsubayashi, Y. Uwatoko, and M. TAKIGAWA, *J. Phys. Soc. Jpn.* **80**, 033705, (2011).
94. L. Ma, G. F. Chen, D.-X. Yao, J. Zhang, S. Zhang, T.-L. Xia, and W. Yu, *Phys. Rev. B* **83**, 132501, (2011).
95. S. Li, C. de la Cruz, Q. Huang, G. F. Chen, T.-L. Xia, J. L. Luo, N. L. Wang, and P. Dai, *Phys. Rev. B.* **80**, 020504, (2009).
96. Y. Mizuguchi and Y. Takano, *J. Phys. Soc. Jpn.* **79**, 102001, (2010).
97. T. Imai, K. Ahilan, F. L. Ning, T. M. McQueen, and R. J. Cava, *Phys. Rev. Lett.* **102**, 177005, (2009).
98. H. Kotegawa, S. M. andY. Hara, H. Tou, Y. Mizuguchi, and Y. Takano, *J. Phys. Soc. Jpn.* **77**, 113703, (2008).
99. Y. Shimizu, T. Yamada, T. Takami, S. Niitaka, H. Takagi, and M. Itoh, *J. Phys. Soc. Jpn.* **78**, 123709, (2009).
100. C. Michioka, H. Ohta, M. Matsui, J. Yang, K. Yoshimura, and M. Fang, *Phys. Rev. B.* **82**, 064506 (Aug, 2010).
101. J. Guo, S. Jin, G. Wang, S. Wang, K. Zhu, T. Zhou, M. He, and X. Chen, *Phys. Rev. B* **82**, 180520, (2010).
102. Y. Zhang, L. X. Yang, M. Xu, Z. R. Ye, F. Chen, C. He, H. C. Xu, J. Jiang, B. P. Xie, J. J. Ying, X. F. Wang, X. H. Chen, J. P. Hu, M. Matsunami, S. Kimura, and D. L. Feng, *Nat. Mater.* **10**, 273, (2011).
103. T. Qian, X.-P. Wang, W.-C. Jin, P. Zhang, P. Richard, G. Xu, X. Dai, Z. Fang, J.-G. Guo, X.-L. Chen, and H. Ding, *Phys. Rev. Lett.* **106**, 187001, (2011).
104. W. Yu, L. Ma, J. B. He, D. M. Wang, T.-L. Xia, G. F. Chen, and W. Bao, *Phys. Rev. Lett.* **106**, 197001, (2011).
105. H. Kotegawa, Y. Hara, H. Nohara, H. Tou, Y. Mizuguchi, H. Takeya, and Y. Takano, *J. Phys. Soc. Jpn.* **80**, 043708, (2011).
106. D. A. Torchetti, M. Fu, D. C. Christensen, K. J. Nelson, T. Imai, H. C. Lei, and C. Petrovic, *Phys. Rev. B* **83**, 104508, (2011).

Chapter 8

Material Specific Model Hamiltonians and Analysis on the Pairing Mechanism

Kazuhiko Kuroki

Department of Engineering Science, The University of Electro-Communications, Chofu, Tokyo 182-8585, Japan, and
JST, TRIP, Chiyoda, Tokyo 102-0075, Japan
kuroki@vivace.e-one.uec.ac.jp

In the present chapter, we mainly focus on theoretical studies which adopt material specific model Hamiltonians that take into account the band structure obtained from first principles calculation. In particular, we pay special attention to the studies based on the models that consider all five of the Fe 3*d* bands. We first review the model construction from the first principles band calculation, and then describe the main features of the band structure and the Fermi surfaces. We then discuss how the many body interactions together with these peculiar band structures and Fermi surfaces result in the development of spin fluctuations or antiferromagnetism. As for the mechanism of superconductivity, we describe mainly the spin-fluctuation-mediated pairing, while we also comment on some other possibilities. Finally, we give a review on how the electronic states depend on the materials, affecting both the spin fluctuations and the superconductivity.

Iron-Based Superconductors: Materials, Properties, and Mechanisms
Edited by Nan-Lin Wang, Hideo Hosono, and Pengcheng Dai

ISBN 978-981-4303-22-4 (Hardcover), 978-981-4303-23-1 (eBook)
www.panstanford.com

8.1 Introduction

The discovery of superconductivity at 26 K in the electron doped LaFeAsO by Hosono's group[1] and subsequent increase in the transition temperature (T_c) exceeding 50 K in the same family of compounds[2] have given great impact to the field of condensed matter physics. This is not only because of the high values of T_c, but also because this poses a fundamental question on unconventional mechanisms of superconductivity in a wider class of compounds other than the high T_c cuprates.

The iron based superconductors share some common features with the cuprates, i.e., they are layered materials with 3*d* electrons located around the Fermi level, the parent compounds are antiferromagnetically ordered, and superconductivity appears when magnetism is suppressed upon doping holes or electrons, suggesting a possible relationship between magnetism and superconductivity. In fact, at the early stage of the study, a purely-phonon-mediated pairing mechanism was shown to be unlikely for these materials since the T_c is too high to result from a moderate electron-phonon coupling.[3,4] Therefore, there is a strong possibility that electronic degrees of freedom are playing an important role in the pairing mechanism, and spin-fluctuation-mediated pairing has indeed been proposed right after the discovery.[5–7] On the other hand, there is a large difference compared to the cuprates, in which a single band intersects the Fermi surface. The iron based superconductors are essentially multiband/multiorbital systems, and there are four or five pocket like (or cylinder in the three dimensional Brillouin zone) Fermi surfaces originating from several different orbitals. In the spin-fluctuation-mediated pairing scenario, the interaction between disconnected electron and hole Fermi surfaces induces spin fluctuations, which can give rise to a superconducting gap that is basically *s*-wave but changes sign between electron and hole pockets, hence termed as $s\pm$wave.

In the present chapter, we adopt the viewpoint that this peculiar band structure is playing an important role in the occurrence of superconductivity, and mainly focus on theoretical studies which adopt material specific model Hamiltonians that take into account the band structure obtained from first principles calculation. In particular, we pay special attention to the studies based on the models that consider all five of the Fe $3d$ bands accurately. Alternative approaches can be (i) those that mainly give focus on the first principles band calculation (e.g., refs. 5), (ii) weak coupling theories that adopt more simplified and generic Hamiltonians (ref. 8), or (iii) to give more focus on the strong correlation effects and adopt models with localized spins (refs. 9,10), and these approaches will be reviewed in more detail in other chapters. We first describe the model construction from the first principles band calculation, and then explain the main features of the band structure and the Fermi surfaces. We then describe how the many body interactions together with these peculiar band structures and Fermi surfaces result in the development of spin fluctuations or antiferromagnetism. We also review some of the studies that discuss the relation between the antiferromagnetism and the structural transition. As for the mechanism of superconductivity, we mainly focus on the spin-fluctuation-mediated pairing, and describe how the multiple modes of spin fluctuations can give rise to different types of sign reversing superconducting gap. We will also comment on the possible competition between sign reversing and non-reversing gaps. One of the strong advantages to consider material specific Hamiltonians is that we can analyze the lattice structure and/or the element dependences. In the final section, we give a review on how the electronic states depend on the materials, affecting both the spin fluctuations and the superconductivity.

8.2 Model Hamiltonian Construction

8.2.1 *The Band Structure*

In order to theoretically investigate the electron correlation effects, it is necessary to obtain a many-body model Hamiltonian that correctly describes the physics within the relevant energy scale. The model Hamiltonian consists of the kinetic energy part and the many-body (electron-electron interaction) part. For an investigation of a given particular material, it is often important to take into account the accurate band structure, especially when the material exhibits peculiar band structures or Fermi surfaces. The kinetic energy part in this case can be described by a tightbinding model whose hopping integrals correctly reproduce the bands around the Fermi level obtained from the first principles band calculation. This downfolding procedure can be performed by constructing Wannier orbitals from the band calculations, which can be accomplished by adopting formalisms such as the "Maximally localized Wannier orbital"[11] or the "N-th order muffin-tin orbital"[12] approaches.

One of the tightbinding models (kinetic energy part) of LaFeAsO adopted in various theoretical studies was obtained in ref. 6. Here, we first review the construction of this model. LaFeAsO has a tetragonal layered structure, where Fe atoms form a square lattice in each layer, which is sandwiched by As atoms (Fig. 8.1). Due to the tetrahedral coordination of As, there are two Fe atoms per unit cell. The band structure obtained with the local density approximation (LDA) using a plane-wave basis (Quantum- ESPRESSO package[13]) is shown in Fig. 8.1. In this calculation, the experimentally determined lattice parameters[1] were used. Near the Fermi level, there are ten bands mainly originating from the Fe 3d orbitals. There are ten bands because there exist two iron atoms per unit cell. In ref. 6, ten maximally localized Wannier orbitals were constructed from these bands using the code developed by Mostofi *et al.*[14] These Wannier orbitals, centered at the two Fe sites in the unit cell, have five orbital symmetries ($d_{3Z^2-R^2}$, d_{XZ}, d_{YZ}, $d_{X^2-Y^2}$, and d_{XY}) as shown in Fig. 8.2. (Hereafter $3Z^2 - R^2$ will be abbreviated as Z^2) Here, X, Y, Z refer to the axes of the unit cell with two Fe sites as shown in Fig. 8.1. It should be noted here that although these Wannier orbitals mainly consist of Fe $3d$ orbitals, they also contain some As $4p$ orbital due to the hybridization effect. The two Wannier orbitals with the same symmetry in each unit cell are equivalent in that each Fe atom has the same local arrangement of the surrounding atoms. Therefore, we can take a unit cell that contains only one orbital (for each orbital symmetry) by unfolding the Brillouin zone,[5,6,15] and we end up with an effective five orbital model on a square lattice as shown in Fig. 8.3. Here, x and y axes are rotated by 45 degrees from X-Y, to which we refer for all the wave vectors hereafter. The in-plain hopping integrals $t(\Delta x, \Delta y, \Delta z = 0; \mu, \nu)$ obtained this way are displayed

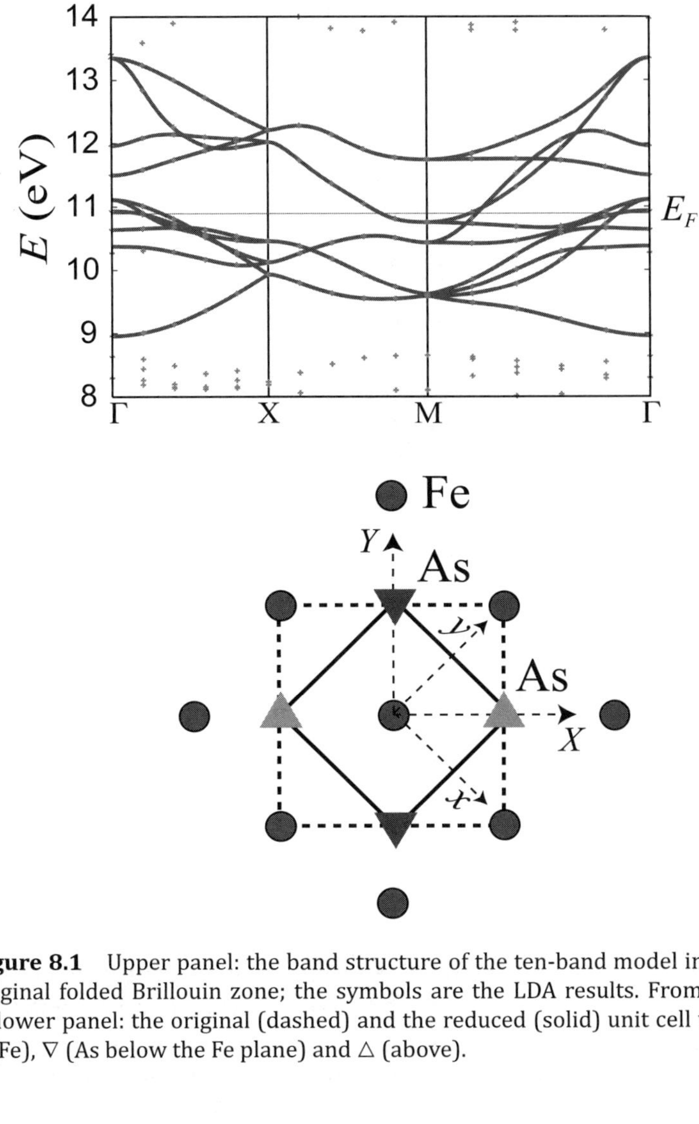

Figure 8.1 Upper panel: the band structure of the ten-band model in the original folded Brillouin zone; the symbols are the LDA results. From ref. 6. lower panel: the original (dashed) and the reduced (solid) unit cell with • (Fe), ∇ (As below the Fe plane) and $\triangle$ (above).

in Table 8.1, where $[\Delta x, \Delta y]$ is the hopping vector, and μ, ν label the five Wannier orbitals (μ, $\mu = 1, 2, 3, 4$ and 5 correspond to Z^2, XZ, YZ, $X^2 - Y^2$, and XY, respectively). The on-site energies for the five orbitals are $(\varepsilon_1, \varepsilon_2, \varepsilon_3, \varepsilon_4, \varepsilon_5) = (10.75, 10.96, 10.96, 11.12, 10.62)$ eV. With these effective hoppings and on-site energies, the in-plane

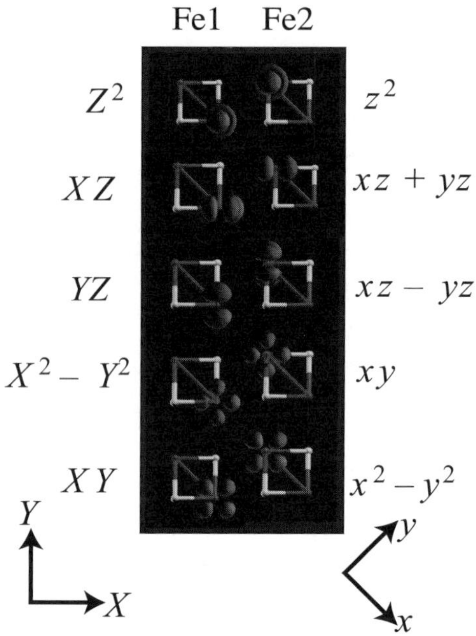

Figure 8.2 The ten 3d Wannier orbitals are shown. The red (yellow) balls indicate the Fe (As) sites.

tightbinding Hamiltonian is given in the form

$$H_0 = \sum_{ij}\sum_{\mu\nu}\sum_{\sigma}\Big[t(x_i - x_j, y_i - y_j; \mu, \nu)c^{\dagger}_{i\mu\sigma}c_{j\nu\sigma} + t(x_j - x_i, y_j - y_i; \nu, \mu)c^{\dagger}_{j\nu\sigma}c_{i\mu\sigma}\Big] + \sum_{i\mu\sigma}\varepsilon_{\mu}n_{i\mu\sigma}, \quad (8.1)$$

where $c^{\dagger}_{i\mu\sigma}$ creates an electron with spin σ on the μ-th orbital at site i, and $n_{i\mu\sigma} = c^{\dagger}_{i\mu\sigma}c_{i\mu\sigma}$. The band filling n is defined as the number of electrons/number of sites (e.g., $n = 10$ for a full filling). There are six electrons per site in the undoped case, and the doping level x in $LaFeAsO_{1-x}F_x$ is related to the band filling as $n = 6 + x$.

The five bands are heavily entangled reflecting the strong hybridization of the five 3*d* orbitals. In Fig. 8.3, the Fermi surface (at the $k_z = 0$ plane) for $n = 6.1$ (corresponding to $x = 0.1$) is shown in the unfolded Brillouin zone. The Fermi surface consists of four pieces (pockets in 2D): two concentric hole pockets (denoted

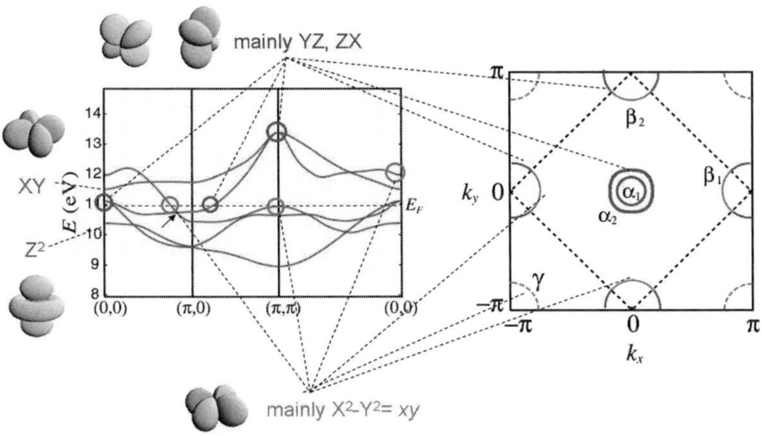

Figure 8.3 Left: the band structure of five-band model A of LaFeAsO in the unfolded Brillouin zone. Orbital characters of the bands are also shown. The dashed line indicates the Fermi level for $n = 6.1$. The short solid arrow denotes the position of the Dirac cone. Right: Fermi surface for $n = 6.1$. The blue (red) portions have strong $X^2 - Y^2$ (XZ/YZ) character. Dashed blue curves represent the portions where the $d_{X^2-Y^2}$ band lies very close to the Fermi level at $n = 6.1$ and actually becomes a Fermi surface for smaller doping. The original (dashed lines) and the unfolded (solid) Brillouin zone are shown. From ref. 16.

as α_1, α_2) around $(k_x, k_y) = (0, 0)$, two electron pockets around $(\pi, 0)$ (β_1) or $(0, \pi)$ (β_2), respectively. Besides these pieces of the Fermi surface, there is a portion of the band around (π, π) that nearly touches the Fermi level at $n = 6.1$, so that this portion acts as a "pseudo Fermi surface" at this doping rate, and will actually form a hole Fermi surface around (π, π) in the undoped case ($n = 6$). We will call this or the pseudo Fermi surface the γ Fermi surface.

As for the orbital character, α and some portions of β near the Brillouin zone edge have mainly d_{XZ} and d_{YZ} character, while the portions of β away from the Brillouin zone edge and γ have mainly $d_{X^2-Y^2}$ orbital character. An interesting feature in the band structure is the presence of Dirac cones, i.e., places where the upper and the lower bands make a conical contact.[15,16] The ones closest to the Fermi level (short arrow in Fig. 8.3) correspond to the crossing

Table 8.1 Hopping integrals $t(\Delta x, \Delta y; \mu, \nu)$ of five orbital model A in units of 0.1 eV. $[\Delta x, \Delta y]$ denotes the in-plain hopping vector, and (μ, ν) the orbitals. σ_y, I, and σ_d corresponds to $t(\Delta x, -\Delta y; \mu, \nu)$, $t(-\Delta x, -\Delta y; \mu, \nu)$, $t(\Delta y, \Delta x; \mu, \nu)$, respectively, where '$\pm$' and '$\pm(\mu', \nu')$' in the row of (μ, ν) mean that the corresponding hopping is equal to $\pm t(\Delta x, \Delta y; \mu, \nu)$ and $\pm t(\Delta x, \Delta y; \mu', \nu')$, respectively. This table, combined with the relation $t(\Delta x, \Delta y; \mu, \nu) = t(-\Delta x, -\Delta y; \nu, \mu)$, gives all the in-plain hoppings ≥ 0.01 eV up to fifth neighbors. From ref. 6

(μ,ν) $[\Delta x, \Delta y]$	[1,0]	[1,1]	[2,0]	[2,1]	[2,2]	σ_y	I	σ_d
(1,1)	−0.7		−0.4	0.2	−0.1	+	+	+
(1,2)	−0.8					−(1,3)	−	−
(1,3)	0.8	−1.5			−0.3	−(1,2)	−	+
(1,4)		1.7			−0.1	−	+	+
(1,5)	−3.0			−0.2		+	+	−
(2,2)	−2.1	1.5				+(3,3)	+	+
(2,3)	1.3		0.2	−0.2		+	+	−
(2,4)	1.7			0.2		+(3,4)	−	−
(2,5)	−2.5	1.4				−(3,5)	−	+
(3,3)	−2.1	3.3		−0.3	0.7	+(2,2)	+	+
(3,4)	1.7	0.2		0.2		+(2,4)	−	+
(3,5)	2.5			0.3		−(2,5)	−	−
(4,4)	1.6	1.2	−0.3	−0.3	−0.3	+	+	+
(4,5)				−0.1		−	+	−
(5,5)	3.1	−0.7	−0.2			+	+	+

points of the $d_{X^2-Y^2}$ and the d_{XZ}/d_{YZ} bands below the β Fermi surface. In the present chapter, we will refer to the five orbital model that adopts the present band structure as "five orbital model A". Five orbital models of LaFeAsO with similar values of the hopping integrals were obtained also in refs. 17,18.

Another five orbital band structure frequently adopted in theoretical studies was obtained by Graser *et al.*[19]. There, the tightbinding hopping integrals were obtained so as to reproduce the first principles calculation in ref. 20, where the lattice parameters determined by theoretical optimization were adopted. The band structure of this model is given in Fig. 8.4. For the iron pnictides, the theoretically optimized lattice structure has somewhat lower position of the pnictogen atoms compared to the experimentally determined position, which, as will be described in detail in

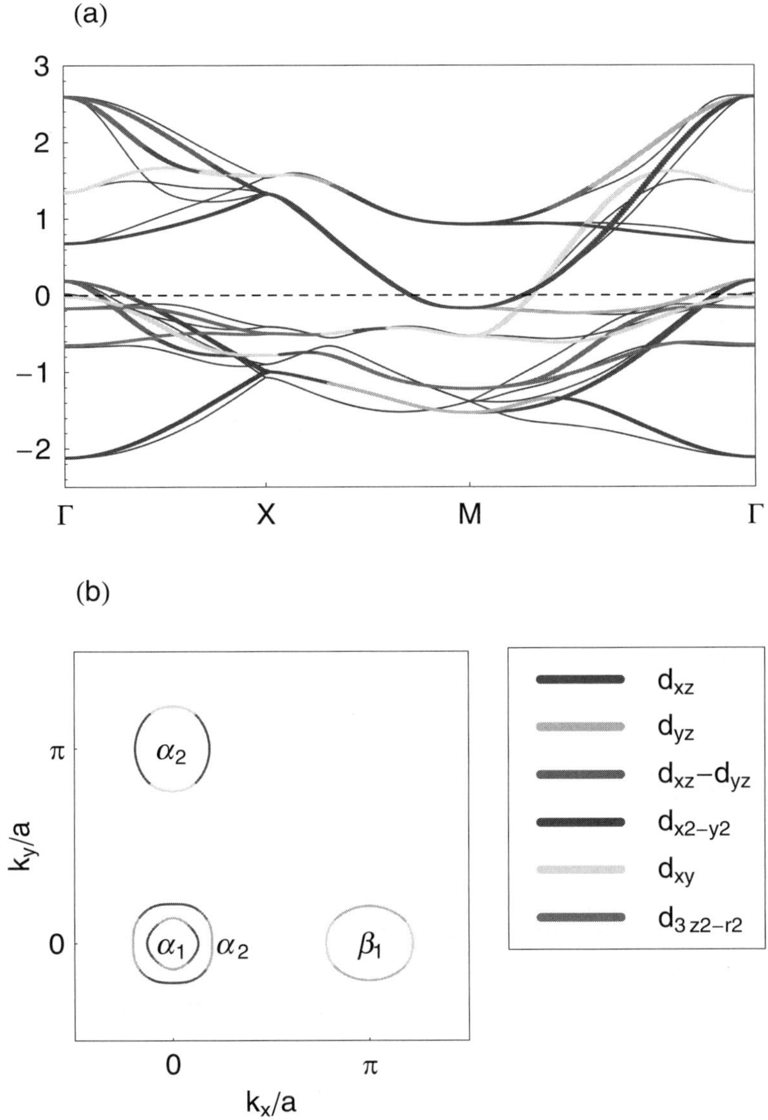

Figure 8.4 (a) The band structure in the folded Brillouin zone and (b) Fermi surfaces in the unfolded Brillouin zone of model B obtained by Graser *et al.* From ref. 19. Here, the Fermi level (dashed line) and the Fermi surfaces are for the undoped case ($n = 6$). Note that the γ Fermi surface around (π, π) is absent, whereas in model A this is present for $n = 6$. See also Color Insert.

section 8.3, results in a lower position of the $d_{X^2-Y^2}$ (or d_{xy} in the notation in ref. 19) band around the wave vector (π,π) (or Γ in the folded Brillouin zone presented in the top of Fig. 8.4). In this case, the γ Fermi surface originating from the $d_{X^2-Y^2}$ band is absent unless holes are doped. We will call this model "five orbital model B".

There are also some models with fewer bands/orbitals. Two band model in refs. 21, three band models in refs. 22,23, and four band models (in the folded Brillouin zone) in refs. 24–26 are some examples. These models have similar shape of the Fermi surface as those in the five orbital model, especially when presented in the folded Brillouin zone, but have different orbital characters on some of the Fermi surfaces. Arita and Ikeda constructed a four orbital model which does not explicitly take into account the d_{Z^2} orbital.[27] Since the entire d_{Z^2} band lies below the Fermi level in LaFeAsO, this four orbital model has the same orbital character of the Fermi surfaces as in the five orbital model, although the band filling is changed from 6 to 4 (half-filled 4 bands) in the undoped case. Alternatively, there are also some models that take into account not only the Fe 3d orbitals, but also the 4p orbitals of the arsenic atoms explicitly.[28,29]

8.2.2 *Electron-Electron Interactions*

For the many body part of the model Hamiltonian of the iron based superconductors, the multiorbital electron-electron interactions are considered. Those are the intra-orbital Coulomb U, the inter-orbital Coulomb U', the Hund's coupling J, and the pair-hopping J' as shown in Fig. 8.5. The many body Hamiltonian of the five orbital model then reads

$$H = \sum_i \sum_\mu \sum_\sigma \varepsilon_\mu n_{i\mu\sigma} + \sum_{ij} \sum_{\mu\nu} \sum_\sigma t_{ij}^{\mu\nu} c_{i\mu\sigma}^\dagger c_{j\nu\sigma}$$
$$+ \sum_i \left(U \sum_\mu n_{i\mu\uparrow} n_{i\mu\downarrow} + U' \sum_{\mu>\nu} \sum_{\sigma,\sigma'} n_{i\mu\sigma} n_{i\nu\sigma'} \right.$$
$$\left. - J \sum_{\mu\neq\nu} \boldsymbol{S}_{i\mu} \cdot \boldsymbol{S}_{i\nu} + J' \sum_{\mu\neq\nu} c_{i\mu\uparrow}^\dagger c_{i\mu\downarrow}^\dagger c_{i\nu\downarrow} c_{i\nu\uparrow} \right), \quad (8.2)$$

where $t_{ij}^{\mu\nu}$ is the hopping integrals described in the previous section.

Figure 8.5 Multiorbital electron-electron interactions.

For the iron based superconductors, there have been some studies that estimate the electron-electron interaction values.[17,29,30,32–35] There are several general methods to evaluate the interaction values from first principles. One of them is the constrained LDA method, where the orbital occupancies in LDA calculations are fixed at certain values, and the interactions are determined by a derivative of the orbital energy over the occupancy.[36] Another approach is the constrained random phase approximation (RPA) scheme, in which the screened Coulomb interactions are calculated within the perturbation theory.[37,38] In Table 8.2, we present the values of the orbital dependent intraorbital and Hund's coupling interactions obtained for LaFeAsO using the constrained RPA method.[17] In the above Hamiltonian formula 8.2, the interactions are given in the orbital independent form for simplicity, but in reality, the strength of the interaction depends on the orbitals as seen in the table. The

Table 8.2 The intra/inter orbital interactions and the Hund's coupling (in units of eV) of the five orbital model of LaFeAsO obtained in ref. 17

"U" or "U'"	XY	YZ	Z^2	ZX	X^2-Y^2
XY	3.31	1.95	1.89	1.95	2.09
YZ	1.95	2.77	2.20	1.78	1.67
Z^2	1.89	2.20	3.27	2.20	1.65
ZX	1.95	1.78	2.20	2.77	1.67
X^2-Y^2	2.09	1.67	1.65	1.67	2.20
"J"	XY	YZ	Z^2	ZX	X^2-Y^2
XY		0.54	0.64	0.54	0.27
YZ	0.54		0.41	0.45	0.43
Z^2	0.64	0.41		0.41	0.50
ZX	0.54	0.45	0.41		0.43
X^2-Y^2	0.27	0.43	0.50	0.43	

$d_{X^2-Y^2}$ Wannier orbital has relatively small intraorbital interaction because this orbital is directed towards the As atom, so that its spread is large due to the hybridization with the $4p$ orbitals. The d_{XY} and d_{Z^2} orbitals have smaller effect of the hybridization, and thus the interaction values are large.

Using the evaluated interaction values, a number of studies, using the dynamical mean field theory (DMFT)[31] or other many-body techniques, have investigated the strength of the electron correlation in iron based superconductors.[33,34,39–43] The strength of the electron correlation is found to be in the regime of moderate to strong, depending on the material family, the values of the interactions, or the orbitals considered in the calculation. The family dependence of the strength of the correlation will be mentioned in section 8.5.2.1.

8.3 Spin Fluctuations and Antiferromagnetism

8.3.1 *Random Phase Approximation*

A number of unusual properties are observed even in the normal state of the iron pnictides. It is natural to consider that these properties are related to the occurrence of high T_c superconductivity. One of the origin for these anomalous normal state properties is considered to be the development of strong spin fluctuations, so it is of great importance to study the spin excitation and related quantities of the theoretical models. Among the first methods that have been applied to the multi orbital models of the iron pnictides is the RPA. In the RPA approach, the spin and charge susceptibilities are obtained by collecting bubble (see Fig. 8.11) and ladder type diagrams consisting of the bare (unrenormalized) Green's function. In the case of the iron based superconductors, the multiorbital version of RPA is adopted, which is described in, e.g., refs. 44, 45. For the five orbital model, the Green's function in the orbital basis $G_{lm}(k)$ ($k \equiv (\boldsymbol{k}, i\omega_n)$) is a 5×5 matrix, where l, m refer to the orbitals. In the RPA, the Green's function matrix is determined solely from the kinetic energy part of the Hamiltonian. The irreducible susceptibility matrix

$$\chi^0_{l_1,l_2,l_3,l_4}(q) = \sum_k G_{l_1 l_3}(k+q) G_{l_4 l_2}(k) \tag{8.3}$$

($l_i = 1, \ldots, 5$) has 5^4 components, and the spin and the charge (orbital) susceptibility matrices are obtained from matrix equations,

$$\hat{\chi}_s(q) = \frac{\hat{\chi}^0(q)}{1 - \hat{S}\hat{\chi}^0(q)}, \tag{8.4}$$

$$\hat{\chi}_c(q) = \frac{\hat{\chi}^0(q)}{1 + \hat{C}\,\hat{\chi}^0(q)}, \tag{8.5}$$

where

$$S_{l_1 l_2, l_3 l_4} = \begin{cases} U, & l_1 = l_2 = l_3 = l_4 \\ U', & l_1 = l_3 \neq l_2 = l_4 \\ J, & l_1 = l_2 \neq l_3 = l_4 \\ J', & l_1 = l_4 \neq l_2 = l_3, \end{cases} \tag{8.6}$$

$$C_{l_1 l_2, l_3 l_4} = \begin{cases} U & l_1 = l_2 = l_3 = l_4 \\ -U' + J & l_1 = l_3 \neq l_2 = l_4 \\ 2U' - J, & l_1 = l_2 \neq l_3 = l_4 \\ J' & l_1 = l_4 \neq l_2 = l_3. \end{cases} \tag{8.7}$$

We denote the largest eigenvalue of the spin (charge) susceptibility matrix for $i\omega_n = 0$ as $\chi_s(\boldsymbol{k})(\chi_c(\boldsymbol{k}))$. We can also monitor the largest eigenvalue of the matrix $\hat{S}\hat{\chi}^0(q)$ for $i\omega_n = 0$ as the strength of the low energy spin fluctuations, namely, the magnetic ordering is signaled by this eigenvalue reaching unity. We will refer to this quantity as the Stoner factor α_S.

8.3.2 *Electron-Hole Interaction*

In Fig. 8.6, we show the RPA spin susceptibility $\chi_s(\boldsymbol{k})$ of five orbital model A in the unfolded Brillouin zone obtained in ref. 6. Here, the electron-electron interactions were taken as orbital independent, and the adopted values are $U = 1.2$, $U' = 0.9$, $J = J' = 0.15$ eV. The band filling is taken as $n = 6.1$, corresponding to 10% electron doping, and the temperature is $T = 0.02$ eV. It can be seen that there are peaks around the wave vector $\boldsymbol{Q} = (\pi, 0)$ and $(0, \pi)$. The static configuration of the spins that correspond to the wave vector $(0, \pi)$ is shown in the bottom of Fig. 8.6. This wave vector of the peak position of the spin susceptibility is consistent with inelastic neutron scattering experiments.[46,47]

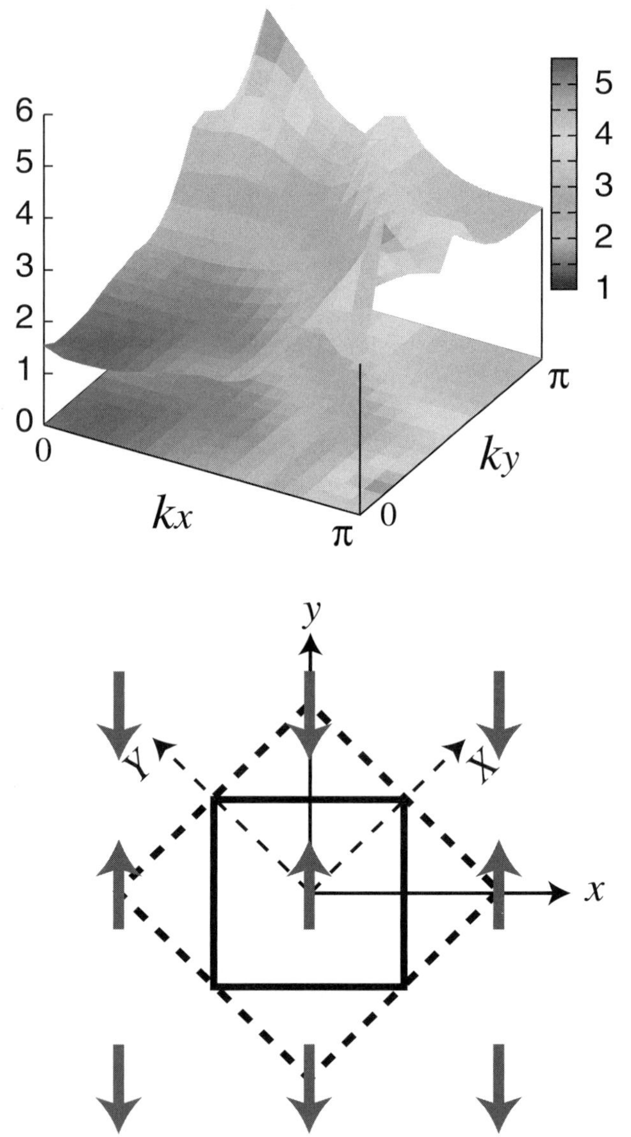

Figure 8.6 Top panel: RPA result of the spin susceptibility χ_s for model A with the parameter values $U = 1.2$, $U' = 0.9$, $J = J' = 0.15$ eV, $n = 6.1$, and $T = 0.02$ eV. From ref. 6. Bottom panel: The spin configuration in real space that corresponds to the wave vector $(0, \pi)$.

Naively, these peaks in the spin susceptibility are considered to be originating from the "nesting" between the electron and hole Fermi surfaces, which are indeed separated by the wave vectors $(\pi, 0)/(0, \pi)$. However, some caution has to be taken here. First of all, the term "good Fermi surface nesting" usually implies that when one Fermi surface is moved by a wave vector $\boldsymbol{Q}$, it almost completely overlaps with another Fermi surface. However, in the present case of iron pnictides, the nesting in that sense is usually not so good. So it is not the shape of the Fermi surface itself that is important for the development of spin fluctuations (as a pairing glue that will be discussed later), but the interaction between electrons around the wave vector $\boldsymbol{k}$ and the holes around the wave vector $\boldsymbol{k} + \boldsymbol{Q}$, where the states $\boldsymbol{k}$ and $\boldsymbol{k} + \boldsymbol{Q}$ can be somewhat away from the Fermi level. In the present chapter, we use the term "nesting at the wave vector $\boldsymbol{Q}$" not in the sense that the Fermi surfaces separated by $\boldsymbol{Q}$ have similar shapes, but in the sense that the spin susceptibility is enhanced at this wave vector due to electron (particle)-hole interactions. Second caution concerns the multiorbital nature of the material. In a single orbital system, the on-site U dominates the spin fluctuations, whereas in multiorbital systems, there are various on-site interactions as described in section 8.2.2, and it is the intraorbital U that mostly governs the spin fluctuations.

To see the origin of the spin fluctuations more carefully, we show in Fig. 8.7 the two largest orbital-diagonal components of the spin susceptibility, χ_{s3333} and χ_{s4444}, of model A with the same values of parameters as in Fig. 8.6.[48] Here, orbitals 3 and 4 stand for d_{YZ} and $d_{X^2-Y^2}$, respectively, as mentioned in section 8.2.1. χ_{s3333} has peaks solely around $(\pi, 0)$ and $(0, \pi)$, which originates from the particle-hole interaction between d_{XZ}, d_{YZ} portions of α and β pockets as shown in the lower panel of Fig. 8.7, where the thickness of the Fermi surface represents the strength of the $d_{X^2-Y^2}$ or d_{XZ}/d_{YZ} characters. On the other hand, χ_{s4444} has peaks around $(\pi, 0)$, $(0, \pi)$ and $(\pi, \pi/2)$, $(\pi/2, \pi)$. The former is due to the particle-hole interaction between the γ pocket and the $d_{X^2-Y^2}$ portion of the β pocket, while the latter originates from the interaction between the $d_{X^2-Y^2}$ portion of the β_1 and β_2.[6,19,48–51] It should be noted here that in model A of LaFeAsO, the γ Fermi surface itself is absent at $n = 6.1$, although the top of the $d_{X^2-Y^2}$ band at (π, π) nearly touches the Fermi level.

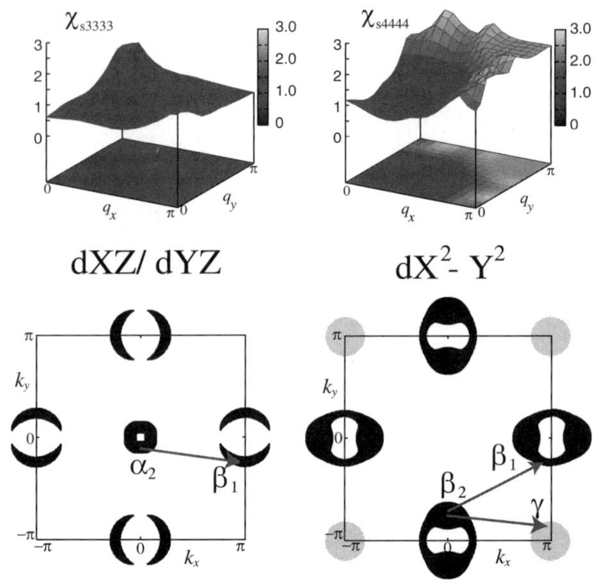

Figure 8.7 Top panels : Diagonal components, χ_{s3333} and χ_{s4444}, of the spin susceptibility matrix in the orbital representation ($3: YZ, 4: X^2 - Y^2$) for model A with the same parameter values as in Fig. 8.6. Bottom: Intraorbital interaction is shown for the portions of the Fermi surface with orbitals XZ, YZ (left) and $X^2 - Y^2$ (right). Here the thickness of the Fermi surface represents the strength of the respective orbital character. From ref. 48. See also Color Insert.

Therefore, the "nesting" in its strict sense between β and γ is not good at all. Still, the interaction between the $d_{X^2-Y^2}$ portions of the bands around (π, π) and $(\pi, 0)/(0, \pi)$ gives rise to large peaks in χ_{s4444}. Hence, we once again stress that the exact shape of the Fermi surface is not directly relevant to what we are focusing on now.

There have also been some theoretical analysis on the dynamics of the normal state spin fluctuations using the five orbital model. In ref. 52, the dynamical spin susceptibility $S(\boldsymbol{q}, \omega)$ has been calculated within RPA as shown in Fig. 8.8. Here, a column like feature was found in $\boldsymbol{q}, \omega$ space, which is in good agreement with inelastic neutron scattering experiments.[53,54] Spin lattice relaxation rate measured in NMR/NQR experiments has also been calculated,[18,52,55] as will be described in section 8.5.3.2.

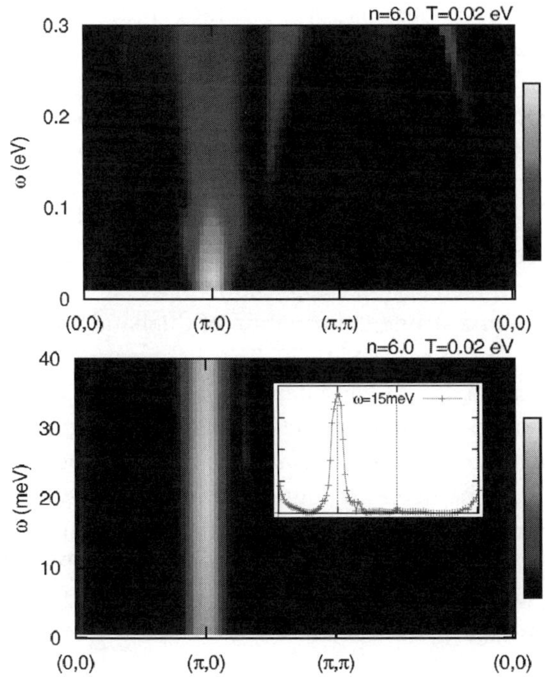

Figure 8.8 The dynamical spin susceptibility of model A in the $(\boldsymbol{q}, \omega)$ plane obtained by RPA. From ref. 52. The lower panel is the low energy blow up of the upper.

Let us now make several comments concerning the RPA approach described above. First, one may note that the values of the electron-electron interactions adopted here are small compared to the values evaluated from the first principles calculation (Table 8.1). This is because in the RPA, due to the neglect of the self energy corrections and the vertex corrections, the magnetic ordering is easily realized (the Stoner factor a_S reaches unity) at very high temperatures if we adopt the interaction values evaluated from first principles. The same problem arises for the mean field approach on the antiferromagnetic state described in section 8.3.3. In ref. 48, RPA calculations with orbital dependent interactions were also performed, but to cope with the problem mentioned above, all the interactions obtained in ref. 17 were multiplied by a constant reduction factor. Within this approach, the orbital dependence of the interaction values was found to have small effect.

Secondly, in the RPA, the band structure determined from the first principles calculation is unaffected by the electron-electron interactions because the self-energy corrections are neglected. In the fluctuation exchange (FLEX) approximation,[56] the self energy correction due to spin and charge fluctuations is taken into account self-consistently, so that the band structure is affected by the interactions. In fact, a direct application of FLEX to the five orbital model results in a strong deformation of the band structure, as has been pointed out in refs. 18,27,55. The deformation is so strong as to push down the $d_{X^2-Y^2}$ band near (π, π) in the unfolded Brillouin zone. This results in a development of the spin fluctuations around (π, π) rather than $(\pi, 0)$ in disagreement with the experiments. In ref. 18, this has been attributed to the double counting of the interaction within the first principles band calculation and FLEX. There, in order to reduce this effect, the static ($\omega = 0$) portion of the self-energy was subtracted in FLEX, thereby re-obtaining the $(\pi, 0)$ spin fluctuations.

8.3.3 *Antiferromagnetism in the Parent Compound*

The wave vector $(\pi, 0)/(0, \pi)$ at which the spin fluctuations are observed experimentally, and also at which the spin susceptibility is peaked in the RPA/FLEX calculations mentioned above, is the same as the wave vector of the static antiferromagnetic ordering in the parent compound.[57] Still, there has been a debate as to whether the origin of the magnetism can be understood as an effect of the band structure and the Fermi surfaces, namely, whether the magnetic state can be regarded as a "Fermi surface nesting induced SDW". Local spin density approximation (LSDA) calculations have often been used to theoretically analyze the antiferromagnetism of the parent compound of the iron pnictides. All those calculations show that the stripe-type (or collinear) antiferromagnetic state with the spin configuration wave vector of (π, π) in the original Brillouin zone (with two irons per unit cell), which corresponds to $(\pi, 0)$ in the unfolded Brillouin zone, has the lowest energy. However, one of the problems raised there is the discrepancy of the magnetic moment, namely, the theoretical calculations give $\sim 2\mu_B$,[58–60] while the experimental values are less than $1\mu_B$.[57] Several proposals have been made regarding the origin of this discrepancy, such as the motion of domains[61] or introduction of negative U.[62]

As for the studies based on multiorbital model Hamiltonians, there have been some mean field approximation approaches for the five orbital model, which show that relatively small values of $U \leq 1$eV and $J \leq 0.2$ eV give small magnetic moments consistent with experiments.[63,64] Ran *et al.* studied five orbital model A of LaFeAsO as well as a simplified two band model, and showed that the antiferromagnetic state cannot acquire a full gap and instead has Dirac nodes in the gap due to the topological feature of the band structure (nodal spin density wave (SDW) state).[63] In Fig. 8.9, the relevant portion of the bands of this state is shown along the k_x or k_y directions. In fact, these Dirac nodes in the antiferromagnetic state is indeed seen in the first principles band calculations[58,65] as well as in an angle resolved photoemission (ARPES) experiment.[66] Kaneshita *et al.* studied the optical conductivity of the nodal SDW state of model A, and showed that the results obtained for the interaction values that give small magnetic moments are consistent with experimental observations.[64]

Regarding the magnitude of the magnetic moment, there are also some model calculations that use larger values of the interaction (closer to those evaluated from first principles), adopting methods that take into account strong correlation effects. Hansmann *et al.* on the other hand adopted the four orbital model that does not consider the d_{Z^2} orbital explicitly.[27] The intraorbital $U = 1.8$ eV was taken there, considering the effect that the electron-electron interactions are reduced from the values evaluated for the five orbital models. This is because the d_{Z^2} orbital is effectively taken into account in the four Wannier orbitals, so that the orbital spread is larger than in the five orbital model, and also the screening of the electron-electron interactions is more efficient. The application of DMFT to this model shows that the local magnetic moment is comparable to those obtained in LSDA calculations, but the experiments usually observe the dynamically screened moment, which is much smaller, resulting in a discrepancy between theory and experiment.[43] Misawa *et al.* applied the variational Monte Carlo (VMC) method to the five orbital model

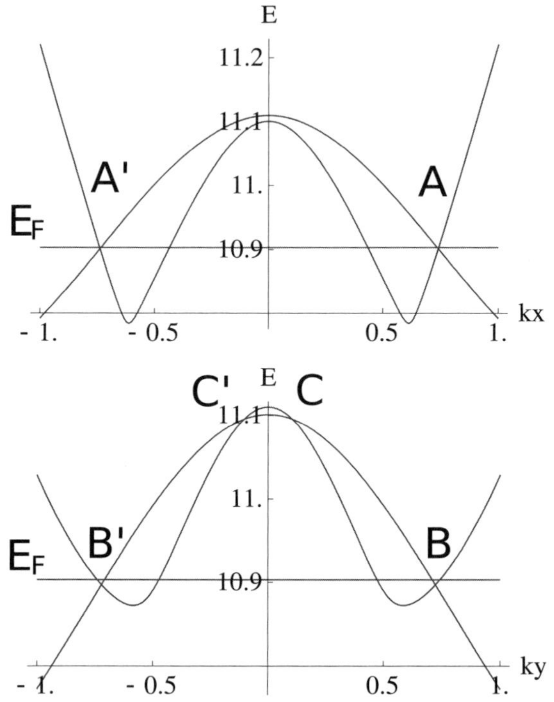

Figure 8.9 Dirac nodes of the nodal SDW state obtained in model A. The Upper and lower panels show the relevant band dispersion along k_x and k_y directions, respectively. From ref. 63.

obtained in ref. 17, and showed that electron correlation effects beyond the mean field level largely suppress the magnetic moment.[67]

In the actual materials, the antiferromagnetic ordering occurs at the Neel temperature which is nearly equal to or slightly lower than the structural transition temperature,[57] where tetragonal to orthorhombic lattice distortion takes place. Later experiments have further revealed possible occurrence of nematic order or fluctuations in the vicinity of these transitions. Namely, anisotropic transport properties in the resistivity and optical conductivity have been observed in detwinned 122 systems,[68,69] and also the lift of the degeneracy between d_{xz} and d_{yz} orbitals as well as the reconstruction of the Fermi surface has been observed in ARPES experiments.[70,71] Also, in refs. 72,73, a softening of the elastic constant was observed in $BaFe_{2-x}Co_xAs_2$.

Theoretically, the relation between the antiferromagnetism, the structural transition, and the possible occurrence of the nematic order has been an issue of hot debate. Yildirim showed using the first principles band calculation that the lattice distortion takes place in the collinear antiferromagnetic state, while not in the normal state.[74] Kubo and Thalmeir considered a simplified $d_{xz} - d_{yz}$ two orbital[75] and showed within the mean field approximation that ferro-orbital ordering, where one of the two originally degenerate orbitals (d_{xz} and d_{yz}) has larger occupancy than the other, coexists with the stripe type antiferromagnetism. The occurrence of the ferro-orbital ordering was proposed as the origin of the lattice distortion transition. An image of the orbital ordering coexisting with the stripe type antiferromagnetism is shown in Fig. 8.10. Yang *et al.* applied VMC method to five orbital model A, and showed that $d_{xz}-d_{yz}$ orbital

Figure 8.10 The ferro-orbital ordering coexisting with the stripe type antiferromagnetic ordering. The orbitals are d_{xz} and d_{yz}. The orbital with larger size has larger occupancy.

ordering is stabilized by the stripe type antiferromagnetism.[76] Alternatively, in ref. 72, the softening of the elastic constant has been interpreted in terms of the spin-nematic fluctuations, where the nematic order parameter of magnetic origin is coupled to the shear strain. There the enhancement of the nematic fluctuations driven by antiferromagnetism reduces the elastic constant through magneto-elastic coupling.

As for the anisotropy of the transport properties, Yin *et al.* used the LDA+DMFT approach. It was proposed there that at low energies, the Pauli's exclusion principle results in the transport anisotropy in the SDW phase, namely, the quasiparticles tend to propagate along the antiferromagnetic direction, which also leads to orbital polarization.[77] Sugimoto *et al.* studied five orbital model A within the mean field theory taking into account both the nodal SDW state and the orbital ordering, and showed that the anisotropy of the optical conductivity can be understood in terms of the orbital characters of the initial and final states contributing to interband transitions, and also considering the linear dispersion of the the Dirac nodes around the Fermi level.[78] There, the orbital ordering was found to be unnecessary to understand the anisotropy of the transport properties.

Another streamline of the theoretical studies on this issue are those that take into account the effect of electron-phonon interactions in the multiorbital models to understand the structural transition originating from ferro-orbital ordering.[79,80] In these studies, the ferro-orbital transition dominates over the antiferro-magnetic transition, and the occurrence of such a situation has been discussed in relation with the orbital fluctuation mediated pairing described in section 8.4.3.

8.4 Superconductivity

8.4.1 *General Theory on Fluctuation Mediated Pairing*

Let us now move on to the superconductivity. As mentioned in the Introduction, there is a strong possibility that bosonic modes other than phonons are responsible as a glue for mediating the Cooper pairing in the iron based superconductors. Although there is a debate concerning the origin and the nature of the antiferromagnetic phase in the parent compound as described in section 8.3.3, it still seems natural to consider that in the doped materials which exhibit superconductivity, the most probable glue for the pairing is the spin fluctuations.[5,6] Hence, we first briefly summarize the spin (and charge) fluctuation pairing in general,[81,82] where we consider a single band system for simplicity. The gap equation for superconductivity can be written in the form,

$$\Delta(\boldsymbol{k}) = -\sum_{\boldsymbol{k}'} V(\boldsymbol{k}-\boldsymbol{k}')\frac{\tanh(E(\boldsymbol{k}')/2k_BT)}{2E(\boldsymbol{k}')}\Delta(\boldsymbol{k}') \tag{8.8}$$

where $E(\boldsymbol{k})$ is the quasiparticle excitation spectrum and $V(\boldsymbol{q})$ is the pairing interaction. When the pairing is mediated by phonons, $V(\boldsymbol{q})$ is negative (attractive) and this requires a constant sign of the gap $\Delta(\boldsymbol{k})$ for the gap equation to have a non-zero solution. When the pairing is mediated by spin and charge fluctuations, on the other hand, the pairing interaction can in general take the form

$$V(\boldsymbol{q}) = \frac{3}{2}V_{\rm sp}(\boldsymbol{q}) - \frac{1}{2}V_{\rm ch}(\boldsymbol{q}) + (\text{ first order terms}), \tag{8.9}$$

where $V_{\rm sp}$ and $V_{\rm ch}$ are contributions from spin and charge fluctuations, respectively. $V_{\rm sp}(\boldsymbol{q})$ and $V_{\rm ch}(\boldsymbol{q})$ are positive and take large values at the wave vectors where the spin and charge fluctuations develop. From the form of the gap equation (8.8), one can see that the positive (negative) pairing interaction at a wave vector $\boldsymbol{q}$ tends to alter (maintain) the gap sign across this wave vector. Thus, the spin and charge fluctuations have opposite effect on the sign of the superconducting gap. Let us consider the case when spin fluctuations strongly dominate over charge fluctuations ($V_{\rm sp} \gg V_{\rm ch}$). When the spin fluctuations develop at a wave vector $\boldsymbol{Q}$, the pairing interaction is repulsive and largely peaked at this wave vector. Then, for the gap equation to have a nontrivial solution, the gap on the Fermi surface has to change its sign across the wave vector $\boldsymbol{Q}$. For example, in the case of the Hubbard model (with on-site repulsive interaction U),

$$V_{\rm sp(ch)}(q) = U^2\chi_{s(c)}(q), \qquad \chi_{s(c)}(q) = \frac{\chi_0(q)}{1-(+)U\,\chi_0(q)}, \tag{8.10}$$

where $\chi_0(q)$ is the irreducible susceptibility.[82] The bubble diagrams responsible for this pairing interaction is shown in the top of Fig. 8.11. On a square lattice near half-filling, $\chi_0(q)$, and thus the repulsive spin-fluctuation-mediated pairing interaction, develop around $\boldsymbol{Q} = (\pi, \pi)$, and this requires sign change of the superconducting gap across this vector, resulting in a d-wave gap as shown in the bottom of Fig. 8.11.

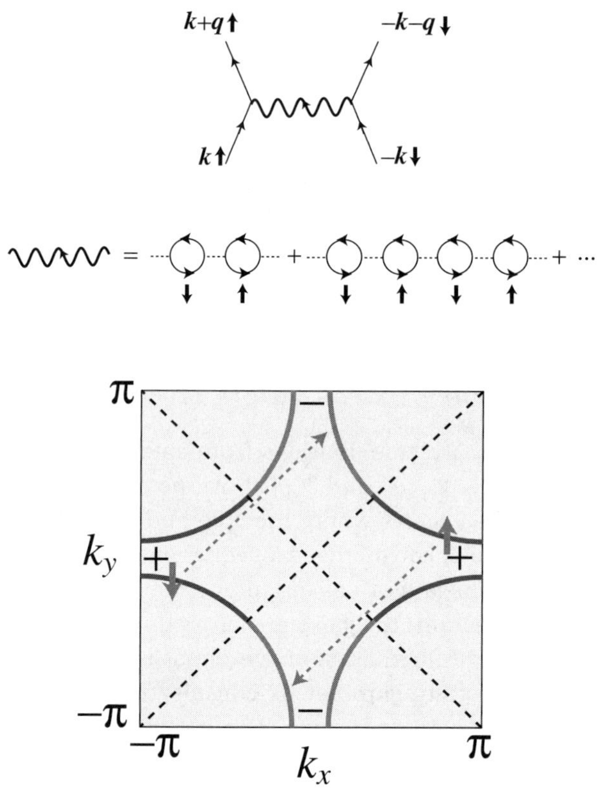

Figure 8.11 Top: the bubble diagrams responsible for the spin-fluctuation-mediated pairing interaction. Bottom: the d-wave superconducting gap. The red (green) lines show portions of the Fermi surface having positive (negative) gap sign. The diagonal dashed lines indicate the nodal lines of the gap. The green short arrows are the electrons (with spin), and the dashed long arrows indicate the pair scattering process with a momentum transfer of (π, π).

Now, in the case of d-wave pairing, the gap has nodes intersecting the Fermi surface. This is necessary in order to have a sign change of the gap, but on the other hand is unfavorable for superconductivity in the sense that the gap is not fully open on the Fermi surface. One of the ways to circumvent this problem is to consider disconnected Fermi surfaces, where spin fluctuations develop at the wave vector that bridge the Fermi surfaces. This can occur for a pair of electron and hole Fermi surfaces with moderate nesting,[83–85] whose examples are shown in Fig. 8.12(a)(b). FLEX studies of the Hubbard

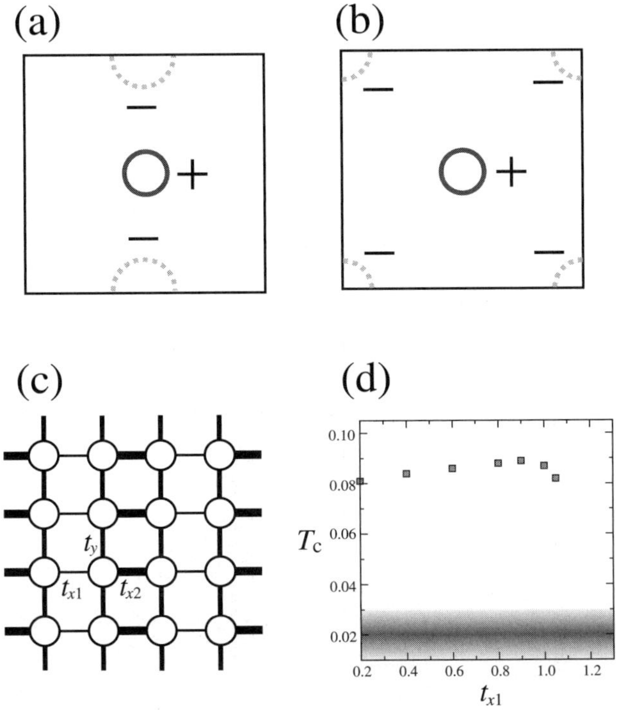

Figure 8.12 (a)(b) Examples of disconnected Fermi surfaces. The ± represents the sign of the superconducting gap induced by spin fluctuations. (c) An example of the lattice that gives the Fermi surface in (a). (d) T_c of the Hubbard model on the lattice shown in (c) obtained by fluctuation exchange method. $U = 8$, $t_y = 1$, $t_{x2} = 1.2$ in units of t_{x1}. The hatched regime indicate the typical value of the T_c of the Hubbard model on the square lattice obtained by the same method. From ref. 84.

model on lattices (for example, Fig. 8.12(c)) that possess this kind of disconnected Fermi surfaces show that T_c can be higher than those obtained on the square lattice using the same method, as shown in Fig. 8.12(d).[84]

8.4.2 *Spin Fluctuation Mediated Pairing*

We now consider applying the spin-fluctuation-mediated pairing theory to the iron based superconductors. When the spin fluctuations play a dominant role in the pairing, the superconducting gap should be determined by the cooperation or competition between the multiple spin fluctuation modes described in section 8.3.2. Specifically, the α-β and γ-β electron-hole interactions tend to favor the fully-gapped, $s\pm$-wave, in which the gap changes sign between electron (β) and hole (α, γ) Fermi surfaces, but has a constant sign on each of them as shown in Fig. 8.13(a).[5] In fact, a number of papers have studied the possibility of this $s\pm$ as one of the most probable candidates of the pairing state.[6,8,21,24–26,86–95] On the other hand, β_1-β_2 interaction tends to change the sign of the gap between these pockets, which can result in either d-wave or an $s\pm$-wave pairing with nodes on the β Fermi surface, as shown schematically in Fig. 8.13(b)(c).[6,19,86]

In the five orbital RPA approach, the gap and T_c of the fluctuation mediated superconductivity can be obtained as follows. Using the spin and charge susceptibilities within the RPA, the pairing interaction is given as

$$\hat{V}^s(q) = \frac{3}{2}\hat{S}\hat{\chi}_s(q)\hat{S} - \frac{1}{2}\hat{C}\,\hat{\chi}_c(q)\hat{C} + \frac{1}{2}(\hat{S} + \hat{C}). \qquad (8.11)$$

This is plugged into the linearized Eliashberg equation,

$$\lambda\phi_{l_1 l_4}(k) = -\frac{T}{N}\sum_q \sum_{l_2 l_3 l_5 l_6} V_{l_1 l_2 l_3 l_4}(q) \times G_{l_2 l_5}(k-q)\phi_{l_5 l_6}(k-q)G_{l_3 l_6}(q-k). \qquad (8.12)$$

The 5×5 matrix gap function ϕ_{lm} in the orbital representation along with the associated eigenvalue λ is obtained by solving this equation. The gap function in the orbital representation can be transformed into those in the band representation by an unitary transformation. The eigenvalue λ monotonically increases upon

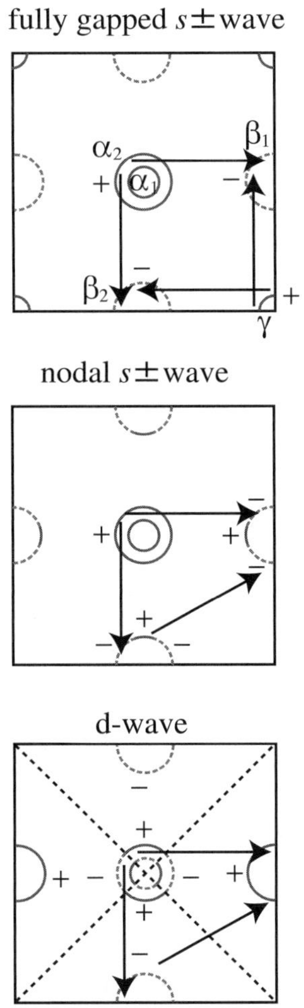

Figure 8.13 The fully-gapped $s\pm$ wave (top panel), the nodal $s\pm$ wave (middle), and the d-wave gap (bottom) are schematically shown. The solid red (dashed blue) curves represent positive (negative) sign of the gap. The arrows indicate the wave vector of the dominating pair scattering processes. γ Fermi surface is present when the pnictogen height is large. From ref. 48.

lowering the temperature, and T_c is the temperature where the largest λ reaches unity. Therefore, the largest eigenvalue λ at a fixed temperature, compared for different cases (pairing symmetries, materials, etc.), can be considered as a qualitative measure for T_c.

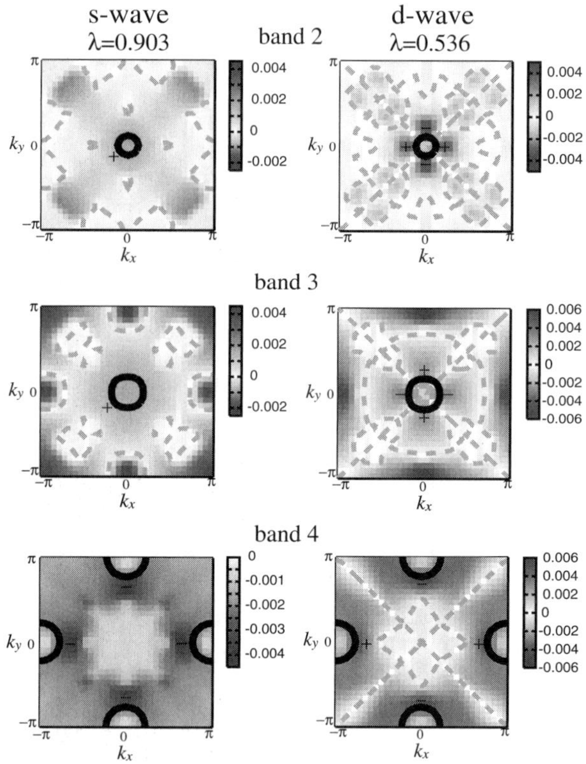

Figure 8.14 The s-wave (left panels) and d-wave (right) gap functions for the 2nd to 4th bands from top to bottom in the band representation for model A of LaFeAsO. The same parameter values as in Fig. 8.6 are adopted. The red (blue) indicates the gap with positive (negative) sign. Solid lines represent the Fermi surface, and green dashed lines the nodes in the gap (the nodes in the s-wave gap do not intersect the Fermi surface). From ref. 48. See also Color Insert.

For model A of LaFeAsO,[6,86] the RPA+Eliashberg equation approach described above gives the fully gapped $s\pm$-wave pairing at the band filling of $n = 6.1$ as shown in Fig. 8.14. The sign of the gap indeed changes between electron and hole Fermi surfaces, and there are no nodes intersecting the Fermi surface. Here, the eigenvalue of the $s\pm$-wave is much larger than d-wave, showing that the former has a higher T_c and thus there is no chance for the latter being realized.

There are other theoretical studies that employ five orbital model A[6] or similar models.[17,18] Nomura applied perturbation theory to calculate the pairing interactions, where diagrams up to third order was taken into account.[88] As mentioned in section 8.3.2, Ikeda *et al.* applied FLEX to the five orbital model, and analyzed superconductivity by solving the Eliashberg equation.[18,55] Functional renormalization group (fRG) was also applied to model A[90,91] (the fRG calculation results will be presented in section 8.5.3.1). These studies all give the fully gapped $s\pm$ state to be the most dominant pairing. A VMC study on model A shows that there is a near degeneracy between the superconducting states with the gap forms $\cos(k_x)\cos(k_y)$ and $\cos(k_x) + \cos(k_y)$.[76] The former gap form is usually considered as the canonical $s\pm$ form (note that the absolute value of $\cos(k_x)\cos(k_y)$ is maximized at $(k_x, k_y) = (0, 0)$ and $(\pi, 0)$ with different signs), but the VMC calculation shows that the latter extended s-wave form also has a significant contribution. In fact, this can be considered as consistent with RPA and other calculations, where there is substantial deviation of the gap from the simple $\cos(k_x)\cos(k_y)$ form (in which the gap should stay nearly constant on each Fermi surface). Alternatively, there have also been some studies that employ five orbital model B, which has stronger tendency of giving nodal pairings.[19,96] We will come back to this point in section 8.5.3.1.

Also, from the early stage, there have been studies on spin fluctuations and superconductivity that employ models with fewer bands as mentioned in section 8.2.1. Two band model in ref. 21 and four band models (in the folded Brillouin zone) in refs. 24–26 are some examples. These models have Fermi surfaces similar to those in the five orbital model when compared in the folded Brillouin zone, but have different orbital characters on some of the Fermi surfaces. Since these models basically have nesting between electron and hole Fermi surfaces, the $(\pi, 0)/(0, \pi)$ spin fluctuations in the unfolded Brillouin (or the (π, π) fluctuation in the folded one) is obtained, and consequently leads to the $s\pm$ pairing state with or without nodes depending on the model or the parameter values. These studies can be regarded as capturing the essence of the spin fluctuation mediated pairing in the iron pnictides, but a more accurate description concerning the material dependence of the superconductivity requires accurate band structures as will be described in section 8.5.

8.4.3 *Orbital Fluctuation Mediated Pairing*

As will be discussed in the next subsection, there is a debate regarding the robustness of the superconductivity against impurities and defects. From this viewpoint, the possibility of the superconducting gap which does not change sign has been proposed.[97,98] This is called the $s{+}{+}$ state as opposed to the $s\pm$. Such a pairing state requires attractive pairing interaction that scatters electron pairs from the electron to the hole Fermi surface and vice versa, rather than the repulsive pair scattering mediated by spin fluctuations. Pairing interaction mediated by phonons, as in conventional superconductors, is one of such interactions, but as mentioned in the Introduction, such an interaction is too weak to account for the high T_c in the iron based superconductors. If we turn to pairing interactions originating from electron-electron interactions, charge fluctuations can mediate such attractive pairing interactions as mentioned in section 8.4.1. In multiorbital systems, the fluctuation of orbital degrees of freedom can also give rise to attractive pairing interactions. In fact, the possible occurrence of orbital ordering in the parent compound was mentioned in section 8.3.3, and the role of the fluctuations that accompany such or related orbital orderings has been discussed by some authors.

Kontani *et al.* took into account Einstein-type iron oscillations in five orbital model A. They showed using RPA and FLEX that for large enough electron-phonon coupling, the attractive pair scattering between electron and hole Fermi surfaces, mediated by antiferro-orbital fluctuations with momentum of $\sim$ $(\pi, 0)/(0, \pi)$, becomes large, and therefore $s{+}{+}$ can dominate over $s\pm$.[97,100,101] Yanagi *et al.* adopted the 16 band d-p model, and showed that for sufficiently large interorbital interaction U', orbital fluctuations can be enhanced resulting in a $s{+}{+}$ superconductivity.[98] They further took into account electron-phonon interactions,[79,99] and proposed in ref. 79 in particular that the ferro-orbital fluctuation induced by the orthorhombic phonon modes of the arsenic atoms is playing an important role in the occurrence of the high T_c superconductivity.

This has been discussed in relation with the possible occurrence of the ferro-orbital ordering and the orthorhombic structural transition described in section 8.3.3. However, it has been pointed out later in refs. 102, 80 that the orthorhombic phonon proposed to be responsible for the ferro-orbital fluctuation in ref. 79 is actually of the acoustic type, and thus has zero frequency in the long wave length limit, so that this phonon should not be responsible for high T_c superconductivity. It remains a future problem to further investigate whether the $(\pi, 0)/(0, \pi)$ antiferro-orbital fluctuations in the iron based superconductors can actually be strong enough to overcome the effect of the antiferromagnetic spin fluctuations so as to make the gap $s{++}$ rather than $s\pm$.

8.4.4 *Theoretical Proposals for the Detection of the Pairing State Based on the Effective Multiorbital Hamiltonian*

In this section, we review some of the issues concerning the determination of the pairing symmetry. A number of experiments for the arsenides, such as ARPES,[103,104] penetration depth measurements,[105] and muon spin relaxation (μSR),[106–109] suggest that the gap is open on the whole Fermi surface, although the magnitude of the gap may vary along the Fermi surface. However, whether the sign change of the superconducting gap between the electron and hole Fermi surfaces exists, as suggested in the spin fluctuation mediated $s\pm$-wave pairing scenario, is still under debate.

There have been some experiments that strongly suggest the presence of sign change in the superconducting gap. The nuclear magnetic relaxation (NMR) rate lacks the coherence peak below T_c,[111–122] which is an indication that the gap must have opposite signs on certain portions of the Fermi surface, where the positive and negative portions have about the same amount of area. In fact, it has also been shown that in heavily electron doped LaFeAsO, the coherence peak does appear, which can be interpreted as due to the shrinking or the disappearance of the hole Fermi surface, resulting in an unbalanced positive and negative portions.[110] In some NMR studies, the relaxation rate is found to decay as T^3 below T_c,[111,114,116–120] which is usually taken as an evidence for the presence of nodes in the superconducting gap, but Nagai *et al.* showed, using five orbital model A, that this temperature dependence can be understood by considering fully open $s\pm$ gap that has some anisotropy on the electron Fermi surface.[123] In ref. 124, half-integer quantum flux transitions was observed in a niobium-poly crystalline NdFeAsO loops, which can be understood by the presence of the sign change in the superconducting gap. Also in ref. 125, STM/STS measurements in a 11 material have detected a magnetic field dependence of the quasiparticle interference in the superconducting state, which suggests the realization of the $s\pm$ gap. Regarding the phase diagram of the iron pnictides, there are suggestions of the coexistence of magnetism and superconductivity, which is considered to be an evidence that there must be a sign change in the superconducting gap.[126]

However, there has been some debate concerning the sensitivity of T_c against impurities. Although some experiments show suppression of T_c by non-magnetic impurities,[127,128] it has been pointed out, e.g., in refs. 129–132 that the suppression of T_c by impurity doping or by particle irradiation is too weak for a pairing state with sign changes in the gap. Onari and Kontani calculated the suppression of T_c by impurities within five orbital model A, which showed that the experimentally observed suppression rate of T_c is small compared to the theoretical expectations of the $s\pm$ state.[133] This robustness of T_c against impurities has motivated the consideration of orbital fluctuation mediated $s{+}{+}$ state as discussed in section 8.4.3.[79,80,97–99]

It has been proposed at the early stage[24,134,135] that one of the ways to determine whether the superconducting gap indeed changes its sign between the electron and hole Fermi surfaces is the observation of neutron scattering resonance at the nesting vector of the electron and hole Fermi surfaces, i.e., $(\pi, 0)$ in the unfolded Brillouin zone. In fact, in the case of the high T_c cuprates, the observation of the resonance has proved to be a powerful tool for detecting the sign change in the d-wave superconducting gap.[136,137] In Fig. 8.15, the spin susceptibility calculated for five orbital model B in various pairing states including $s\pm$ is shown.[135] As seen in this figure, a peak appears around $\omega = 2\Delta$ at the wave vector close to $(\pi, 0)$. In fact, neutron scattering experiments have indeed observed such a peak like structure in the superconducting state around the

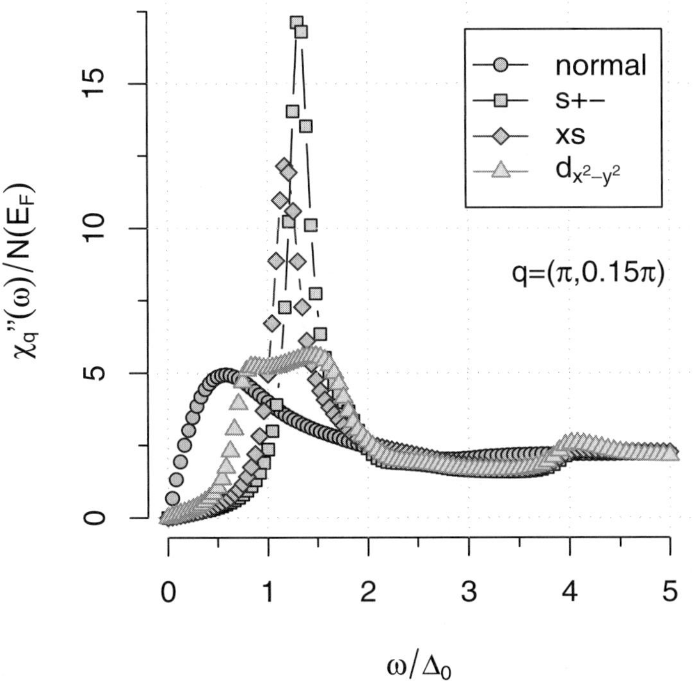

Figure 8.15 The imaginary part of the dynamical spin susceptibility at the wave vector close to $(\pi, 0)$ obtained for various pairing states in five orbital model B. From ref. 135.

nesting vector of the Fermi surface.[46,47,138–142] This has been taken as a strong evidence for the sign change in the superconducting gap.

However, Onari *et al.* later took into account the quasiparticle damping effect in the calculation of the dynamical spin susceptibility using five orbital model A, and showed that a peak like enhancement over the normal state values can be obtained even in the $s{+}{+}$ state, which is due to the suppression of the normal state susceptibility originating from the damping.[143] In a later paper by Nagai and Kuroki, also using five orbital model A, it was shown that the experimental results are close to the calculation results of the $s\pm$ state when the superconducting gap and the quasiparticle damping is $\sim$10 meV, while they are more closer to those of the $s{+}{+}$ state for larger gap and damping.[144] It was proposed there that the two possible candidates can be distinguished by investigating experimentally the wave vector (π, π) along with $(\pi, 0)$. Alternatively, in ref. 145, it was proposed that the realization of $s{++}$ state can be detected by probing the orbital susceptibility through measuring the phonon spectra in the neutron scattering experiments. Also, a Bogoliubov-de Gennes equation approach for the single impurity problem in five orbital model A shows that the bound state peak in the local density of states can be used to distinguish between $s\pm$ and $s{++}$.[146]

8.5 Material Dependence

8.5.1 *Some Experimental Observations*

One of the interesting and important features of the iron pnictides is the unusually strong material dependence of T_c, which ranges from $\simeq$ 5K in LaFePO[147] to 55K in SmFeAsO[2] even within the same group of elements. More systematically, Lee *et al.*[148] pointed out that T_c varies with the Fe-Pn-Fe (Pn: pnictogen) bond angle α (Fig. 8.16), where T_c is maximized around the bond angle ($\simeq$109 degrees) for materials such as SmFeAsO or NdFeAsO at which the pnictogens form a regular tetrahedron, while T_c is low for materials with large α such as LaFePO. The importance of the bond angle has also been pointed out by Zhao *et al.*[149] and others. On the other hand, as will be described later, some of the theoretical studies have focused on the pnictogen (or chalcogen) height h_{Pn} namely, the distance between a pnictogen atom and the Fe layer (Fig. 8.16), and this has motivated also systematic experimental studies on the height dependence of T_c.[150] It should be noted here that in LnFeAsO, the systematic decrease of the bond angle in Ln=La $\rightarrow$ Dy is due to both the increase of the height and the decrease of the in-plane lattice constant.[151] Further experimental investigations have shown that T_c tends to increase with the increase of the height to about $h_{\mathrm{Pn}} = 1.4$ Å, but materials having very large h_{Pn} of $>$1.5 Å do not have very high T_c.[150] For example, $Ca_4Al_2O_6Fe_2As_2$,[152] which is a variation of the materials with perovskite layers sandwiched between the FeAs layers,[153,154] has a very large h_{Pn} ($\simeq$1.5 Å), small

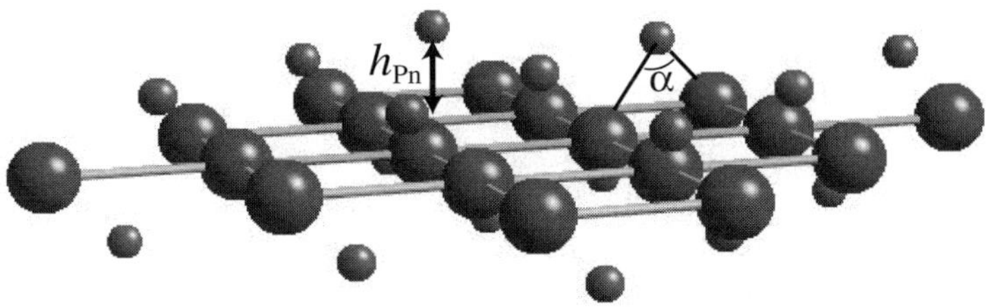

Figure 8.16 Lattice structure of one Fe-Pn layer, with the pnictogen height h_{Pn} and the Fe-Pn-Fe bond angle α are indicated.

lattice constant a ($\simeq$3.71 Å), and consequently a very small bond angle ($\sim$102 degrees). This material have been found to exhibit superconductivity of 28.3K,[152] which is certainly lower than the T_c of SmFeAsO or NdFeAsO.

Pressure effects have also been experimentally elaborated. For LaFeAsO for example, T_c first increases with pressure, but then decreases when the pressure becomes too large,[155–157] which is contrasted with materials having $T_c > 50$ K at ambient pressure such as NdFeAsO for which T_c monotonically and rapidly decreases with pressure.[158] On the other hand, T_c of FeSe is strongly enhanced by more than 20K by applying pressure.[159–161] All these experimental results indicate that T_c is unusually sensitive to the lattice structure in the iron-based compounds.

There also seems to be strong material dependence of the superconducting gap. While a number of experiments on arsenides suggest that the gap is fully open on the Fermi surface as mentioned in section 8.4.4, experiments on LaFePO show that there are nodes in the superconducting gap.[162–164] Arsenides and LaFePO also exhibit sharp contrast in nuclear magnetic resonance (NMR) experiments. In the arsenides, some experiments show that the NMR relaxation rate decays as T^3 below T_c,[111,114,116–120] while some experiments[121] indicates a more rapid decay. In LaFePO, $1/T_1$ below T_c is shown to decay even more slowly than above T_c.[122] So the low T_c and the presence of the nodes in the superconducting gap seems to be correlated in LaFePO. On the other hand, in a 122 material $BaFe_2(As_{1-x}P_x)_2$,[165,166] in which a transition from antiferromagnetism to superconductivity occurs by partially substituting As by P (isovalent doping), a number of experiments suggest presence of line nodes in the superconducting gap,[167,168] but T_c is relatively high (about 30K)[166] in contrast to LaFePO.

If we now focus on the amount of doped electrons/holes, the spin fluctuation and superconductivity is known to exhibit strong doping dependence as well. The NMR/NQR experiments show that there is strong low energy spin fluctuation in the hole doped materials, while in the electron doped materials, especially in the electron doped LaFeAsO, the spin fluctuation becomes weaker.[111,169–171] Also, neutron scattering experiments have revealed that the wave vector at which the spin fluctuations develop exhibits doping dependence. In the electron doped materials or 11 materials, the spin fluctuation wave vector tends to deviate from $(\pi, 0)$ toward (π, π) in the unfolded Brillouin zone,[172–175] while in KFe_2As_2, a heavily hole doped end material, the spin fluctuations develop at incommensurate wave vectors that deviate from $(\pi, 0)$ toward $(0, 0)$.[176] Interestingly, various experiments on KFe_2As_2 show presence of nodes in the superconducting gap.[177–180] In this material, the electron Fermi surface is absent,[181,182] so it is unlikely to have gap nodes there. Also, an experiment on the magnetic vortex lattice of KFe_2As_2 indicates that the superconducting gap does not exhibit large anisotropy around the c-axis,[183] ruling out the possibility of vertical nodes.

8.5.2 *Lattice Structure Dependence of the Band Structure and the Electron-Electron Interactions*

8.5.2.1 Pnictogen height

From the early days of the study on the iron pnictide superconductors, it has been shown in various theoretical studies that the pnictogen height h_{Pn} affects the band structure. For the 1111 materials, $h_{\mathrm{Pn}} = (z_{\mathrm{Pn}} - 0.5) \times c$, where z_{Pn} is the internal coordinate of the pnictogen atom and c the c-axis lattice constant. For the original lattice structure of LaFeAsO, $z_{\mathrm{As}} = 0.651$[1] corresponds to the As height of $h_{\mathrm{As}} = 1.32$ Å. It has been shown in refs. 15, 48, 42, 185, 186 that z_{Pn} controls the relative position of $d_{X^2-Y^2}(= d_{xy})$ and d_{Z^2} bands near the Γ point of the folded (original) Brillouin zone. In

Table 8.3 Materials and lattice parameters, and the nearest and second-nearest neighbor hopping integrals (in eV) in the corresponding tightbinding models. From ref. 48

	a(Å)	c(Å)	z_{Pn}	h_{Pn} (Å)	α	$t_{X^2-Y^2}$	$t'_{X^2-Y^2}$	t_{XZ}	t'_{XZ}
LaFeAsO[1]	4.04	8.74	0.6512	1.32	113.6	0.163	0.124	−0.210	0.329
NdFeAsO[148]	3.94	8.51	0.6624	1.38	109.9	0.135	0.123	−0.202	0.332
LaFePO[147]	3.96	8.51	0.6339	1.14	120.2	0.253	0.156	−0.234	0.377

the unfolded Brillouin zone, these bands appear near (π, π). In Fig. 8.17, we show the band structure in the unfolded Brillouin zone for hypothetical lattice structures with $h_{\mathrm{Pn}} = 1.38$ Å and 1.14 Å with the lattice constants and z_{La} fixed at the original values of LaFeAsO.[48] $h_{\mathrm{Pn}} = 1.38$ Å and 1.14 Å are the same as in the optimally doped NdFeAsO and LaFePO, respectively (see Table 8.3). It can be seen that the $d_{X^2-Y^2}$ band which forms the γ Fermi surface around (π, π) rises as h_{As} is increased, while the d_{Z^2} band sinks below the Fermi level.

The band structure of the five orbital model of LaFePO is shown in Fig. 8.17. The lattice constants (both the in-plane a and the out-of-plane c) of LaFePO are small compared to LaFeAsO as shown in Table 8.3. Nonetheless, the band structure is very similar to that of the hypothetical lattice structure of LaFeAsO with $h_{\mathrm{Pn}} = 1.14$ Å. Thus, the main difference from LaFeAsO is caused by the pnictogen height, and the top of the $d_{X^2-Y^2}$ band at (π, π) is indeed pushed below the Fermi level. The band structure of the optimally doped NdFeAsO is shown in Fig. 8.17. The $d_{X^2-Y^2}$ band at (π, π) is seen to cross the Fermi level even at $n = 6.1$ as expected.

The reason for these shifts in the band positions can be understood in terms of the hopping integrals. As h_{As} increases, the nearest-neighbor hopping for the $d_{X^2-Y^2}$ orbital is reduced as shown in Table 8.3. If we approximate the $d_{X^2-Y^2}$ portion of the bands by

$$\begin{aligned}\varepsilon_{X^2-Y^2}(\boldsymbol{k}) = &-2t_{X^2-Y^2}[\cos(k_x) + \cos(k_y)]\\ &-4t'_{X^2-Y^2}\cos(k_x)\cos(k_y), \qquad (8.13)\end{aligned}$$

where $t_{X^2-Y^2}$ and $t'_{X^2-Y^2}$ stand for the nearest and second-nearest neighbor hoppings, respectively, the energy difference between $(0, 0)$ and (π, π) is proportional to $t_{X^2-Y^2}$, so that the reduction in $t_{X^2-Y^2}$ acts to push up the $d_{X^2-Y^2}$ band at (π, π). Besides the

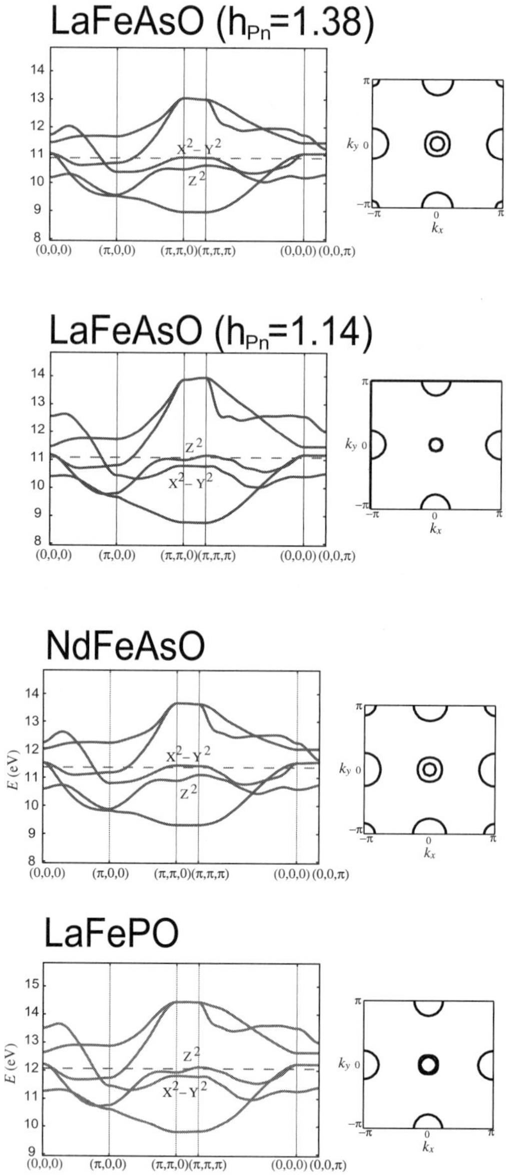

Figure 8.17 From top to bottom: The band structure and the Fermi surface of the hypothetical lattice structures of LaFeAsO with h_{Pn} = 1.38 Å, 1.14 Å. and those for the actual (experimentally determined) lattice structure of NdFeAsO and LaFePO. From ref. 48.

Table 8.4 The intraorbital interaction and Hund's coupling of the five orbital model of various materials obtained in ref. 30 (in units of eV). $\bar{U}$ is the average intraorbital interaction. The values for LaFeAsO somewhat differ from those presented in Table 8.1,[17] which is mainly because in ref. 17, where the pseudopotential calculation with plane-wave basis set was adopted, the screening effect of the La-4f electrons was neglected, whereas in ref. 30, this screening effect was taken into account in the full-potential LMTO calculation

Material(family)	h_{Pn} (Å)	$U_{X^2-Y^2}$	U_{XZ}	$\bar{U}$	J_{XZ-YZ}
LaFePO (1111)	1.14	1.68	2.42	2.47	0.36
LaFeAsO (1111)	1.32	1.91	2.43	2.53	0.37
$BaFe_2As_2$ (122)	1.36	2.29	2.64	2.80	0.40
LiFeAs (111)	1.51	2.85	2.96	3.15	0.39
FeSe (11)	1.47	3.78	4.11	4.24	0.48
FeTe (11)	1.77	3.59	2.88	3.41	0.37

variation in the $d_{X^2-Y^2}$ hoppings, the increase in h_{As} results in an overall reduction in the hopping integrals of other orbitals because the effective hopping path Fe→ As→ Fe becomes less effective.

The material dependence of the electron-electron interactions is also present. Miyake *et al.* studied various families of the iron based superconductors, and found a correlation between the anion height and the strength of the electron correlation.[30] Namely, the electron-electron interaction tends to be large in the 11 materials, where the chalcogen atom is located at high positions, while in the 1111 materials, where the pnictogen height is relatively small, the electron-electron interaction is weaker. The interaction values of some of the materials are summarized in Table 8.4. In the 11 family, the large electronegativity of the chalogen elements compared to the pnictogen as well as the large chalcogen height leads to a strong ionic character of the material, thereby resulting in a small Wannier orbital radius (spread) and thus strong electron-electron interaction.

8.5.2.2 Bond angle

As mentioned in section 8.5.1, the bond angle dependence of T_c has been widely examined experimentally, and in this sense, it is also

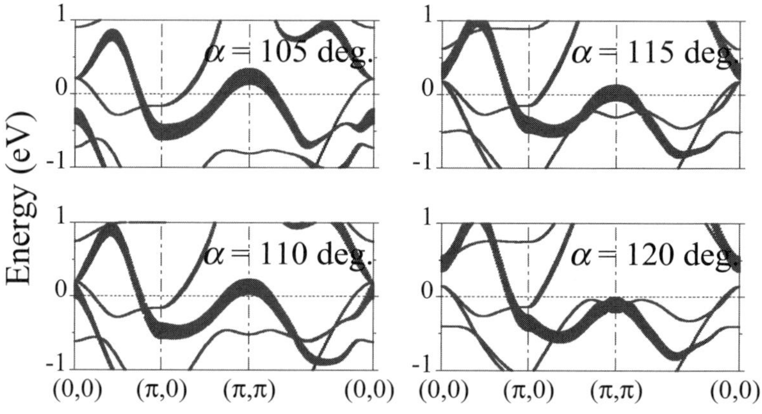

Figure 8.18 Band structure of $Ca_4Al_2O_6Fe_2As_2$ for hypothetical lattice structures, where the Fe-As-Fe bond angle α is varied while fixing the bond length. The thickness represents the strength of the $d_{X^2-Y^2}$ (or d_{xy}) orbital character. From ref. 187.

important to understand how the bond angle affects the electronic structure. The effect of the bond angle is of course included in the effect of the height, but the small angle regime which turns out to be of importance can be reached (within the realistic parameter regime) only when the height and the in-plane lattice constant are varied simultaneously (which was not done in ref. 48). In this context, $Ca_4Al_2O_6Fe_2As_2$[152] mentioned in section 8.5.1 is of special interest in that it has a very small bond angle of 102 degrees due to both the large h_{As} and the small in-plane lattice constant. In fact, Miyake *et al.* examined the band structure of this material and found that the α_1 hole Fermi surface disappears due to the reconstruction of the band structure near the Γ point.[187] They also showed how the band structure changes when the bond angle is varied hypothetically with the Fe-As bond length fixed. In Fig. 8.18, the band structure in the unfolded Brillouin zone for four bond angles are shown, where the thickness represents the strength of the $d_{X^2-Y^2}$ orbital character. As the bond angle is decreased from 120 degrees, the $d_{X^2-Y^2}$ band near the wave vector $(0, 0)$ lowers and approaches the two-fold degenerate bands that originate from the $d_{XZ/YZ}$ orbitals. At 110 degrees, where the As form nearly a regular tetrahedron, the $d_{X^2-Y^2}$ and $d_{XZ/YZ}$ (namely, d_{xy}, d_{yz}, d_{zx}) bands are nearly degenerate at $(0, 0)$.

For smaller bond angle, the $d_{X^2-Y^2}$ portion at (0, 0) sinks below the Fermi level, and only one hole Fermi surface is left around (0, 0).

In fact, this disappearance of the α_1 hole Fermi surface for small bond angle is found to be universal in the iron based superconductors. In Fig. 8.19, the Fermi surface of LaFeAsO in the unfolded Brillouin zone is shown for hypothetical lattice structures with various bond angles (with the Fe-As bond length fixed at its original value). When the bond angle is large, two hole Fermi surfaces, α_1 and α_2 are present around the wave vector (0,0). As the bond angle decreases, the γ Fermi surface appears around (π, π), and there are now three hole Fermi surfaces. This is what was described in section 8.3 as an effect of increasing the pnictogen height.[42,48,185,186] As the bond angle is decreased even further, the α_1 Fermi surface disappears, and again there are only two hole Fermi surfaces but in this case one around (0, 0) and another around (π, π). So the tendency found in $Ca_4Al_2O_6Fe_2As_2$ holds also for LaFeAsO. However, the range of the bond angle in which there are three hole Fermi surfaces (three hole Fermi surface zone in Fig. 8.19) turns out to be much larger for LaFeAsO than for $Ca_4Al_2O_6Fe_2As_2$.

8.5.2.3 Three dimensionality

So far we have mainly concentrated on materials which have small three dimensionality. The 122 materials such as $BaFe_2As_2$[188] have relatively large three dimensionality. A characteristic feature of the band structure is that the Brillouin zone unfolding procedure[6] that adopts the reduced unit cell (with one iron) cannot be performed strictly.[189] Namely, Fe1 and Fe2 shown in the top of Fig. 8.20 are not completely equivalent. Thus, a strictly accurate Hamiltonian is a ten orbital model given in the Brillouin zone of the body centered tetragonal lattice. Nonetheless, there are some five orbital models of the 122 materials obtained by approximately unfolding the Brillouin zone.[176,184] This corresponds to considering the simple tetragonal lattice structure shown in the top right of Fig. 8.20. The five orbital models of the 122 systems, despite the approximation, can reproduce a large portion of the original band structure, but caution has to be taken when these models are used since the

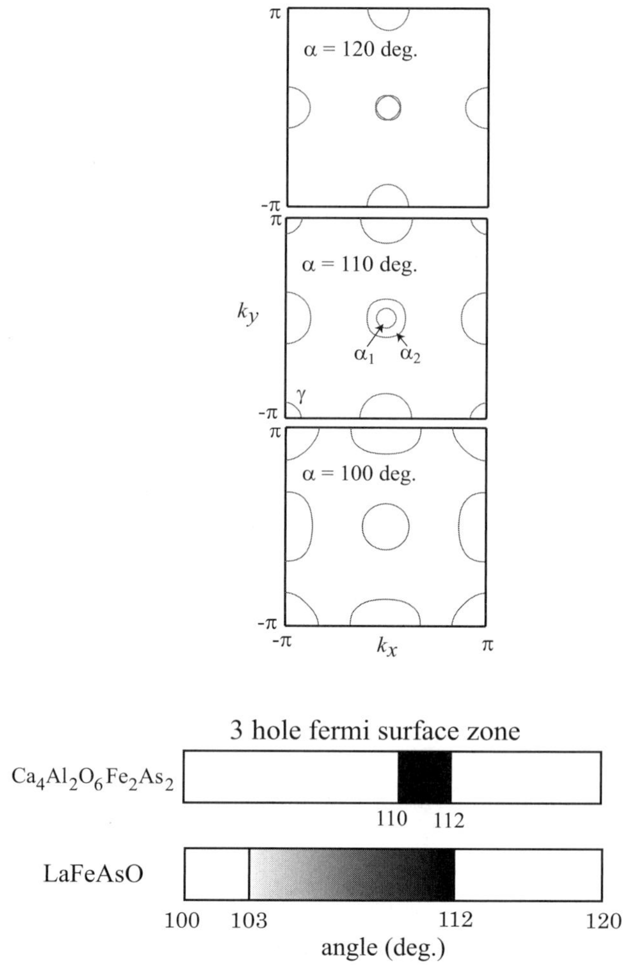

Figure 8.19 Top: the Fermi surface of the hypothetical lattice structures of LaFeAsO with varying the Fe-As-Fe bond angle while fixing the Fe-As bond length. Bottom: The bond angle range where three hole Fermi surfaces coexist is shown for LaFeAsO and $Ca_4Al_2O_6Fe_2As_2$. From ref. 197.

deviation from the original ten orbital band structure can be crucial in some cases.

As an example of the ten orbital model, we show in Fig. 8.20 the band structure of the model of $BaFe_2(As_{1-x}P_x)_2$ ($x = 0.64$ here). In

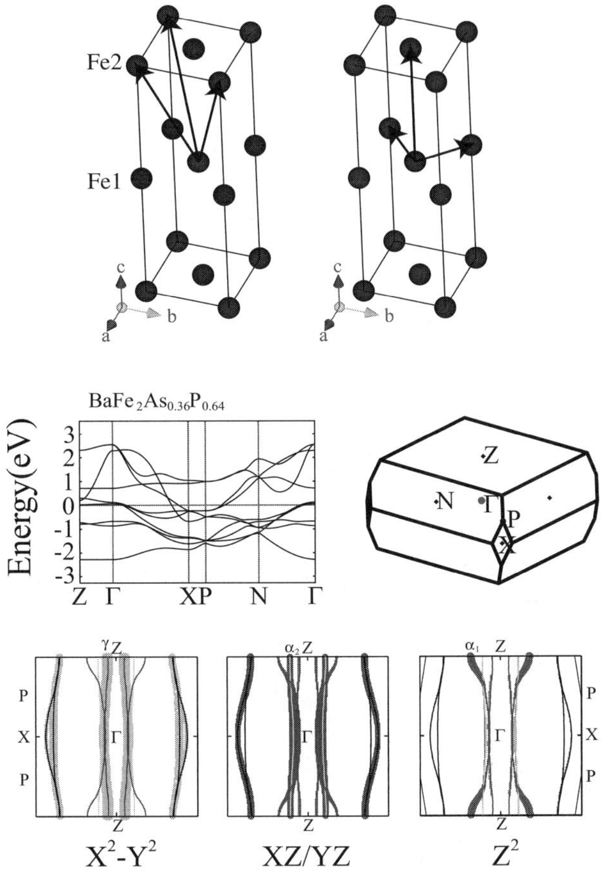

Figure 8.20 Top: the body centered tetragonal lattice structure of the 122 systems is shown in the left, where Fe1 and Fe2 are inequivalent. When the two Fe atoms can be considered as equivalent, the unit cell is reduced to result in a simple tetragonal lattice structure shown in the right. Middle: the band structure of the ten orbital model for $BaFe_2(As_{0.36}P_{0.64})_2$ along with the Brillouin zone of the body centered tetragonal lattice structure. Bottom: vertical cut of the Fermi surface. The thickness represents the strength of the orbital character. From ref. 190.

the bottom of Fig. 8.20, we show the vertical cut of the Fermi surface of $BaFe_2(As_{0.36}P_{0.64})_2$ for each orbital character, where the thickness represents the strength of the character.[190] The 122 materials share common features with the 1111 materials in that they have three hole (around Γ-Z) and two electron Fermi (around X-P) surfaces. There are, however, some differences. One is that the portion of the Fermi surface around the Z point having strong d_{Z^2} character is continuously connected to the d_{XZ}/d_{YZ} portion of the Fermi surface around the Γ point in 122, while in 1111 the d_{Z^2} Fermi surface, when present, is an isolated three-dimensional pocket. Another difference of 122 from 1111 is that the hole Fermi surface having strong $d_{X^2-Y^2}$ orbital character coexists with the d_{Z^2} Fermi surface; in 1111, either the three-dimensional d_{Z^2} or the cylindrical $d_{X^2-Y^2}$ (γ) Fermi surfaces exist depending on the pnictogen height.[42,48,185,186]

In the case of $BaFe_2(As_{1-x}P_x)_2$, the main effect of replacing As by P is to reduce the pnictogen height. The band structure calculations using experimentally determined lattice parameters show that lowering the height leads to a higher d_{Z^2} orbital level as in 1111, resulting in a larger warped Fermi surface around the Z point.[166] Such a strong warping of the Fermi surface has been directly observed in an ARPES experiment.[191] At the same time, the $d_{X^2-Y^2}$ level is lowered also like in 1111, so that the $d_{X^2-Y^2}$ hole Fermi surface shrinks, but it exists up to a P content of about $x \sim 0.7$ within the LDA calculation.

8.5.3 *Material Dependence of the Spin Fluctuations and Superconductivity*

8.5.3.1 Lattice structure dependence

Let us now see how the band structure modified by the lattice structure in turn affects the spin fluctuations and superconductivity. Although the strength of the interactions are also material dependent as mentioned in section 8.5.2.1, its variation is not so large within each families like 1111, 122, 11 etc., so here we concentrate on the variation of the band structure.

The effect of varying the height on the spin fluctuations and the superconductivity was studied in detail in ref. 48. As seen in Fig. 8.21, varying h_{As} (while fixing the lattice constants) affects the spin susceptibility in several ways. The d_{YZ} orbital component χ_{s3333} (the same notation as in section 8.3.2 is used) always has peaks around $(\pi, 0)$,$(0, \pi)$ reflecting the α-β particle-hole interaction.

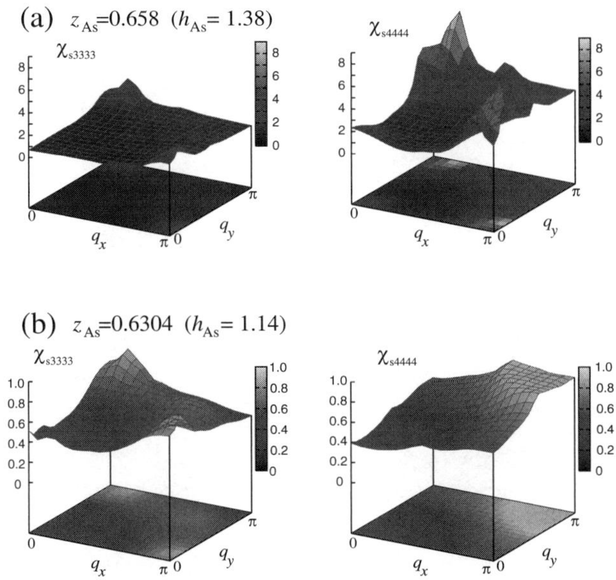

Figure 8.21 χ_{s3333} (left panels) and χ_{s4444} (right) for the model with (a) $h_{As} = 1.38$ Å and (b) $h_{As} = 1.14$ Å. The same band filling and the interaction values as in Fig. 8.6 are adopted. From ref. 48.

On the other hand, the $d_{X^2-Y^2}$ orbital component of the spin susceptibility χ_{s4444} exhibits a strong variation with h_{As}: when h_{As} is large and the γ Fermi surface is present, χ_{s4444} (Fig. 8.21(a), right) is strongly peaked at $(\pi, 0)$, owing to the γ-β interaction. However, as h_{As} is reduced, the structure around $(\pi, \pi/2)$, which arises from the β_1-β_2 interaction, dominates (Fig. 8.21(b), right).

The lattice structure effect on the spin susceptibility in turn affects spin-fluctuation-mediated superconductivity. When h_{As} is large, the two different $(\pi, 0)$ spin fluctuations, namely those from χ_{s4444} and χ_{s3333}, cooperate to mediate the fully-gapped, $s\pm$-wave superconductivity. In Fig. 8.22, the $s\pm$ gap obtained by RPA for $h_{Pn} = 1.38$ Å is shown along with the less dominating d-wave gap. Note that the $s\pm$ gap is fully consistent with the requirement that the gap has to change the sign across the wave vector at which the spin fluctuations develop (see also top of Fig. 8.13). When h_{As} is small, by contrast, the development of the $(\pi, \pi/2)$ spin fluctuations

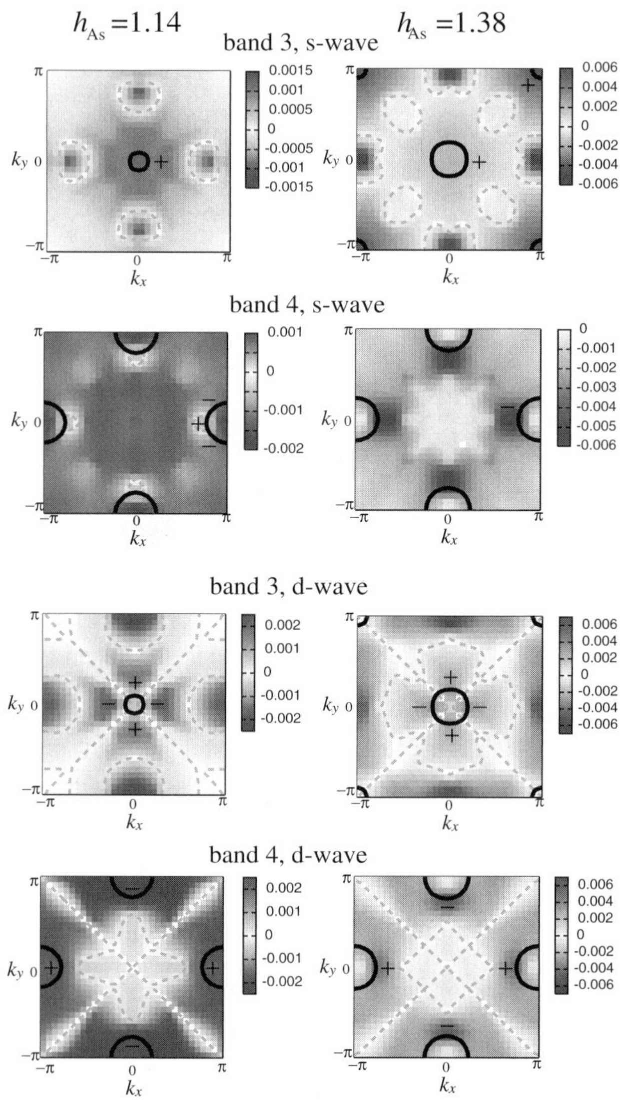

Figure 8.22 The gap functions for the model with h_{As} = 1.14 Å (left panels) or h_{As} = 1.38 Å (right). From top to bottom: *s*-wave in band 3 (with hole Fermi surface α_1), *s*-wave in band 4 (with electron Fermi surfaces), *d*-wave in band 3, *d*-wave in band 4. The band filling and the interaction values as well as the lines and the color codings are the same as in Fig. 8.14. From ref. 48. See also Color Insert.

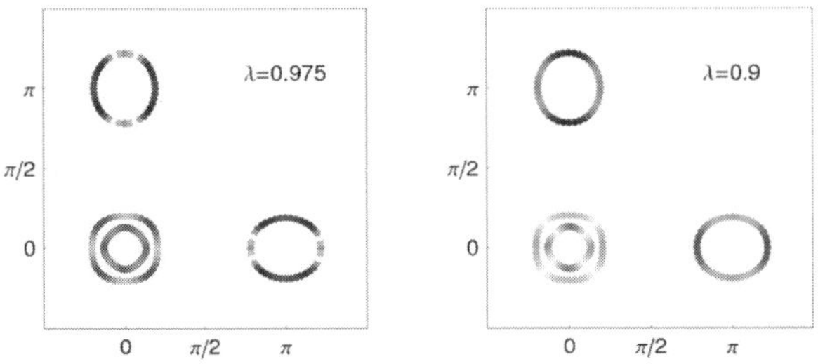

Figure 8.23 The nodal $s\pm$-wave (left) and d-wave (right) superconducting gap of model B on the Fermi surfaces obtained by RPA for $n = 6$ and $U = 1.25$. The red (blue) indicates positive (negative) sign. From ref. 19.

favor superconducting gap that changes the sign across this wave vector. The coexistence of such a $(\pi, \pi/2)$ interaction with the $(\pi, 0)$ interaction will induce a frustration in momentum space regarding the sign change of the gap (see the schematic figures in Fig. 8.24), and can result either in the d-wave or the nodal $s\pm$-wave gap as was described schematically in Fig. 8.13. In Fig. 8.22, the nodal $s\pm$-wave and d-wave gap functions obtained for $h_{\rm Pn} = 1.14$ Å by RPA are shown.[48] In fact, the nodal $s\pm$-wave pairing was first obtained in ref. 19 using five orbital model B, which has a lower position of the $d_{X^2-Y^2}$ band around (π, π) than in model A. In Fig. 8.23, a typical nodal superconducting gap obtained in ref. 19 is shown. In the description above, the effect of the height on the gap function is mainly presented in the momentum space. On the other hand, Kariyado and Ogata provided a intuitive picture of the present effect in real space.[192]

In Fig. 8.24, the eigenvalue λ of the Eliashberg equation (qualitative measure for T_c) for the s-wave and d-wave pairings at a fixed temperature of $T = 0.02$ eV is plotted against $z_{\rm As}$ (lower scale) or against $h_{\rm As}$ (upper scale). For large $h_{\rm As}$ where the fully-gapped, $s\pm$-wave dominates, λ is large because the strong spin fluctuations arising from α-β and γ-β interactions cooperate. In this regime, d-wave pairing has much smaller λ, so that it is not realized. This is contrasted with the case of small $h_{\rm As}$, where d-wave or nodal $s\pm$-wave begin to dominate. In this region, λ is small because the γ-β interaction is no longer effective, or to be more precise, its remaining

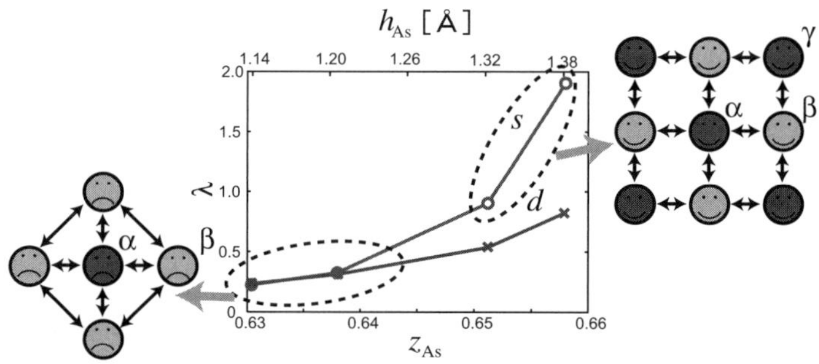

Figure 8.24 s-wave and d-wave eigenvalues of the Eliashberg equation plotted against z_{As} (lower scale) or h_{As} (upper scale) for $n = 6.1$ and $T = 0.02$ eV. The interaction values are the same as in Fig. 8.6. The lattice constants are fixed at the original values for LaFeAsO. For the s-wave, the open (solid) circles indicate that the gap is nodeless (nodal). From ref. 48. The figures in the left and right insets show schematically the presence (left) or absence (right) of frustration in momentum space.

effect *competes* with the effect of the β_1-β_2 interaction. These results show that the pnictogen height can act as a "switch" between the fully-gapped, high-T_c $s\pm$-wave and the low-T_c gapless (either d-wave or nodal $s\pm$-wave) superconductivity.

For the band structure of the actual LaFePO, the RPA calculation gives a result similar to that for the hypothetical lattice structure of LaFeAsO with $h = 1.14$ Å (the same height as in the actual LaFePO), namely, a close competition between the nodal $s\pm$-wave and d-wave with small λ.[48] fRG studies for the model of LaFePO[91,193] also show that the $s\pm$ gap exhibits nodes in the electron Fermi surface as shown in Fig. 8.25. However, in the case of fRG, the d-wave pairing is found to be less competitive against nodal $s\pm$-wave compared to RPA.[91] In any case, the band structure of LaFePO gives nodal superconductivity, and this is consistent with experiments on LaFePO that suggest the presence of nodes in the superconducting gap.[162–164] Since there is a possibility of a close competition between s and d-wave pairings, some authors have studied the possibility of $s + id$ pairing.[194]

It should also be noted here that apart from the above mentioned mechanism of nodeless-nodal gap competition due to the presence/absence of the γ Fermi surface around (π, π), various

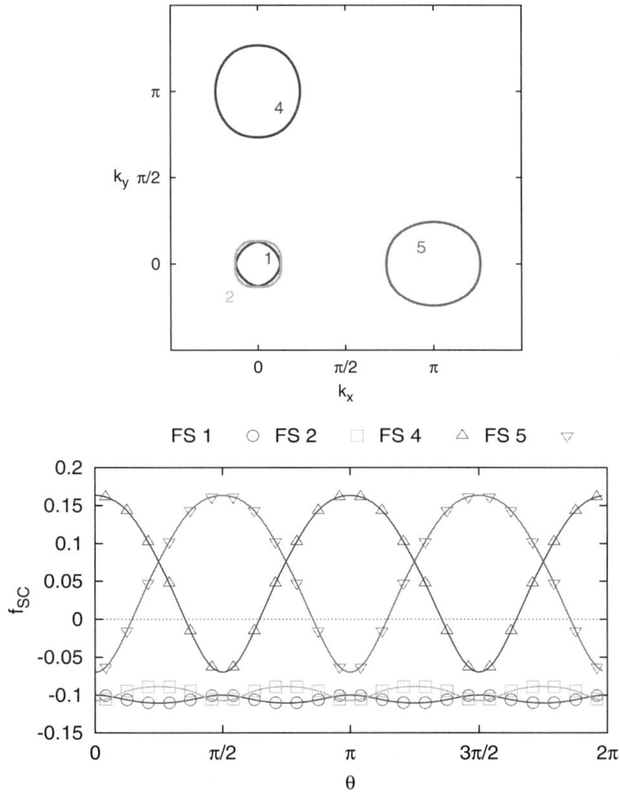

Figure 8.25 The gap function on the Fermi surfaces of LaFePO obtained by functional renormalization group method. The horizontal axis θ is the polar angle of the Fermi surface points with respect to the center of each Fermi surface. $\theta = 0$ corresponds to the $+k_x$ direction. From ref. 193.

mechanisms for the competition between nodeless and nodal pairings have also been discussed. The effect of the intra-Fermi-surface repulsive scattering induced by the Coulomb interaction was discussed by Chubukov *et al.*[195] The effect is relatively strong when the inter-Fermi-surface scattering mediated by the $(\pi, 0)$ spin fluctuations is weak, so that nodes tend to emerge in systems with weaker tendency towards magnetism. On the other hand, the lifting of the nodes by impurities was discussed in ref. 196. The nodal-nodeless gap competition has also been a two-orbital exchange coupling model (J_1-J_2 model),[10] and also in the orbital fluctuation pairing mechanism.[100] The doping dependence of the superconducting gap will be discussed in section 8.5.3.2.

Let us now look more closely into the problem of T_c vs. lattice structure. In the theory that focuses on the γ Fermi surface around (π, π), higher pnictogen position results in a higher T_c, so this alone cannot account for the experimental observations that T_c is maximized around the bond angle where the pnictogen atoms form a regular tetrahedron.[148] In ref. 48, the saturation of T_c in LnFeAsO in the bond angle regime of less than 109 degrees was attributed to the decrease of the in-plane lattice constant a,[151] which is unfavorable for superconductivity due to the increase of the band width and thus the reduction of the density of states. However, as mentioned in section 8.5.1, later experiments have revealed that T_c of the materials having very large h_{Pn} of > 1.5 Å is not so high,[150] which seemed to be difficult to understand from the viewpoint of considering the effect of the height and the in-plane lattice constant *independently* as was done in ref. 48.

Here, the bond angle dependence of the band structure and the Fermi surface described in section 8.5.2.2 has turned out to be of great importance. In Fig. 8.26, the eigenvalue of the Eliashberg equation λ for the s$\pm$-wave superconductivity and the antiferromagnetic stoner factor α_S at $(\pi,0)$ obtained by FLEX are shown as functions of the bond angle while fixing the bond length at its original value.[197] In the large angle regime, the eigenvalue of the Eliashberg equation remains relatively small. As the angle is decreased, the γ Fermi surface around (π, π) becomes effective and the eigenvalue becomes large, as mentioned in the previous section.[19,48,91,193] As the angle is decreased further, superconductivity tends to be suppressed, correlated with the disappearance of the α_1 hole Fermi surface. Hence the superconductivity is optimized around 110 degrees, in agreement with the experimental observations.[148] Here, the number of Fermi surfaces is maximized around the bond angle where the symmetry of the lattice structure is good, and there, superconductivity is optimized since larger number of the Fermi surfaces gives rise to larger number of pair scattering channels. On the other hand, the strength of the low energy spin fluctuations, as measured by the stoner factor α_S, is not necessarily correlated

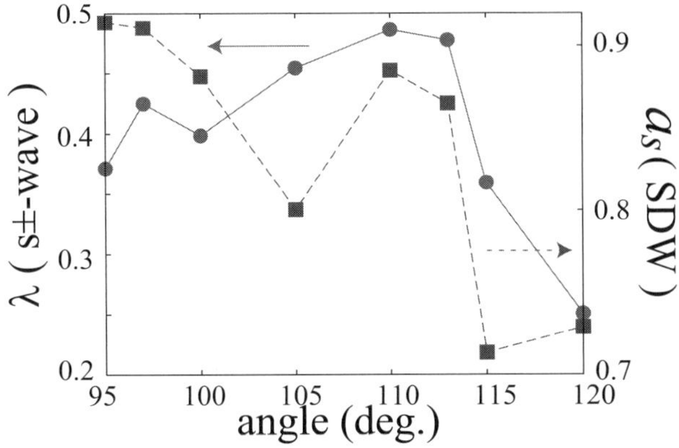

Figure 8.26 The Eliashberg equation eigenvalue obtained by FLEX for s±-wave pairing (circle, left axis) and the stoner factor at $(\pi,0)$ (square, right) against the bond angle for temperature $T = 0.005$ eV. Here, the interaction values of LaFeAsO in ref. 30 was used by multiplying a factor of 0.6 to all of the interactions so as to reduce the double counting effect discussed in section 8.3.2. From ref. 197.

with the eigenvalue of the Eliashberg equation. For instance, in the relatively large angle regime of $115 < \alpha < 120$ degrees, superconductivity begins to grow as the bond angle is decreased, but the change in the stoner factor is small. This angle regime is close to that of LaFeAsO, so that this can be related to the NMR/NQR experiments in LaFeAsO,[111,169–171] which do not find strong enhancement of the low energy spin fluctuations for the optimally doped systems despite T_c being not so low (much higher than in LaFePO having larger bond angle). Also, in the small angle regime, the spin fluctuations tend to increase with the decrease of the bond angle, while the superconductivity is degraded. In fact, an NQR experiment for $Ca_4Al_2O_6Fe_2As_2$ has found that the low energy spin fluctuations of this material is very strong, although T_c is not so high (~28K).[198]

8.5.3.2 Doping dependence

As for theoretical studies on the doping dependence, Ikeda *et al.* performed FLEX calculation of the dynamical spin susceptibility and the nuclear spin-lattice relaxation rate $1/T_1$[55,199] using five orbital

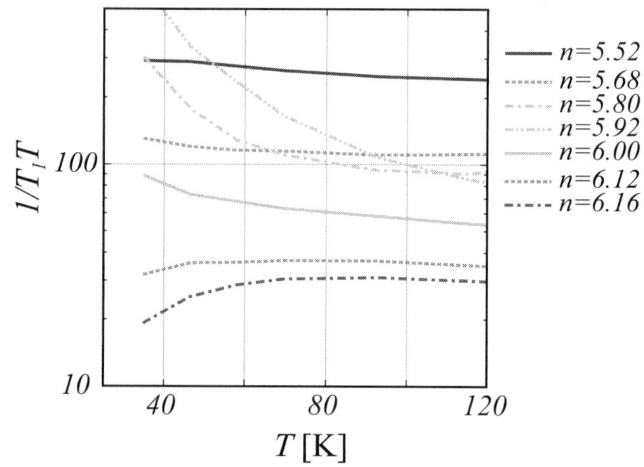

Figure 8.27 $1/T_1T$ as functions of temperature obtained using FLEX for various band fillings. The five orbital model obtained in ref. 18 is used. Original results are in ref. 199. Here, calculation results for some of the doping levels have been omitted for clarity with the permission of the authors.

model A of LaFeAsO or a similar band dispersion obtained in ref. 18. In Fig. 8.27, calculation results of $1/T_1T$ for various doping levels are shown. It can be seen that $1/T_1T$ is enhanced at low temperatures in the lightly hole doped regime, while the enhancement is absent in the electron doped regime. This calculation at least qualitatively reproduces the tendency that hole doping strengthen the low energy spin fluctuations. This is because the γ Fermi surface around (π, π) tends to become robust in the hole doped case. On the other hand, within the FLEX calculation, T_c (eigenvalue of the Eliashberg equation to be precise) of the spin fluctuation mediated superconductivity is not higher in the hole doped regime than in the electron doped regime,[18] namely, T_c is not necessarily correlated with the strength of the low lying spin excitations. This is consistent with the FLEX calculation on the bond angle dependence presented in Fig. 8.26, where T_c and the strength of the low energy spin fluctuations are not directly correlated.

The doping dependence of the incommensurability of the spin fluctuation has also been understood theoretically. A number of studies show that the spin fluctuations tend to deviate in the direction of (π, π) in the unfolded Brillouin zone in the electron doped case,[6,19,199–203] in agreement with the experiments. As for the electron-hole asymmetry of the spin fluctuations, an approximate five orbital model of the 122 system in the unfolded Brillouin zone was used in ref. 176 to show that the wave vector of the spin fluctuations is strongly governed by the orbital character of the Fermi surfaces. Namely, as shown in Fig. 8.28, in the electron doped regime, the portion of the Fermi surfaces having the same character on electron and hole Fermi surfaces are connected by a wave vector that deviates from $(\pi, 0)$ in the (π, π) direction, while in the hole

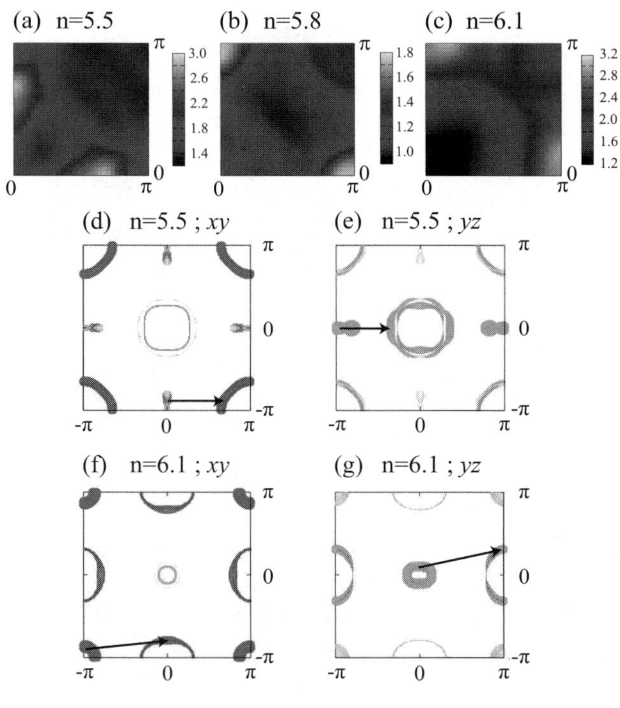

Figure 8.28 The contour plot of the RPA spin susceptibility for the approximate five orbital model of KFe_2As_2 ($n = 5.5$) (a), with the band filling changed to $n = 5.8$ (b), and BFe_2As_2 with the band filling $n = 6.1$ (c). The strength of the xy and yz orbital character on the Fermi surface for $n = 5.5$ (d,e) and $n = 6.1$ (f,g). The arrows connect Fermi surface portions with similar orbital character. Here, the unfolded Brillouin zone ($k_z = 0$ plane) is adopted. From ref. 176. See also Color Insert.

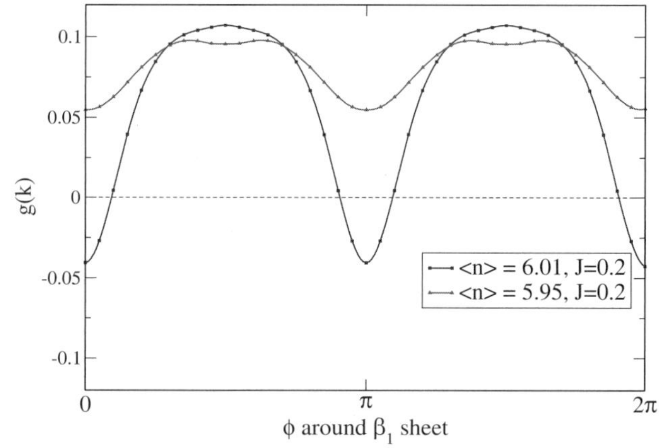

Figure 8.29 The doping dependence of the superconducting gap on the electron (β) Fermi surface obtained by RPA for five band model model B. From ref. 50. The horizontal axis ϕ is the polar angle measured from the k_x axis.

doped case the wave vector deviates toward (0, 0). Consequently, the wave vector at which the spin fluctuations develop in the five orbital model is consistent with the experimental observations.

As for superconductivity mediated by spin fluctuations, the presence/absence of the γ Fermi surface around (π, π) affects the form of the superconducting gap as was discussed in section 8.5.3.1, so the gap is also affected by the doping level. In refs. 48 and, 50 this was discussed in detail using five orbital models A and B, respectively. Fig. 8.29 shows a typical form of the superconducting gap in the hole doped ($n = 5.95$) and electron doped ($n = 6.01$) regimes, obtained in the RPA calculation using five orbital model B. It is found that the superconducting gap tends to have nodes on the electron Fermi surface when electrons are doped due to the disappearance of the γ Fermi surface.

8.5.3.3 Effect of the three dimensionality

As described in section 8.5.2.3, the band structure and the Fermi surface of the 122 materials exhibit significant three dimensionality. There have been several studies on the superconducting gap that take into account this three dimensionality of the band structure.

In Fig. 8.30, we show the superconducting gap obtained by RPA calculation for the ten orbital model of $BaFe_2(As_{0.36}P_{0.64})_2$.[190] The electron and hole Fermi surfaces have different signs of the gap for most portions, namely, the gap is basically $s\pm$, mediated by the spin fluctuation originating from the electron-hole interaction. The gap on the electron Fermi surfaces is fully open since the $d_{X^2-Y^2}$ (γ) Fermi surface is present, as is the case for the 1111 systems with high pnictogen positions.[48] On the other hand, as the volume of the α_1 Fermi surface around the Z point grows upon lowering the pnictogen height by isovalent doping, a three-dimensional sign change of the gap takes place within this Fermi surface, as shown in Fig. 8.30. The gap function on the α_1 Fermi surface is schematically shown at the bottom of Fig. 8.30. This sign change can be considered to be due to the repulsive intraband interaction within the α_1 Fermi surface (shown by the arrow in the schematic figure), which becomes more effective as the volume of the Fermi surface around the Z point grows. A similar three-dimensional sign change of the superconducting gap has been found in a five orbital study of Ba122 in refs. 184, 204, although in the model of ref. 184. the α_1 Fermi surface around the Z point has an $d_{X^2-Y^2}$ character (or xy) rather than d_{Z^2}.

Since the three-dimensional nodes here occur at portions of the Fermi surface having strong d_{Z^2} character, the presence of the nodes does not strongly reduce T_c. This is in contrast to the case of LaFePO, where nodes of the gap intersect the $d_{X^2-Y^2}$ and/or d_{XZ}/d_{YZ} Fermi surfaces,[18,48,91,192,193] which play an important role in the spin-fluctuation-mediated superconductivity. This view may explain why T_c is relatively high in $BaFe_2(As_{1-x}P_x)_2$ despite the presence of nodes in the superconducting gap. It is worth mentioning here that a similar RPA study for KFe_2As_2 also shows that three-dimensional nodes go into the d_{Z^2} hole Fermi surface[205] in the $s\pm$ state, which can be consistent with experimental observations.[177–180,183] Thus, regardless of the doping level, there is a general tendency of having three dimensional horizontal-like nodes (or dips) around the d_{Z^2} portion of the hole Fermi surface, at least within the RPA calculation.

Within the LDA band + RPA approach, the nodes of the superconducting gap do not intersect the electron Fermi surface even for the end composition of $x = 1$ in $BaFe_2(As_{1-x}P_x)_2$, where

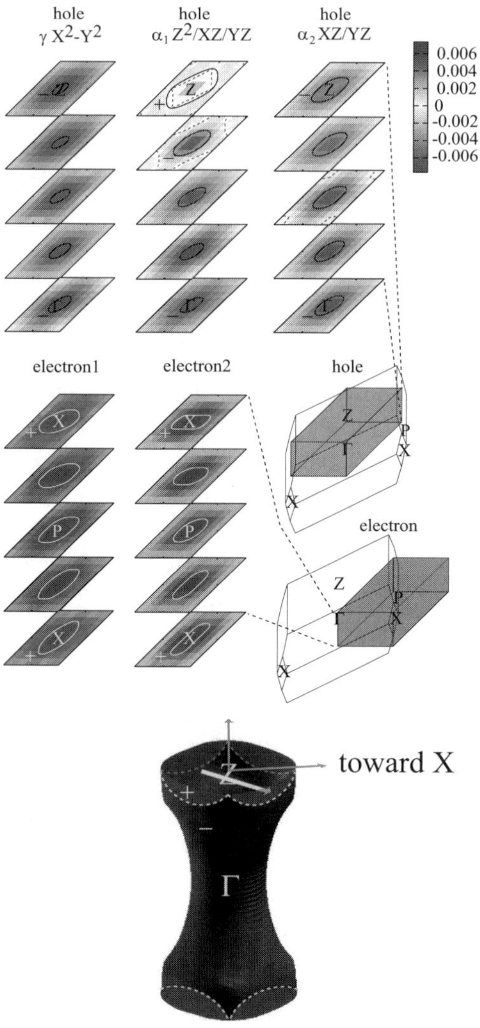

Figure 8.30 Contour plots of the gap function for the ten orbital model of $BaFe_2(As_{0.36}P_{0.64})_2$ obtained by RPA. The gap is shown for the five Fermi surfaces with five horizontal cuts around Γ−Z (hole) or X−P−X (electron). The solid lines represent the Fermi surfaces, while the dashed lines are the nodes of the gap. The Brillouin zone is shown along with the regions where the contour plots of the gap are presented. In the bottom, a schematic view of the gap function on the $\alpha_1 = Z^2/XZ/YZ$ hole Fermi surface is shown. The arrow bridging the Fermi surface indicates the intraband interaction that induces the sign change of the gap. From ref. 190.

the pnictogen height is substantially reduced. This is because as far as the LDA band calculation is concerned, the top of the $d_{X^2-Y^2}$ band (the band that gives the γ Fermi surface) still lies close to the Fermi level even for $x = 1$, in contrast to the case of LaFePO. On the other hand, an analysis of the Raman experiment for the electron doped 122 material $Ba(Fe_{1-x}Co_x)_2As_2$ shows that the nodes or gap minima occur on loops stretching along the c-direction on the electron Fermi surface. The discrepancy may be due to the difference of the material (doping), or due to the neglect of the self energy in the RPA.

11[206] and 111[207] materials also have relatively large three dimensionality in the band structure, although not as large as in 122. In ref. 208, an RPA study was performed for the three dimensional five orbital model of FeSe. Since the height of Se is relatively large (see Table 8.4), the γ Fermi surface is present within the LDA calculation, which results in the fully gapped $s\pm$ gap. In ref. 193, fRG was applied to the two dimensional five orbital model of FeSe, and also obtained the fully gapped $s\pm$ gap. FRG was applied to the five orbital model of LiFeAs, where $k_z = 0$ and $k_z = \pi$ cuts of the band structure was considered.[209] There again, the $s\pm$ state is found to be the dominant pairing.

8.6 Concluding Remarks and Perspectives

In recent years, there have been a number of studies on correlated electron systems which are based on material specific Hamiltonians that take into account the accurate band structure obtained from first principles. Iron based superconductors have been discovered in a period where such a streamline of theoretical studies has been acceleratingly growing in the field of condensed matter physics. As for the theoretical analysis of unconventional superconductivity in particular, there is still an ambiguity concerning the mechanism of the pairing, etc., but the reliability of the studies based on accurate model Hamiltonians is rapidly increasing. In fact, in the days before the study of the iron pnictides started, it was hard to imagine an analysis on the mechanism of unconventional superconductivity based on model Hamiltonians that consider five, ten or even 16 orbitals with very accurate band structures. This owes to both the substantial growth of the computational resource and the development of the theoretical formalisms, including the methods to obtain the model Hamiltonians from first principles using, e.g., maximally localized Wannier orbitals[11] as has been described in this chapter. It is also worth noting that those theorists (including the present author himself) who were mainly working on simplified model Hamiltonians can now perform band calculations and construct accurate model Hamiltonians by themselves due to the wide spread use of the band structure calculation and model construction packages.[13,210,211] Also, the development of the methods to evaluate the electron-electron interactions[36–38] has further enabled the *ab-initio* construction of the low energy model Hamiltonians.

The accurate construction of material specific model Hamiltonians has a great advantage in that we can analyze theoretically the lattice structure and/or the element dependence of the material properties in a more reliable and non-empirical way. Particularly for the iron based superconductors, the Fermi surface consists of small pockets (or thin cylinders to be precise), so a small change in the lattice structure and/or the elements constructing the material leads to a large change in the volume or the presence/absence of the Fermi surface. As described in the present chapter, this leads to a large difference in the strength or even the wave vector of the spin fluctuations, which in turn affects the spin-fluctuation-mediated pairing. The material dependence of the strength of the spin fluctuations and T_c as well as the superconducting gap form obtained within the spin-fluctuation-mediated pairing scenario seems to be in agreement at least qualitatively with the experimental observations.

In this context, it is worth mentioning that such a lattice structure sensitivity is not peculiar to materials having subtle band structures as in the iron based superconductors. In fact, even in the high T_c cuprates, a change in the lattice structure can affect the band structure, which can have significant influence on superconductivity as has been shown in ref. 212. Let us briefly comment on this issue. The LDA band calculation of the cuprates gives a large Fermi surface originating from a single band that has strong Cu $3d_{x^2-y^2}$ orbital character, hybridized with the oxygen $2p_\sigma$ orbitals. Therefore, a lattice structure sensitivity of the band structure like the one seen in the iron based superconductors is at first glance not expected. However, it has been noticed that there actually is a material dependence of the Fermi surface shape, where relatively low T_c materials have a square shaped Fermi surface, while the high T_c ones have more rounded shape.[214,215] In many of the theoretical studies of the cuprates, the model Hamiltonian that considers only the $3d_{x^2-y^2}$ orbital (plus the hybridizing oxygen $2p$ orbitals) has been adopted, and the material dependence has been taken into account by varying the ratio of the second/third and nearest neighbor hopping integrals, which controls the shape of the Fermi surface. However, this Fermi surface variation is actually due to the effect of the $3d_{z^2}$ orbitals,[213,214] whose energy level with respect to $d_{x^2-y^2}$ depends on the materials. In ref. 212, a two orbital model that explicitly considers maximally localized $d_{x^2-y^2}$ and d_{z^2} Wannier orbitals was introduced to analyze the material dependence of superconductivity. There, it was shown that T_c of the d-wave superconductivity is low when the level offset ΔE between the two orbitals is small so that the d_{z^2} orbital character mixes into the Fermi surface around the wave vector $(\pi, 0)/(0, \pi)$, making the Fermi surface more square shaped. Thus the explicit consideration of the two orbitals, despite the Fermi surface consisting of a single band, explains the the material dependence of T_c. This example shows that the consideration of the accurate kinetic energy term of the model Hamiltonian can lead to a better understanding of the experimental observations, even in materials with seemingly simple band structures.

Back to the iron based superconductors, there still remain various theoretical issues to be resolved. One of them is the form of the superconducting gap function. As described in this chapter, the most probable candidate up to date is the $s\pm$ gap, with nodes at certain portions of the Fermi surface depending on the materials. However, there is still a debate going on concerning whether there actually is a sign change of the superconducting gap between the electron and hole Fermi surfaces. Although various experiments do suggest the presence of this sign change, further experiments are necessary to confirm this point. Theoretically, it remains a future problem to investigate non-empirically whether the attractive pairing interactions (such as those mediated by the orbital fluctuations) can actually be large enough to overcome the effect of the repulsive pairing interaction mediated by the antiferromagnetic spin fluctuations. Regarding this issue, it should also be important to thoroughly examine to what extent the sign reversing gap states such as nodeless or nodal $s\pm$ can explain the robustness of T_c against the impurities/defects observed in some experiments. In fact, a possible origin of the robustness peculiar to multiorbital systems has been pointed out in ref. 216. Further investigations are required on issues such as the strength of the impurities as has been done in ref. 217, refined formalisms to analyze how the impurities/defects affect T_c, and also the effect of the impurities/defects other than pair breaking (such as the Fermi surface variation).

Another topic related to the issue of the superconducting gap form is the discovery of superconductivity in a heavily electron doped 122 Selenide $K_xFe_{2-y}Se_2$.[218] In the ARPES experiments of this material,[219,220] the hole Fermi surface was found to be missing, which is in agreement with band calculation for the stoichiometric composition KFe_2Se_2 (which has a band filling of $n = 6.5$ in the sense of the five orbital model).[221] The possible realization of nodeless d-wave superconductivity in the absence of the hole Fermi surface was pointed out in ref. 6 right after the discovery of superconductivity in doped LaFeAsO, and in fact, studies that specifically focus on this material have proposed the possibility of d-wave superconductivity adopting models in the unfolded Brillouin zone.[222–224] On the other hand, it has been pointed out that d-wave gap is not possible as far as the ARPES observations[219,220] are concerned considering the fact that the unfolding procedure cannot be done strictly in 122 systems.[225,226] Moreover, it has become clear that the composition of the superconducting materials are away from stoichiometry (there is a large amount of Fe defects), so that the absence/presence of the hole Fermi surface might have to be reconsidered more carefully. More experiments are necessary to clarify the nature of this material.

Looking at how we have understood various properties of the iron based superconductors using material specific Hamiltonians, it may seem that we are somewhat close to the stage where we can theoretically predict unknown exotic materials such as high T_c superconductors. Still, many breakthroughs have to be accomplished for such a purpose. One is the further development of techniques to treat the electron correlation effects. As for unconventional superconductivity in particular, formulations for a more non-empirical evaluation of T_c would be a great theoretical challenge. Another problem is the way to link the ideas for realizing exotic materials to the prediction of an unknown material. As mentioned in this chapter, the possible realization of high T_c superconductivity in materials having disconnected Fermi surfaces was predicted in theoretical studies using some "toy models."[84,85] A more recent dynamical cluster approximation calculation shows that T_c of such system is indeed very high.[227] Those studies suggest that T_c of systems with disconnected Fermi surfaces consisting of a *single orbital*, as contrasted to multiorbital in iron-based compounds, may give T_c even higher than the cuprates. The problem is that although we can say from band structure calculations whether a certain material has such kind of Fermi surfaces *once the material is given*, it is still difficult to go the other way around, namely, to predict unknown materials that have peculiar band structures. The development of theoretical approaches have enabled us to analyze the physical properties in a material specific way, but further effort is necessary for us to approach the theoretical stage where the *ab-initio* prediction of exotic superconductors becomes more feasible.

Acknowledgments

The author would like to acknowledge H. Aoki, R. Arita, S. Onari, H. Usui, K. Suzuki, Y. Nagai, M. Machida, H. Kontani, Y. Tanaka, and H. Sakakibara for collaboration in the theoretical studies described in the present chapter. He is also grateful to D. J. Scalapino, P. J. Hirschfeld, D. H. Lee, T. Miyake, H. Ikeda, K. Nakamura, T. Kariyado and M. Ogata for the permission to use the figures/tables from their paper and also for many fruitful discussions. The author also thanks many of the theorists and experimentalists for valuable discussions; H. Hosono, Y. Kamihara, M. Hirano, H. Fukuyama, I. I. Mazin, D. Singh, M. Imada, S. Sachdev, T. Nomura, Y. Ōno, A. V. Chubukov, I. Eremin, K. Terakura, A. V. Balatsky, C. H. Lee, H. Eisaki, T. Ito, N. Takeshita,

R. Kumai, T. Hanaguri, H.-H. Wen, S. Shamoto, A. Fujimori, T. Yoshida, Y. Matsuda, T. Shibauchi, S. Kasahara, Y. Takano, K. Prassides, K. Ishida, G.-q. Zheng, H. Mukuda, H. Fukazawa, H. Takahashi, H. Ogino, J. Shimoyama, K. Yamada, N. Fujiwara, A. Maeda, E. Takayama-Muromachi, T. Shimojima, S. Shin, M. Sato, Y. Kobayashi, and many others.

References

1. Y. Kamihara, T. Watanabe, M. Hirano, and H. Hosono, J. Am. Chem. Soc. **130**, 3296 (2008).
2. Z.-A. Ren, W. Lu, J. Yang, W. Yi, X.-L. Shen, Z.-C. Li, G.-C. Che, X.-L Dong, L.-L. Sun, F. Zhou, and Z.-X. Zhao, Chin. Phys. Lett. **25**, 2215 (2008).
3. L. Boeri, O. V. Dolgov, and A. A. Golubov, Phys. Rev. Lett. **101**, 026403 (2008).
4. L. Boeri, M. Calandra, I. I. Mazin, O. V. Dolgov, and F. Mauri, Phys. Rev. B **82**, 020506 (2010).
5. I. I. Mazin, D. J. Singh, M. D. Johannes, and M. H. Du, Phys. Rev. Lett. **101**, 057003 (2008).
6. K. Kuroki, S. Onari, R. Arita, H. Usui, Y. Tanaka, H. Kontani, and H. Aoki, Phys. Rev. Lett. **101**, 087004 (2008).
7. For a review, see, I. I. Mazin and J. Schmalian, Physica C **469**, 614 (2009).
8. A. V. Chubukov, V. Efremov, and I. Eremin, Phys. Rev. B **78** (2008) 134512.
9. Q. Si and E. Abrahams, Phys. Rev. Lett. **101**, 076401 (2008).
10. K. Seo, B. A. Bernevig, and J. Hu, Phys. Rev. Lett. **101**, 206404 (2008).
11. N. Marzari and D. Vanderbilt, Phys. Rev. B **56** (1997) 12847; I. Souza, N. Marzari and D. Vanderbilt, Phys. Rev. B **65**, 035109 (2002).
12. O. K. Andersen and T. Saha-Dasgupta, Phys. Rev. B **62**, 16219(R) (2000).
13. S. Baroni, A. Dal Corso, S. de Gironcoli, P. Giannozzi, C. Cavazzoni, G. Ballabio, S. Scandolo, G. Chiarotti, P. Focher, A. Pasquarello, K. Laasonen, A. Trave, R. Car, N. Marzari, A. Kokalj, http://www.pwscf.org/.
14. A. A. Mostofi, J. R. Yates, N. Marzari, I. Souza and D. Vanderbilt, (http://www.wannier.org/).
15. O. K. Andersen and L. Boeri, Annalen der Physik, **1** 8 (2011).

16. K. Kuroki, S. Onari, R. Arita, H. Usui, Y. Tanaka, H. Kontani, and H. Aoki, New J. Phys. **11**, 025017 (2009).
17. K. Nakamura, R. Arita, and M. Imada, J. Phys. Soc. Jpn. **77**, 093711 (2008).
18. H. Ikeda, R. Arita, and J. Kuneš, Phys. Rev. B **81**, 054502 (2010).
19. S. Graser, T. A. Maier, P. J. Hirschfeld, and D. J. Scalapino, New J. Phys. **11**, 025016 (2009)
20. C. Cao, P. J. Hirschfeld, and H.-P. Cheng, Phys. Rev. B **77**, 220506 (R) (2008).
21. S. Raghu, X.-L. Qi, C.-X. Liu, D. J. Scalapino, and S.-C. Zhang, Phys. Rev. B **77**, 220503(R) (2008).
22. P. A. Lee and X.-G. Wen, Phys. Rev. B **78**, 144517 (2008)
23. M. Daghofer, A. Nicholson, A. Moreo, and E. Dagotto, Phys. Rev. B **81**, 014511 (2010).
24. M. M. Korshunov and I. Eremin, Phys. Rev. B **78**, 140509(R) (2008).
25. C. Platt, C. Honerkamp, and W. Hanke, New J. Phys. **11**, 055058 (2009).
26. R. Thomale, C. Platt, J. Hu, C. Honerkamp, B. A. Bernevig, Phys. Rev. B **80**, 180505(R) (2009).
27. R. Arita and H. Ikeda, J. Phys. Soc. Jpn. **78**, 113707 (2009).
28. Y. Yanagi, Y. Yamakawa, and Y. Ōno, J. Phys. Soc. Jpn. **77**, 123701 (2008).
29. T. Miyake, L. Pourovskii, V. Vildosola, S. Biermann, and A. Georges, J. Phys. Soc. Jpn. **77**, Suppl. C 99 (2008).
30. T. Miyake, K. Nakamura, R. Arita, and M. Imada, J. Phys. Soc. Jpn. **79**, 044705 (2010).
31. For a review, see A. Georges, G. Kotliar, W. Krauth and M. J. Rosenberg, Rev. Mod. Phys. **68**, 13 (1996), G. Kotliar, S. Y. Savrasov, K. Haule, V. S. Oudovenko, O. Parcollet, and C. A. Marianetti, Rev. Mod. Phys. **78**, 865 (2006).
32. K. Nakamura, Y. Yoshimoto, Y. Nohara, and M. Imada, J. Phys. Soc. Jpn. **79**, 123708 (2010).
33. V. I. Anisimov, E. Z. Kurmaev, A. Moewes, and I. A. Izyumov, Physica C **469**, 442 (2009).
34. M. Aichhorn, L. Pourovskii, V. Vildosola, M. Ferrero, O. Parcollet, T. Miyake, A. Georges, and S. Biermann, Phys. Rev. B **80**, 085101 (2009).
35. A. Kutepov, K. Haule, S. Y. Savrasov, and G. Kotliar Phys. Rev. B **82**, 045105 (2010)
36. O. Gunnarsson, O. K. Andersen, O. Jepsen, and J. Zaanen, Phys. Rev. B **39**, 1708 (1989).

37. F. Aryasetiawan, M. Imada, A. Georges, G. Kotliar, S. Biermann, and A. I. Lichtenstein, Phys. Rev. B **70**, 195104 (2004).
38. I. V. Solovyev and M. Imada, Phys. Rev. B **71**, 045103 (2005).
39. K. Haule, J. H. Shim, and G. Kotliar, Phys. Rev. Lett. **100**, 226402 (2008).
40. L. Craco, M. S. Laad, S. Leoni, and H. Rosner, Phys. Rev. B **78**, 134511 (2008).
41. H. Ishida and A. Liebsch, Phys. Rev. B **81**, 054513 (2010).
42. V. Vildosola, L. Pourovskii, R. Arita, S. Biermann, and A. Georges, Phys. Rev. B **78**, 064518 (2008).
43. P. Hansmann, R. Arita, A. Toschi, S. Sakai, G. Sangiovanni, and K. Held, Phys. Rev. Lett. **104**, 197002 (2010).
44. K. Yada and H. Kontani J. Phys. Soc. Jpn. **74**, 2161 (2005).
45. T. Takimoto, T. Hotta, and K. Ueda, Phys. Rev. B **69**, 104504 (2004).
46. M. D. Lumsden, A. D. Christianson, D. Parshall, M. B. Stone, S. E. Nagler, G. J. MacDougall, H. A. Mook, K. Lokshin, T. Egami, D. L. Abernathy, E. A. Goremychkin, R. Osborn, M. A. McGuire, A. S. Sefat, R. Jin, B. C. Sales, and D. Mandrus, Phys. Rev. Lett. **102**, 107005 (2009).
47. S. Chi, A. Schneidewind, J. Zhao, L. W. Harriger, L. Li, Y. Luo, G. Cao, Z. Xu, M. Loewenhaupt, J. Hu, and P. Dai, Phys. Rev. Lett. **102**, 107006 (2009).
48. K. Kuroki, H. Usui, S. Onari, R. Arita, and H. Aoki, Phys. Rev. B **79**, 224511 (2009).
49. K. Kuroki and H. Aoki, Physica C **469**, 635 (2009).
50. A. F. Kemper, T. A. Maier, S. Graser, H.-P. Cheng, P. J. Hirschfeld, and D. J. Scalapino, New J. Phys. **12**, 073030 (2010).
51. T. A. Maier, S. Graser, D. J. Scalapino, and P. J. Hirschfeld, Phys. Rev. B **79**, 224510 (2009).
52. T. Kariyado and M. Ogata, J. Phys. Soc. Jpn. **78**, 043708 (2009).
53. M. Ishikado, R. Kajimoto, S. Shamoto, M. Arai, A. Iyo, K. Miyazawa, P. M. Shirage, H. Kito, H. Eisaki, S. Kim, H. Hosono, T. Guidi, R. Bewley, and S. M. Bennington, J. Phys. Soc. Jpn. **78**, 043705 (2009).
54. K. Matan, R. Morinaga, K. Iida, and T. J. Sato, Phys. Rev. B **79**, 054526 (2009).
55. H. Ikeda, J. Phys. Soc. Jpn. **77**, 123707 (2008).
56. N. E. Bickers, D. J. Scalapino, and S. R. White, Phys. Rev. Lett. **62**, 961 (1989).
57. C. de la Cruz, Q. Huang, J. W. Lynn, Jiying Li, W. Ratcliff II, J. L. Zarestky, H. A. Mook, G. F. Chen, J. L. Luo, N. L. Wang, and P. Dai, Nature **453**, 899 (2008).

58. S. Ishibashi, K. Terakura, and H. Hosono, J. Phys. Soc. Jpn. **77**, 053709 (2008).
59. I. I. Mazin, M. D. Johannes, L. Boeri, K. Koepernik, and D. J. Singh, Phys. Rev. B **78**, 085104 (2008).
60. For a review, see D. J. Singh, Physica C **469**, 418 (2009).
61. I. I. Mazin and M. D. Johannes, Nature Phys. **5**, 141 (2009).
62. H. Nakamura, N. Hayashi, N. Nakai, M. Okumura, and M. Machida, Physica C **469**, 908 (2009).
63. Y. Ran, F. Wang, H. Zhai, A. Vishwanath, and D.-H. Lee, Phys. Rev. B **79**, 014505 (2009).
64. E. Kaneshita, T. Morinari, and T. Tohyama, Phys. Rev. Lett. **103**, 247202 (2009).
65. H. Fukuyama, J. Phys. Soc. Jpn. Online-News and Comments [May 12, 2008]
66. P. Richard, K. Nakayama, T. Sato, M. Neupane, Y.-M. Xu, J. H. Bowen, G. F. Chen, J. L. Luo, N. L. Wang, X. Dai, Z. Fang, H. Ding, and T. Takahashi, Phys. Rev. Lett. **104**, 137001 (2010).
67. T. Misawa, K. Nakamura, and M. Imada, J. Phys. Soc. Jpn. **80**, 023704 (2011).
68. J.-H. Chu, J. G. Analytis, K. De Greve, P. L. McMahon, Z. Islam, Y. Yamamoto and I. R. Fisher, Science **329**, 824 (2010).
69. M. Nakajima, S. Ishida, T. Liang, K. Kihou, Y. Tomioka, T. Ito, C. H. Lee, H. Kito, A. Iyo, H, Eisaki, K. M. Kojima, T. Kakeshita, and S. Uchida, Phys. Rev. B **81**, 104528 (2010).
70. T. Shimojima, K. Ishizaka, Y. Ishida, N. Katayama, K. Ohgushi, T. Kiss, M. Okawa, T. Togashi, X. -Y. Wang, C. -T. Chen, S. Watanabe, R. Kadota, T. Oguchi, A. Chainani, S. Shin, Phys. Rev. Lett. **104**, 057002 (2010).
71. M. Yi, D. H. Lu, J.-H. Chu, J. G. Analytis, A. P. Sorini, A. F. Kemper, B. Moritz, S.-K. Mo, R. G. Moore, M. Hashimoto, W. S. Lee, Z. Hussain, T. P. Devereaux, I. R. Fisher, Z.-X. Shen, PNAS **108**, 6878–6883 (2011).
72. R. M. Fernandes, L. H. VanBebber, S. Bhattacharya, P. Chandra, V. Keppens, D. Mandrus, M. A. McGuire, B. C. Sales, A. S. Sefat, and J. Schmalian, Phys. Rev. Lett. **105**, 157003 (2010).
73. M. Yoshizawa, R. Kamiya, R. Onodera, Y. Nakanishi, K. Kihou, H. Eisaki, and C. H. Lee, arXiv:1008.1479.
74. T. Yildirim, Phys. Rev. Lett. **101**, 057010 (2008).
75. K. Kubo and P. Thalmeir, J. Phys. Soc. Jpn. **78**, 083704 (2009).

76. F. Yang, H. Zhai, F. Wang, and D.-H. Lee, Phys. Rev. B **83**, 134502 (2011).
77. Z. P. Yin, K. Haule, and G. Kotliar, Nat. Phys. **7**, 294 (2011).
78. K. Sugimoto, E. Kaneshita, and T. Tohyama, J. Phys. Soc. Jpn. **80**, 033706 (2011).
79. Y. Yanagi, Y. Yamakawa, N. Adachi, and Y. Ōno, J. Phys. Soc. Jpn. **79**, 123707 (2010).
80. H. Kontani, T. Saito, and S. Onari, Phys. Rev. B **84**, 024528 (2011).
81. K. Miyake, S. Schmitt-Rink, and C. M. Varma, Phys. Rev. B **34**, 6554 (1986).
82. D. J. Scalapino, E. Loh, and J. E. Hirsch, Phys. Rev. B **34**, 8190 (1986).
83. N. Bulut, D. J. Scalapino, and R. T. Scalletar, Phys. Rev. B **45**, 5577 (1992).
84. K. Kuroki and R. Arita, Phys. Rev. B **64**, 024501 (2001).
85. K. Kuroki, T. Kimura nad R. Arita, Phys. Rev. B **66**, 184508 (2002).
86. K. Kuroki, S. Onari, R. Arita, H. Usui, Y. Tanaka, H. Kontani, and H. Aoki, Phys. Rev. Lett. **102**, 109902 (Erratum) (2009).
87. M. Daghofer, A. Moreo, J. A. Riera, E. Arrigoni, D. J. Scalapino, and E. Dagotto, Phys. Rev. Lett. **101**, 237004 (2008).
88. T. Nomura, J. Phys. Soc. Jpn. **77**, Suppl. C 123 (2008).
89. H. Ikeda, J. Phys. Soc. Jpn. **77**, 123707 (2008).
90. F. Wang, H. Zhai, Y. Ran, A. Vishwanath, and D.-H. Lee, Phys. Rev. Lett. **102**, 047005 (2009).
91. R. Thomale, C. Platt, W. Hanke, B. A. Bernevig, Phys. Rev. Lett. **106**, 187003 (2011).
92. Z.-J. Yao, J.-X. Li, and Z. D. Wang, New. J. Phys. **11** 025009 (2009).
93. S.-L. Yu, J. Kang, and J.-X. Li, Phys. Rev. B **79**, 064517 (2009).
94. Y. Yanagi, Y. Yamakawa, and Y. Ōno, J. Phys. Soc. Jpn. **77**, 123701 (2008).
95. V. Cvetkovic and Z. Tesanovic, Europhys. Lett. **85**, 37002 (2009).
96. Q. Luo, G. Martins, D.-X. Yao, M. Daghofer, R. Yu, A. Moreo, and Elbio Dagotto, Phys. Rev. B **82**, 104508 (2010).
97. H. Kontani and S. Onari, Phys. Rev. Lett. **104**, 157001 (2010).
98. Y. Yanagi, Y. Yamakawa, and Y. Ōno, Phys. Rev. B **81**, 054518 (2010).
99. Y. Yanagi, Y. Yamakawa, N. Adachi, and Y. Ōno, Phys. Rev. B **82**, 064518 (2010).
100. S. Onari and H. Kontani, arXiv: 1009.3882.
101. T. Saito, S. Onari, and H. Kontani, Phys. Rev. B **82**, 144510 (2010).
102. H. Kontani and S. Onari, J. Phys. Soc. Jpn. **80**, 056001 (2011).

103. H. Ding, P. Richard, K. Nakayama, T. Sugawara, T. Arakane, Y. Sekiba, A. Takayama, S. Souma, T. Sato, T. Takahashi, Z. Wang, X. Dai, Z. Fang, G. F. Chen, J. L. Luo, N. L. Wang, Europhys. Lett. **83**, 47001 (2008).

104. K. Nakayama, T. Sato, P. Richard, Y.-M. Xu, Y. Sekiba, S. Souma, G. F. Chen, J. L. Luo, N. L. Wang, H. Ding, and T. Takahashi, Europhys. Lett. **85**, 67002 (2009).

105. K. Hashimoto, T. Shibauchi, T. Kato, K. Ikada, R. Okazaki, H. Shishido, M. Ishikado, H. Kito, A. Iyo, H. Eisaki, S. Shamoto, and Y. Matsuda, Phys. Rev. Lett. **102**, 017002 (2009).

106. H. Luetkens, H.-H. Klauss, R. Khasanov, A. Amato, R. Klingeler, K. Hellmann, N. Leps, A. Kondrat, C. Hess, A. Kohler, G. Behr, J. Werner, and B. Buchner, Phys. Rev. Lett. **101**, 097009 (2008).

107. A. A. Aczel, E. Baggio-Saitovitch, S. L. Budko, P. C. Canfield, J. P. Carlo, G. F. Chen, P. Dai, T. Goko, W. Z. Hu, G. M. Luke, J. L. Luo, N. Ni, D. R. Sanchez-Candela, F. F. Tafti, N. L. Wang, T. J. Williams, W. Yu, and Y. J. Uemura, Phys. Rev. B **78**, 214503 (2008).

108. T. Goko, A. A. Aczel, E. Baggio-Saitovitch, S. L. Bud'ko, P. C. Canfield, J. P. Carlo, G. F. Chen, P. Dai, A. C. Hamann, W. Z. Hu, H. Kageyama, G. M. Luke, J. L. Luo, B. Nachumi, N. Ni, D. Reznik, D. R. Sanchez-Candela, A. T. Savici, K. J. Sikes, N. L. Wang, C. R. Wiebe, T. J. Williams, T. Yamamoto, W. Yu, and Y. J. Uemura, arXiv:0808.1425.

109. M. Hiraishi, R. Kadono, S. Takeshita, M. Miyazaki, A. Koda, H. Okabe, and J. Akimitsu, J. Phys. Soc. Jpn. **78**, 023710 (2009).

110. H. Mukuda, M. Nitta, M. Yashima, Y. Kitaoka, P. M. Shirage, H. Eisaki, and A. Iyo, J. Phys. Soc. Jpn. **79**, 113701 (2010).

111. For a review, see e.g., K. Ishida, Y. Nakai, and H. Hosono, J. Phys. Soc. Jpn. **78**, 062001 (2009) and references therein.

112. Y. Nakai, K. Ishida, Y. Kamihara, M. Hirano, and H. Hosono, J. Phys. Soc. Jpn. **77** 073701 (2008).

113. H.-J. Grafe, D. Paar, G. Lang, N. J. Curro, G. Behr, J. Werner, J. Hamann-Borrero, C. Hess, N. Leps, R. Klingeler, and B. Büchner, Phys. Rev. Lett. **101**, 047003 (2008).

114. H. Mukuda, N. Terasaki, H. Kinouchi, M. Yashima, Y. Kitaoka, S. Suzuki, S. Miyasaka, S. Tajima, K. Miyazawa, P. Shirage, H. Kito, H. Eisaki, and A. Iyo, J. Phys. Soc. Jpn. **77**, 093704 (2008).

115. N. Terasaki, H. Mukuda, M. Yashima, Y. Kitaoka, K. Miyazawa, P. M. Shirage, H. Kito, H. Eisaki, and A. Iyo, J. Phys. Soc. Jpn. **78**, 013701 (2009).

116. K. Ahilan, F. L. Ning, T. Imai, A. S. Sefat, R. Jin, M. A. McGuire, B. C. Sales, D. Mandrus, Phys. Rev. B **78**, 100501(R) (2008).
117. K. Matano, Z. A. Ren, X. L. Dong, L. L. Sun, Z. X. Zhao, and G.-Q. Zheng, Europhys. Lett. **83**, 57001 (2008).
118. H. Kotegawa, S. Masaki, Y. Awai, H. Tou, Y. Mizuguchi, and Y. Takano, J. Phys. Soc. Jpn. **77**, 113703 (2008).
119. H. Fukazawa, T. Yamazaki, K. Kondo, Y. Kohori, N. Takeshita, P. M. Shirage, K Kihou, K. Miyazawa, H. Kito, H. Eisaki, and A. Iyo, J. Phys. Soc. Jpn. **78**, 033704 (2009).
120. M. Yashima, H. Nishimura, H. Mukuda, Y. Kitaoka, K. Miyazawa, P. M. Shirage, K. Kihou, H. Kito, H. Eisaki, and Akira Iyo, J. Phys. Soc. Jpn. **78**, 103702 (2009).
121. Y. Kobayashi, A. Kawabata, S. C. Lee, T. Moyoshi, and M. Sato, J. Phys. Soc. Jpn. **78**, 073704 (2009).
122. Y. Nakai, K. Ishida, Y. Kamihara, M. Hirano, and H. Hosono, Phys. Rev. Lett. **101**, 077006 (2008).
123. Y. Nagai, N. Hayashi, N. Nakai, H. Nakamura, M. Okumura, and M. Machida, New J. Phys. **10**, 103026 (2008).
124. C.-T. Chen, C. C. Tsuei, M. B. Ketchen, Z.-A. Ren and Z. X. Zhao, Nat. Phys. **6**, 260 (2010).
125. T. Hanaguri, S. Niitaka, K. Kuroki, and H. Takagi, Science **328**, 474 (2010).
126. R. M. Fernandes, and J. Schmalian, Phys. Rev. B **82**, 014520 (2010).
127. Y. F. Guo, Y. G. Shi, S. Yu, A. A. Belik, Y. Matsushita, M. Tanaka, Y. Katsuya, K. Kobayashi, I. Nowik, I. Felner, V. P. S. Awana, K. Yamaura, and E. Takayama-Muromachi, Phys. Rev. B **82** 054506 (2010).
128. S. Kitagawa, Y. Nakai, T. Iye, K. Ishida, Y. F. Guo, Y. G. Shi, K. Yamaura, and E. Takayama-Muromachi, Phys. Rev. B **83**, 180501(R) (2011).
129. M. Sato, Y. Kobayashi, S. C. Lee, H. Takahashi, E. Satomi, and Y. Miura, J. Phys. Soc. Jpn. **79** (2010) 014710.
130. Y. K. Li, X. Lin, Q. Tao, C. Wang, T. Zhou, L. J. Li, Q. B. Wang, M. He, G. H. Cao, and Z. A. Xu, New J. Phys. **11** 053008 (2009).
131. C. Tarantini, M. Putti, A. Gurevich, Y. Shen, R. K. Singh, J. M. Rowell, N. Newman, D. C. Larbalestier, P. Cheng, Y. Jia, and H.-H. Wen, Phys. Rev. Lett. **104**, 087002 (2010).
132. Y. Nakajima, T. Taen, Y. Tsuchiya, T. Tamegai, H. Kitamura, and T. Murakami, Phys. Rev. B **82**, 220504 (2010).
133. S. Onari and H. Kontani, Phys. Rev. Lett. **103** (2009) 177001.

134. T. A. Maier and D. J. Scalapino, Phys. Rev. B **78**, 020514(R) (2008).
135. T. A. Maier, S. Graser, D. J. Scalapino, and P. Hirschfeld, Phys. Rev. B **79**, 134520 (2009).
136. P. Monthoux and D. J. Scalapino, Phys. Rev. B **72**, 1874 (1994).
137. J. Rossat-Mignod, L. P. Regnault, C. Vettier, P. Bourges, P. Burlet, J. Bossy, J. Y. Henry, and G. Lapertot, Physica C **185**, 86 (1991).
138. A. D. Christianson, E. A. Goremychkin, R. Osborn, S. Rosenkranz, M. D. Lumsden, C. D. Malliakas, I. S. Todorov, H. Claus, D. Y. Chung, M. G. Kanatzidis, R. I. Bewley and T. Guidi, Nature **456**, 930 (2008).
139. Y. Qiu, W. Bao, Y. Zhao, C. Broholm, V. Stanev, Z. Tesanovic, Y. C. Gasparovic, S. Chang, J. Hu, B. Qian, M. Fang, and Z. Mao, Phys. Rev. Lett. **103**, 067008 (2009).
140. D. S. Inosov, J. T. Park, P. Bourges, D. L. Sun, Y. Sidis, A. Schneidewind, K. Hradil, D. Haug, C. T. Lin, B. Keimer and V. Hinkov, Nat. Phys. **6**, 178 (2010).
141. J. Zhao, L.-P. Regnault, C. Zhang, M. Wang, Z. Li, F. Zhou, Z. Zhao, C. Fang, J. Hu and P. Dai, Phys. Rev. B **81**, 180505(R) (2010).
142. M. Ishikado, Y. Nagai, K. Kodama, R. Kajimoto, M. Nakamura, Y. Inamura, S. Wakimoto, H. Nakamura, M. Machida, K. Suzuki, H. Usui, K. Kuroki, A. Iyo, H. Eisaki, M. Arai, and S. Shamoto, Phys. Rev. B **84**, 144517 (2011)
143. S. Onari, H. Kontani, and M. Sato, Phys. Rev. B **81**, 060504(R) (2010).
144. Y. Nagai and K. Kuroki, Phys. Rev. B **83**, 220516 (2011).
145. T. A. Maier, S. Graser, P. J. Hirschfeld, and D. J. Scalapino, Phys. Rev. B **83**, 220505 (2011).
146. T. Kariyado and M. Ogata, J. Phys. Soc. Jpn. **79**, 083704 (2010).
147. Y. Kamihara, H. Hiramatsu, M. Hirano, R. Kawamura, H. Yanagi, T. Kamiya, and H. Hosono, J. Am. Chem. Soc. **128**, 10012 (2006).
148. C. H. Lee, A. Iyo, H. Eisaki, H. Kito, M. T. Fernandez-Diaz, T. Ito, K. Kihou, H. Matsuhata, M. Braden, and K. Yamada, J. Phys. Soc. Jpn. **77**, 083704 (2008).
149. J. Zhao, Q. Huang, C. de la Cruz, S. Li, J. W. Lynn, Y. Chen, M. A. Green, G. F. Chen, G. Li, Z. Li, J. L. Luo, N. L. Wang, and P. Dai, Nat. Mater. **7**, 953 (2008).
150. Y. Mizuguchi and Y. Takano, J. Phys. Soc. Jpn. **79**, 102001 (2010) and references therein.
151. K. Miyazawa, K. Kihou, P. M. Shirage, C.-H. Lee, H. Kito, H. Eisaki, and A. Iyo, J. Phys. Soc. Jpn. **78**, 034712 (2009).

152. P. M. Shirage, K. Kihou, C. H. Lee, H. Kito, H. Eisaki, A. Iyo, Appl. Phys. Lett. **97**, 172506 (2010).
153. H. Ogino, Y. Matsumura, Y. Katsura, K. Ushiyama, S. Horii, K. Kishio, J. Shimoyama, Supercond. Sci. Technol. **22**, 075008 (2009);
154. H. Ogino, K. Machida, A. Yamamoto, K. Kishio, J. Shimoyama, T. Tohei, Y. Ikuhara, Supercond. Sci. Technol. **23**, 115005 (2010).
155. H. Takahashi, K. Igawa, K. Arii, Y. Kamihara, M. Hirano, and H. Hosono, Nature (London) **453**, 376 (2008).
156. K. Tatsumi, N. Fujiwara, H. Okada, H. Takahashi, Y. Kamihara, M. Hirano, and H. Hosono, J. Phys. Soc. Jpn. **78**, 023709 (2009).
157. H. Okada, K. Igawa, H. Takahashi, Y. Kamihara, M. Hirano, H. Hosono, K. Matsubayashi, and Y. Uwatoko, J. Phys. Soc. Jpn. **77**, 113712 (2008).
158. N. Takeshita, A. Iyo, H. Eisaki, H. Kito, and T. Ito, J. Phys. Soc. Jpn. **77**, 075003 (2008).
159. S. Medvedev, T. M. McQueen, I. Trojan, T. Palasyuk, M. I. Eremets, R. J. Cava, S. Naghavi, F. Casper, V. Ksenofontov, G. Wortmann, C. Felser, Nat. Mater. **8**, 630 (2009).
160. S. Margadonna, Y. Takabayashi, Y. Ohishi, Y. Mizuguchi, Y. Takano, T. Kagayama, T. Nakagawa, M. Takata, K. Prassides, Phys. Rev. B **80**, 064506 (2009).
161. G. Garbarino, A. Sow, P. Lejay, A. Sulpice, P. Toulemonde, M. Mezouar, M. Nunez-Regueiro, Europhys. Lett. **86** (2009) 27001
162. J. D. Fletcher, A. Serafin, L. Malone, J. Analytis, J-H Chu, A. S. Erickson, I. R. Fisher, and A. Carrington, Phys. Rev. Lett. **102**, 147001 (2009).
163. C. W. Hicks, T. M. Lippman, M. E. Huber, J. G. Analytis, J.-H. Chu, A. S. Erickson, I. R. Fisher, and K. A. Moler, Phys. Rev. Lett. **103**, 127003 (2009).
164. M. Yamashita, N. Nakata, Y. Senshu, S. Tonegawa, K. Ikada, K. Hashimoto, H. Sugawara, T. Shibauchi, and Y. Matsuda, Phys. Rev. B **80**, 220509 (2009)
165. S. Jiang , H. Xing, G. Xuan, C. Wang, Z. Ren, C. Feng, J. Dai, Z. Xu, and G. Cao, J. Phys.: Condens. Matter **21**, 382203 (2009).
166. S. Kasahara , T. Shibauchi, K. Hashimoto, K. Ikada, S. Tonegawa, R. Okazaki, H. Ikeda, H. Takeya, K. Hirata, T. Terashima, and Y. Matsuda, Phys. Rev. B **81**, 184519 (2010).
167. K. Hashimoto, M. Yamashita, S. Kasahara, Y. Senshu, N. Nakata, S. Tonegawa, K. Ikada, A. Serafin, A. Carrington, T. Terashima, H. Ikeda, T. Shibauchi and Y. Matsuda, Phys. Rev. B **81**, 220501(R) (2010).

168. Y. Nakai, T. Iye, S. Kitagawa, K. Ishida, H. Ikeda, S. Kasahara, H. Shishido, T. Shibauchi, Y. Matsuda and T. Terashima, Phys. Rev. B. **81**, 020503(R) (2010).

169. Y. Nakai, S. Kitagawa, K. Ishida, Y. Kamihara, M. Hirano, and H. Hosono, New J. Phys. **11**, 045004 (2009).

170. H. Mukuda, N. Terasaki, N. Tamura, H. Kinouchi, M. Yashima, Y. Kitaoka, K. Miyazawa, P. M. Shirage, S. Suzuki, S. Miyasaka, S. Tajima, H. Kito, H. Eisaki, and A. Iyo, J. Phys. Soc. Jpn. **78**, 084717 (2009).

171. T. Nakano, N. Fujiwara, K. Tatsumi, H. Okada, H. Takahashi, Y. Kamihara, M. Hirano, and H. Hosono, Phys. Rev. B **81**, 100510 (2010).

172. C. Lester, J.-H. Chu, J. G. Analytis, T. G. Perring, I. R. Fisher, and S. M. Hayden, Phys. Rev. B **81**, 064505 (2010).

173. H.-F. Li, C. Broholm, D. Vaknin, R. M. Fernandes, D. L. Abernathy, M. B. Stone, D. K. Pratt, W. Tian, Y. Qiu, N. Ni, S. O. Diallo, J. L. Zarestky, S. L. Bud'ko, P. C. Canfield, and R. J. McQueeney, Phys. Rev. B **82**, 140503 (2010).

174. M. D. Lumsden, A. D. Christianson, E. A. Goremychkin, S. E. Nagler, H. A. Mook, M. B. Stone, D. L. Abernathy, T. Guidi, G. J. MacDougall, C. de la Cruz, A. S. Sefat, M. A. McGuire, B. C. Sales, D. Mandrus, Nat. Phys. **6**, 182 (2010).

175. S.-H. Lee, G. Xu, W. Ku, J. S. Wen, C. C. Lee, N. Katayama, Z. J. Xu, S. Ji, Z. W. Lin, G. D. Gu, H.-B. Yang, P. D. Johnson, Z.-H. Pan, T. Valla, M. Fujita, T. J. Sato, S. Chang, K. Yamada, and J. M. Tranquada, Phys. Rev. B **81**, 220502(R) (2010).

176. C. H. Lee, K. Kihou, H. Kawano-Furukawa, T. Saito, A. Iyo, H. Eisaki, H. Fukazawa, Y. Kohori, K. Suzuki, H. Usui, K. Kuroki, and K. Yamada, Phys. Rev. Lett. **106**, 067003 (2011).

177. H. Fukazawa, Y. Yamada, K. Kondo, T. Saito, Y. Kohori, K. Kuga, Y. Matsumoto, S. Nakatsuji, H. Kito, P. M. Shirage, K. Kihou, N. Takeshita, C. H. Lee, A. Iyo and H. Eisaki, J. Phys. Soc. Jpn. **78**, 083712 (2009).

178. J. K. Dong, S. Y. Zhou, T. Y. Guan, H. Zhang, Y. F. Dai, X. Qiu, X. F. Wang, Y. He, X. H. Chen and S. Y. Li, Phys. Rev. Lett. **104**, 087005 (2010).

179. K. Hashimoto, , A. Serafin, S. Tonegawa, R. Katsumata, R. Okazaki, T. Saito, H. Fukazawa, Y. Kohori, K. Kihou, C. H. Lee, A. Iyo, H. Eisaki, H. Ikeda, Y. Matsuda, A. Carrington and T. Shibauchi, Phys. Rev. B **82**, 014526 (2010).

180. S. W. Zhang, L. Ma, Y. D. Hou, J. Zhang, T.-L. Xia, G. F. Chen, J. P. Hu, G. M. Luke and W. Yu, Phys. Rev. B **81**, 012503 (2010).

181. T. Terashima, M. Kimata, N. Kurita, H. Satsukawa, A. Harada, K. Hazama, M. Imai, A. Sato, K. Kihou, C. H. Lee, H. Kito, H. Eisaki, A. Iyo, T. Saito, H. Fukazawa, Y. Kohori, H. Harima, and S. Uji, J. Phys. Soc. Jpn. **79**, 053702 (2010).
182. T. Sato, K. Nakayama, Y. Sekiba, P. Richard, Y.-M. Xu, S. Souma, T. Takahashi, G. F. Chen, J. L. Luo, N. L. Wang, and H. Ding, Phys. Rev. Lett. **103**, 047002 (2009).
183. H. Kawano-Furukawa, C. J. Bowell, J. S. White, R. W. Heslop, A. S. Cameron, E. M. Forgan, K. Kihou, C. H. Lee, A. Iyo, H. Eisaki, T. Saito, H. Fukazawa, Y. Kohori, R. Cubitt, C. D. Dewhurst, J. L. Gavilano and M. Zolliker, Phys. Rev. B **84**, 024507 (2011).
184. S. Graser, A. F. Kemper, T. A. Maier, H.-P. Cheng, P. J. Hirschfeld, and D. J. Scalapino, Phys. Rev. B **81**, 214503 (2010).
185. D. J. Singh and M. H. Du, Phys. Rev. Lett. **100**, 237003 (2008).
186. S. Lebegue, Z. P. Yin, and W. E. Pickett, New J. Phys. **11**, 025004 (2009).
187. T. Miyake, T. Kosugi, S. Ishibashi, and K. Terakura, J. Phys. Soc. Jpn. **79**, 123713 (2010).
188. M. Rotter, M. Tegel, and D. Johrendt, Phys. Rev. Lett. **101**, 107006 (2008).
189. I. I. Mazin, T. P. Devereaux, J. G. Analytis, Jiun-Haw Chu, I. R. Fisher, B. Muschler and R. Hackl, Phys. Rev. B **82**, 180502 (2010).
190. K. Suzuki, H. Usui and K. Kuroki, J. Phys. Soc. Jpn. **80**, 103710 (2011).
191. T. Yoshida, I. Nishi, S. Ideta, A. Fujimori, M. Kubota, K. Ono, S. Kasahara, T. Shibauchi, T. Terashima, Y. Matsuda, H. Ikeda and R. Arita, Phys. Rev. Lett. **106**, 117001 (2011).
192. T. Kariyado and M. Ogata, J. Phys. Soc. Jpn. **79**, 033703 (2010).
193. F. Wang, H. Zhai, and D.-H. Lee, Phys. Rev. B **81**, 184512 (2010).
194. W.-C. Lee, S.-C. Zhang, and C. Wu, Phys. Rev. Lett. **102**, 217002 (2009).
195. A. V. Chubukov, M. G. Vavilov, and A. B. Vorontsov, Phys. Rev. B **80**, 140515(R) (2009).
196. V. Mishra, G. Boyd, S. Graser, T. Maier, P. J. Hirschfeld, and D. J. Scalapino, Phys. Rev. B **79**, 094512 (2009).
197. H. Usui and K. Kuroki, Phys. Rev. B **84**, 024505 (2011).
198. H. Kinouchi, H. Mukuda, M. Yashima, Y. Kitaoka, P. M. Shirage, H. Eisaki, and A. Iyo, Phys. Rev. Lett. **107**, 047002 (2011).
199. H. Ikeda, R. Arita, and J. Kuneš, Phys. Rev. B **82**, 024508 (2010).
200. A. N. Yaresko, G.-Q. Liu, V. N. Antonov, and O. K. Andersen, Phys. Rev. B **79**, 144421 (2009).

201. J. Knolle, I. Eremin, A. V. Chubukov, and R. Moessner, Phys. Rev. B **81**, 140506(R) (2010).

202. J. T. Park, D. S. Inosov, A. Yaresko, S. Graser, D. L. Sun, Ph. Bourges, Y. Sidis, Y. Li, J.-H. Kim, D. Haug, A. Ivanov, K. Hradil, A. Schneidewind, P. Link, E. Faulhaber, I. Glavatskyy, C. T. Lin, B. Keimer, and V. Hinkov, Phys. Rev. B **82**, 134503 (2010).

203. J. Zhang, R. Sknepnek, and J. Schmalian, Phys. Rev. B **82**, 134527 (2010).

204. P. J. Hirschfeld and D. J. Scalapino, Physics **3** (2010) 64.

205. K. Suzuki, H. Usui, and K. Kuroki, Phys. Rev. B 84, 144514 (2011). Here $s\pm$ state is in close competition with the d-wave state. On the other hand, an fRG study on KFe_2As_2 in R. Thomale *et al.*, Phys. Rev. Lett. **107**, 117001 (2011) gives d-wave pairing dominating over $s\pm$.

206. F.-C. Hsu, J.-Y. Luo, K.-W. Yeh, T.-K. Chen, T.-W. Huang, P. M. Wu, Y.-C. Lee, Y.-L. Huang, Y.-Y. Chu, D.-C. Yan, and M.-K. Wu, Proc. Nat. Acad. Sci. **105**, 14262 (2008).

207. X. C. Wang, Q. Q. Liu, Y. X. Lv, W. B. Gao, L. X. Yang, R. C. Yu, F. Y. Li, C. Q. Jin, Solid State Comm. **148**, 538 (2008).

208. H. Usui and K. Kuroki, Physica C **470**, S382 (2010).

209. C. Platt, R. Thomale, and W. Hanke, Phys. Rev. B **84**, 235121 (2011).

210. P. Blaha, K. Schwarz, G. K. H. Madsen, D. Kvasnicka, J. Luitz, WIEN2K, An Augmented Plane Wave and Local Orbitals Program for Calculating Crystal Properties, Vienna University of Technology, Austria, 2001; P. Blaha, K. Schwarz, P. Sorantin, S. B. Trickey, Comput. Phys. Commun. 59 (1990) 399.

211. J. Kuneš, R. Arita, P. Wissgott, A. Toschi, H. Ikeda, and K. Held, Comp. Phys. Commun. **181**, 1888 (2010).

212. H. Sakakibara, H. Usui, K. Kuroki, R. Arita, and H. Aoki, Phys. Rev. Lett. **105**, 057003 (2010).

213. O. K. Andersen, A. I. Liechtenstein, O. Jepsen, and F. Paulsen, J. Phys. Chem. Solids **56**, 1573 (1995).

214. E. Pavarini, I. Dasgupta, T. Saha-Dasgupta, O. Jepsen, and O. K. Andersen, Phys. Rev. Lett. **87**, 047003 (2001).

215. K. Tanaka, T. Yoshida, A. Fujimori, D. H. Lu, Z.-X. Shen, X.-J. Zhou, H. Eisaki, Z. Hussain, S. Uchida, Y. Aiura, K. Ono, T. Sugaya, T. Mizuno, and I. Terasaki, Phys. Rev. B **70**, 092503 (2004).

216. Y. Nagai, K. Kuroki, M. Machida, and H. Aoki, arXiv: 1012.5565.

217. K. Nakamura, R. Arita, and H. Ikeda, Phys. Rev. B **83**, 144512 (2011).

218. J. Guo, S. Jin, G. Wang, S. Wang, K. Zhu, T. Zhou, M. He, and X. Chen, Phys. Rev. B **82**, 180520(R) (2010).
219. T. Qian, X.-P. Wang, W.-C. Jin, P. Zhang, P. Richard, G. Xu, X. Dai, Z. Fang, J.-G. Guo, X.-L. Chen, H. Ding, Phys. Rev. Lett. **106**, 187001 (2011).
220. Y. Zhang, L. X. Yang, M. Xu, Z. R. Ye, F. Chen, C. He, J. Jiang, B. P. Xie, J. J. Ying, X. F. Wang, X. H. Chen, J. P. Hu, D. L. Feng, Nat. Mater. **10**, 273 (2011).
221. I. R. Shein and A. L. Ivanovskii, Phys. Lett. A **375**, 1028 (2011).
222. T. A. Maier, S. Graser, P. J. Hirschfeld, D. J. Scalapino, Phys. Rev. B **83**, 100515(R) (2011).
223. F. Wang, F. Yang, M. Gao, Z.-Y. Lu, T. Xiang, and D.-H. Lee, Europhys. Lett. **93**, 57003 (2011).
224. T. Das and A. V. Balatsky, Phys. Rev. B **84**, 014521 (2011).
225. I. I. Mazin, Phys. Rev. B **84**, 024529 (2011).
226. T. Saito, S. Onari, and H. Kontani, Phys. Rev. B **83**, 140512 (2011).
227. T. A. Maier and D. J. Scalapino, Phys. Rev. B **84**, 180513 (2011).

Chapter 9

The Antiferromagnetic Phase of Iron-Based Superconductors: An Itinerant Approach

Johannes Knolle[a] and Ilya Eremin[b,*]
[a] *Max Planck Institute for the Physics of Complex Systems, D-01187 Dresden, Germany*
[b] *Institut für Theoretische Physik III, Ruhr-Universität Bochum, D-44801 Bochum, Germany*
*jknolle@pks.mpg.de, ieremin@tp3.rub.de

9.1 Introduction

A comprehensive understanding of the relationship between magnetism and superconductivity in the Fe-based superconductors, discovered in 2008 by Hideo Hosono and collaborators [Kamihara *et al.* (2008)], ultimately requires an analysis of the magnetic ground states in these compounds and their evolution with doping. In particular, the origin of magnetism in the FeSC parent compounds is hotly debated especially because it is believed that the same magnetic interactions that drive the magnetic ordering also produce the Cooper-pairing [Hirschfeld *et al.* (2011)]. The

Iron-Based Superconductors: Materials, Properties, and Mechanisms
Edited by Nan-Lin Wang, Hideo Hosono, and Pengcheng Dai

ISBN 978-981-4303-22-4 (Hardcover), 978-981-4303-23-1 (eBook)
www.panstanford.com

phase diagram of ferropnictides (FPs) is similar to high-T_c cuprates and contains an antiferromagnetic (AF) phase in close proximity to the superconducting (SC) one. Most iron-based superconductors exhibit an AF state at low carrier concentrations which is suppressed with doping, pressure, or disorder allowing for the emergence of superconductivity. This shows strong similarities to the generic cuprate phase diagram and is evidence for the interplay of magnetism and superconductivity in the Fe-based materials. There are two important distinctions, however. First, parent compounds of iron-based superconductors are antiferromagnetic *metals*, and second, the superconducting pairing symmetry in most of the materials is, most likely, an extended *s*-wave, with or without nodes [Hirschfeld *et al.* (2011)]. The electronic structure of parent FPs in the normal state has been measured by angle-resolved photoemission (ARPES) [Liu *et al.* (2008); Terashima *et al.* (2009); Zabolotnyy *et al.* (2009); Yang *et al.* (2009); Lu *et al.* (2008); Ding *et al.* (2011)] and by magneto-oscillations [Coldea *et al.* (2008); Carrington (2011)]. Both agree largely with *ab-initio* band structure calculations [Singh and Du (2008); Boeri *et al.* (2008)]. It consists of two quasi-two-dimensional near-circular hole pockets of unequal size, centered around the Γ-point (0,0), and two quasi-2D elliptic electron pockets centered around $(0,\pi)$ and $(\pi, 0)$ points in the unfolded Brillouin zone (BZ) which includes only Fe atoms. For tetragonal symmetry, the two electron pockets transform into each other under rotation by 90°. In the folded BZ, which is used for experimental measurements because of two nonequivalent As positions with respect to an Fe plane, both electron pockets are centered around (π, π). The dispersions near electron pockets and near hole pockets are reasonably close to each other apart from the sign change, i.e., there is a substantial degree of nesting between hole and electron bands. One has to mention that nesting of electron and hole bands is not always present in iron-based superconductors and we comment on these systems at the end of this chapter.

Here we analyze theoretically the formation of antiferromagnetic order in parent FPs and its consequences for the orbital and electronic structures, as well as, the appearance of Ising nematic order above the magnetic transition. In addition, we address the spin excitations in the magnetically ordered state and compare them critically to the experimental data. We will only focus on FeAs materials where the weak-coupling analysis seems to be applicable. Note, neutron scattering measurements on parent FeAs pnictides have revealed the ordering momentum in the unfolded BZ to be either $(0, \pi)$, or $(\pi, 0)$, i.e., magnetic order consists of ferromagnetic chains along one crystallographic direction in a Fe plane which alternate antiferromagnetically along the other direction. Such magnetic order emerges in the $J_1 - J_2$ model of localized spins with interactions between nearest and next-nearest neighbors J_1 and J_2, respectively, for $J_2 > 0.5J_1$ [Chandra *et al.* (1990); Si and Abrahams (2008); Xu *et al.* (2008); Yildirim (2008); Uhrig *et al.* (2009)]. However, here we review an alternative scenario which assumes that parent FPs are good metals made of itinerant electrons, and antiferromagnetic order is of SDW type. The nesting-driven mechanism is known to give rise to the incommensurate AF in Cr [Rice (1970); Keldysh and Kopaev (1965)]. Given the electronic structure of FPs, it is natural to assume that AF order emerges, at least partly, due to near nesting between the dispersions of holes and electrons. This is confirmed by *ab-initio* analysis of the total energy in the antiferromagnetic state which shows that the main energy gain with respect to the paramagnetic state comes from regions of the BZ where electron and hole pockets reside. [Andersen and Boeri (2011)]. Several groups followed this scenario over the last three years [Cvetkovic and Tesanovic (2009); Chubukov *et al.* (2008); Brydon and Timm (2009); Wang *et al.* (2009); Platt *et al.* (2009)].

The chapter is organized as follows: First, we review the main idea of the weak-coupling RPA formalism for the single-band Hubbard model in the AF state and present a general recipe to compute spin excitations below the antiferromagnetic transition temperature. Then we move to the situation of multiband FPs. There, due to certain degrees of frustration associated with multiple nesting wave vectors we first address the question on the selection of the experimentally observed $(\pi, 0)$ or $(0, \pi)$ order out of the degenerate manifold. In addition, we discuss the possibility of an Ising nematic order to form above the actual Neel temperature and its coupling to orbital and lattice degrees of freedom. Finally, we extend the RPA spin-wave formalism to the multiband (multiorbital) case and compare our results to available experimental data including the coexistence of AF and s^{+-}-wave superconductivity. At the end of the chapter we critically review the materials which do not show $(\pi, 0)$ or $(0, \pi)$ order and discuss open questions and limitations of the itinerant perspective.

9.2 A Primer: Single-Band Hubbard Model

As a primer of the spin wave calculations in the itinerant weak-coupling limit let us start with the example of a single-band Hubbard model at half-filling which was analyzed in the context of parent high-T_c cuprates about two decades ago [Schrieffer *et al.* (1989); Singh and Tesanovic (1990); Chubukov and Frenkel (1992)]:

$$H_1 = \sum_{\mathrm{k},\sigma} \epsilon_{\mathrm{k}} c^{\dagger}_{\mathrm{k},\sigma} c_{\mathrm{k},\sigma} + \frac{U}{V} \sum_{\mathrm{k},\mathrm{k}',\mathrm{q}} c^{\dagger}_{\mathrm{k+q},\uparrow} c_{\mathrm{k},\uparrow} c^{\dagger}_{\mathrm{k'-q},\downarrow} c_{\mathrm{k}',\downarrow}. \tag{9.1}$$

Here, $c^{\dagger}_{\mathrm{k},\sigma}$ $(c_{\mathrm{k},\sigma})$ refers to the creation (annihilation) of a fermionic quasiparticle with momentum $\mathbf{k}$ and spin σ. U denotes the on-site Coulomb repulsion and ϵ_{k} is the energy dispersion based on the tight-binding approximation with hopping between nearest neighbors. At half filling ($n = 1$) the Fermi surface is perfectly nested, i.e. $\epsilon_{\mathrm{k}} = -\epsilon_{\mathrm{k+Q}}$ where $\mathbf{Q} = (\pi, \pi)$. Similar to the one-dimensional case the non-interacting susceptibility $\chi_0(\mathrm{Q}, \omega)$ diverges at the nesting wave vector and the paramagnetic ground state is unstable with respect to the formation of density waves. For repulsive Coulomb interaction the spin density wave (SDW) instability occurs first and the system undergoes the paramagnetic $|PM\rangle$ to antiferromagnetic $|AF\rangle$ transition for any weak value of U. In a mean-field approximation one can decouple the interaction with respect to a finite magnetization. Putting the spin quantization axis along the z-direction we define the SDW gap

$$\Delta = \frac{U}{V} \sum_{\mathrm{k},\sigma}{}' sgn(\sigma) \langle AF | c^{\dagger}_{\mathrm{k+Q},\sigma} c_{\mathrm{k},\sigma} | AF \rangle. \tag{9.2}$$

and the quadratic mean-field Hamiltonian can be diagonalized leading to two new Eigenenergies $E_{1/2} = \pm\sqrt{\epsilon_{\mathrm{k}}^2 + \Delta^2}$. They look formally the same as in a superconductor with two bands separated by an energy gap 2Δ. Therefore, the system is an insulator for any small value of U at half-filling and complete nesting. If the nesting is not complete, there are two important points to bear in mind. First, the SDW transition will occur only at some critical value of U_{cr}, and second, for $U > U_{cr}$ the system is an insulator at half-filling.

In order to calculate the spin excitations of the AF state one looks at the dynamical spin susceptibility tensor

$$\chi^{st}(\mathrm{q}, \mathrm{q}', t) \propto \langle AF|T_\tau S^s_{\mathrm{q}}(t) S^t_{-\mathrm{q}'}(0)|AF\rangle \tag{9.3}$$

where $S^s_{\mathrm{q}} = \sum_{\mathrm{k},\alpha,\beta} c^\dagger_{\mathrm{k+q},\alpha} \sigma^s_{\alpha,\beta} c_{\mathrm{k},\beta}$. It is important to calculate the fluctuations with respect to the true AF ground state with finite Δ. The SU(2) symmetry of the Hubbard Hamiltonian is explicitly broken in the AF state and longitudinal(zz) and transverse ($+-$) components of the susceptibility tensor are inequivalent. In particular, the zz component is gapped by 2Δ and is less interesting. The gapless Goldstone modes and corresponding spin waves appear in the transverse channel and govern the low energy behavior of the system. The transverse susceptibility $\chi^{\pm}(\mathrm{q}, \mathrm{q}', t)$ can be computed within a random phase approximation (RPA) in the AF ground state. This amounts to summing all ladder diagramms in a Dyson equation:

$$\chi^{\pm}_{RPA}(\mathrm{q}, \mathrm{q}', \omega) = \sum_{\mathrm{q}_1} \chi^{\pm}_0(\mathrm{q}, \mathrm{q}_1, \omega)[1 - U\chi^{\pm}_0(\mathrm{q}_1, \mathrm{q}', \omega)]^{-1} \tag{9.4}$$

The breaking of translational symmetry requires a careful treatment of momentum conservation. Due to a reduction of the BZ off-diagonal components of the susceptibility appear in momentum space (Umklapp susceptibility). Their presence is crucial for the correct description of spin excitations. Moreover, in order to treat correctly gapless spin waves in the charge gap set by 2Δ, it is necessary to compute the magnitude of the gap, Eq. 9.2, and RPA spin susceptibility self-consistently. It is easy to show that the condition for the pole formation in the RPA transverse susceptibility for $\mathrm{q} = \mathrm{q}' = \mathrm{Q}$ and $\omega \to 0$ (Goldstone mode) coincides with the SDW mean-field gap equation [Schrieffer *et al.* (1989)].

Although the formalism developed above assumes well-defined quasiparticles (weak-coupling) in the AF state the calculations of the spin waves were carried out in great detail in application to the un-doped cuprate high-T_c superconductors together with a thourough comparison to experimental data. For example, in Fig. 9.1 we show

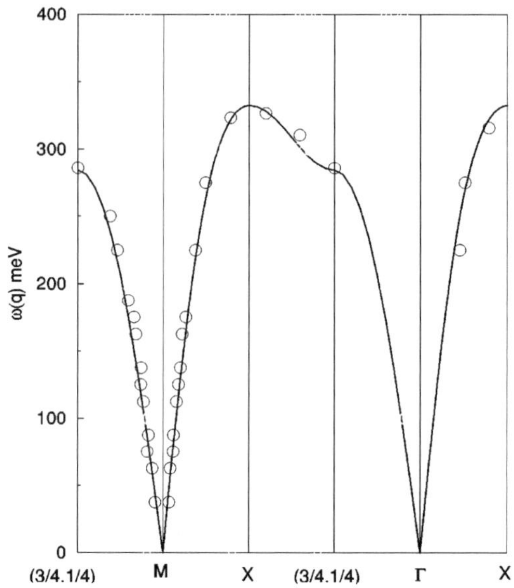

Figure 9.1 Calculated spin wave dispersion using the self-consistent RPA type formalism for the single band Hubbard model in comparison to experimental data for cuprates of [Coldea *et al.* (2001)]. Taken from [Peres and Araujo (2002)].

the results of a theoretical calculation within the formalism described above [Peres and Araujo (2002)] together with experimental data [Coldea *et al.* (2001)]. The agreement is remarkable given the strongly correlated nature of cuprates. It is interesting to notice that the RPA formalism taken in the large U limit gives the spin wave velocity $v = \frac{1}{\sqrt{2}}J$ with $J = -\frac{4t^2}{U}$ in agreement with strong-coupling result of the corresponding Heisenberg model. This indicates that at least at half-filling the spin excitations can be consistently described by both weak- and strong-coupling approaches. From the weak-coupling perspective, an inclusion of self-energy corrections [Singh and Tesanovic (1990)] further improves the agreement between Monte-Carlo calculations of the sublattice magnetization and that obtained within the self-consistent RPA type theory. It points to the idea that there is indeed a smooth crossover between weak and strong coupling limits [Frenkel and Hanke (1990)], with the proper inclusion of higher order corrections.

The weak-coupling RPA formalism gives not only immediately information about spin excitations but also knowledge on the Stoner continuum and its interaction with the spin waves. In the opposite strong-coupling limit this requires some extra care. A disadvantage of the weak-coupling theory is the assumption that the transition between strong and weak-coupling is continuous which is not always the case especially at finite doping.

Having introduced some key ideas with a single-band system, let us now move to FPs. The aim of the next sections is to generalize the single-band theory to the multiband, or multiorbital, case in order to gain insights into the magnetic phase of iron-based superconductors. An additional motivation is a desire to understand the magnetic properties of models which are most commonly used to describe the superconducting phase. This is required in order to obtain a coherent understanding of all different phases within a single approach. Moreover, in contrast to cuprates there is a principal difficulty to derive the analog of an effective low energy spin Hamiltonian for multiorbital systems such as FPs in a controlled way.

9.3 Magnetic Order in Ferropnictides

9.3.1 *Magnetic Frustration*

As pointed out in the introduction the magnetic order in the FeAs materials was originally detected by neutron scattering [Cruz *et al.* (2008)] and μsR experiments [Klauss *et al.* (2008)]. The magnetic transition temperature varies slightly from compound to compound and is of the order of $T_N \sim 150$K. In real space the magnetic ordering consists of ferromagnetic chains along one crystallographic direction in the Fe square lattice which are coupled antiferromagnetically. In momentum space the order can be characterized either by $Q_1 = (\pi, 0)$ or $Q_2 = (0, \pi)$ wavevector. Within the localized scenario this order emerges in the context of the $J_1 - J_2$ model [Chandra *et al.* (1990)] for $J_2 > 0.5 J_1$ once quantum fluctuations are taken into account. In the following we review how this order appears in the itinerant picture.

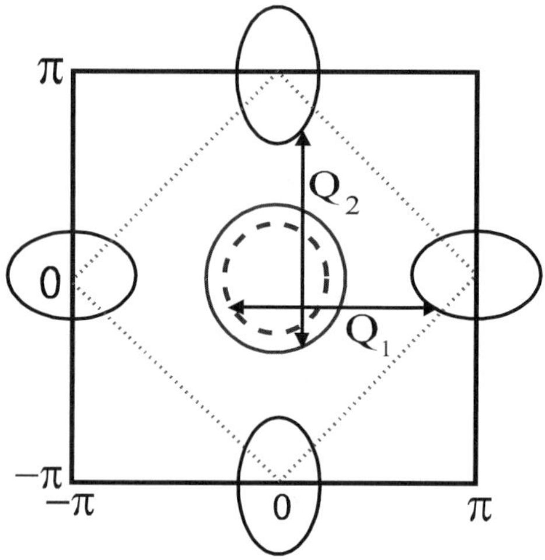

Figure 9.2 Schematic Fermi surface of ferropnictides in the unfolded Fe-based Brillouin zone with two circular hole-like pockets centered around the Γ-point and two elliptical electron-like pockets located around the $(\pi, 0)$ and $(0, \pi)$ point of the BZ, respectively. Q_1 and Q_2 represent two nesting wave vectors. Taken from [Eremin and Chubukov (2010)].

The schematic Fermi surface (FS) in the normal state of FPs is reproduced in Fig. 9.2 for the unfolded iron-based BZ. As we pointed out in the introduction the elliptic electron bands and nearly circular hole bands are nearly nested. For the Fermi surface topology of FPs it means that there are two nesting wave vectors $Q_1 = (\pi, 0)$ and $Q_2 = (0, \pi)$ between the hole- and electron-like pockets. For the idealized case of zero ellipticity of the electron pockets and equal masses for the electron and hole bands the situation is similar to the half-filled Hubbard model with nearest neighbor hopping. In particular, the susceptibility in the particle-hole channel diverges logarithmically as it usually does in the particle-particle Cooper-channel. A renormalization group analysis shows that the leading instability is magnetic. [Chubukov *et al.* (2008); Cvetkovic and Tesanovic (2009)] However, in contrast to the Hubbard model, here it occurs at two wave vectors Q_1 and Q_2 leaving the question open how the experimentally observed state is realized.

To formulate the problem in a formal way let us start with a generic spin configuration described by two mean-field SDW order

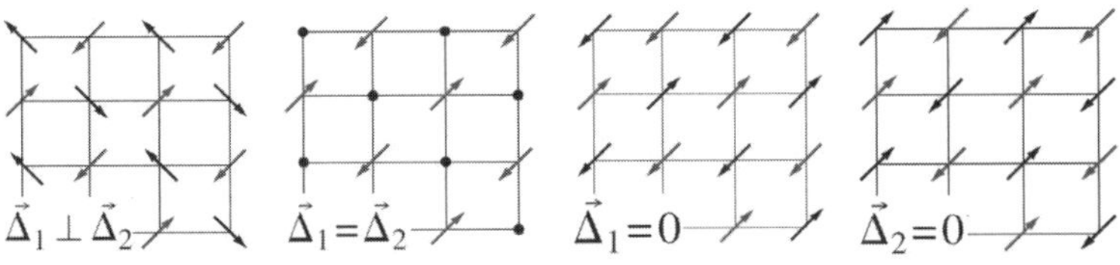

Figure 9.3 Possible real space orderings of the magnetic state. Either one of the last two corresponds to the experimentally realized one. Taken from [Eremin and Chubukov (2010)].

parameters $\vec{\Delta}_i$ for each of the wave vector $\mathrm{Q_i}$ in the form

$$\vec{S}(\mathrm{R}) = \vec{\Delta}_1 e^{i\mathrm{Q}_1\mathrm{R}} + \vec{\Delta}_2 e^{i\mathrm{Q}_2\mathrm{R}}. \qquad (9.5)$$

For such a configuration the Fe lattice decouples into two antiferromagnetically ordered sublattices with magnetizations $\vec{\Delta}_1 + \vec{\Delta}_2$ and $\vec{\Delta}_1 - \vec{\Delta}_2$, respectively, which interpenetrate each other. However, neither the angle between two Neel lattices is fixed nor the magnitude. For example, in Fig. 9.3 we show four possible orderings for a generic $\vec{S}(\mathrm{R})$ out of many possibilities. The last two configurations with one of the $\vec{\Delta}_i$ going to zero refer to the experimentally observed one.

Without loss of generality one can assume that one of the hole pockets is interacting stronger with the two electron-like pockets. Therefore it is useful to consider a model consisting of a single circular hole FS centered around the Γ-point (α-band) and two elliptical electron Fermi surface pockets centered around $(\pm\pi, 0)$ and $(0, \pm\pi)$ points in the unfolded BZ (β-bands):

$$H_2 = \sum_{\mathrm{p},\sigma} \left[\varepsilon_{\mathrm{p}}^{\alpha_1} \alpha_{1\mathrm{p}\sigma}^{\dagger} \alpha_{1\mathrm{p}\sigma} + \varepsilon_{\mathrm{p}}^{\beta_1} \beta_{1\mathrm{p}\sigma}^{\dagger} \beta_{1\mathrm{p}\sigma} + \varepsilon_{\mathrm{p}}^{\beta_2} \beta_{2\mathrm{p}\sigma}^{\dagger} \beta_{2\mathrm{p}\sigma} \right]. \qquad (9.6)$$

Here, $\varepsilon_{\mathrm{p}}^{\alpha_1} = -\frac{\hbar^2 p^2}{2m_1} + \mu$ and $\varepsilon_{\mathrm{p}}^{\beta_1} = \frac{\hbar^2 p_x^2}{2m_x} + \frac{\hbar^2 p_y^2}{2m_y} - \mu$, $\varepsilon_{\mathrm{p}}^{\beta_2} = \frac{\hbar^2 p_x^2}{2m_y} + \frac{\hbar^2 p_y^2}{2m_x} - \mu$ are the dispersions of hole and electron bands. The momenta of $\alpha-$fermions are counted from $(0, 0)$, the momenta of β_1- and β_2-fermions are counted from $(0, \pi)$ and $(\pi, 0)$.

One assumes that the most important interactions are those which give rise to the SDW order. According to the terminology of Ref. [Chubukov *et al.* (2008)] these are density-density interactions between α and β fermions in the form

$$H_4 = U_1 \sum \alpha^\dagger_{1p_3\sigma} \beta^\dagger_{jp_4\sigma'} \beta_{jp_2\sigma'} \alpha_{1p_1\sigma} + \frac{U_3}{2} \sum \left[\beta^\dagger_{jp_3\sigma} \beta^\dagger_{jp_4\sigma'} \alpha_{1p_2\sigma'} \alpha_{1p_1\sigma} + h.c \right]. \tag{9.7}$$

At this level we neglect the potential angular dependencies of U_1 and U_3 along the FSs which arise due to hybridization of different orbitals which we will discuss later.

The two SDW order parameters are expressed as $\vec{\Delta}_1 \propto \sum_p \langle \alpha^\dagger_{1p\delta} \beta_{1p\gamma} \sigma_{\delta\gamma} \rangle$ with momentum Q_1 and $\vec{\Delta}_2 \propto \sum_p \langle \alpha^\dagger_{1p\delta} \beta_{2p\gamma} \sigma_{\delta\gamma} \rangle$ with momentum Q_2. Without loss of generality we can set $\vec{\Delta}_1$ along z-axis and $\vec{\Delta}_2$ in the xz-plane. In explicit form, we introduce

$$\Delta_1^z = -U_{SDW} \sum_p \langle \alpha^\dagger_{1p\uparrow} \beta_{1p\uparrow} \rangle$$
$$\Delta_2^{z(x)} = -U_{SDW} \sum_p \langle \alpha^\dagger_{1p\uparrow} \beta_{2p\uparrow(\downarrow)} \rangle. \tag{9.8}$$

where $U_{SDW} = U_1 + U_3$.

In the simplest case all masses are equal, i.e., $m_x = m_y$ and $\varepsilon_p^{\beta_1} = \varepsilon_p^{\beta_2} = \varepsilon_p^{\beta}$, such that we have perfect nesting between all circular pockets. Again similar to the single-band case we can treat the interaction term, Eq. 9.7, with the SDW order parameters, Eq. 9.8 in mean-field. With two subsequent Bogolyubov transformations [Eremin and Chubukov (2010)] the quadratic Hamiltonian can be written as

$$H_2^{eff} = \sum_{a,p} \varepsilon_p^\beta d^\dagger_{ap} d_{ap} + \sum_p E_p \left(e^\dagger_{ap} e_{ap} + p^\dagger_{bp} p_{bp} - e^\dagger_{bp} e_{bp} - p^\dagger_{ap} p_{ap} \right), \tag{9.9}$$

where $E_p = \pm\sqrt{(\varepsilon_p)^2 + |\Delta|^2}$ and $\Delta = \sqrt{(\Delta_1^z)^2 + (\Delta_2)^2}$, $\Delta_2 = \sqrt{(\Delta_2^z)^2 + (\Delta_2^x)^2}$ and there is only one single self-consistent equation for the total gap magnitude

$$1 = \frac{U_{SDW}}{2N} \sum_p \frac{1}{\sqrt{(\varepsilon_p)^2 + \Delta^2}}. \tag{9.10}$$

As one clearly sees the self-consistency equation (9.10) only fixes the total magnitude of $(\Delta_1^z)^2 + (\Delta_2)^2$ but not the magnitude and the direction of individual Δ_1 and Δ_2. This implies a huge groundstate degeneracy at the mean-field level where SDW ordering corresponds to spontaneous breaking of the $O(6)$ symmetry. The experimentally realized states with either $\Delta_i = 0$ are just two of infinitely many possibilities. Moreover, the degeneracy of the itinerant model is even larger than that in the localized $J_1 - J_2$ model where the moments are fixed. In the itinerant picture the magnitude of each sublattice magnetization can vary by the cost of the other.

In the following we will show that within mean-field the degeneracy can still be lifted by the ellipticity of the electron pockets or interaction within electron or within hole pockets. This explains why the particular order is realized and then due to magnetostriction the structural order is imposed on the lattice. However, it does not explain why the structural transition occurs sometimes at higher temperature than T_N. To cover this aspect we also briefly discuss how the wave vector can be selected prior to the onset of long-range magnetic ordering. This refers to the notion of Ising nematic order. We also discuss how it couples to the lattice and orbitals.

9.3.2 *Lifting the Magnetic Ground State Degeneracy at T_N*

So far, the analysis assumed the idealized situation of fully nested circular electron and hole pockets with interactions only between holes and electrons. If one takes into account the ellipticity of the electron pockets and additional interactions the degeneracy may be lifted. One includes, in addition, the four other possible $\beta - \beta$ interactions:

$$\begin{aligned} H_4^{ex} &= U_6 \sum \beta^\dagger_{1p_3\sigma}\beta^\dagger_{2p_4\sigma'}\beta_{2p_2\sigma'}\beta_{1p_1\sigma} + U_7 \sum \beta^\dagger_{2p_3\sigma}\beta^\dagger_{1p_4\sigma'}\beta_{2p_2\sigma'}\beta_{1p_1\sigma} \\ &+ \frac{U_8}{2} \sum \left[\beta^\dagger_{2p_3\sigma}\beta^\dagger_{2p_4\sigma'}\beta_{1p_2\sigma'}\beta_{1p_1\sigma} + h.c\right] \\ &+ \frac{U_4}{2} \sum \left[\beta^\dagger_{1p_3\sigma}\beta^\dagger_{1p_4\sigma'}\beta_{1p_2\sigma'}\beta_{1p_1\sigma} + \beta^\dagger_{2p_3\sigma}\beta^\dagger_{2p_4\sigma'}\beta_{2p_2\sigma'}\beta_{2p_1\sigma}\right] \end{aligned} \tag{9.11}$$

In the AF state one again applies the sequence of Bogolyubov transformations, takes averages $\langle \cdots \rangle$, to obtain the contribution to the ground state energy. The final correction to the ground state energy was obtained in [Eremin and Chubukov (2010)] and has the form:

$$E_{gr}^{ex} = 2A^2\left[(U_6 + U_8 - U_7 - U_4)\right]\frac{|\Delta_1|^2|\Delta_2|^2}{\Delta^4} + 4A^2U_7\frac{(\Delta_1\cdot\Delta_2)^2}{\Delta^4} \tag{9.12}$$

Observe that E_{gr}^{ex} depends on $|\Delta_1|^2|\Delta_2|^2$ and on $(\Delta_1\cdot\Delta_2)^2$, i.e., it is sensitive to both, relative values and relative directions of Δ_1 and Δ_2. When all interactions are of equal strength, the first term vanishes, and the last term favors $\Delta_1 \perp \Delta_2$. In this situation, the $O(6)$ degeneracy of the ideal model is broken, but only down to $O(3)\times O(2)$, i.e., the magnitude of the order parameter at each site is now the same because $\left(\vec{\Delta}_1+\vec{\Delta}_2\right)^2 = \left(\vec{\Delta}_1-\vec{\Delta}_2\right)^2$, but the angle between the directions of the SDW order in the two sublattices (i.e., between $\vec{\Delta}_1+\vec{\Delta}_2$ and $\vec{\Delta}_1-\vec{\Delta}_2$) is still arbitrary. This is exactly the same situation as in the classical $J_1 - J_2$ model. However, once $U_6+U_8-U_7-U_4$ is nonzero, the degeneracy is broken down to a conventional $O(3)$ already at the mean-field level. Because U_4 is reduced and even changes sign under RG [Chubukov *et al.* (2008)], while other U_i do not flow, the most likely situation is that $U_6+U_8-U_7-U_4>0$, in which case E_{gr}^{ex} is minimized when either $\vec{\Delta}_1=0$, or $\vec{\Delta}_2=0$, i.e., SDW order is either $(0,\pi)$ or $(\pi,0)$. This is exactly the same SDW order as observed in experiments. If $U_6+U_8-U_7-U_4$ was negative, E_{gr}^{ex} would be minimized when $|\vec{\Delta}_1| = |\vec{\Delta}_2|$, in which case the SDW OPs of the two sublattices would align orthogonal to each other. The spin configuration for such state is shown the left panel of Fig. 9.3.

The other effect concerns a distortion of the electron Fermi pockets, i.e. the electron pockets are actually ellipses rather than circles. The effective masses m_x and m_y are not equal, and $\varepsilon_{\mathrm{k}}^{\beta_1} \neq \varepsilon_{\mathrm{k}}^{\beta_2}$. To continue with the analytical analysis, one assumes that the ellipticity is small, and introduces $m_x=(1+\delta)m$ and $m_y=(1-\delta)m$, where $\delta << 1$, and computes the correction to the ground state energy to second order in δ. The contribution to the ground state energy coming from the ellipticity is

$$E_{gr}^{ellipt} = C\,|\vec{\Delta}_1|^2|\vec{\Delta}_2|^2,\;\; C = \delta^2\frac{m\mu^2}{4\pi\Delta^4} \tag{9.13}$$

where the coefficient C is positive, i.e., the correction due to ellipticity of electron pockets again breaks the degeneracy and selects either $(0, \pi)$ or $(\pi, 0)$ state. This is again the same selection as one needs for consistency with experiments. It is remarkable that ellipticity leads to a term in the ground state energy that is similar to an effective interaction between two SDW OPs which, for $\Delta << \mu$, leads to the same selection of the ground state SDW order as the direct interaction between the two electron pockets.

It follows from the above analysis that it is indeed possible to understand in a simple and transparent way the connection between the peculiar FS topology and the magnetic groundstate in FPs. Before we continue to calculate the most striking manifestation of symmetry breaking in the AF state, the Goldstone modes, we shortly discuss different aspects. Since the magnetic $(\pi, 0)$ state not only breaks spin rotational symmetry but also translational symmetry it leads to a reconstruction of the Fermi surface when going from the paramagnetic to the antiferromagnetic phase. In Fig. 9.4 we show the data of this reconstruction (upper panel) from [Liu *et al.* (2010)] together with a theoretical modeling of an itinerant five orbital model (lower panel). The agreement is remarkably good.

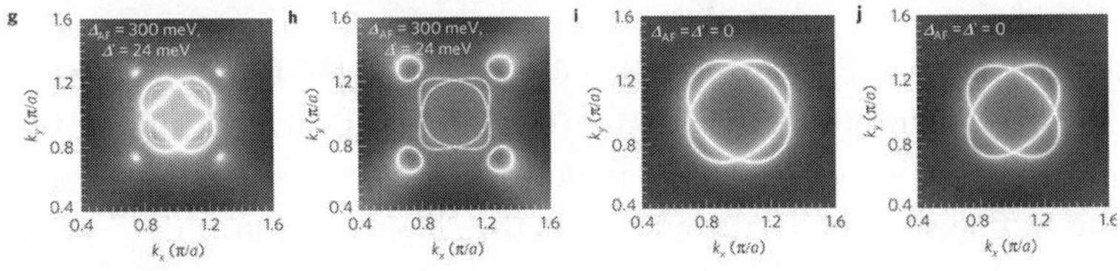

Figure 9.4 Reconstruction of the Fermi surface pockets from the paramagnetic state, panels e,f and i,j, to the antiferromagnetic state, panels c,d and g, h. The ARPES data (upper panels) are nicely explained by the theoretical calculation of an itinernat five orbital model (lower panel). Taken from Liu *et al.* (2010).

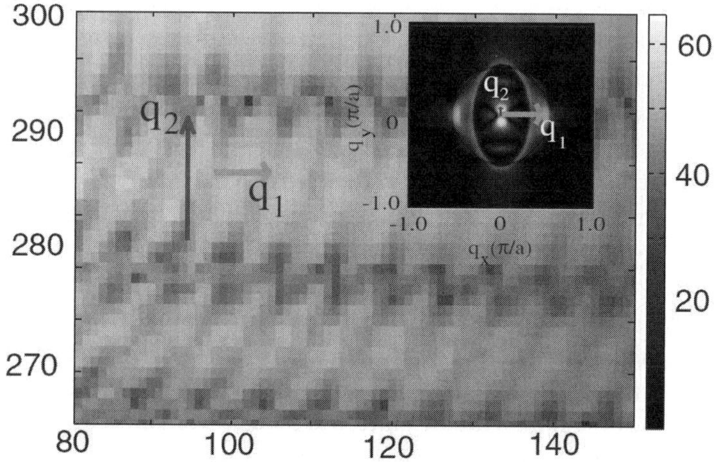

Figure 9.5 Real space image of the QPI at −7 meV away from the impurity site. The inset shows the corresponding QPI image. The QPI interference originating from the structure in the spectral density is reflected in the strong $10 \div 15a$ modulation along the FM (y) direction and a weak $5 \div 7a$ modulation along the AF (x) direction. The impurity potential is a mixture of magnetic (20%) and nonmagnetic (80%) parts. Intensity refers to states/eV. Taken from Knolle *et al.* (2010b). See also Color Insert.

Note that a reconstruction of the electronic structure in the AF state leads to a corresponding modification of the local density of states. The latter shows a pronounced C_2 anisotropy in the AF state which is consistent with STM experiments [Chuang *et al.* (2010)]. Fig. 9.5 shows a calculated real space image of the quasiparticle intereference pattern (QPI) for a non-magnetic impurity computed within a T-matrix approach [Knolle *et al.* (2010b)] away from the impurity position and at −7meV. Observe that the QPI modulation shown in the inset is reflected in periodic structures seen along the x and y crystallographic directions. In the AF x direction with q_1 structure, one finds a weak stripe pattern with a periodicity of the order of $7a$ where a is a lattice (Fe-Fe) constant. In addition, there is a stronger modulation of $10 \div 15a$ along the FM y direction, consistent with the structure described as nematic in [Chuang *et al.* (2010)]. These extra modulations arise purely due to scattering between bands involved in the SDW, no extra folding of the bands occurs.

In addition, two remarks are necessary. First, once the linear coupling term between lattice distortion and magnetization which breaks C(4) symmetry is included in the theory, it forces the lattice to follow the magnetic ordering [Cano *et al.* (2010)], and thus, leads to a structural transition. Experimentally such a structural transition at T_s is observed but depending on the material it either coincides or precedes the magnetic transition at T_N.

9.3.3 *Ising Nematic Order Above T_N*

The emergence of a phase with broken rotational symmetry above or at T_N is experimentally not only manifested by a tetragonal to orthorhombic transition at T_s, but also by the onset of significant anisotropies in several quantities [Fisher *et al.* (2011)], such as dc resistivity [Chu *et al.* (2010); Tanatar *et al.* (1990)], optical conductivity [Dusza *et al.* (2011); Nakajima *et al.* (2011)], and orbital occupancy [Yi *et al.* (2011)].

Early theoretical proposals explored two alternative scenarios for the nematic order. On the one hand orbital ordering might induce the structural transition and trigger the magnetic transition at a lower temperature by renormalizing the exchange constants [Kruger *et al.* (2009)]. On the other hand, a purely magnetic scenario based on a localized $J_1 - J_2$ model [Chandra *et al.* (1990)], for FPs see [Si and Abrahams (2008); Fang *et al.* (2008); Xu *et al.* (2008)], is build upon the assumption that the order parameter manifold for two possible stripe magnetic orders is $O(3) \times Z_2$ (compare to the actual degeneracy which has O(6) symmetry), with the Z_2 Ising symmetry differentiating between the two possible ordering vectors $(0, \pi)$ and $(\pi, 0)$. This scenario explores the possibility that the Z_2 Ising variable orders before the O(3) rotational symmetry is broken, leading to an intermediate phase with a broken tetragonal symmetry but no long-range magnetic order, called Ising nematic. This magnetic scenario was later extended to itinerant systems in a phenomenological way [Qi *et al.* (2009); Fernandes *et al.* (2010)].

Here, we review how the Ising nematic order can be understood within a magnetic scenario for an itinerant fermionic model [Fernandes *et al.* (2011)]. From the microscopic Hamiltonian Eq. 9.6

and Eq. 9.7 one can derive the partition function

$$Z \propto \int d\vec{\Delta}_1 d\vec{\Delta}_2 e^{-S_{\text{eff}}[\vec{\Delta}_1,\vec{\Delta}_2]} \tag{9.14}$$

in terms of the effective action for the two magnetic order parameters

$$S_{\text{eff}}\left[\vec{\Delta}_1, \vec{\Delta}_2\right] = r_0\left(\vec{\Delta}_1^2 + \vec{\Delta}_2^2\right) + \frac{u}{2}\left(\vec{\Delta}_1^2 + \vec{\Delta}_2^2\right)^2 - \frac{g}{2}\left(\vec{\Delta}_1^2 - \vec{\Delta}_2^2\right)^2 + v\left(\vec{\Delta}_1 \cdot \vec{\Delta}_2\right)^2 . \tag{9.15}$$

The coefficients can be computed in terms of the bare fermionic propagators. The Z_2 symmetry breaking without magnetic order would imply that fluctuations associated with one of the bosonic fields are larger than the fluctuations associated with the other one, e.g., $\langle\vec{\Delta}_1^2\rangle > \langle\vec{\Delta}_2^2\rangle$ while $\langle\vec{\Delta}_1\rangle = \langle\vec{\Delta}_1\rangle = 0$. We can now introduce scalar fields $\phi = \langle\vec{\Delta}_1^2\rangle - \langle\vec{\Delta}_2^2\rangle$ and $\psi = \langle\vec{\Delta}_1^2\rangle + \langle\vec{\Delta}_2^2\rangle$. Note that in contrast to the previous subsection we do not treat $\vec{\Delta}$ as a mean-field order but rather as a fluctuating quantity. Here the field ϕ is associated with the appearance of Ising nematic order. It is now possible via a second Hubbard-Stratonovich transformation to derive an effective action in terms of the new scalar fields only:

$$S_{\text{eff}}\left[\phi, \psi\right] = \int_q \left\{\frac{\phi^2}{2g} - \frac{\psi^2}{2u} + \frac{3}{2}\log\left[\left(\chi_q^{-1} + \psi\right)^2 - \phi^2\right]\right\} \tag{9.16}$$

The action can now be analyzed in a mean-field and renormalization group treatment [Fernandes *et al.* (2011)] as a function of the dimensionless *inverse* nematic coupling $\alpha = \frac{u}{g}$. Its magnitude can readily be computed from the microscopic Hamiltonian. Similar to the previous mean-field analysis it decreases with increasing ellipticity of the electron pockets. In terms of physical parameters it decreases with pressure but increases with increasing lattice stiffness and electron doping. In Fig. 9.6 we show one of the phase diagrams obtained in [Fernandes *et al.* (2011)].

Depending on the strength of the nematic coupling different types of system behavior can be found with two distinct multi-critical points in the phase diagram, namely a magnetic tri-critical point and a nematic tri-critical point. In the case of strongly anisotropic, quasi-2D systems, the nematic tri-critical point precedes the magnetic one. For the largest nematic coupling the

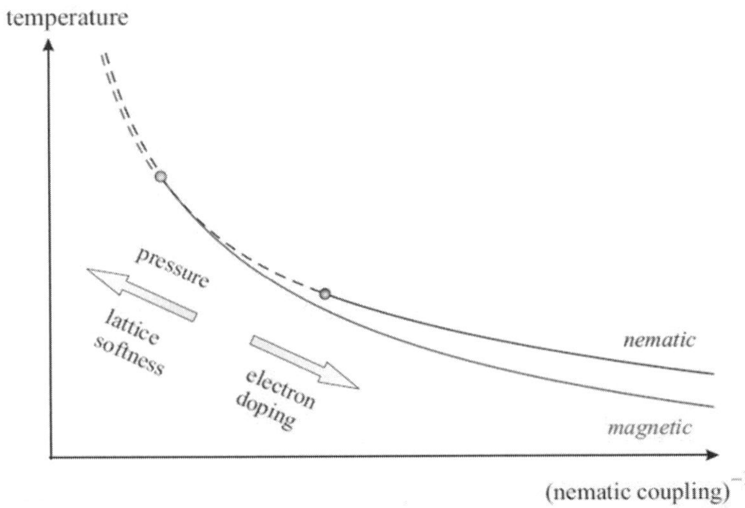

Figure 9.6 Characteristic phase diagram as function of temperature and inverse nematic coupling for strongly anisotropic, quasi-two dimensional systems. Red and blue curves represent magnetic and Ising-nematic transitions, respectively. The arrows show how the inverse nematic coupling changes as function of doping, pressure, and lattice stiffness. Here a solid (dashed) line denotes a second-order (first-order) transition, and a double-dashed line indicates a simultaneous first-order transition. The two solid points mark the positions of the nematic and magnetic tri-critical points. Taken from [Fernandes *et al.* (2011)].

stripe magnetic order emerges simultaneously with the Ising-nematic order, via a first- order magnetic transition ($T_N = T_s$). For intermediate couplings a first-order preemptive Ising-nematic transition accompanied by a discontinuous increase of the magnetic correlation length to a larger but still finite value is found, such that the stripe magnetic order does not develop at T_s and emerges only at a smaller T. In this case, the magnetic transition is second-order. For small nematic couplings a second-order preemptive Ising-nematic transition is followed by a second-order magnetic stripe transition at a smaller temperature.

In addition, one can also show that the nematic order parameter couples linearly to the orbital polarization and lattice distortion [Fernandes *et al.* (2011)], and hence nematic order generates orbital and structural order. In principal all these orders break the same C_4 symmetry and, therefore, appear simultaneously. From the experimental point of view this fact presents a difficulty to answer the question what is the actual driving mechanism for the structural transition. Nevertheless, so far the itinerant scenario gives a natural understanding for a variety of phenomena in the FPs phase diagrams.

9.4 Spin Waves in Itinerant Multiorbital Systems

Now, let us move to a discussion of the spin excitations below T_N in iron-based superconductors. We first generalize the self-consistent random phase approximation formalism for the calculation of spin waves to multiorbital systems (MOS) as the FPs.

9.4.1 *Multiorbital Models — Spin Wave Theory*

The main ideas are similar as for the single band model, but with additional complications due to orbital indices. In the following we follow closely Knolle *et al.* (2011). We define the spin operator as

$$\mathrm{S}_s(\mathrm{q}) = \sum_{\mathrm{k},\sigma,\sigma'} \psi^{\dagger}_{s\sigma}(\mathrm{k}+\mathrm{q}) \frac{\sigma_{\sigma\sigma'}}{2} \psi_{s\sigma'}(\mathrm{k}), \tag{9.17}$$

where $\psi^{\dagger}_{s\sigma}$ is the creation operator of the orbital s with spin σ in the MOS. The transverse spin susceptibility in the paramagnetic state can be calculated from the spin-spin correlation function in Matsubara frequency space:

$$\chi^{\pm}_{st}(\mathrm{q}, i\Omega) = \int_0^{\beta} d\tau e^{i\Omega\tau} \left\langle T_\tau S^+_s(\mathrm{q}, \tau) S^-_t(-\mathrm{q}, 0) \right\rangle . \tag{9.18}$$

In MOS it is convenient to introduce a dynamical spin susceptibility tensor in a more general way using four orbital indices [Kemper *et al.* (2010); Brydon and Timm (2009)]

$$\chi^{\pm}_{spqt}(\mathrm{q}, i\Omega) = \int_0^{\beta} d\tau e^{i\Omega\tau} \sum_{\mathrm{k},\mathrm{k}',\gamma,\delta,\gamma',\delta'} \sigma^+_{\gamma\delta} \sigma^-_{\gamma'\delta'} \tag{9.19}$$

$$\left\langle T_\tau \psi^{\dagger}_{p\gamma}(\mathrm{k}, \tau) \psi_{q\delta}(\mathrm{k}+\mathrm{q}, \tau) \psi^{\dagger}_{t\gamma'}(\mathrm{k}', 0) \psi_{s\delta'}(\mathrm{k}'-\mathrm{q}, 0) \right\rangle . \tag{9.20}$$

With the Green's function (GF)

$$G^{\sigma}_{st}(\mathrm{k}, i\omega_n) = -\int_0^{\beta} d\tau e^{i\omega\tau}\left\langle T_\tau \psi_{s\sigma}(\mathrm{k}, \tau)\psi^{\dagger}_{t\sigma}(\mathrm{k}, 0)\right\rangle \quad (9.21)$$

the susceptibility matrix can be written as

$$\chi^{\pm}_{spqt}(\mathrm{q}, i\Omega) = -\frac{1}{N\beta}\sum_{\mathrm{k}, i\omega_n} G^{\uparrow}_{sp}(\mathrm{k}, i\omega_n) G^{\downarrow}_{qt}(\mathrm{k}+\mathrm{q}, i\omega_n + i\Omega) \quad (9.22)$$

The susceptibility in the paramagnetic phase within the RPA can be written in a matrix form $\hat{\chi}$, see for example Kemper *et al.* (2010)

$$\chi^{RPA}_{bsta} = \chi^{0}_{bsta} + \chi^{RPA}_{b'sta'} U_{c'b'a'd'} \chi^{0}_{bc'd'a}, \quad (9.23)$$

and the solution of the RPA in matrix form is straightforward.

$$[\hat{\chi}]^{RPA} = \hat{\chi}^{0}(1 - \hat{U}\hat{\chi}^{0})^{-1} \quad (9.24)$$

The physical susceptibility which is measured in experiment is then obtained

$$\chi^{RPA}_{phys}(\mathrm{q}, \omega) = \frac{1}{2}\sum_{st} \chi^{RPA}_{sstt}(\mathrm{q}, \omega). \quad (9.25)$$

However, important information is also contained in the individual components of the susceptibility tensor χ_{spqt} which can be used to distinguish between predominantly intra- or interorbital fluctuations. The RPA in diagrammatic form is illustrated in Fig. 9.7.

Generally, for MOS, the kinetic or tight-binding part H_0 of the Hamiltonian $H = H_0 + H_{int}$, including the intra- and inter-orbital hoppings, can be diagonalized via a unitary transformations which gives the eigenergies $E_\nu(\mathrm{k})$ of the system. The operators in the

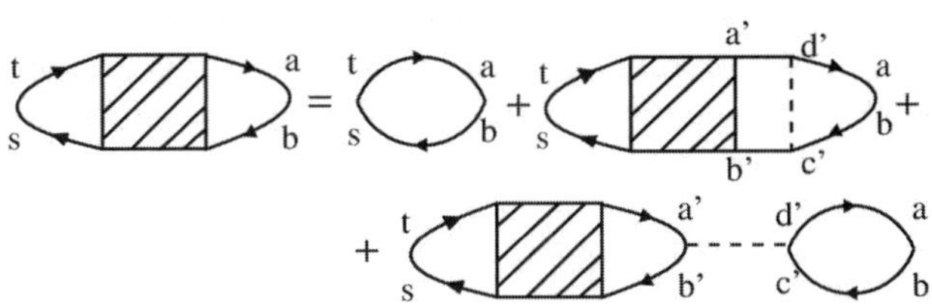

Figure 9.7 Diagrammtic form of the RPA Dyson equation for the multiorbital susceptibility. Depending on the form of the interactions ladder and bubble diagramms contribute to the longitudinal and transverse component.

orbital and the band basis are connected via the coefficients of the unitary transformations $a_\nu^s(\mathrm{k},\sigma)$

$$\psi_{s\sigma}(\mathrm{k}) = \sum_\nu a_\nu^s(\mathrm{k},\sigma)\gamma_{\nu\sigma}(\mathrm{k}) \tag{9.26}$$

Without interactions the GFs are easily obtained as

$$G_{st}^{\sigma}(\mathrm{k}, i\omega_n) = \sum_\nu \frac{a_\nu^s(\mathrm{k},\sigma)a_\nu^{t*}(\mathrm{k},\sigma)}{i\omega - E_\nu(\mathrm{k})} \tag{9.27}$$

and the transverse spin susceptibility becomes:

$$\chi_{spqt}^{\pm}(\mathrm{q}, i\Omega) = -\frac{1}{N\beta}\sum_{\mathrm{k},i\omega,\nu,\mu} \tag{9.28}$$

$$\frac{a_\mu^s(\mathrm{k},\uparrow)a_\mu^{p*}(\mathrm{k},\uparrow)a_\nu^q(\mathrm{k}+\mathrm{q},\downarrow)a_\nu^{t*}(\mathrm{k}+\mathrm{q},\downarrow)}{\left[i\omega - E_\mu(\mathrm{k})\right]\left[i\omega + i\Omega - E_\nu(\mathrm{k}+\mathrm{q})\right]}. \tag{9.29}$$

The Matsubara sum can be performed via a standard contour integration and by analytic continuation $i\Omega \rightarrow \Omega + i\Gamma$. Observe that for large Hamiltonian matrices the coefficients a_ν have to be calculated numerically by diagonalizing the Hamiltonian matrix written in **k**-space.

In a mean-field treatment of the SDW phase, the interaction part of the hamiltonian H_{int} is decoupled with respect to a groundstate with a finite magnetization at wave vector Q, e.g. $(\pi, 0)$ or $(0, \pi)$ for iron-based superconductors. Here, we assume that only one of them is present below T_N. The order parameters read

$$\Delta_{\nu\nu'} = \sum_\mathrm{k} sgn(\sigma)\left\langle \psi_{\nu\sigma}^\dagger(\mathrm{k})\psi_{\nu'\sigma}(\mathrm{k}+\mathrm{Q})\right\rangle. \tag{9.30}$$

In the magnetic state, SU(2) spin symmetry is broken which yields $\chi^{\pm} \neq \frac{1}{2}\chi^{zz}$. In addition, the AF state also breaks translational symmetry and the mean-field Hamiltonian in the reduced BZ takes the form

$$H_{SDW} = \sum_\mathrm{k}{}' \hat{\psi}_\sigma^\dagger(\mathrm{k})\begin{pmatrix} H_0(\mathrm{k}) & \sigma M \\ \sigma M & H_0(\mathrm{k}+\mathrm{Q})\end{pmatrix}\hat{\psi}_\sigma(\mathrm{k}) \tag{9.31}$$

where we introduce the new spinor

$$\hat{\psi}_\sigma(\mathrm{k}) = \begin{pmatrix} \psi_\sigma(\mathrm{k}) \\ \psi_\sigma(\mathrm{k}+\mathrm{Q})\end{pmatrix}. \tag{9.32}$$

Note for MOS each component of the spinor may acquire a spinor structure in orbital space as well. Then the GF acquires off-diagonal elements in momentum space and becomes a matrix. In short-hand notation, we write only the orbital index with a tilde to indicate that the momentum is shifted by the AFM wave vector. We thus obtain

$$\hat{G}_{sp} = \left\langle \begin{pmatrix} s \\ \tilde{s} \end{pmatrix} (p^\dagger \; \tilde{p}^\dagger) \right\rangle = \begin{pmatrix} G_{sp} & G_{s\tilde{p}} \\ G_{\tilde{s}p} & G_{\tilde{s}\tilde{p}} \end{pmatrix} \tag{9.33}$$

Analogously, the susceptibility matrix becomes a matrix in q-space [Schrieffer *et al.* (1989); Knolle *et al.* (2010a); Brydon and Timm (2009)]:

$$\hat{\chi}^{\pm}_{spqt} = \begin{pmatrix} \chi^{\pm}_{spqt}(\mathrm{q}, \mathrm{q}) & \chi^{\pm}_{spqt}(\mathrm{q}, \mathrm{q}+\mathrm{Q}) \\ \chi^{\pm}_{spqt}(\mathrm{q}+\mathrm{Q}, \mathrm{q}) & \chi^{\pm}_{spqt}(\mathrm{q}+\mathrm{Q}, \mathrm{q}+\mathrm{Q}) \end{pmatrix} \tag{9.34}$$

where the off-diagonal components, in short hand notation $\chi^{\pm}_{spqt}(\mathrm{q}, \mathrm{q}+\mathrm{Q}) = \left[G_{sp}G_{q\tilde{t}} + G_{s\tilde{p}}G_{\tilde{q}\tilde{t}} + G_{\tilde{s}\tilde{p}}G_{\tilde{q}t} + G_{\tilde{s}p}G_{qt}\right]$ are Umklapp susceptibilities as they connect spins with momentum shifted by the AFM wave vector, which is a reciprocal wave vector of the magnetic BZ. The detailed expressions can be found in Knolle *et al.* (2011).

The form of the RPA in the SDW state remains the same, see Eq. (9.23), with the only extension that the interaction matrix also doubles in size with $\hat{U}$ matrices on the diagonal and zeros on the off-diagonal. To obtain the physical susceptibility we perform an RPA with Eq. (9.34) and trace over the orbital indices of the first entry $\chi^{RPA}_{spqt}(\mathrm{q}, \mathrm{q})$ of the susceptibility matrix. The imaginary part is directly related to the cross section of INS measurements. We also recall that self-consistency enters via taking the same set of interaction parameters in the RPA and the Hartree-Fock mean-field equations.

9.4.2 *Accidental Collective Modes in Itinerant Frustrated Antiferromagnets*

Before computing spin waves for realistic parameters of iron-based superconductors we again come back to the idealized model of fully nested electron and hole pockets discussed in the previous section. At a mean-field level there is a large ground state degeneracy. In

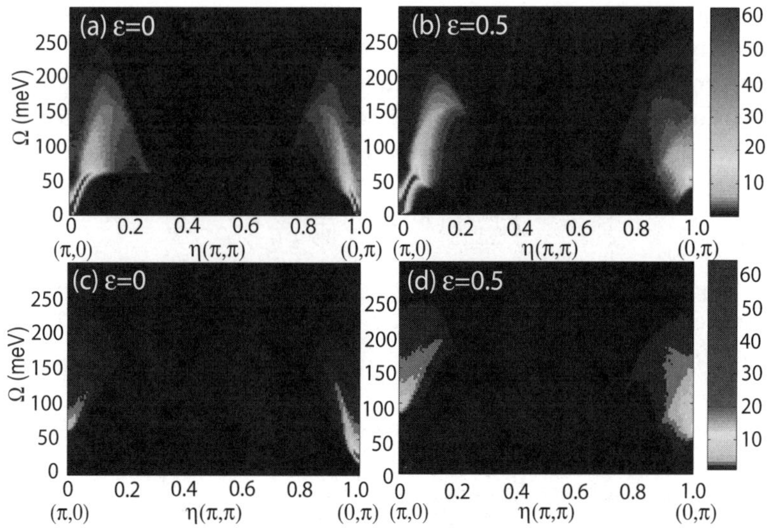

Figure 9.8 Imaginary part of the transvers (upper panel) and longitudinal (lower panel) susceptibility along $(0, \pi) \rightarrow (\pi, 0)$ points of the BZ for two different ellipticities of the elctron pockets parameterized by ϵ. Taken from [Knolle *et al.* (2010a)]. See also Color Insert.

a localized model this would be reflected by additional soft modes in the spin wave spectra. It is interesting to check whether signs of frustration in the magnetic excitation spectrum also show up in the itinerant description. In Fig. 9.8 we show the transverse (upper panel) and longitudinal (lower panel) components of the imaginary part of the spin susceptibility for zero and finite ellipticities. Observe that the magnetic order is fixed to be along $(\pi, 0)$ wave vector. Consider first the idealized case of circular electron pockets, $\epsilon = 0$ (left panels). In this case two out of three FS are completely gapped. In Fig. 9.8(a) we show the calculated imaginary part of the transverse component of the susceptibility, Im $[\hat{\chi}_{+-}(\mathrm{q}, \Omega)]_{RPA}$ along the direction $(0, \pi)$ to $(\pi, 0)$. Because the corresponding ground state is degenerate, the excitation spectrum has a Goldstone mode at the ordering momentum $\mathrm{Q}_1 = (\pi, 0)$ and another accidental gapless mode at $\mathrm{Q}_2 = (0, \pi)$. The excitations near the ordering momentum are propagating up to $\Omega_c(\mathrm{Q}_1) = 2\Delta_1 \sim 60$ meV which is expected because both pockets separated by Q_1 are gapped.

Excitations near Q_2 are also propagating, but up to smaller energies, $\Omega_c(Q_2) = \Delta_1$, that is consistent with the fact that only one of the two pockets separated by Q_2 is gapped. At higher energies spin-waves enter the continuum and become overdamped Stoner-like excitations, however, still with well defined peaks. The difference between $\Omega_c(Q_1)$ and $\Omega_c(Q_2)$ is also seen in Fig. 9.8(c) where the longitudinal component of the susceptibility is plotted. As expected, $\text{Im}[\hat{\chi}_{+-}(q, \Omega)]_{RPA}$ vanishes below $\Omega_c(Q)$.

For finite $\epsilon = 0.5$ the results are presented in panels (b) and (d). There are two key effects introduced by ellipticity. First, the degeneracy is now lifted [Eremin and Chubukov (2010)] and the transverse excitations near Q_2 acquire a finite gap clearly visible in Fig. 9.8(b). Second, spin-wave excitations near Q_1 have a finite Landau damping down to $\Omega = 0$ because the SDW order no longer completely gaps the hole and electron FSs separated by Q_1. At the same time, spin-wave excitations are still clearly visible at low energies and do not become overdamped. The reason is that SDW coherence factors suppress the scattering between electron and hole like pockets involved in the formation of magnetic order and reduce the Landau damping around the ordering momentum Q_1 from $\Omega/|q - Q_1|$ to $\Omega|q - Q_1|$ what for linear dispersion is of the same order as Ω^2 and $(q - Q_1)^2$. Once Ω exceeds $\Omega_c(Q_2)$, the coherence factors no longer screen Landau damping, and spin-waves become Stoner-like, overdamped excitations. The scattering between between the pockets not involeved in magnetism is not suppressed by the SDW coherence factors because Q_2 is not the ordering momentum, therefore, Landau damping makes the transverse excitations near Q_2 overdamped immediately above the gap. Note, that there is also a tendency towards incommensuration near Q_2. Longitudinal excitations are still gapped up to $\Omega_c(Q)$ and become Stoner-like at higher energies.

Overall, it is interesting to observe that in our itinerant description the groundstate degeneracy for the fully nested situation which arises due to frustration between the two wave vectors Q_1 and Q_2 shows up in the magnetic excitation spectrum as an additional soft mode.

In the following we briefly review how signs of itinerant frustration can show up in the coexistence regime of AF and s^{+-}

superconductivity, as well. In particular, the spin excitation spectrum of the superconducting state with s^{+-}-symmetry is dominated by the so-called magnetic resonance mode which arises due to the condition $\Delta_{\mathbf{k+Q}} = -\Delta_{\mathbf{k}}$ for the SC gap where $\mathbf{Q}$ is either $\mathbf{Q}_1$ or $\mathbf{Q}_2$ [Korshunov and Eremin (2008); Maier and Scalapino (2008)]. It is an open question how the presence of superconductivity affects the magnetic excitation spectrum of the AF phase. Note that the existence of a gapless Goldstone mode (assuming continuous spin symmetry and singlet SC) is guaranteed based on general symmetry principles. It seems plausible to speculate that the resonance mode is simply absorbed by the Goldstone mode. Nevertheless, the enhanced Z_2-symmetry with two potential ordering vectors does allow for a strong resonance mode around $\mathbf{Q}_2 = (0, \pi)$ in the coexistence SC+AF phase if the magnetic order corresponds to $\mathbf{Q}_1 = (\pi, 0)$ and vice versa.

To include superconductivity and AF at the mean-field level, one decouples the interaction part of the Hamiltonian, Eq. (1.7), with respect to the AF and superconducting (SC) order. Without any loss of generality the mean-field matrix can be diagonalized by a sequence of unitary transformations provided that the symmetry of the SC is s^{+-}-type and the gaps on the electron and hole pockets are of similar magnitudes [Vorontsov *et al.* (2010); Fernandes and Schmalian (2010)]. Therefore, one first applies a unitary transformations with respect to the AF state and introduces new quasiparticle operators. In the AF state with ordering at $\mathbf{Q}_1 = (\pi, 0)$ [which should refer in three dimensions to the actual $(\pi, 0, \pi)$ ordering wave vector] the system remains a metal as the pocket at $(0, \pi)$ remains intact. In addition, because of the finite doping and ellipticity of the electron pockets, tiny pockets of electron and hole character remain around the Γ point of the BZ at intermediate values of the AF order as shown in Fig. 9.9. As we neglected the underlying orbital structure of the electronic wave functions, these small FS pockets are generally gapped for large sizes of the SDW gap. In reality, they are more stable against an increase of the SDW gap due to the formation of Dirac nodes driven by the symmetry between xz and yz orbitals [Ran *et al.* (2009)]. However, as we will show later, their presence does not influence spin excitations close to the high

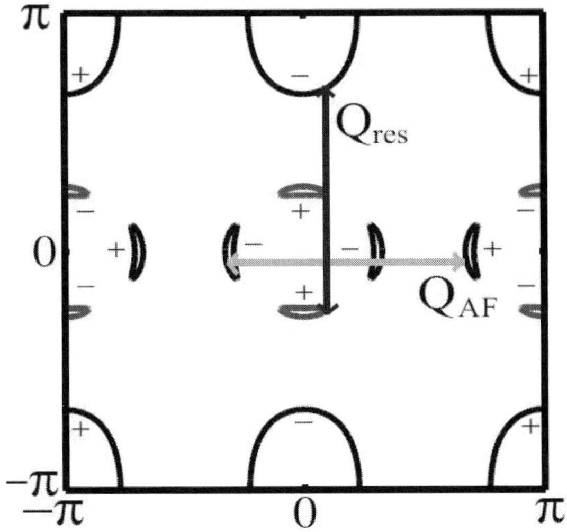

Figure 9.9 Calculated Fermi surface topology in the AF state with $Q_1 = (\pi, 0)$ ordering momentum. The + and − signs refer to the phase of the superconducting s^{+-}-wave order parameter in the coexistence phase. The arrows refer to the AF ordering wave vector and the wavevector of the incommensurate resonant excitations below T_c. Taken from [Knolle *et al.* (2011b)].

symmetry points of the BZ. Thus, the physics induced by Dirac nodes can be safely neglected for our purposes.

In the next step one rewrites the SC pairing interaction in the s^{+-} channel with the unitary AF transformation, and subsequently performs a mean-field (MF) decoupling in the particle-particle channel [Knolle *et al.* (2011b)]. The resulting MF Hamiltonian is then diagonalized by two independent Bogolyubov transformations. The new mean-field equations for the superconducting and SDW gaps have to be solved self-consistently.

Extending the spin susceptibility calculations to the coexistence SC+AF phase is straightforward. The only difference is that anomalous Green's functions, which arise due to superconductivity, have to be taken into account [Knolle *et al.* (2011b)]. Fig. 9.10(a) shows the evolution of the transverse spin excitations around the AF ordering momentum $\mathbf{Q}_1$. Due to the finite magnetization, well-defined anisotropic spin waves exist around the ordering

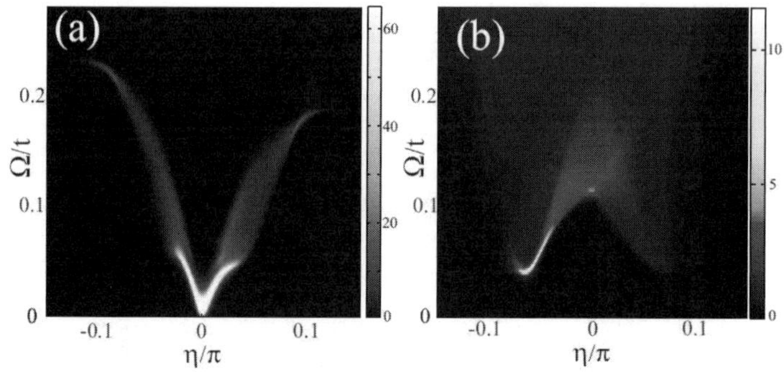

Figure 9.10 The (q, Ω) intensity map (in states/t) of the imaginary part of the transverse spin susceptibility around the AFM wave vector $Q_1 = (\pi, 0)$ along $(\pi, 0.15\pi) \rightarrow (\pi, 0) \rightarrow (1.15\pi, 0)$ direction of the BZ (a) and around the $Q_2 = (0, \pi)$ along $(0, 1.15\pi) \rightarrow (0, \pi) \rightarrow (0.15\pi, \pi)$ direction in the coexistence SC+AF phase. Taken from [Knolle *et al.* (2011b)].

momentum similar to the pure AF state. This is expected as the Goldstone mode in the AF phase reflects the breaking of the spin-rotational symmetry of the system and as such is not affected by the presence of the additional spin-singlet superconducting state. At the same time, and in contrast to the pure AF state, spin waves are damped away from Q_1 at energies well below twice the AF gap $2W$. In particular, additional damping starts at energies of twice the superconducting gap. This new feature of the coexistence phase arises due to the renormalization of the particle-hole (p-h) continuum. In the AF phase, the p-h response is determined by transitions between α- and β-bands as well as intraband transitions. The latter are suppressed by vanishing AF matrix elements around the ordering wave vector. This also guarantees the stability of the Goldstone mode even in the situation when AF does not gap the entire FS and remnant electron and hole pockets are still present. Damping of spin waves arises from interband transitions. They are gapped exactly by $2W$ at $\mathbf{Q}_1$; however, this gap becomes progressively smaller away from it. In the coexistence phase, the interband excitations at $\mathbf{Q}_1$ are gapped by the sum of twice SC and AF gaps. Away from $\mathbf{Q}_1$ the effect of AF quickly vanishes. However, the p-h continuum is still gapped by approximately twice the SC gap, $2\Delta_0$. Above this energy it experiences a discontinuous jump due to $\text{sgn}(\Delta_k^\alpha) = -\text{sgn}(\Delta_k^\beta)$. As a result, the spin wave excitations are damped at energies associated with twice the SC gap as seen from Fig. 9.10(a). The 'stronger' damping of the spin waves in the coexistence region originates from a smaller value of the AF gap in the coexistence phase. The discontinuous jump of the particle-hole continuum at approximately $2\Delta_0$ is absent in the pure AF state. Note that this feature may be used in inelastic neutron scattering to distinguish between the microscopic coexistence of AF and SC orders and the formation of macroscopic phase separated regions between them.

Fig. 9.10(b) shows the spin excitations in the coexistence phase around $\mathbf{Q}_2$. As discussed above, in the pure AF phase the excitations around $\mathbf{Q}_2$ are paramagnons centered at frequency Ω_2. In the coexistence phase, however, the SC gap modifies the behavior of the p-h continuum around $\mathbf{Q}_2$. Due to a sign change of the SC gaps between the pockets, the p-h continuum experiences a discontinuous jump at $2\Delta_0$ and the necessary condition for the formation of an excitonic resonance [Korshunov and Eremin (2008); Maier and Scalapino (2008)] at energies below $2\Delta_0$ forms. It is remarkable that this resonance exist at energies which are much lower than the energy position of the original mode at Ω_2 in the AF state. The incommensuration of the resonance in the coexistence phase is closely related to the reconstructed FS for which hole pocket states that would be connected by $\mathbf{Q}_2$ to an electron pocket are gapped for sufficiently large magnetic order. Note also that contrary to the ordering momentum, the excitations around $\mathbf{Q}_2$ remain isotropic in spin space, i.e. the excitations in the longitudinal and transverse channel show qualitatively similar behavior.

Overall the results in the coexistence region show the peculiarity of spin excitations in the ferropnictides: while around the ordering momentum $\mathbf{Q}_1$ the excitations are determined by the presence of the AF order, the excitations around $\mathbf{Q}_2$ are dominated by SC and the formation of the incommensurate spin resonance. The formation of the resonance effectively lowers the energy splitting between the different magnetic states in the coexistence region. Correspondingly the system tends to increase its magnetic degeneracy in the presence of s^{+-} superconductivity. The collective excitations associated with the presence of the incommensurate spin resonance close to $\mathbf{Q}_2$ may occur in the coexistence region, a phenomenon which would be absent in the pure AF state. This possibility would be interesting to check experimentally.

A direct probe to see the excitations around $\mathbf{Q}_1$ and $\mathbf{Q}_2$ separately in INS experiments is to look at untwinned crystals which have been grown recently [Blomberg *et al.* (2011); Chu *et al.* (2010)]. However, existing experiments on twinned crystals already indicate that both spin waves and spin resonance are present in the coexistence phase. Ref. [Pratt *et al.* (2010)] found that spin excitations in the coexistence AF+SC phase show strong anisotropy along the c-crystallographic direction, absent in the pure SC state. The difference in energy of the spin excitations at (π, π, q_z) is $3 - 4$ meV between $q_z = 0$ and $q_z = \pi$. At the same time, it was argued [Park *et al.* (2010)] that due to the peculiar crystallographic structure of 122 compounds, the excitations at (π, π, π) and $(\pi, \pi, 0)$ in the folded BZ zone correspond to excitations at $(\pi, 0, \pi)$ and $(0, \pi, 0)$ wave vectors in the unfolded BZ, respectively. In other words, while excitations around (π, π, π) refer to the true spin waves at $\mathbf{Q}_1$ (gapped in the experimental system by spin-orbit coupling we have omitted) the excitations around $(\pi, \pi, 0)$ originate from the incommensurate resonance at $\mathbf{Q}_2$. The crucial test to see whether this is true is to study the spin rotational symmetry of these excitations and their possible incommensurability.

9.4.3 *Two Orbital Model: Orbital versus Excitonic Scenario*

So far we have not discussed the role of orbitals in the formation of the AF order. At the same time, there is no doubt that the orbitals do play an important role. For example they introduce a significant angular dependence for the intraband and interband interactions resulting in a change of important details in the superconducting and antiferromagnetic states. It was argued that the originally nodeless s^{+-}-wave superconducting state may acquire nodes on the electron pockets [Maiti *et al.* (2011); Thomale *et al.* (2011); Kemper *et al.* (2010)]. For the AF state it was predicted that iron-based superconductors show nodal elementary excitations with a linear Dirac spectrum at low energies as a result of orbital effects

[Ran *et al.* (2009); Yang and Kee (2010); Morinari *et al.* (2010)]. In addition, due to the fact that the antiferromagnetic order is either $(\pi, 0)$ or $(0, \pi)$, the magnetic anisotropy induces orbital anisotropies which in turn affect the electronic properties along a and b directions [Sugimoto *et al.* (2011); Daghofer *et al.* (2010); Valenzuela *et al.* (2010)]. They may also introduce novel features such as ferro-orbital or antiferro-orbital order [Kubo *et al.* (2008); Lv *et al.* (2010)] associated with intra and inter-orbital electron-hole pairing. Therefore, band and orbital lattice descriptions are not necessarily equivalent even in the limit of interactions being much smaller than the bandwidth. Furthermore, nesting features do not play an essential role in the formation of AF order starting from the orbital description.

In order to get a basic feeling for the formalism and especially to see the effect of the orbital degrees of freedom we concentrate on the simplest multioribital model that was introduce in the context of FPs [Raghu *et al.* (2008)]. It takes into account only the d_{xz} and d_{yz} orbitals which, however, have the largest weight around the Fermi energy.

The kinetic part of the Hamiltonian is written in the two component spinor basis in orbital space with the help of the Pauli matrices τ_i

$$H_0 = \sum_{\mathbf{k},\sigma} \psi_\sigma^\dagger \left[(\epsilon_+ - \mu)\,\tau_0 + \epsilon_- \tau_3 + \epsilon_{xy} \tau_1 \right] \psi_\sigma \tag{9.35}$$

where

$$\psi_\sigma(\mathbf{k}) = \begin{pmatrix} d_{xz\sigma}(\mathbf{k}) \\ d_{yz\sigma}(\mathbf{k}) \end{pmatrix}. \tag{9.36}$$

Here, $\varepsilon_\pm(k) = \frac{\varepsilon_x(k) \pm \varepsilon_y(k)}{2}$, $\varepsilon_x(k) = -2t_1 \cos k_x - 2t_2 \cos k_y - 4t_3 \cos k_x \cos k_y$, $\varepsilon_y(k) = -2t_2 \cos k_x - 2t_1 \cos k_y - 4t_3 \cos k_x \cos k_y$, and $\varepsilon_{xy}(k) = -4t_4 \sin k_x \sin k_y$.

In general, the interacting part of the Hamiltonian consists of different on-site electron-electron interactions:

$$H_{int}^{orb} = U \sum_i \sum_\nu n_{i\nu\uparrow} n_{i\nu\downarrow} + V \sum_{\nu \neq \mu, \sigma, \sigma'} n_{i\nu\sigma} n_{i\mu\sigma'} \tag{9.37}$$

$$-J \sum_{\nu \neq \mu} \mathbf{S}_{i\nu} \cdot \mathbf{S}_{i\mu} + J' \sum_{\nu \neq \mu} d_{i\nu\uparrow}^\dagger d_{i\nu\downarrow}^\dagger d_{i\mu\downarrow} d_{i\mu\uparrow}, \tag{9.38}$$

where U and V refer to the intra- and inter-orbital Coulomb repulsion, J and $J' = J/2$ denote the Hund and the pair hopping terms. The susceptibility matrix, with labels 1 for xz and 2 for yz orbitals, reads

$$\hat{\chi} = \begin{bmatrix} \chi_{1111} & \chi_{2111} & \chi_{1112} & \chi_{2112} \\ \chi_{1211} & \chi_{2211} & \chi_{1212} & \chi_{2212} \\ \chi_{1121} & \chi_{2121} & \chi_{1122} & \chi_{2122} \\ \chi_{1221} & \chi_{2221} & \chi_{1222} & \chi_{2222} \end{bmatrix} \tag{9.39}$$

and the interaction matrix for the transverse component of the spin susceptibility is given by:

$$\hat{U} = \begin{bmatrix} U & 0 & 0 & J \\ 0 & J & U-2J & 0 \\ 0 & U-2J & J & 0 \\ J & 0 & 0 & U \end{bmatrix}. \tag{9.40}$$

Following the procedure described above it is straightforward to compute the physical susceptibility in the paramagnetic state. The response is governed by antiferromagnetic fluctuations centered at $(\pi, 0)$ and $(0, \pi)$ resulting from scattering that connects the hole pocket around $(0, 0)$ and the two electron pockets at the BZ boundaries.

In the next step we introduce self-consistent mean-field averages associated with the AF state. We consider two scenarios. In the first case, which we denote as *orbital*, one introduces the magnetic state in the orbital basis using the local interactions. The mean-field phase diagram of the two orbital model was studied by several groups [Ran *et al.* (2009); Kubo *et al.* (2008); Moreo *et al.* (2009)] and the main conclusion is that AF ordering involves only very weakly interorbital components. Thus we assume in the following that the mean-field AF order involves only states within the same orbital, separated by $\mathbf{Q}_{AF}$ ($\Delta_{\nu\nu'} = \delta_{\nu,\nu'}\Delta_\nu$), and decouple the on-site local interaction, Eq. (9.37), with respect to it. In addition, we set $\mathbf{Q}_{AF} = (0, \pi)$ and do not consider the problem of its selection out of the manifold of degenerate magnetic states as it was discussed in the first section. We obtain two coupled gap equations

$$M_x = -\left[U\,\Delta_{xz} + J\,\Delta_{yz}\right] \tag{9.41}$$

$$M_y = -\left[U\,\Delta_{yz} + J\,\Delta_{xz}\right] \tag{9.42}$$

and the Hamiltonian matrix becomes

$$\hat{h}^{Orbital} = \begin{pmatrix} \varepsilon_x & \varepsilon_{xy} & \sigma M_x & 0 \\ \varepsilon_{xy} & \varepsilon_y & 0 & \sigma M_y \\ \sigma M_x & 0 & \tilde{\varepsilon}_x & \tilde{\varepsilon}_{xy} \\ 0 & \sigma M_y & \tilde{\varepsilon}_{xy} & \tilde{\varepsilon}_y \end{pmatrix} \tag{9.43}$$

where the tilde again means that the momentum is shifted by $\mathbf{Q}_{AF}$ The matrix can be inverted and with the new eigenvectors one can solve the coupled gap equations and carry out the spin wave calculations as outlined above.

The second, *excitonic*, scenario is built on the belief that in the iron-based superconductors, the antiferromagnetic instability, at least partially, is connected to nesting of electron and hole bands [Chubukov *et al.* (2008); Mazin *et al.* (2008)]. This scenario we largely explored in the previous subsection. Within this scenario the tight-binding hamiltonian, Eq. (9.35), is first diagonalized to obtain the band structure. One further assumes, that the bands are the relevant degrees of freedom and ignores the role of the underlying orbital structure. The interactions are then also re-written in the band representation. Moreover, within the RG flow they are strongly renormalized, to such an extent that their values at the Fermi energy are not the same as the initial ones determined by the local on-site interactions [Chubukov *et al.* (2008)]. This allows us to approximate the band interactions as Fermi-liquid like density-density type interactions given in Eq. (9.7), which at the lowest order are angular independent. By decoupling the band interaction with respect to Δ_{12} (we assume now only one wave vector for the order parameter) one finds a single gap equation

$$M = -\frac{u_1 + u_3}{2}\Delta_{12} \tag{9.44}$$

and the Hamiltonian matrix becomes

$$\hat{h}^{Band} = \begin{pmatrix} E_1 & 0 & 0 & \sigma M \\ 0 & E_2 & \sigma M & 0 \\ 0 & \sigma M & \tilde{E}_1 & 0 \\ \sigma M & 0 & 0 & \tilde{E}_2 \end{pmatrix}. \tag{9.45}$$

When comparing the orbital and the excitonic scenario one realizes that for small magnetic moments and low energies the spin

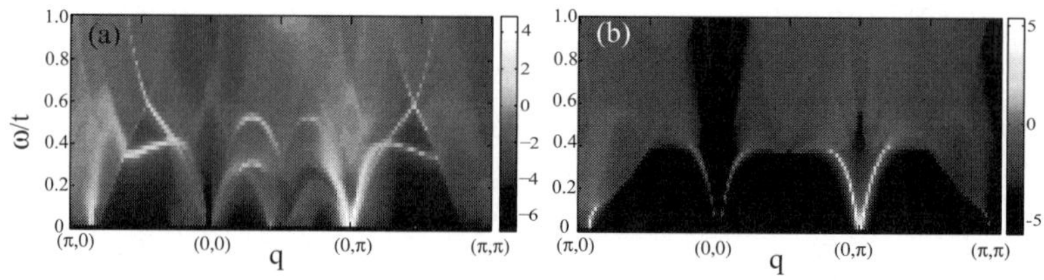

Figure 9.11 Imaginary part of the transverse spin susceptibility in the orbital (a) and excitonic (b) scenarios for a magnetic moment of $m = 0.38\mu_B$. Taken from [Knolle *et al.* (2011)].

excitations look very similar [Knolle *et al.* (2011)]. The Fermi surface has tiny pockets in both cases, however, in the excitonic scenario they are due to imperfect nesting and disappear quickly for larger moments, whereas, in the orbital picture the pockets are coming from Dirac cones away from the Fermi energy and are thus protected even at larger interactions [Ran *et al.* (2009)]. In Fig. 9.11 we show the transverse susceptibility for a magnetic moment of $m = 0.38\mu_B$ in both scenarios along the route $(\pi, 0) \to (0, 0) \to (0, \pi) \to (\pi, \pi)$ of the first BZ. The peculiar feature of the spin excitations in the AF state in both scenarios are well-defined spin waves around the ordering wave vector. They disperse until they hit the particle-hole continuum and become overdamped.

As we pointed out in the previous subsection there is an $O(6)$ degeneracy of the magnetic ground state [even in the orbital model] which manifests itself in additional zero energy modes at the non-ordered momenta of $(\pi, 0)$ and (π, π) for larger magnetic moments (not shown, see [Knolle *et al.* (2011)]). These modes, however, are again accidental in the sense that they are not stable with regard to perturbations such as the finite Landau damping introduced by remnant Fermi surfaces. This is why they are not seen for small magnetic moments. However, as soon as the FSs are completely gapped they become visible. In the orbital scenario, however, the FSs will be gapped completely only for very large interactions for symmetry reasons and the particle-hole continuum remains gapless. Thus, the spin waves remain well-defined only around the ordering momentum. Away from it they are overdamped because the Dirac cones in the electronic excitation spectra lead to remnant FS pockets.

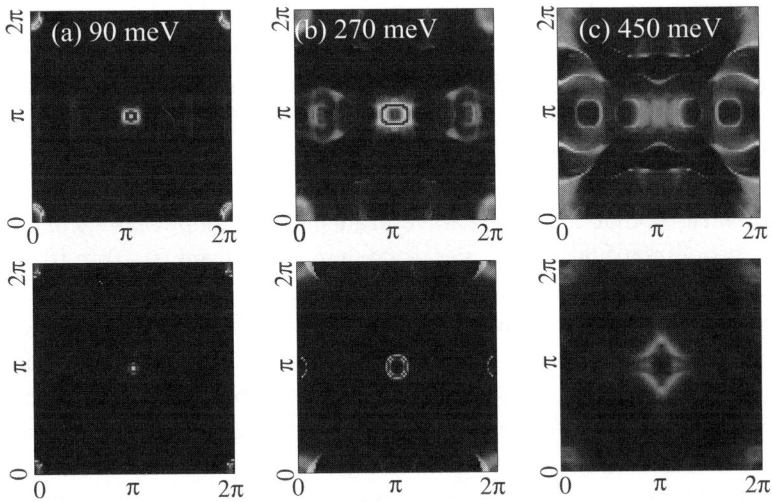

Figure 9.12 Constant energy cuts of the imaginary part of the total transverse susceptibilities in the orbital (upper) and excitonic (lower) scenarios for a magnetic moment of $\mu = 0.38\mu_B$ and three different energies. See also Color Insert.

We show in Fig. 9.12 constant energy cuts for both scenarios on an absolute scale. In both cases the elliptical shaped ring around the ordering momentum (center of the plots) disperses at low energies until it enters the particle-hole continuum. Due to the orbital structure of the wave functions the response in the orbital scenario has a lot more structure which is supported by recent inelastic neutron scattering experiments [Ewings *et al.* (2011)].

A peculiarity of the spin waves within the orbital scenario is the relatively rich structure of the spin excitations away from the ordering momentum. Here the complication is that each orbital has its own magnetic gap which correspondingly affects the magnetic excitations within each orbital. It is possible to analyze separately the intra and inter-orbital components of the transverse spin susceptibility. Within the orbital scenario, the Goldstone modes originate from the intra-orbital xz and yz susceptibilities, while in the excitonic scenario the entire low energy excitations are in the inter-band susceptibility by construction. The AF magnetic order breaks the symmetry between xz and yz orbitals which leads to different behavior of the spin excitations away from the ordering momentum and at higher energies in the corresponding intra-orbital susceptibilities. Given the fact that the spin waves show some orbital dependencies we believe that this feature is measurable in RIXS experiments [Knolle *et al.* (2011); Kaneshita *et al.* (2011)].

Another interesting feature of the spin excitations in the orbital scenario is that their spin wave branches can reappear out of the continuum at higher energies, e.g. when going from $(0, 0)$ to $(0, \pi)$ as shown in Fig. 9.11. Here, orbital coherence factors yielding the larger continuum (damping) are small at these higher energies, thus allowing the spin waves to reappear. As soon as the itinerant magnetism appears in MOS with multiple order parameters this reappearance of spin waves at higher energies away from the ordering momentum should be a general feature not restricted to iron-based superconductors. Spin excitations within the orbital scenario show a rich structure at energies which exceed well the onset of the particle-hole continuum. Note that at these energies the spin excitations should not any longer be interpreted as spin waves originating from breaking of the spin-rotational symmetry in the antiferromagnetic state but rather should be attributed to a Stoner continuum. These excitations are almost unchanged across the Neel temperature and as such can be well separated in the experimental data.

A natural next step would be to carry out spin wave calculations with more accurate multiorbital models. [Kaneshita and Tohyama (2010)] have done this for a more realistic five orbital model and found good agreement with experiments [Ewings *et al.* (2011)]. However, there are various models used in the literature with slightly varying parameters. Depending on the choice of model and dispersion input the results may actually differ [Brydon *et al.* (2011)].

9.4.4 *Comparison to Experiments*

After having introduced the general theory how to calculate spin waves in itinerant magnets with multiple orbitals (bands) contributing at the Fermi energy we presented results for a two orbital (band) model. In the following, we want to compare the itinerant description of magnetism in FPs with actual experiments. As mentioned above the two orbital model can give qualitative insights especially to elucidate the role of orbital degrees of freedom. However, in addition to the wrong electronic struture in the unfolded zone it is not possible to fit the dispersion parameters of the model quantitatively to experiments such as ARPES [Raghu *et al.* (2008)]. Therefore, we will show the results for a phenomenological three band model which can be fitted to experiments [Knolle *et al.* (2010a)].

Neutron scattering measurements on parent $FeAs$ compounds revealed that magnetic order consists of ferromagnetic chains along one crystallographic direction in an Fe plane and antiferromagnetic chains along the other [Cruz *et al.* (2008); Klauss *et al.* (2008)], i.e., the system selects $\mathbf{Q}_{AF}$ to be either $\mathbf{Q}_1 = (\pi, 0)$ *or* $\mathbf{Q}_2 = (0, \pi)$. Sharp propagating spin-wave like excitations have been observed near the ordering momentum (e.g., $\mathbf{Q}_1$) up to energies of around 100 meV, with different velocities along the two crystallographic directions [Zhao *et al.* (2009); Diallo *et al.* (2009)]. At larger energies, excitations become overdamped [Diallo *et al.* (2009)].

Magnetism with $(\pi, 0)$ stripe order arises also in the $J_1 - J_2$ model of localized spins for $J_2 < 0.5J_1$ [Chandra *et al.* (1990)], and it was argued that such a model can successfully describe some of the experimental data on the spin wave spectra, although, an unusually large in-plane anisotropy of the antiferromagnetic exchange between nearest neighbor spins has to be assumed [Zhao *et al.* (2009)] in order to account for the anisotropic spin wave velocities. We have seen in the first section of this review that it is possible to understand the order from an itinerant perspective, which takes into account the peculiar electronic structure of the FPs. The ellipticity of the electron pockets and interactions between them played a key role. The issue we address now is whether inelastic neutron scattering (INS) measurements of magnetic excitations can be explained within the itinerant approach, as well. This is a crucial test of the itinerant description of magnetism in FPs.

In Fig. 9.13 we compare the calculations performed in [Knolle *et al.* (2010a)] with inelastic neutron scattering experiments. Measured excitations near $\mathbf{Q}_1$ are propagating spin-waves up to about 100 meV, with a finite but small damping. In the left panel

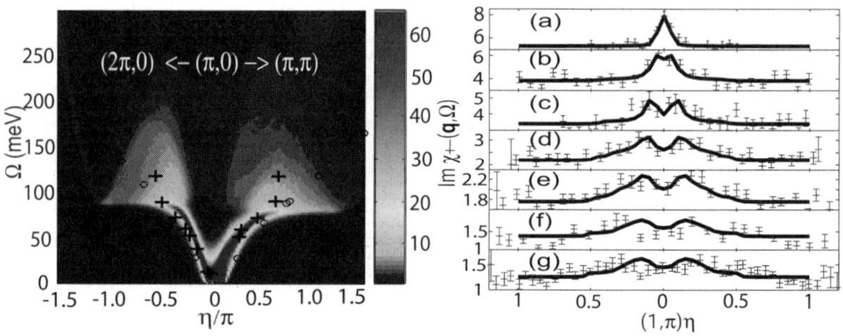

Figure 9.13 Comparison of the calculated spin excitations to experimental data of [Diallo *et al.* (2009); Zhao *et al.* (2009)]. Taken from [Knolle *et al.* (2010a)].

of Fig. 9.13 we compare the dispersion with the experimental data [Zhao *et al.* (2009); Diallo *et al.* (2009)]. We emphasize that both the measured and the calculated spectra have different velocities along different crystallographic directions. In the itinerant theory, it is a consequence of the non-zero ellipticity *and* the fact that only one electron pocket is involved in the SDW formation. Remarkably, for $\epsilon = 0.5$ ellipticity which agrees with the ARPES data the anisotropy of the spin wave velocities is the same as in the experimental INS data. Another verifiable theory prediction for this range is that the width of the spin-wave peak should scale as Ω^2. Additionally, the excitations are still visible even when the spin-waves enter the continuum. In the right panel of Fig. 9.13 we compare the calculated and the measured $Im\chi^{+-}(q,\Omega)$ along $(0,0) \to (\pi,0) \to (2\pi,0)$ direction for different Ω. We see that the agreement is quite good even for higher frequencies.

9.5 Discussion and Conclusion

In this review we presented the itinerant description of antiferromagnetism in parent materials of the iron-based superconductors. In contrast to a purely localized scenario this theory allows for a coherent understanding of the full phase diagram of these materials as a function of doping, disorder and pressure. All the magnetic properties such as magnitude of the magnetic moment, selection of the order from the degenerate manifold of the possible ground states, appearance of the Ising nematic order above T_N and finally the anisotropic spin waves can be well understood and connected to the basic electronic structure of these materials. A particular advantage is the possibility to compute charge and spin response functions in a consistent fashion within a single formalism and for the same set of parameters.

In principle one can draw a phenomenological connection between the itinerant model and the $J_1 - J_2$ model. First, the relationship between the real-space order parameters M_1 and M_2 and the momentum-space order parameters Δ_1 and Δ_2 is $\Delta_1 = \mathrm{M}_1 + \mathrm{M}_2$ and $\Delta_2 = \mathrm{M}_1 - \mathrm{M}_2$. The scalar product $\mathrm{M}_1 \cdot \mathrm{M}_2$ is then the same as $\Delta_1^2 - \Delta_2^2$ and the nematic coupling is given by $g = \zeta J_1^2/J_2$. Second, in the itinerant approach, the hard constraint $\mathrm{M}_1^2 = \mathrm{M}_2^2 = 1$ is replaced by the quartic terms, which play the role of soft constraints. We see that the itinerant and the $J_1 - J_2$ models are indeed quite similar, and in both models the Ising-nematic order results from the Z_2 degeneracy of the stripe magnetic ground state. However, one has to be careful with a description of the spin waves in iron-based superconductors via $J_1 - J_2$ like models. The original derivation of the $J_1 - J_2$ model goes back to the single-band Hubbard model with hopping between nearest and next-nearest neighbors. Iron-based superconductors are certainly multi-orbital materials and their localized strong-coupling limit is not exactly known at present. Therefore, spin wave calculations based on the $J_1 - J_2$ model even taking into account biquadratic couplings have limitations especially at higher energies [Ewings *et al.* (2011)].

Nevertheless, there are also limitations of the itinerant description at present. This concerns the iron chalcogenides $\mathrm{FeTe_{1-x}Se_x}$. The itinerant scenario is applicable in the regime of intermediary Se doping, near the superconducting dome of the (x, T) phase diagram. In this region, the electronic structure is similar to the one considered in the itinerant model and neutron scattering [Liu *et al.* (2010b)] shows that magnetic fluctuations are peaked at the ordering vectors $\mathrm{Q}_1 = (\pi, 0)$ and $\mathrm{Q}_2 = (0, \pi)$. On the other hand, the itinerant model fails for undoped samples where nesting might be absent and the magnetic order has $(\pi/2, \pi/2)$ ordering wave vector. The same problem concerns $A_x\mathrm{Fe}_{2-x/2}\mathrm{Se}_2$ [A =Cs, K, (Tl,Rb),

(Tl,K)] iron-selenide compounds where the electronic structure is still discussed.

Further open issues on the theoretical side which have to be studied are the possible role of spin-orbit coupling, interaction with magnetic and non-magnetic impurities, and the doping dependence. This promises interesting perspectives for future research in the field of iron-based superconductors.

We acknowledge helpful discussions and collaborations with A. Akbari, Ph. Brydon, A. V. Chubukov, R. Fernandes, P. Hirschfeld, S. Maiti, R. Moessner, J. Schmalian, P. Thalmeier, R. Thomale, M. Vavilov, and W. Wang. IE acknowledges financial support of the SFB Transregio 12, and German Academic Exchange Service (DAAD PPP USA No. 50750339). JK acknowledges support from the Studienstiftung des Deutschen Volkes; the IMPRS Dynamical Processes in Atoms, Molecules and Solids and the GRK 1621 Itinerant magnetism and superconductivity in intermetallic compounds. IE is thankful to Kazan Federal University (Grant RNP-31) for partial support.

References

O. K. Andersen, and L. Boeri, Annalen der Physik **1**, 8 (2011).

E. C. Blomberg, M. A. Tanatar, A. Kreyssig, N. Ni, A. Thaler, Rongwei Hu, S. L. Bud'ko, P. C. Canfield, A. I. Goldman, and R. Prozorov, Phys. Rev. B **83**, 134505 (2011).

L. Boeri, O. V. Dolgov, and A. A. Golubov, Phys. Rev. Lett. **101**, 026403 (2008).

P. M. R. Brydon and C. Timm, Phys. Rev. B **80**, 174401 (2009); **79**, 180504(R) (2009).

P. M. R. Brydon, Maria Daghofer, Carsten Timm, J. Phys.: Condens. Matter **23**, 246001 (2011).

A. Cano, M. Civelli, I. Eremin, and I. Paul, Phys. Rev. B **82**, 020408(R) (2010).

A Carrington, Rep. Prog. Phys. **74**, 124507 (2011).

P. Chandra, P. Coleman and A. I. Larkin, Phys. Rev. Lett. **64**, 88 (1990).

J.-H. Chu, J. G. Analytis, K. De Greve, P. L. McMahon, Z. Islam, Y. Yamamoto, and I. R. Fisher, Science **329**, 824 (2010).

T.-M. Chuang *et al.*, Science **327**, 181 (2010).

A. V. Chubukov and D. M. Frenkel, Phys. Rev. B **46**, 11884 (1992).

A. V. Chubukov, D. V. Efremov, and I. Eremin, Phys. Rev. B **78**, 134512 (2008).

R. Coldea, S. M. Hayden, G. Aeppli, T. G. Perring, C. D. Frost, T. E. Mason, S.-W. Cheong, and Z. Fisk, Phys. Rev. Lett. **86**, 5377 (2001).

A. I. Coldea, J. D. Fletcher, A. Carrington, J. G. Analytis, A. F. Bangura, J.-H. Chu, A. S. Erickson, I. R. Fisher, N. E. Hussey, and R. D. McDonald, Phys. Rev. Lett. **101**, 216402 (2008); H. Shishido, A. F. Bangura, A. I. Coldea, S. Tonegawa, K. Hashimoto, S. Kasahara, P. M. C. Rourke, H. Ikeda, T. Terashima, R. Settai, Y. Onuki, D. Vignolles, C. Proust, B. Vignolle, A. McCollam, Y. Matsuda, T. Shibauchi, A. Carrington, Phys. Rev. Lett. **104**, 057008 (2010)

Clarina de la Cruz, Q. Huang, J. W. Lynn, J. Li, W. Ratcliff II, J. L. Zarestky, H. A. Mook, G. F. Chen, J. L. Luo, N. L. Wang, and P. Dai, Nature **453**, 899 (2008).

V. Cvetkovic and Z. Tesanovic, Europhys. Lett. **85**, 37002 (2009).

M. Daghofer, Q. Luo, R. Yu, D. Yao, A. Moreo, and E. Dagotto, Phys. Rev. B **81**, 180514(R) (2010).

S. O. Diallo, V. P. Antropov, T. G. Perring, C. Broholm, J. J. Pulikkotil, N. Ni, S. L. Budko, P. C. Canfield, A. Kreyssig, A. I. Goldman, and R. J. McQueeney, Phys. Rev. Lett. **102**, 187206 (2009).

H. Ding, K. Nakayama, P. Richard, S. Souma, T. Sato, T. Takahashi, M. Neupane, Y.-M. Xu, Z.-H. Pan, A. V. Federov, Z. Wang, X. Dai, Z. Fang, G. F. Chen, J. L. Luo, N. L. Wang, J. Phys.: Condens. Matter **23**, 135701 (2011); P. Richard, T. Sato, K. Nakayama, T. Takahashi, H. Ding, Rep. Prog. Phys. **74**, 124512 (2011).

A. Dusza, A. Lucarelli, F. Pfuner, J.-H. Chu, I. R. Fisher, and L. Degiorgi, EPL 93, 37002 (2011).

I. Eremin and A. V. Chubukov, Phys. Rev. B **81**, 024511 (2010).

R. A. Ewings, T. G. Perring, J. Gillett, S. D. Das, S. E. Sebastian, A. E. Taylor, T. Guidi, and A. T. Boothroyd, Phys. Rev. B **83**, 214519 (2011).

C. Fang, H. Yao, W.-F. Tsai, J. P. Hu, and S. A. Kivelson, Phys. Rev. B **77**, 224509 (2008).

R. M. Fernandes, L. H. VanBebber, S. Bhattacharya, P. Chandra, V. Keppens, D. Mandrus, M. A. McGuire, B. C. Sales, A. S. Sefat, and J. Schmalian, Phys. Rev. Lett. **105**, 157003 (2010).

R. M. Fernandes, and J. Schmalian, Phys. Rev. B **82**, 014521 (2010).

R. M. Fernandes, A. V. Chubukov, J. Knolle, I. Eremin, J. Schmalian, Phys. Rev. B **85**, 024534 (2012).

I. R. Fisher, L. Degiorgi, and Z. X. Shen, Rep. Prog. Phys. **74**, 124506 (2011).

David M. Frenkel and W. Hanke, Phys. Rev. B **42**, 6711 (1990).

P. J. Hirschfeld, M. M. Korshunov, I. I. Mazin, Rep. Prog. Phys. **74**, 124508 (2011).

Y. Kamihara, T. Watanabe, M. Hirano, and H. Hosono, J. Am. Chem. Soc. **130**, 3296 (2008).

E. Kaneshita and T. Tohyama, Phys. Rev. B **82**, 094441 (2010).

E. Kaneshita, K. Tsutsui, T. Tohyama, Phys. Rev. B **84**, 020511(R) (2011).

L. V. Keldysh and Yu. V. Kopaev, Sov. Phys. Solid State **6**, 2219 (1965).

A. F. Kemper, T. A. Maier, S. Graser, H.-P. Cheng, P. J. Hirschfeld, and D. J. Scalapino, New J. Phys. **12**, 073030 (2010); T. A. Maier, S. Graser, D. J. Scalapino, and P. J. Hirschfeld, Phys. Rev. B **79**, 224510 (2009).

H.-H. Klauss, H. Luetkens, R. Klingeler, C. Hess, F. J. Litterst, M. Kraken, M. M. Korshunov, I. Eremin, S.-L. Drechsler, R. Khasanov, A. Amato, J. Hamann-Borreo, N. Leps, A. Kondrat, G. Behr, J. Werner, B. Büchner, Phys. Rev. Lett. **101**, 077005 (2008).

J. Knolle, I. Eremin, A. V. Chubukov, and R. Moessner, Phys. Rev. B **81**, 140506 (2010).

J. Knolle, I. Eremin, A. Akbari, and R. Moessner, Phys. Rev. Lett. **104**, 257001 (2010).

J. Knolle, I. Eremin, and R. Moessner, Phys. Rev. B **83**, 224503 (2011).

J. Knolle, I. Eremin, J. Schmalian, and R. Moessner, Phys. Rev. B **84**, 180510(R) (2011).

M. M. Korshunov and I. Eremin, Phys. Rev. B **78**, 140509(R) (2008).

F. Kruger, S. Kumar, J. Zaanen, J. van den Brink, Phys. Rev. B **79**, 054504 (2009).

Katsunori Kubo, Peter Thalmeier, J. Phys. Soc. Jpn. **78**, 083704 (2009).

C. Liu, G. D. Samolyuk, Y. Lee, N. Ni, T. Kondo, A. F. Santander-Syro, S. L. Bud'ko, J. L. McChesney, E. Rotenberg, T. Valla, A. V. Fedorov, P. C. Canfield, B. N. Harmon, A. Kaminski, Phys. Rev. Lett. **101**, 177005 (2008).

D. H. Lu, M. Yi, S.-K. Mo, A. S. Erickson, J. Analytis, J.-H. Chu, D. J. Singh, Z. Hussain, T. H. Geballe, I. R. Fisher, Z.-X. Shen, Nature **455**, 81 (2008).

C. Liu, T. Kondo, R. M. Fernandes, A. D. Palczewski, E. D. Mun, N. Ni, A. N. Thaler, A. Bostwick, E. Rotenberg, J. Schmalian, S. L. Bud'ko, P. C. Canfield and A. Kaminski, Nat. Phys. **6**, 419 (2010).

T. J. Liu, J. Hu, B. Qian, D. Fobes, Z. Q. Mao, W. Bao, M. Reehuis, S. A. J. Kimber, K. Proke, S. Matas, D. N. Argyriou, A. Hiess, A. Rotaru, H. Pham, L. Spinu, Y. Qiu, V. Thampy, A. T. Savici, J. A. Rodriguez, and C. Broholm, Nat. Mater. **9**, 718 (2010)

W. Lv, F. Krüger, and Ph. Phillips, Phys. Rev. B **82**, 045125 (2010).

T. A. Maier and D. J. Scalapino, Phys. Rev. B **78**, 020514(R) (2008).

Saurabh Maiti and Andrey V. Chubukov, Phys. Rev. B **82**, 214515 (2010).

S. Maiti, M. M. Korshunov, T. A. Maier, P. J. Hirschfeld, and A. V. Chubukov Phys. Rev. B **84**, 224505 (2011).

I. I. Mazin, D. J. Singh, M. D. Johannes, and M. H. Du, Phys. Rev. Lett. **101**, 057003 (2008).

A. Moreo, M. Daghofer, J. A. Riera, and E. Dagotto, Phys. Rev. B **79**, 134502 (2009).

T. Morinari, E. Kaneshita, and T. Tohyama, Phys. Rev. Lett. **105**, 037203 (2010).

M. Nakajima, T. Liang, S. Ishida, Y. Tomioka, K. Kihou, C. H. Lee, A. Iyo, H. Eisaki, T. Kakeshita, T. Ito, and S. Uchida, PNAS **108**, 12238 (2011).

J. T. Park, D. S. Inosov, A. Yaresko, S. Graser, D. L. Sun, Ph. Bourges, Y. Sidis, Y. Li, J.-H. Kim, D. Haug, A. Ivanov, K. Hradil, A. Schneidewind, P. Link, E. Faulhaber, I. Glavatskyy, C. T. Lin, B. Keimer, and V. Hinkov, Phys. Rev. B **82**, 134503 (2010).

C. Platt, C. Honerkamp, and W. Hanke, New J. Phys. **11**, 055058 (2009).

D. K. Pratt, A. Kreyssig, S. Nandi, N. Ni, A. Thaler, M. D. Lumsden, W. Tian, J. L. Zarestky, S. L. Budko, P. C. Canfield, A. I. Goldman, and R. J. McQueeney, Phys. Rev. B **81**, 140510(R) (2010).

Y. Qi and C. Xu, Phys. Rev. B **80**, 094402 (2009).

N. M. R. Peres and M. A. N. Arajo, Phys. Rev. B **65**, 132404 (2002).

S. Raghu, Xiao-Liang Qi, Chao-Xing Liu, D. J. Scalapino, and Shou-Cheng Zhang, Phys. Rev. B **77**, 220503 (2008).

Ying Ran, Fa Wang, Hui Zhai, Ashvin Vishwanath, Dung-Hai Lee, Phys. Rev. B **79**, 014505 (2009).

M. T. Rice, Phys. Rev. B **2**, 3619 (1970).

J. R. Schrieffer, X. G. Wen, and S. C. Zhang, Phys. Rev. B **39**, 11663 (1989).

Q. Si and E. Abrahams, Phys. Rev. Lett. **101**, 076401 (2008).

D. J. Singh and M.-H. Du, Phys. Rev. Lett. **100**, 237003 (2008).

A. Singh and Z. Tesanovic, Phys. Rev. B **41**, 614 (1990).

K. Sugimoto, E. Kaneshita, T. Tohyama, J. Phys. Soc. Jpn. **80**, 033706 (2011).

M. A. Tanatar, E. C. Blomberg, A. Kreyssig, M. G. Kim, N. Ni, A. Thaler, S. L. Budko, P. C. Canfield, A. I. Goldman, I. I. Mazin, and R. Prozorov, Phys. Rev. B **81**, 184508 (2010).

K. Terashima, Y. Sekiba, J. H. Bowen, K. Nakayama, T. Kawahara, T. Sato, P. Richard, Y.-M. Xu, L. J. Li, G. H. Cao, Z.-A. Xu, H. Ding, T. Takahashi, Proc. Natl. Acad. Sci. **106**, 7330 (2009).

R. Thomale, Ch. Platt, W. Hanke, B. A. Bernevig, Phys. Rev. Lett. **106**, 187003 (2011).

Goetz S. Uhrig, Michael Holt, Jaan Oitmaa, Oleg P. Sushkov, and Rajiv R. P. Singh, Phys. Rev. B **79**, 092416 (2009).

B. Valenzuela, E. Bascones, M. J. Calderon, Phys. Rev. Lett. **105**, 207202 (2010).

A. B.Vorontsov, M. G. Vavilov, A. V. Chubukov, Phys. Rev. B **81**, 174538 (2010).

F. Wang, H. Zhai, Y. Ran, A. Vishwanath, and D.-H. Lee, Phys. Rev. Lett. **102**, 047005 (2009).

Cenke Xu, Markus Muller, and Subir Sachdev, Phys. Rev. B **78**, 020501(R) (2008).

L. X. Yang, Y. Zhang, H. W. Ou, J. F. Zhao, D. W. Shen, B. Zhou, J. Wei, F. Chen, M. Xu, C. He, Y. Chen, Z. D. Wang, X. F. Wang, T. Wu, G. Wu, X. H. Chen, M. Arita, K. Shimada, M. Taniguchi, Z. Y. Lu, T. Xiang, D. L. Feng, Phys. Rev. Lett. **102**, 107002 (2009).

B.-J. Yang and H.-Y. Kee, Phys. Rev. B **82**, 195126 (2010).

M. Yi, D. Lu, J.-H Chu, J. G. Analytis, A. P. Sorini, A. F. Kemper, B. Moritz, S.-K. Mo, R. G. Moore, M. Hashimoto, W.-S. Lee, Z. Hussain, T. P. Devereaux, I. R. Fisher, and Z.-X. Shen, PNAS **108**, 6878 (2011).

T. Yildirim, Phys. Rev. Lett. **101**, 057010 (2008).

V. B. Zabolotnyy, D. S. Inosov, D. V. Evtushinsky, A. Koitzsch, A. A. Kordyuk, G. L. Sun, J. T. Park, D. Haug, V. Hinkov, A. V. Boris, C. T. Lin, M. Knupfer, A. N. Yaresko, B. Buechner, A. Varykhalov, R. Follath, S. V. Borisenko, Nature **457**, 569 (2009).

J. Zhao, D. T. Adroja, D.-X. Yao, R. Bewley, S. Li, X. F. Wang, G. Wu, X. H. Chen, J. Hu, and P. Dai, Nat. Phys. **5**, 555 (2009).

Chapter 10

Magnetism in Parent Compounds of Iron-Based Superconductors

Jiangping Hu
Department of Physics, Purdue University, West Lafayette, Indiana 47907, USA
Beijing National Laboratory for Condensed Matter Physics, and Institute of Physics, Chinese Academy of Sciences, Beijing 100190, China
jphu@iphy.ac.cn, hu4@purdue.edu

Magnetism plays an important role in high temperature superconductors. Compared with cuprates, iron-based superconductors exhibit even much richer magnetic properties. Different families of iron-based sueprconductors can approach different magnetically ordered ground states with mixed local and itinerant characters. Here, we review the magnetic properties of the parent compounds of iron-based superconductors, including iron-pnictides and iron-chalcogenides. The essential message of this review is that the magnetism in all iron-based superconductors is well described by models with short-range effective magnetic exchange couplings, and among the couplings, the next nearest neighbour antiferromagnetic coupling J_2 is almost universal in all iron-based superconductors, which suggests it is the key magnetic exchange coupling that controls superconductivity pairing.

Iron-Based Superconductors: Materials, Properties, and Mechanisms
Edited by Nan-Lin Wang, Hideo Hosono, and Pengcheng Dai

ISBN 978-981-4303-22-4 (Hardcover), 978-981-4303-23-1 (eBook)
www.panstanford.com

10.1 Introduction

The discovery of iron-based superconductors in 2008 [1] rekindles intensive interest in high transition temperature (high-T_c) superconductors (SC). Up to date, many families of iron-based superconductors have been found. They can be divided into two basic categories: iron-pnictides and iron-chalcogenides. The iron-pnictides include the quaternary family (or 1111-family) [1–3], such as LaOFeAs, the tertiary family which includes 111-family, such as NaFeAs, and 122p-family, such as $BaFe_2As_2$ [4], and more complex structures such as $Sr_4V_2O_6Fe_2As_2$ [5]. The iron-chalcogenides include the binary-family Fe/Se(Te) (or 11-family) [6], the alkali-doped $A_{1-x}Fe_{2-y}Se_2$ (A=K,Cs,Rb) [7–10] (or 122c-family) and others, such as $BaFe_2Se_3$ [11, 12] and $(CeO)_2FeSe_2$ [13] whose superconductivity needs further experimental investigation. The transition temperatures of these new superconductors reach as high as 56 K [2] for iron-pnictides and 48 K [14] for iron-chalcogenides.

Soon after the discovery, similar to copper oxide superconductors, cuprates, magnetism was identified to be entwined with superconductivity in these compounds [15–17]. Superconductivity is developed by introducing doping or applying pressure while long-range magnetic orders are suppressed. These facts suggest that magnetism is a competitor as well as a predecessor to superconductivity. Understanding the magnetism is therefore pivotal to understand superconducting mechanism as proposed in many theories (for a review, [18]).

Unlike all parent compounds of cuprates that share a common antiferromagnetically (AF) ordered ground state, those of iron-based superconductors exhibit many different AF orders in their ground states, including collinear-AF (C-AF) order in iron-pnictides [15], bicollinear-AF (B-CAF) order in the 11 family of iron-chalcogenide, FeTe [19–22], and block-AF order together with vacancy order (B-AF_v) in the 122c-family of iron-chalcogenide $K_{0.8}Fe_{1.6}Se_2$ [23, 24]. While the universal presence of antiferromagnetism suggests that superconductivity is strongly interrelated with magnetism, the diversity of the AF ordered states obscures their interplay.

The AF ordered states in iron based superconductors can either be metallic or insulating. More specifically, the parent compounds of iron-pnictides are bad metal (for a review [25]) while the newly discovered 122c-iron-chalcogenide, $K_{0.8}Fe_{1.6}Se_2$ is a block AF insulator [7–10, 23, 24, 26]. The fact of the existence of both magnetically ordered insulating and metallic ground states has led to a strong debate on the origin of magnetism.

Theoretically, it has been extremely difficult to find a consistent magnetic model to describe the magnetism of iron-based superconductors because of the dichotomic behavior featured with aspects of both local moments and itinerant electrons. It is well known that magnetism can be explained by either local moment models where local spins interact with each other or itinerant electron models where a magnetic order arises from nested Fermi surfaces. The former is appropriate for insulating materials such as the parent compounds of cuprates while the latter is suitable for metallic systems, such as chromium. However, in iron-based superconductors, it was shown that even in the insulating parent compounds, itinerant electron aspects have to be included [27] and vise versa, in the metallic parent compounds, local moment aspects are manifested [28]. Such a dichotomy of the magnetic properties [27–33] leads to many diversified viewpoints on the origin of the magnetism [22, 34–42].

In this paper, we review both experimental and theoretical results regarding the magnetism in the parent compounds of iron based superconductors. The essential message of this review is that *the magnetic states of the parent compounds can be described by local magnetic exchange models in all families of iron based superconductors, which suggests that magnetism stems from local physics, such as superexchange mechanism mediated by As or Se(Te), rather than nesting between Fermi surfaces.* In the first section, we review magnetic orders and magnetic excitations reported by neutron scattering experiments and electronic structures of magnetic states revealed by angle resolved photoemission spectroscopy (ARPES) and other experimental techniques. In the second section, we discuss theoretical calculations and models that explain the observed magnetic properties. In the third section, we discuss the interplay of spin, lattice and orbital degrees of freedom and electronic nematic states observed in iron-based superconductors. Finally we briefly discuss the relationship between magnetism and superconductivity. Our presentation is ordered to reflect the chronological order in the study of iron-based superconductors.

10.2 Experimental Results on the Parent Compounds of Iron-Based Superconductors

10.2.1 *Results on Iron-Pnictides*

Iron-pnictides, in particular, the 122p family, have been intensively investigated by neutron scattering experiments(for a review, [43]). The experiments show that all the parent compounds of iron-pinctides share a common C-AF state as a ground state shown in Fig. 10.1. The C-AF phase is also accompanied by a small orthorhombic lattice distortion which is much less than 1% of the original tetragonal lattice [44, 45]. For example, in $SrFe_2As_2$, the lattice constants in the orthorhombic phase are $a = 5.5695(9)$ Å and $b = 5.512(1)$ Å [44]. The ordered magnetic moment in the C-AF phase is parallel to the a-axis–the axis with relative larger lattice constant in the orthorhombic structure. However, phase transitions and magnetic moments are rather different in different families. In the quaternary family (or the 1111-family) of iron-based SC's [1], the parent compounds generally have two second-order phase transitions as temperature decreases [15, 16, 46, 47]. At high temperature, the system is a paramagnet (PM), and undergoes a tetragonal-orthorhombic lattice transition at the temperature T_s around 150 k. As temperature further decreases to T_N slightly lower than T_s, there is a magnetic phase transition into the C-AF state, characterized by the commensurate propagation vector $Q = (\pi, 0, \pi)$ [15, 16]. In the PM phase, RFeAsO is a bad metal, where 'bad' means that it is not cleanly insulating but has a resistivity much larger than most metals. For example, a typical resistivity in these materials, $\rho \sim 7\text{m}\Omega\cdot\text{cm}$ [1] at room temperature, which is very large compared with the resistivity of iron metal ($10^{-2}\text{m}\Omega\cdot\text{cm}$). In the C-AF state, the ordered magnetic moment is about $0.4\mu_B$ [15], a value comparable to cuprates.

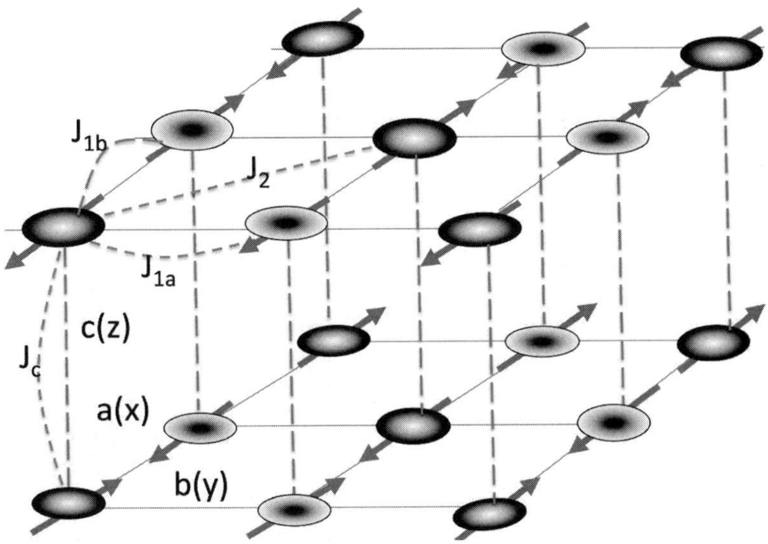

Figure 10.1 Schematic lattice and magnetic structure of the C-AF state in iron-pnictides. The effective magnetic exchange couplings are indicated.

For the 122p-family of iron-pnictides, the parent compounds generally have only one phase transition as temperature decreases, namely $T_s = T_N$ [44], the lattice distortion and C-AF transition take place at the same temperature. This transition is a first-order type. Upon doping, however, the phase transition can be separated into two phase transitions, which can be either a two-second-order type or an one-first-order and one-second order type as shown in experiments [48, 49]. The transition temperature, T_s or T_N, in the 122p-family reaches around 220 k, a value larger than T_s in the 1111-family. Meanwhile, the ordered magnetic moment in the 122p-family with a typical value, $0.8\mu_B$ [44, 45], is also typically much larger than those measured in the 1111-family.

For the 111-family of iron-pnictides, the ground states of the compounds with a nominal formula A(Na, Li) FeAs are superconducting. In NaFeAs, the superconductivity coexists with C-AF magnetic order [50]. The magnetic and structural transitions in NaFeAs are separated in a way similar to those in the 1111 family. T_N and T_s are much lower in the 111-family than in the previous two families. The ordered magnetic moment is around $0.32\mu_B$ [51] in NaFeAs. No static magnetic order has been observed in $Li_{1-x}FeAs$. Nevertheless, non-superconducting $Li_{1-0.06}FeAs$ has strong C-AF magnetic fluctuations at low energy. The magnetic fluctuating moment, $< m^2 >= 2.1 \pm 0.6\mu_B^2$ [52], is rather large. Overall, all these results on the 111-family are consistent with the possibility that these materials are self-doped even no additional carriers have been introduced.

An important fact is that from the lattice structures of the three families of iron-pnictides, it is clear that the 122p is much more three dimensional than the other two. The magnetic exchange coupling between layers along c-axis in the 122p can be as large as one third of in-plane magnetic exchange couplings [28]. This may explain the fact that the parent compounds of the 122p has the largest magnetic ordered moment and the highest magnetic transition temperature. Upon doping, the effective magnetic coupling along c-axis can decrease quickly and the magnetic excitations become much more two-dimensional [53]. Moreover, upon doping, non-commensurate magnetic fluctuations can also be observed in the 122p family [43, 54–58].

10.2.2 *Results on Iron-Chalcogenides*

Iron-chalcogenides and iron-pnictides share many common properties [6]. The iron-selenide layer in iron-chalcogenides is almost identical to the iron-arsenic layer in iron-pnictides. The band structures and Fermi surfaces of the 11 family of iron-chalcogenides are also similar to those of iron-pnictides. Moreover, the superconducting transition temperature, T_c, of iron-chalcogenides is also comparable to iron-pnictides. Therefore, before the 122c family of iron-chalcogenides was discovered, much less research attention has been paid to iron-chalcogenides than to iron-pnictides.

However, the discovery of new alkali-doped iron selenides $R_{1-x}Fe_{2-y}Se_2$ [7–10] made a complete twist. While the new superconductors can still achieve a transition temperature as high as 48 K [14], they have several distinct characters that are noticeably absent in other iron-based superconductors, (1) the absence of hole pockets at Γ point of Brillouin zone in their superconducting

(SC) phases [59–61]; (2) AF ordered insulating phases [23, 24, 26] with very high Néel transition temperatures in their parental compounds [9]; (3) intrinsic vacancy orders; (4) theoretical models based on different mechanisms have suggested different pairing symmetries for $R_{1-x}Fe_{2-y}Se_2$: weak coupling approaches based on spin-excitation mediated pairing predict a d-wave pairing symmetry [62–64], strong coupling approaches [65–67] which emphasize the importance of next nearest neighbor (NNN) AF local exchange coupling suggest the pairing symmetry is a robust s-wave, not different from the $S^{\pm}$-wave symmetry obtained in their pnictide counterparts and models with orbital fluctuation mediated pairing suggest a S^{++}-wave pairing for both iron selenide and pnictide materials [68]. Currently, research on iron-chalcogenides is still in the incipient stage. We only focus on confirmed experimental results on magnetism.

Unlike iron-pnictides whose different families share a common C-AF state, different families of iron-chalcogenides exhibit different magnetically ordered states. The FeTe compounds of the 11 familiy has a B-CAF ground state as shown in Fig. 10.2. The B-CAF state is accompanied by a monoclinic lattice distortion. The distortion can produce 1.3% change of the tetragonal lattice constant [20], a value much larger than the orthorhombic distortion in iron-pnictides. The large lattice distortion indicates a strong coupling between spin and lattice. It is also consistent with the fact there is only one strong first-order transition in FeTe, namely, the B-CAF and monoclinic lattice distortion take place at the same transition temperature which is about 70 K [69]. The B-CAF order can also be changed by introducing additional iron atoms in $Fe_{1+y}Te$. When $y > 0.1$, a static incommensurate magnetic state with an ordered wavevector $(\frac{\pi}{2} + \delta, \frac{\pi}{2} + \delta)$ is developed. The incommensurability δ increases as y increases [19]. The ordered magnetic moment in FeTe, which is about $2.3\mu_B$, is also significantly larger than the values in iron-pnictides. Upon replacing Te by Se, namely increasing x in $FeSe_xTe_{1-x}$, the static B-CAF order vanishes and incommensurate magnetic excitations develop quickly [55, 70].

For the 122c family, $R_{1-x}Fe_{2-y}Se_2$, one magnetically ordered state with iron vacancy order, called B-AF_v has been identified when $x = 0.8$ and $y = 0.4$ as shown in Fig. 10.3. Iron vacancies form

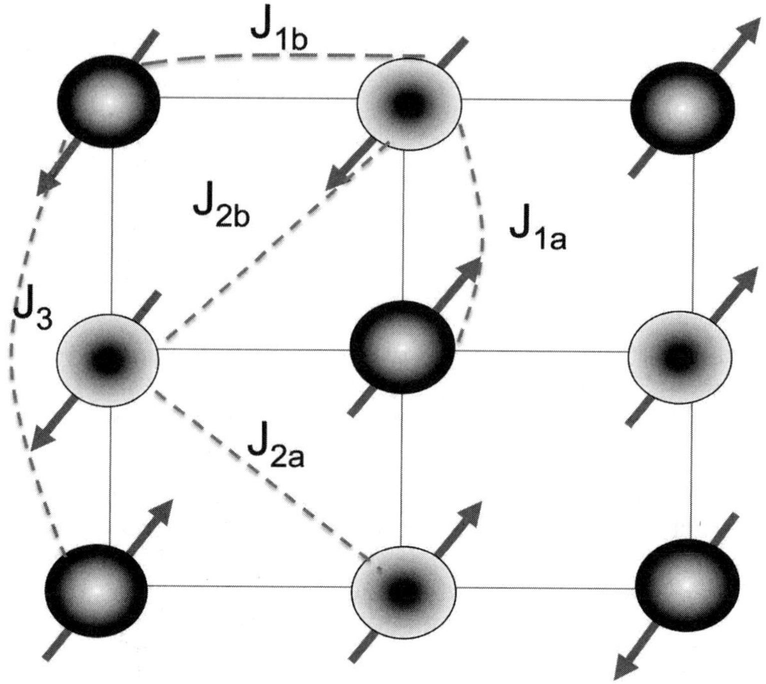

Figure 10.2 Schematic magnetic structure of the B-CAF state in FeTe. The effective magnetic exchange couplings are indicated.

a $\sqrt{5} \times \sqrt{5}$ pattern and four iron spins group together to form a super-site with ferromagnetic(FM) alignment. In the lattice of super-sites, it is AF ordered. The ordered moment per iron site in the B-AF_v state reaches $3.3\mu_B$ and the transition temperature is about 500 k [23, 24]. The B-AF_v is also accompanied with lattice distortion. The lattice distortion in the B-AF_v state can reach 2% of the original lattice constant. More importantly, the B-AF_v state is insulating with a gap larger than 0.3 ev, which is the first insulating state observed in iron-based superconductors.

Two additional iron selenide materials $Ce_2O_2FeSe_2$ [13] and $BaFe_2Se_3$ [11, 12] have also been reported. $Ce_2O_2FeSe_2$ is effectively a system with weakly coupled chains and $BaFe_2Se_3$ is a system with weakly coupled two-leg ladders as shown in Fig. 10.4. In $Ce_2O_2FeSe_2$, the spins are FM along each chain and are AF between

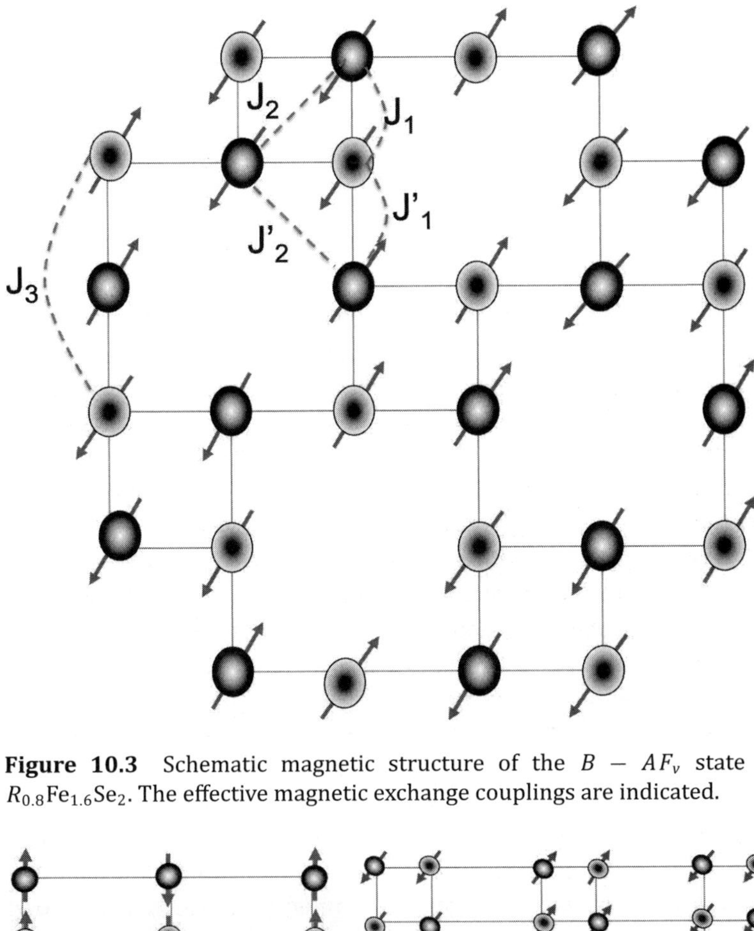

Figure 10.3 Schematic magnetic structure of the $B - AF_v$ state in $R_{0.8}Fe_{1.6}Se_2$. The effective magnetic exchange couplings are indicated.

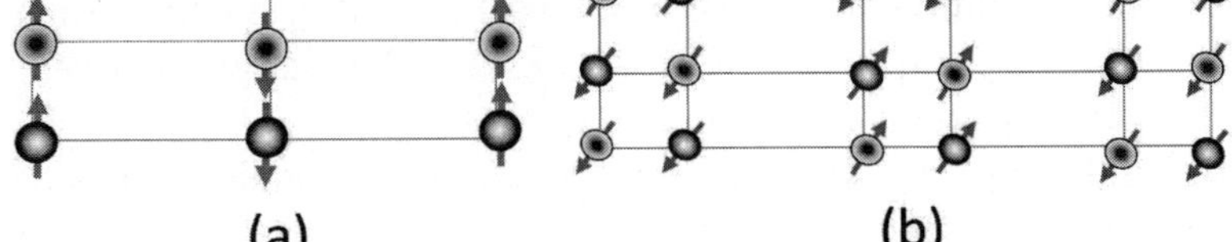

Figure 10.4 Schematic magnetic structures of (a) $Ce_2O_2FeSe_2$ and (b) $BaFe_2Se_3$.

two nearest neighbor chains. The ordered magnetic moment is around $3.3\mu_B$. The magnetic order in $BaFe_2Se_3$ has not been fully resolved.

10.2.3 *Electronic Structures and Resistivity Anisotropy*

While we focus on results from neutron scattering measurements, the most powerful probe to determine static magnetic orders and collective magnetic excitations, it is also necessary to briefly mention other electronic properties in magnetically ordered state.

First, the gaps of single particle excitations are expected in a magnetically ordered state. Iron-pnictides are metallic even in the C-AF state. The gap can only be opened partially on Fermi surfaces. Gap opening associated with magnetic transition was observed in momentum unresolved optical measurements [71–73] in iron-pnictides. Continuous spectra weight suppression was also observed in large energy scale (larger than 0.6ev at least) starting at a temperature much higher than the magnetic transition temperature [74]. In angle resolved photoemission spectra (ARPES), no clear gap opening is detected, instead a strong band reconstruction in a large energy window below Fermi surfaces are observed [69, 75]. It suggests that electrons in entire bands, rather than electrons nearby Fermi surfaces, are contributed to magnetic transitions.

Second, in FeTe, no clear gap opening was observed in optical measurements even if the magnetic transition is a strong first-order type accompanied by large lattice distortion. Large spectra weight transfer was observed as temperature decreases. Strong band reconstruction takes place at the transition temperature as well [50]. The B-AF_v state in $K_{0.8}Fe_{1.6}Se_2$ is insulating with a large gap, larger than 0.3 ev. Many experiments also indicated the existence of additional insulating states [10, 76–81]. However, their exact structure and electronic properties, and the interplay between different states remain to be investigated [82].

Finally, the interplay of spin, lattice and orbital degrees of freedom is very intriguing. In iron-pnictides, the C-AF magnetic ordering, orthorhombic lattice distortion and ferrorbital ordering all break the C^4 symmetry of a tetragonal lattice. Experimentally, the symmetry breaking is shown in the intrinsic in-plane electronic anisotropy as revealed in transport measurements [83, 84] and ARPES measurements [85] of detwinned single crystals of under-doped compounds in the 122p family. Theoretically, it has been debated that which degree of freedom is the dominant player that drives the transition. Here, instead of reviewing all the proposals, we discuss the magnetically-driven mechanism [35, 36, 41].

10.3 Theoretical Models

10.3.1 *Results of First Principle Electronic Structure Calculation*

The first principle electronic structure calculation has provided substantial understanding of iron based superconductors, not only band structures but magnetism as well [21, 86–90].

The earlier calculations showed that LaFeAsO is semimetal with a C-AF order [87]. The magnetic order is attributed to the pnicogen-bridged antiferromagnetic superexchange interactions between two next nearest neighbour iron spins, rather than the nesting between Fermi surfaces as suggested in the band structure of the non-magnetic ordered state [86, 90]. More interestingly, the calculations also predicted values of magnetic exchange couplings [91] in the C-AF state which are consistent with experimental results measured by neutron scattering experiments [28]. However, the magnetic moment obtained in the LDA calculations for LaFeAsO is consistently much larger than experimental values. So far, all the calculations show that the ordered moment in iron-pnictides is around $2\mu_B$ [87], which is much larger than the largest experimental value ($\sim 0.8\mu_B$) in the 122p family. In order to reduce the ordered magnetic moment, the pnicogen height with respect to Fe-plane has to be reduced significantly.

The first principle calculation also captures magnetically ordered state in iron-chalcogenides including the B-CAF state in FeTe [21] and the B-AF$_v$ state in $K_{0.8}Fe_{1.6}Se_2$ [92–94]. The results appear to be consistent with experiments as well. The ordered moments in iron-chalcogenides are more or less consistent between theoretical calculations and experimental measurements. The calculations also correctly suggested that the B-AF$_v$ state in $K_{0.8}Fe_{1.6}Se_2$ is insulating. However, the results from these calculations on the magnetic exchange values are not consistent with experiments. So far there are two ways to obtain the magnetic exchange couplings in the calculation. First, in order to derive the exchange interactions between the Fe-Fe moments, one has to design different magnetic structures and calculates the energy difference among them. Then the energy difference is used to extract magnetic exchange couplings in an effective magnetic exchange model. In this way, the result highly depends on the designed magnetic structures. Second, the exchange couplings are extracted as linear response parameters to the magnetically ordered state. A result from such a method on FeTe strongly contradicts to experimental results [95].

Analytically, there have been two different approaches to understand magnetism. The first approach is based on weak coupling approach. In iron-pnictides, hole pockets locate at Γ point and electron pockets locate at M point in reduced Brillouin zone [17]. This feature suggests that a simple itinerant mechanism can produce the C-AF order as a standard spin density wave order through nesting the electron and hole pockets. The nesting mechanism is somewhat successful in explaining magnetic properties including spin excitations and the development of incommensurate spin fluctuations upon hole or electron doping. However, the pure nesting mechanism from weak coupling is not sufficient. A pure nesting mechanism is not enough to produce the large ordered magnetic moment because the Fermi pockets are rather small so that only electrons near Fermi surfaces can not provide a significant magnetic moment measured experimentally. For example, in the case of the 122p family, the ordered moment roughly equates to the moment expected in a fully ordered spin-one system. We thus expect two electrons per iron are fully polarized in the C-AF state. If we view it within a picture of band structure, it would suggest the gap generated by magnetic ordering is comparable to band width.

The nesting mechanism is more severely challenged in iron-chalcogenides. In fact, the nesting mechanism can not even be considered as a starting point to address the magnetism in iron-chalcogenides. The ordering wavevectors in the magnetic states of iron chalcogenides do not match the nesting wavevector of Fermi surfaces [89]. In fact, iron-chalcogenides exhibit so many different magnetically ordered states. There is no direct correlation between Fermi surface topologies and magnetic orders.

The second approach is to construct effective models with short-range magnetic exchange couplings. This approach requires the formulation of local spin moments. In a strongly correlated electron system, such as cuprates, there is no debate about the existence of local spin moments. While iron-pnictides are not as strongly correlated as cuprates, in the parent compounds of iron-pnictides, the correlation effect is very important as we have witnessed that the entire band structure is reconstructed in the magnetic state [50, 75]. This suggests all electrons, not just electrons near Fermi surface, participate in forming the magnetic order, an implication of the formation of local spin moments. We will show that the models with short-range magnetic exchange couplings indeed capture entire spin-wave excitation spectra. Moreover, the models provide a clear picture of microscopic interactions and allow to identify the existence of universal characteristics shared by all iron based superconductors.

10.3.2 *Effective Magnetic Exchange Models for Iron-Pnictides*

A simple model to capture the C-AF state is so called $J_1 - J_2 - J_z$ Heisenberg model [36]. The model can be intuitively argued in the following way. If we adopt a local moment picture and consider a single FeAs layer, as all spins are assumed to be localized on iron sites, the layer can be modeled by a simple square lattice (see Fig. 10.1). Considering the magnetic exchange coupling mediated by As, since As locates over the center of four nearest neighbor iron sites, we can naturally conclude that magnetic exchange couplings must include those between two nearest neighbor(NN) and two next nearest neighbor (NNN) sites. The Hamiltonian of a single layer thus is given by,

$$H_{n,ab} = J_1 \sum_{\langle ij \rangle} \mathbf{S}_{i,n} \cdot \mathbf{S}_{j,n} + J_2 \sum_{\langle\langle ij \rangle\rangle} \mathbf{S}_{i,n} \cdot \mathbf{S}_{j,n}, \qquad (10.1)$$

where n denotes the layer number. Including the magnetic exchange coupling between layers, we obtain a 3D J_1-J_2-J_z model defined as

$$H = \sum_n H_{n,ab} + J_z \sum_{n,i} \mathbf{S}_{i,n} \cdot \mathbf{S}_{i,n+1}, . \qquad (10.2)$$

If the superexchange mechanism dominates, we would expect all the exchange couplings are AF, namely, $J_1, J_2 > 0$.

To be more specific, the Hamiltonian in Eq. 10.1 has been proposed from both perspectives of first principle calculations and of strong coupling analysis [34, 36, 86, 87, 90]. Both approaches have agreed on that the NNN coupling J_2 stems mostly from superexchange mechanism. In the paper [34], the exchanges are expressed as $J_2 \sim \frac{2t^4}{(\epsilon_p - \epsilon_d)^3}$, where t is the hopping between the $3d$ orbital of Fe and $4p$ orbital of As, $\epsilon_{p,d}$ are onsite energies of $4p$ and $3d$ orbitals. The NN exchange consists of both ferromagnetic and antiferromagnetic parts, $J_1 = J_1^{FM} + J_1^{AF}$, where $J_1^{FM} < 0$ and $J_1^{AF} > 0$. The explicit formula of the two parts depend on the details of the hybridization of different d-p orbital pairs.

The model in Eq. 10.2 in the classical limit or the $S = \infty$ limit has two different ground states in the case of both J_1 and J_2 are AF. To see this, we view the square lattice as two interpenetrating sublattices as shown in Fig. 10.1 where irons in the two sublattices are shown with dark and light colors respectively. Each of the two sublattices is a square lattice. The NNN exchange J_2 in the original lattice becomes NN in the sublattice. Therefore if J_2 is dominant, the ground state of each sublattice is AF in which every spin is antiparallel to four adjacent spins; the relative spin direction between the two sublattices is, however, not necessarily parallel or antiparallel. In such a classical model, the AF orders in two sublattices are completely independent. The ground state energy per spin is $E = -2J_2S^2$. On the other hand, if J_1 is very large, the ground state should be a familiar AF in the original lattice (not sublattice), the ground state of undoped cuprates, with the energy per site $E_{(\pi,\pi)} = -2(J_1 - J_2)S^2$. Comparing the two possible ground states, we obtain if $J_1 > 2J_2$, (π, π) AF is the real ground state and if $J_1 < 2J_2$, the state with two decoupled AF ordered sublattices is the ground state.

In the case of $J_1 < 2J_2$, the ground state is infinitely degenerate, as the relative angle between the spin directions on the two sublattices is entirely arbitrary. If we consider quantum fluctuation, such a degeneracy will be broken. To show this, we use Holstein-Primakoff transform for spin operators:

$$S_i^+ = \sqrt{2S}a_i\sqrt{1-\frac{a^\dagger a_i}{2S}},\ S_i^- = \sqrt{2S}a_i^\dagger\sqrt{1-\frac{a_i^\dagger a_i}{2S}},$$
$$S_i^z = S - a_i^\dagger a_i, \tag{10.3}$$

where a_i's are annihilation operators of spin wave bosons satisfying $[a_i, a_j] = [a_i^\dagger, a_j^\dagger] = 0$ and $[a_i, a_j^\dagger] = \delta_{ij}$. We consider the following spin configuration in the two sublattices: if a site (i_x, i_y) belongs to the first sublattice, its spin moment is $\mathbf{S}_{1,i} = S(0, 0, (-1)^{i_x})$; if it belongs to the second sublattice, its spin moment is $\mathbf{S}_{2,i} = S(-1)^{i_x}(\sin\theta, 0, \cos\theta)$. Taking linear spin wave approximation in the Holstein-Primakoff transform, we have

$$S_{1,i}^+ = \sqrt{2S}\left(\frac{1+(-1)^{i_x}}{2}a_i - \frac{1-(-1)^{i_x}}{2}\right)a_i^\dagger,$$
$$S_{1,i}^- = \sqrt{2S}\left(\frac{1+(-1)^{i_x}}{2}a_i^\dagger - \frac{1-(-1)^{i_x}}{2}\right)a_i,$$
$$S_{1,i}^z = (-1)^{i_x}(S - a_i^\dagger a_i),$$
$$S_{2,i}^+ = (-1)^{i_x}\left(\sqrt{\frac{S}{2}}(a_i + a_i^\dagger)\cos\theta + (S - a_i^\dagger a_i)\sin\theta\right) + \sqrt{\frac{S}{2}}(a_i - a_i^\dagger),$$
$$S_{2,i}^- = (-1)^{i_x}\left(\sqrt{\frac{S}{2}}(a_i + a_i^\dagger)\cos\theta + (S - a_i^\dagger a_i)\sin\theta\right) - \sqrt{\frac{S}{2}}(a_i - a_i^\dagger),$$
$$S_{2,i}^z = (-1)^{i_x}\left(-\sqrt{\frac{S}{2}}(a_i + a_i^\dagger)\sin\theta + (S - a_i^\dagger a_i)\cos\theta\right). \tag{10.4}$$

Substituting these expressions into Eq. (10.1), and keeping up to the quadratic order of spin wave operators, we obtain

$$H_{ab}^{LSW} = \sum_k \left[A_k a_k^\dagger a_k + \frac{B_k}{2}(a_k a_{-k} + h.c.)\right], \tag{10.5}$$

where

$$A_k = 4J_1S + 4J_2S + 2J_1S[(1+\cos\theta)\cos(k_x) + (1-\cos\theta)\cos(k_y)],$$
$$B_k = -\{2J_1S[(1-\cos\theta)\cos(k_x) + (1+\cos\theta)\cos(k_y)]$$
$$+2J_2S(\cos(k_x+k_y) + \cos(k_x-k_y))\}. \tag{10.6}$$

Taking the Bougliubov transform for bosons, we obtain

$$H_{ab}^{LSW} = \sum_k \left(\gamma_k^\dagger \gamma_k + \frac{1}{2} \right) \omega_k, \tag{10.7}$$

where

$$\omega_k = \sqrt{A_k^2 - B_k^2}, \tag{10.8}$$

and $\gamma_k = \cosh \psi_k a_k + \sinh \psi_k a_{-k}^\dagger$, $\gamma_k^\dagger = \cosh \psi_k a_k^\dagger + \sinh \psi_k a_{-k}$, and $\tanh(2\psi_k) = -\frac{B_k}{A_k}$. At zero temperature, the quantum fluctuation contributes a zero-point energy for bosons

$$E_0^Q = \frac{1}{2} \sum_k \omega_k = -\frac{\gamma_Q S J_1^2}{4 J_2} \cos^2\theta + J_2 O\left(\left(\frac{J_1}{J_2} \right)^4 \right), \tag{10.9}$$

where $\gamma_Q = 0.13$. The ground state, therefore, is either $\theta = 0$ or $\theta = \pi$, which means that spins in two sublattice are collinearly aligned, namely, the C-AF state. The collinear configuration is an effect of spin fluctuation. This effect is called "order by disorder" [96]. We will discuss this further next section.

The energy in Eq. 10.9 can be taken into account if we add a quadratic spin-spin interaction term between two sublattices, namely,

$$H_q = - \sum_{<ij>_{NN},n} K (\mathbf{S}_{i,n} \cdot \mathbf{S}_{j,n})^2 \tag{10.10}$$

with $K = \frac{\gamma_Q J_1^2}{8 S^3 J_2}$. By adding such a term into the Hamiltonian in Eq. 10.2, we obtain a model called $J_1 - J_2 - J_z - K$ model. The C-AF state is the ground state of the model when $J_1 < 2J_2$ and $K > 0$. This extension has been recently considered by [38, 97]. The important thing here is that the $J_1 - J_2 - J_z - K$ model is symmetric with respect to tetragonal lattice. In the C-AF state where the spin configuration breaks the tetragonal symmetry, the H_q can be decoupled as

$$H_q = -2 \sum_{<ij>_{NN},n} K \sigma_{ij,n} (\mathbf{S}_{i,n} \cdot \mathbf{S}_{j,n}) + cons. \tag{10.11}$$

where $\sigma_{ij,n} =< \mathbf{S}_{i,n} \cdot \mathbf{S}_{j,n} >= \pm\sigma$ is the expectation value of the spin-spin interaction in the C-AF state. The sign is chosen to be positive if the spins are FM along one axis (b-axis) and negative if they are AF along the other axis (a-axis) in the C-AF state. Combining this term with J_1 term in the original Hamiltonain, the effective nearest neighbour exchange cuplings in the C-AF State becomes

$$J_{1a} = J_1 + 2K\sigma, J_{1b} = J_1 2K\sigma. \quad (10.12)$$

Thus, the spin wave excitations in the C-AF state are effectively described by a $J_{1a} - J_{1b} - J_2 - J_z$ model.

Experimentally, the ordered magnetic moment in iron-pnictides is along the a-axis in the C-AF state. This also suggests that there is a small spin anisotropy due to small spin orbital coupling. This anisotropy can be taken into account by adding $H_a = -J_s \sum_{i,n}(S^a_{i,n})^2$ to the model. The full explanation of the spin wave excitations is still given by Eq. 10.8 with the following replacement of A_k and B_k,

$$\begin{aligned} A_k &= 2S(J_{1b}\cos(k_y) + J_{1a} - J_{1b} + J_z + 2J_2 + J_s), \\ B_k &= 2S(J_{1a}\cos(k_x) + 2J_2\cos(k_x)\cos(k_y) + J_z cos(k_z)) \end{aligned} \quad (10.13)$$

where k_x and k_y are wavevectors along a-axis and b-axis, respectively so that the ordering wavevector in the C-AF state is $(\pi, 0)$. The band width of the spin wave excitations is roughly determined by $2S(2J_2 + J_{1a} + J_z)$. It is very helpful for us to gain intuitive understanding of the spin wave spectra by marking a few energy values at some specific wavevectors of the dispersion of the spin wave excitations,

$$\begin{aligned} \omega(\pi,0,\pi) &= 2S\sqrt{J_s(2J_{1a} + 4J_2 + 2J_z)} \\ \omega(\pi,0,0) &= 2S\sqrt{(J_s + 2J_z)(2J_{1a} + 4J_2 + J_s)} \\ \omega(\pi,\pi,\pi) &= 2S\sqrt{(4J_2 - 2J_{1b} + J_s)(2J_{1a} - 2J_{1b} + 2J_z + J_s)} \\ \omega(0,\pi,\pi) &= 2S\sqrt{(2J_{1a} - 2J_{1b} + J_s)(-2J_{1b} + 4J_2 + 2J_z + J_s)}. \end{aligned} \quad (10.14)$$

These values can help to quickly determine the coupling parameters from experimental data. The energy gap of the spin wave excitations is proportional to $\sqrt{J_s}$, which means a very small anisotropy can produce a sizable gap. The energy gap along $(k_x, k_y, 0)$ is proportional to $\sqrt{J_s + 2J_z}$, which also suggests that a small inter-layer coupling can produce a large gap. The energy at (π, π, π) is very sensitive to the difference between J_{1a} and J_{1b}. For small J_z and J_s, the energy is proportional to $\sqrt{J_{1a} - J_{1b}}$. Experimentally,

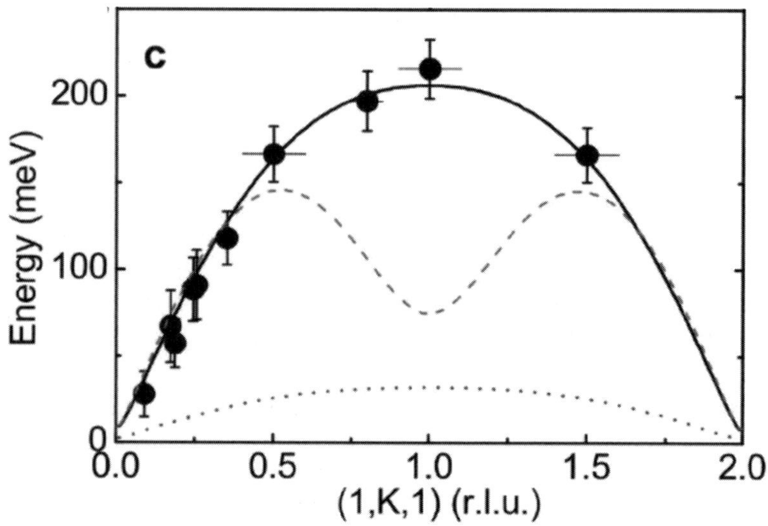

Figure 10.5 The spin wave dispersion along (π, k, π) direction reported in [28]: the solid line is fitted when $J_{1a} >> J_{1b}$ and the dashed line is fitted if $J_{1a} \sim J_{1b}$.

The gap $\omega(\pi, \pi, \pi)$ measured in the 122p family has been very large as shown in Fig. 10.5, namely, the difference of J_{1a} and J_{1b} is very large. Such a large difference can not be fully generated by quantum fluctuation. This means the value of K must include contributions from other couplings. We will discuss this later.

A summary of the coupling values measured by neutron scattering [28, 44, 98] for different iron-pnictides are shown in Table. 10.1. In general, the anisotropy J_s is very small. J_z can be sizable in the parent compound of the 122p family such as $CaFe_2As_2$ [28] where $J_z \sim J_2/3$. It suggests that the 122p family is indeed very 3D.

10.3.3 *Effective Magnetic Exchange Models for Iron-Chalcogenides*

As mentioned earlier, different families of iron-chalcogenides do not share a common magnetically ordered state in their parent compounds. Nevertheless, the effective magnetic exchange models that describe different ordered states can be unified into a $J_{1a}-J_{1b}-J_2-J_3-J_z$ model [22, 99],

$$H=\sum_{n,i} J_{ij}\mathbf{S}_i^n\cdot\mathbf{S}_j^n+\sum_{n,i}\mathbf{S}_i^n\cdot\mathbf{S}_i^{n+1}, \tag{10.15}$$

where J_{1a} is coupling between two AF ordered NN sites, J_{1b} is coupling between two FM ordered NN sites and J_3 is coupling between two next next nearest neighbour (NNNN) sites. The model has been shown to capture magnetically ordered states and their spin-wave excitations in both the 11 and 122c families of iron-chalcogenides.

10.3.3.1 FeTe

In the 11 family, the parent compound is usually considered to be FeTe, or $Fe_{1+y}Te$, where y denotes the concentration of excess iron atoms. Isovalent doping of Se replacing Te atoms induces superconductivity at $\sim 14K$ [6]. The magnetic structure of FeTe is a B-CAF phase if $y<0.08$ (see Fig. 10.2) characterized by the propagation vector $Q=(\pi/2,\pi/2)$ [20, 27, 100] in the unfolded BZ, but enters an incommensurate magnetic phase at $y\geq 0.08$ with the propagation vector $Q=(\pi/2+q,\pi/2+q)$, where q, the incommensurate part, depends on the excess iron concentration y [100].

The most general model to describe the B-CAF phase is a $J_{1a}-J_{1b}-J_{2a}-, J_{2b}-J_3-J_z$ model as proposed in [22] since the spin configurations along two NNN coupling directions are also different. The exact classical phase diagram of the model is shown in Fig. 10.6 [22]. This model captures both commensurate and incommensurate phases measured in $Fe_{1+y}Te_{1-x}Se_x$.

It is worth to note that the phase diagram given by the model remains completely unchanged if all NN exchanges $J_{1a,1b}$ flip signs and become FM. This is because, in this *classical* spin model, one can always do the transform that flips all spins on one sublattice, that is, transforming $\mathbf{S}_i$ to $-\mathbf{S}_i$ if $i_x+i_y=odd$. This maps a ground state for positive J_1's to the one for negative J_1's without changing the energy.

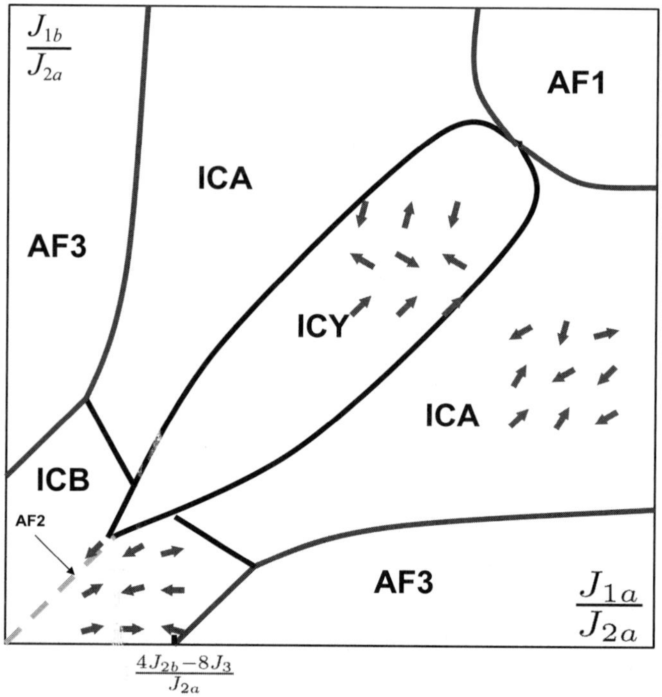

Figure 10.6 The phase diagram of the $J_1(J_{1a}, J_{1b}) - J_2(J_{2a}, J_{2b}) - J_3$ model: (1) AFM1 is the familiar (π, π) antiferromagnetic order in a square lattice; (2) AFM2 is the $(0, \pi)$ C-AF order along the x-axis observed in iron-pnictides; (3) AFM3 is the B-CAF phase observed in $Fe_{1+y}Te_{1-x}Se_x$;(4) ICB is an incommensurate phase with the incommensurate vector along the b-axis; (5) ICA is an incommensurate phase with the incommensurate vector along the a-axis which has been observed in $Fe_{1+y}Te$ [19]; (6) ICY is an incommensurate phase that has a wavevector of the form (π, q) which has been observed in $Fe_{1+y}Te_{1-x}Se_x$ [55, 70].

The spin wave excitations in the B-CAF state are given by

$$H = \sum_k \Psi_k^\dagger \mathcal{H}_k \Psi_k = \sum_k \Psi_k^\dagger \begin{pmatrix} A & J_{1b}D & B & J_{1a}C \\ J_{1b}C & A & J_{1a}D & B \\ B & J_{1a}C & A & J_{1b}D \\ J_{1a}D & B & J_{1b}C & A \end{pmatrix} \Psi_k, \quad (10.16)$$

where

$$A = 2J_{2b}\cos(k_x - k_y) + 2(J_{1a} + J_{2a} - J_{1b} - J_{2b} + 2J_3 + J_z),$$
$$B = 2J_{2a}\cos(k_x + k_y) + 2J_3(\cos 2k_x + \cos 2k_y) + 2J_z\cos k_z, \tag{10.17}$$

and $\Psi_k = (a_k, b_k, a^\dagger_{-k}, b^\dagger_{-k})$, where a_k and b_k are spinwave operators on the two sublattices as given in the previous section. Diagonalizing the spin wave Hamiltonian gives us the two branches spin wave dispersion $\epsilon_{1,2}(k)$,

$$H = \sum_{i=1,2;k} (\gamma_i^\dagger(k)\gamma_i(k) + 1/2)\omega_i(k), \tag{10.18}$$

and

$$a_i(k) = \sum_j U_{ij}(k)\gamma_j(k) + V_{ij}(k)\gamma_j^\dagger(-k). \tag{10.19}$$

where U and V are 2×2 matrices diagonalizing the Hamiltonian obtained by the scheme introduced in the previous section. The spin wave excitations are plotted in Fig. 10.7. Regardless of the specific parameters, a common feature of the spin wave dispersion is an almost dispersionless line along the [1, 1] direction through $(\pm\pi, 0)$ in the unfolded BZ, with an energy $\sim 4S\sqrt{(J_{1a} - J_{1b} + 2J_{2a} - 2J_{2b})(J_{1a} - J_{1b} - 2J_{2b} + 4J_3)}$. It is interesting to compare spin wave dispersions for different J_3. In Fig. 10.7, dispersions along the $[1, -1]$ direction for $J_3/J_{2a} = 0.1, 0.4, 0.8$ are shown respectively. The major different lies in the higher branch, which turns from a convex to a concave-shaped line.

The directly observable quantity in inelastic neutron experiment is the differential scattering cross section:

$$\frac{k_i}{k_f}\frac{d^2\sigma}{d\Omega dE} = \frac{2(\gamma r_e)^2}{\pi g^2\mu_B^2}|F(Q)|^2\frac{\chi''(q,\omega)}{1 - \exp(-\hbar\omega/kT)}, \tag{10.20}$$

where the dynamic local susceptibility $\chi(\omega)$ is defined as

$$\chi(i\omega_n) = \frac{1}{N}\sum_q \int d\tau e^{i\omega_n\tau}[\langle S^+(q,\tau)S^-(-q,0)\rangle$$
$$+\langle S^-(q,\tau)S^+(-q,0)\rangle]. \tag{10.21}$$

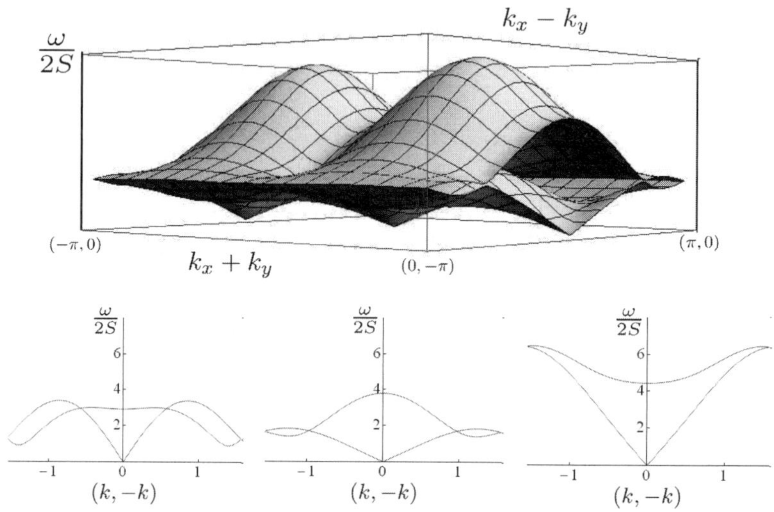

Figure 10.7 3D spinwave dispersion for the parameter set of $J_{2a} = J_{2b} = 1$, $J_{1a} = 2$, $J_{1b} = 0.94$ and $J_3 = 0.4$, which gives a point on the AFM3-ICA phase boundary on the phase diagram. Also note that along the BZ boundary $(k \pm \pi, k)$, the two sheets are degenerate and almost dispersionless. Below: Different dispersion at different parameter regions with common parameters $J_{2a} = J_{2b} = 1$. All the three dispersions are drawn at AFM3-ICA phase boundary given by (from left to right) $(J_{1a}, J_{1b}, J_3) = (1, 0.72, 0.8)$, $(J_{1a}, J_{1b}, J_3) = (2, 0.94, 0.4)$ and $(J_{1a}, J_{1b}, J_3) = (5, 0.9, 0.1)$, respectively.

After taking the Fourier transform, we obtain

$$\chi(\omega) = \frac{1}{N}\sum_{k,i,j}(|V_{i,j}(k)|^2 + |U_{i,j}(k)|^2)\left(\frac{1}{\omega - E_j(k) + i0} + \frac{1}{\omega + E_j(k) + i0}\right). \tag{10.22}$$

By taking $J_{1a}S = -17.5 \pm 5.7$, $J_{1b}S = -51.0 \pm 3.4$, $J_2S = J_{2a}S = J_{2b}S = 21.7 \pm 3.5$, $J_3S = 6.8 \pm 2.8$ all in unit of meV [27], the theoretical result $\chi(\omega, k)$ as shown in Fig. 10.8 gives a good fit to the experimental result in Fig. 10.9 for the 11-family FeTe.

For these experimental results, we emphasize that first the NN exchange are both ferromagnetic, in sharp contrast against the case of the 122p-family, where $J_{1a} > 0$ and $J_{1b} < 0$. Second, there is

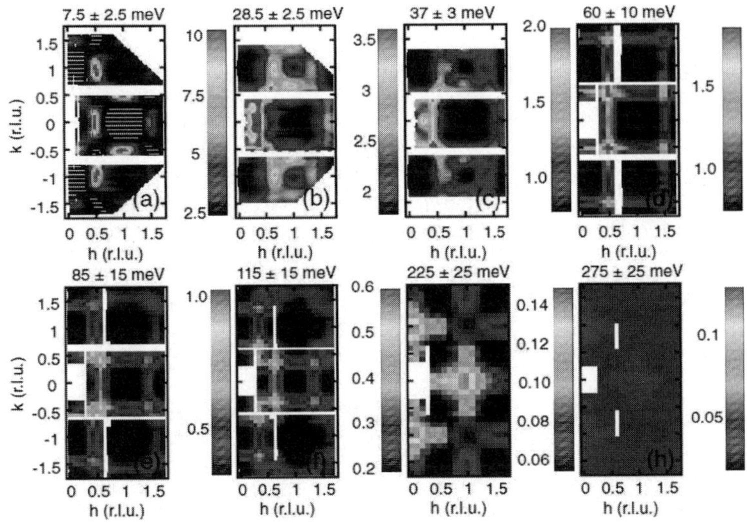

Figure 10.8 Resolution convolved simulation (using TOBYFIT at http://tobyfit.isis.rl.ac.uk) of the Heisenberg model using the best fit parameters in the text plus an out-of-plane coupling of $J_z = 1$ meV [27]. See also Color Insert.

little difference between J_{2a} and J_{2b} even the spin configurations of the two NNN couplings are opposite. Finally, the NNNN exchange J_3 is found necessary to match the experimental data. These results suggest that the B-CAF state in FeTe is described by $J_{1a}-J_{1b}-J_2-J_3$ model.

10.3.3.2 $A_{0.8}Fe_{1.6}Se_2$

$A_{0.8}Fe_{1.6}Se_2$ is unique because of the presence of vacancy ordering. The vacancy pattern is shown in Fig. 10.3 with a block-AF ordering. The magnetism in 122 iron-chacogenides can be understood within the previous model obtained for the 11-family [99].

If we consider the tetragonal lattice, the model describing FeTe is essentially the $J_1 - J_2 - J_3$ model with J_1 being FM. In a B-CAF state, both J_1 and J_2 couplings are frustrated. The only energy saving is from AF J_3. We can thus create iron vacancies to reduce the magnetic frustration. The $\sqrt{5} \times \sqrt{5}$ pattern of the vacancy ordering can be obtained by maximizing the magnetic

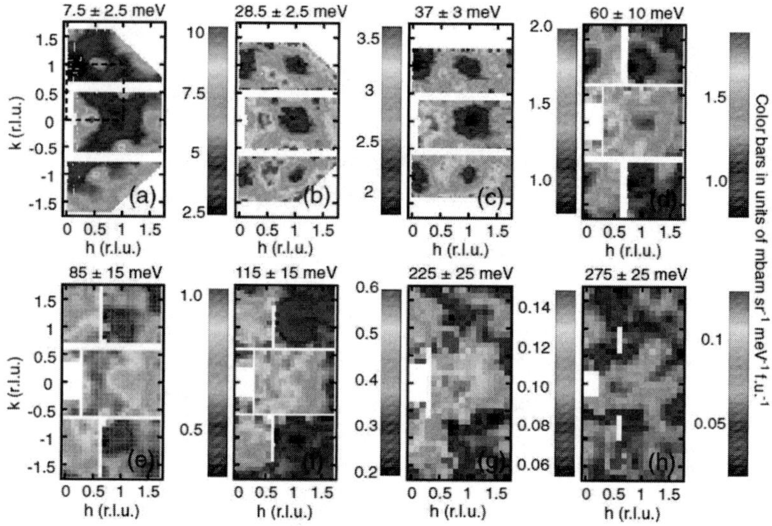

Figure 10.9 Constant energy slices of the spin waves as a function of increasing energy at 10 K for $Fe_{1.05}Te$ [27]. All data are normalized to absolute units with a vanadium standard. (a-c) Collected with incident neutron energy $E_i = 90$ meV, (d-f) $E_i = 350$ meV, (g),(h) $Ei = 500$meV. The dashed line in (a) shows a crystallographic BZ. In these plots (h, k, l) denotes the wave vector $k = (h, k, l)\pi$. See also Color Insert.

energy saving through reducing the magnetic frustration. In order to create a vacancy pattern that minimizes the magnetic exchange energy, the pattern must sufficiently suppress magnetic frustration and maximize energy saving from each magnetic exchange coupling simultaneously. It is straightforward to show that such a pattern is exactly the $\sqrt{5} \times \sqrt{5}$ pattern. With this vacancy order, each site has three NN and NNN couplings to its neighbors. Moreover one spin in each connected five spins (5-1 pattern) along Fe-Fe direction is removed. The odd number of links for both NN and NNN couplings reduce the magnetic frustration completely. The 5-1 pattern saves energy from J_3 as well. Therefore, the state maximizes the energy saving from all magnetic exchange couplings. The magnetic energy saving per spin $E = (-J_1 + J_2 + 2J_3)S^2/2$ (we set $J_3 = 0$ between two NNN sites if there is a vacancy between them). If we ignore the lattice distortion and use the exchange parameters derived from FeTe, $E \sim 70mev$, in contrast to the energy of the B-CAF state which is $2J_3 \sim 14mev$. The vacancy pattern is hence a result of the combination of energy saving from three magnetic couplings including the large contribution from the NN ferromagnetic exchange coupling. The reduction of frustration and the magnetic energy saving also explains why the magnetic transition in the 122c chalcogenides can be as high as 500 K [23], much higher than ~70 K in the 11 chalcogenides.

In the B-AF$_v$ state, there is a strong lattice distortion. Taking the lattice symmetry breaking into account, the general model which describes the spin wave excitations is a $J_1 - J_1' - J_2 - J_2' - J_3 - J_s$ model as shown in Fig. 10.3. J_s is easy axis anisotropy along c-axis since the spin moments in these materials align along the c-axis. The total energy saving in the B-AF$_v$ state, can be written as $E_0 = (-2J_1 - J_1' + 2J_2' - J_2 + 2J_3 - J_3' - J_s)S^2/2$ where the parameters are defined in Fig. 10.3. The spin wave excitations in the $B - AF_v$ state is given by

$$H = \frac{S}{2}\sum_k \psi^\dagger(k)\begin{pmatrix} A(k) & B(k) \\ B(k) & A(k) \end{pmatrix}\psi(k), \qquad (10.23)$$

where $A(k)$ and $B(k)$ are four-by-four matrices,

$$A(k) = S\begin{pmatrix} E_0 & J_1 e^{ik_x} & J_2 e^{ik_x + ik_y} + J_3' e^{-i2k_x} & J_1 e^{ik_y} \\ \cdot & E_0 & J_1 e^{ik_y} & J_2^{-ik_x + ik_y} + J_3' e^{-2ik_y} \\ \cdot & \cdot & E_0 & J_1 e^{-ik_x} \\ \cdot & \cdot & \cdot & E_0 \end{pmatrix}, \qquad (10.24)$$

$$B(k) = S\begin{pmatrix} 2J_c\cos(k_z) & J_2' e^{-ik_x + ik_y} + J_3 e^{-2ik_y} & J_1' e^{-ik_y} & J_2' e^{-ik_x - ik_y} + J_3 e^{2ik_x} \\ \cdot & 2J_c\cos(k_z) & J_2' e^{-ik_x - ik_y} + J_3 e^{2ik_x} & J_1' e^{ik_x} \\ \cdot & \cdot & 2J_c\cos(k_z) & J_2' e^{ik_x - ik_y} + J_3 e^{2ik_y} \\ \cdot & \cdot & \cdot & 2J_c\cos(k_z) \end{pmatrix}. \qquad (10.25)$$

The spinwave has four branches. The fit to experimental results produced the following exchange coupling values for $Rs_{0.8}Fe_{1.6}Se_2$: $SJ_1 = -36$, $SJ_1' = 15$, $SJ_2 = 12$, $SJ_2' = 16$, $SJ_3 = 9.5$, $SJ_3' = 0$, $SJ_c = 1.4$, $SJ_s = 0.44$ meV [26]. Once again, similar to the values in FeTe, here J_1, which is corresponding to J_{1b} in FeTe, is strongly FM, $J_2 \sim J_2'$ and J_3 is AF and necessary.

10.3.4 *A Unified Minimum Magnetic Exchange Model for Iron-Based Superconductors*

If we compare the above effective magnetic exchange couplings in different magnetically ordered states of iron-based superconductors, an intriguing pattern emerges. First, comparing iron-chalcogenides to iron-pnictides, they share similar (i) AF NNN coupling J_2 and (ii) large anistropy of NN exchange couplings ($J_{1a} >> J_{1b}$). They differ from each other by the large FM J_{1a} and significant AF J_3 in iron-chalcogenides vs the large AF J_{1a} and negligible J_3 in iron-pnictides. Second, comparing $Rb_{0.89}Fe_{1.58}Se_2$ with $Fe_{1.05}Te$, even if their static AF orders appear to be different, these two iron chalcogenides are very similar in terms of the values of their effective exchange couplings [26, 27]. Both of them have: (i) large FM J_1 (or J_{1a}), (ii) large anisotropy between the two NN couplings $J_1(J_{1a})$ and J_1' (or J_{1b}), (iii) AF NNN couplings and small anisotropy between two NNN couplings J_2(or, J_{2a}) and J_2' (or J_{2b}), and (iv) significant AF NNNN couplings J_3. Therefore, in all these families, a common difference between effective NN exchange couplings and NNN exchange couplings is that while the NN exchange couplings vary significantly according to the spin configurations between the corresponding two NN sites in the magnetically ordered states, the AF NNN exchange coupling remains almost uniform amongst different classes of materials even though their AF structures and Néel temperatures can be quite different.

The similarities allow us to propose the following general Hamiltonian $J_1 - J_2 - J_3 - K - J_z$ that can unifiedly describe the magnetic states in all families of iron based superconductors [97],

$$H = \sum_{ij,n} [J_{ij}\vec{S}_{i,n} \cdot \vec{S}_{j,n} - K_{ij}(\vec{S}_{i,n} \cdot \vec{S}_{j,n})^2] + J_z \sum_{i,n} \vec{S}_{i,n} \cdot \vec{S}_{i,n+1} \quad (10.26)$$

where J_{ij} is limited to J_1, J_2 and J_3 between two NN, NNN, or NNNN sites respectively and $K_{ij} = K$ if and only if i, j are two NN sites. This model preserves the tetragonal lattice symmetry. The classical phase diagram of the model can be obtained exactly. In Fig. 10.10, we draw a typical phase diagram in the $J_3/|J_1|$-$J_2/|J_1|$ plane by taking $KS^2/|J_1| = 0.2$. The phase diagram is almost symmetric between $J_1 > 0$ (the right part of Fig. 10.2) and $J_1 < 0$ (the left part of Fig. 10.10). For $J_1 > 0$, there are three regions for

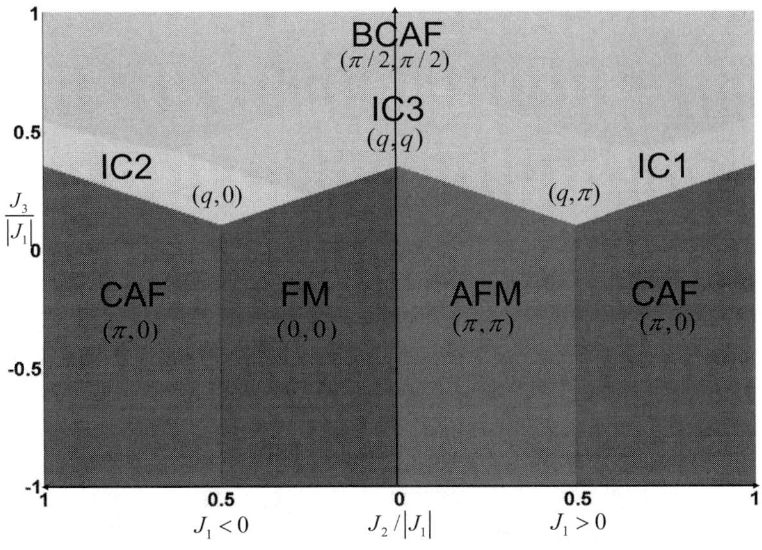

Figure 10.10 The classical phase diagram of the $J_1 - J_2 - J_3 - K$ model at $K = 0.2J_1$ [97]. There are total four commensurate and three incommensurate magnetic phases. Their labels and the ordered wavevectors are specified in the figure. See also Color Insert.

commensurate phases labeled as AFM, C-AF, B-CAF in Fig. 10.10, which exactly describe the static magnetic states of the parent compounds of curpates, ferropnictides and 11-ferrochalcogenide FeTe, respectively. Intriguingly, a B-AF commensurate state as shown in Fig. 10.11 is degenerate with the B-CAF state. The difference between them is that the B-CAF state breaks the C^4 rotation symmetry of the tetragonal lattice while the B-AF does not. The spin configuration of the B-AF state is the same as the B-AF$_v$ state except that there is no vacancy. If J_1 is switched to negative, namely ferromagnetic (FM), the AFM phase becomes a FM phase. In these commensurate phases, the biquadratic interaction term effectively creates the anisotropy of the NN magnetic exchange interactions by taking a meanfield decoupling as we discussed earlier. Therefore, effectively, this model becomes a $J_{1a} - J_{1b} - J_2 - J_3 - J_c$ model which fits the spin-wave excitations in all magnetically ordered phases reviewed in the previous sections.

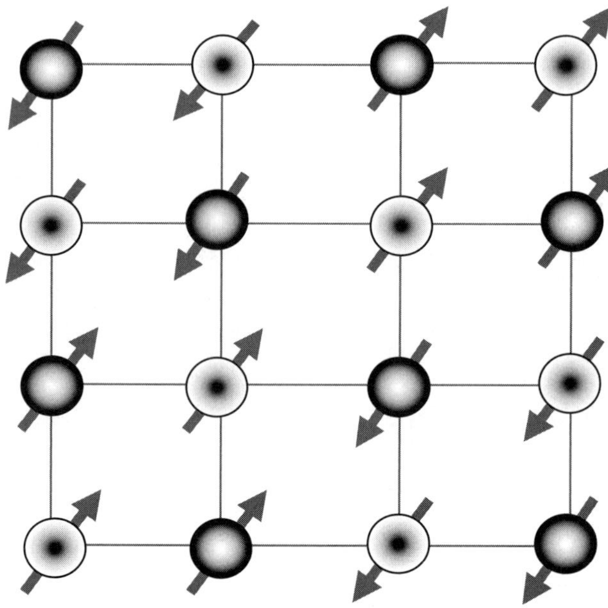

Figure 10.11 The sketch of the block-antiferromagnetic state (B-AF).

The degeneracy of the B-AF and B-CAF in the classical limit is important. Even if we consider the model as a quantum spin model, the energy difference between two phases due to quantum fluctuations are extremely small. Therefore, the degeneracy between them can be strongly broken only when the couplings to other degrees of freedom are included. For FeTe, the development of the B-CAF order is strongly tied to a monoclinic lattice distortion [20]. This monoclinic distortion breaks the same rotational symmetry of the B-CAF. Therefore, it lowers the energy of the B-CAF state. This explains why the magnetic order transition in FeTe is a strong first order type [20]. In the presence of $\sqrt{5} \times \sqrt{5}$ vacancy ordering, the magnetic frustration is strongly reduced. As the vacancy ordering does not break rotational symmetry, the B-AF order becomes the B-AF$_v$ phase. Without vacancy ordering and monoclinic lattice distortion, B-AF state should be the magnetic ground state. Recently, such state has proposed as the possible parent state of KFe_2Se_2 [82] to explain

Table 10.1 The values of the magnetic exchange interactions in $J_1 - J_2 - J_3 - K$ model obtained from the experimental results of different parent compounds of iron-based superconductors [26–28, 98]

Material	Phases	(Q_x, Q_y)	J_1S	$J_{1a}S$	$J_{1b}S$	J_2S	J_3S	KS^2	J_cS
$CaFe_2As_2$	CAF	$(0, 1)\pi$	22	50	-4	19	-	14	5
$BaFe_2As_2$	CAF	$(0, 1)\pi$	25	59	-9	14	-	17	2
FeTe	BCAF	$(\frac{1}{2}, \frac{1}{2})\pi$	-34	-15	-53	18.5	9.5	9	-
$K_{0.8}Fe_{1.6}Se_2$	BAF_v	$(\frac{3}{5}, \frac{1}{5})\pi$	-10	14	-34	16	9	12	1.4

mysterious 2×2 ordered patterns and hidden orders observed in various different experiments [77–79].

From the experimental results of spin wave excitations [26–28, 98], we can extract the magnetic exchange parameters. The results are summarized in Table 10.1. This table displays a central message that all iron based superconductors share a similar AF NNN exchange interaction J_2. However the sign of J_1 is different between ferropnictides and ferrochalcogenides and a significant AF J_3 exists in ferrochalcogenides but not in ferropnictides.

The model can also explains the incommensurate fluctuations observed in neutron experiments. There are also three incommensurate phases sandwiched between the commensurate phases with ordered incommensurate wavevectors (q, π) or (π, q) (labeled as IC1), $(0, q)$ or $(q, 0)$(labeled as IC2) and (q, q) (labeled as IC3) respectively. The static (q, q) phase was observed in $Fe_{1+y}Te$ when $y > 0.1$ [100]. Although no static (q, π) and $(0, q)$ phase has been detected, the (q, π) incommensurate spin fluctuations have been observed in $FeTe_{1-x}Se_x$ [55, 70] and electron-overdoped $Ba(Fe_{1-x}Co_x)_2As_2$ [101]. We can calculate the distinct features of spin excitations in these incommensurate phases. The typical spin excitation spectra in the three incommensurate phases as the function of wavevectors δ are shown in Fig. (10.12a–10.12f). In general, spin excitations include two branches, which can be identified as an acoustic mode and an optical mode. In the (π, q) phase, the two modes connect with each other at two new incommensurate positions $\delta = \pm q_0$ when the wavevector (π, δ) varies as shown Fig. (10.12a). If we calculate the intensity of these spin wave excitations measured by neutron scattering, an hour-glass

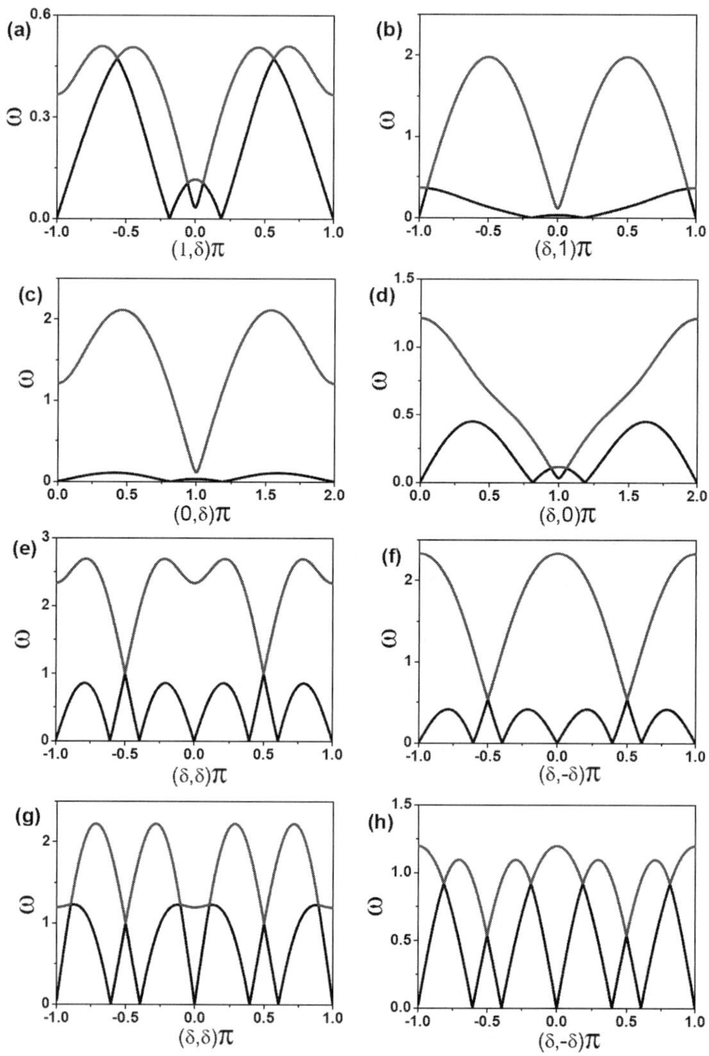

Figure 10.12 The spin waves in incommensurate phases: (a) along $(1,\delta)\pi$ and (b) along $(\delta,1)\pi$ in the (π,q) phase, (c) along $(0,\delta)\pi$ and (d) along $(\delta,0)\pi$ in the $(0,q)$ phase, and (e)/(g) along $(\delta,\delta)\pi$ and (f)/(h) along $(\delta,-\delta)\pi$ in the (q,q)-phase. The parameters are fitted to be $S = 1$, $(J_1, J_2/|J_1|, J_3/|J_1|, KS^2/|J_1|) = (1, 0.6, 0.06, 0)$ in (a) and (b), $(-1, 0.6, 0.06, 0)$ in (c) and (d), $(-1, 0.6, 0.5, 0.05)$ in (e) and (f), and $(1, 0.6, 0.5, 0.05)$ in (g) and (h).

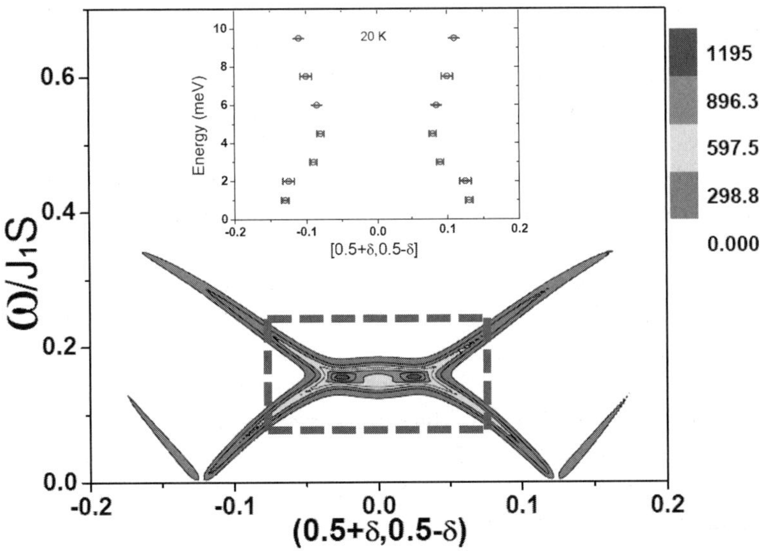

Figure 10.13 The hour-glass-like spin waves along $(1/2+\delta, 1/2-\delta)_{\mathrm{T}}$ in the (π, q) phase. The parameters are set to $(J_1S, J_2/J_1, J_3/J_1, KS^2/J_1) = (30, 0.55, 0.045, 0.02)$. The inset depicts the experimental results released in Ref. [55]. The observed region is labeled by the dashed lines.

like behavior along this direction becomes prominent at low energy as shown in Fig. 10.13. This behavior has been recently reported in $FeSe_{0.4}Te_{0.6}$ [55] and a clear explanation was not given before.

10.4 Electronic Nematism and the Interplay Between Lattice, Spin and Orbital

In the previous section, we have mentioned that the C-AF state in iron-pnictides breaks C^4 tetragonal symmetry. This symmetry breaking is shared by the structural transition which takes place above or at the C-AF magnetic transition temperature. Such a C^4 symmetry breaking can be described by an additional order parameter in the C-AF state, a nematic order, $\sigma_i = \mathbf{S}_i\mathbf{S}_{i+\hat{x}} - \mathbf{S}_i\mathbf{S}_{i+\hat{y}}$, where $\hat{x}$ and $\hat{y}$ are lattice unit vectors in the tetragonal lattice of iron atoms. A nematic phase is developed when $<\sigma_i> \neq 0$. Such a phase is naturally included in the effective model $J_1 - J_2 - K - J_z$ model. The nematic phase has also been formulated in the context of spin density wave order with two degenerate order wavevectors $(0, \pi)$ and $(\pi, 0)$ [41], as well as a ferro-orbital order [83]. Regardless of the difference among different proposals, the effective theories resulted from these proposals are quite similar. Here we will focus on the effective magnetic model, $J_1 - J_2 - K - J_z$, to discuss the nematism based on spin interactions [35, 36, 97].

Before we proceed quantitative analysis, we would like to mention the different sources that can contribute the value of K effectively. In the previous section, we have shown that magnetic fluctuations in $J_1 - J_2$ model contributes a small K value effectively. The second contribution is the spin-lattice coupling. The coupling parameters can also be a function of the distance between spins. In particular, the NN coupling $J_1(r)$ can couple to the lattice distortion. Let's consider a small orthorhombic lattice distortion δr, the coupling,

$$\begin{aligned} J_1(r)(\mathbf{S}_i\mathbf{S}_{i+\hat{x}} + \mathbf{S}_i\mathbf{S}_{i+\hat{y}}) \to\, & J_1(r_0 + \delta r/2)\mathbf{S}_i\mathbf{S}_{i+\hat{x}} \\ & + J_1(r_0 - \delta r/2)\mathbf{S}_i\mathbf{S}_{i+\hat{y}}. \end{aligned} \tag{10.27}$$

We can expand the above term and obtain a coupling between the lattice distortion and the nematic order, $\lambda_{sl}(r_0)\delta r\sigma_i$, where $\lambda_{sl}(r) = \frac{1}{2}\frac{\partial J_1(r)}{\partial r}$. There is also coupling between the nematic order and an onsite orbital order between d_{xz} and d_{yz} orbitals, namely, $\lambda_{so}(n_{xz} - n_{yz})\delta r\sigma_i$, where $n_{xz(yz)}$ is electron density at the corresponding orbitals. If we consider that the magnetism is the dominant player in the system, we can integrate out the above lattice and orbital fluctuations in the tetragonal lattice, which contributes to the value of K.

For convenience in analytic and quantitative analysis, we can derive an effective Hamiltonian in the continuum limit. This derivation has been done in [36, 97]. The low energy effective Hamiltonian reads

$$\begin{aligned} H_{eff} = \int d^2r \sum_{n,\alpha} [\tilde{J}_2|\nabla \mathbf{n}_{\alpha,n}|^2 - \tilde{J}_z \mathbf{n}_{\alpha,n} \cdot \mathbf{n}_{n+1,\alpha}] \\ -\tilde{K}\sum_n [\mathbf{n}_{1,n} \cdot \mathbf{n}_{2,n}]^2 \end{aligned} \tag{10.28}$$

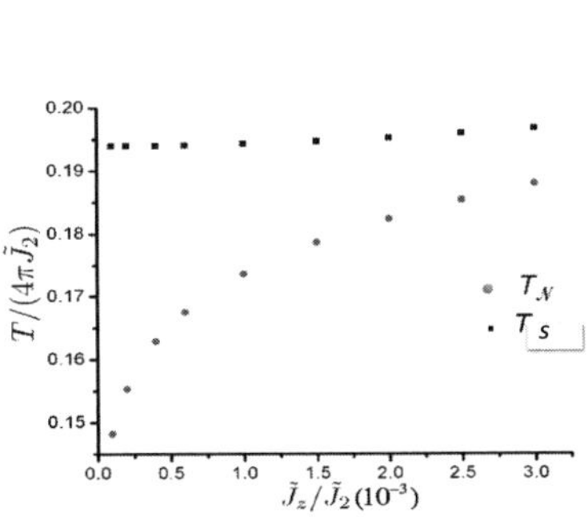

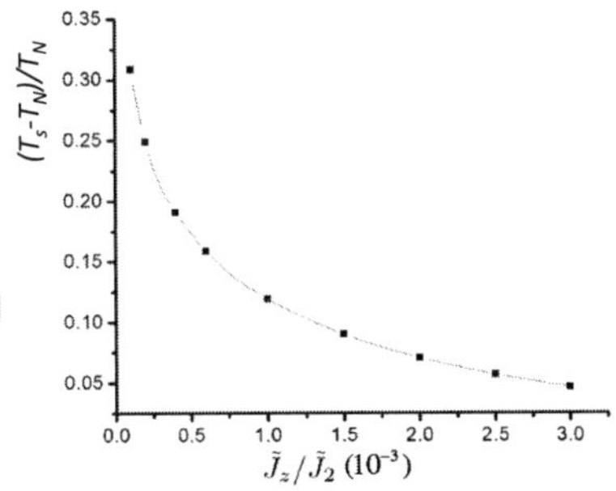

Figure 10.14 Left: T_s and T_N as the function of $\tilde{J}_z$ for $\tilde{J}_2 = 2\tilde{J}_1$ and $N = 3$. Right: $\frac{T_s - T_N}{T_N}$ as the function of $\tilde{J}_z$ for $\tilde{J}_1 = 0.5\tilde{J}_2$.

where $\tilde{J}_i = J_i S^2$, $\tilde{K} = K S^4/2$ where $\mathbf{n}_{1(2),n}$ are AF orders in the two sublattices respectively as shown in Fig. 10.1 and the nematic order in the continuum limit becomes $\sigma = \mathbf{n}_{1,n} \cdot \mathbf{n}_{2,n}$.

We can solve the effective model in the large-N limit analytically. In general, there are two transition temperatures. The nematic transition temperature T_s is given by

$$\frac{4\pi\tilde{J}_2}{NT_s} = \ln\frac{\tilde{J}_2/(NT_s)}{\sqrt{\left(\frac{\tilde{K}}{4\pi\tilde{J}_2}\right)^2 + \left(\frac{\tilde{J}_z}{NT_s}\right)^2} + \frac{\tilde{K}}{4\pi\tilde{J}_2}}, \tag{10.29}$$

and the C-AF transition temperatures, T_N is determined by,

$$\begin{aligned}\frac{\sigma_c}{2\tilde{K}} &= \frac{1}{8\pi\tilde{J}_2}\ln\frac{\sigma_c + \frac{\tilde{J}_z}{NT_N} + 2\sqrt{\frac{\sigma_c^2}{4} + \frac{\sigma_c\tilde{J}_z}{NT_N}}}{\tilde{J}_z/(NT_N)}\\ \frac{\sigma_c}{2\tilde{K}} + \frac{1}{NT_N} &= \frac{1}{4\pi\tilde{J}_2}\ln\frac{\tilde{J}_2}{\tilde{J}_z}.\end{aligned} \tag{10.30}$$

T_N is always lower than T_s in the solutions of above equations. Moreover, $T_N = 0$ if $J_z = 0$, and is finite otherwise. For small J_z, one can approximately have $T_N \sim \tilde{J}_2/\ln(\frac{\tilde{J}_z}{\tilde{J}_z})$. In Fig. 10.14, we plot the two transition temperatures and their difference as the function of J_z/J_2. The difference decreases as J_z increases. This qualitative prediction is confirmed in the 1111 family [47] as shown in Fig. 10.15.

This calculation shows that as soon as J_z is much larger than $10^{-3}J_z$, the two transitions are close to each other so that

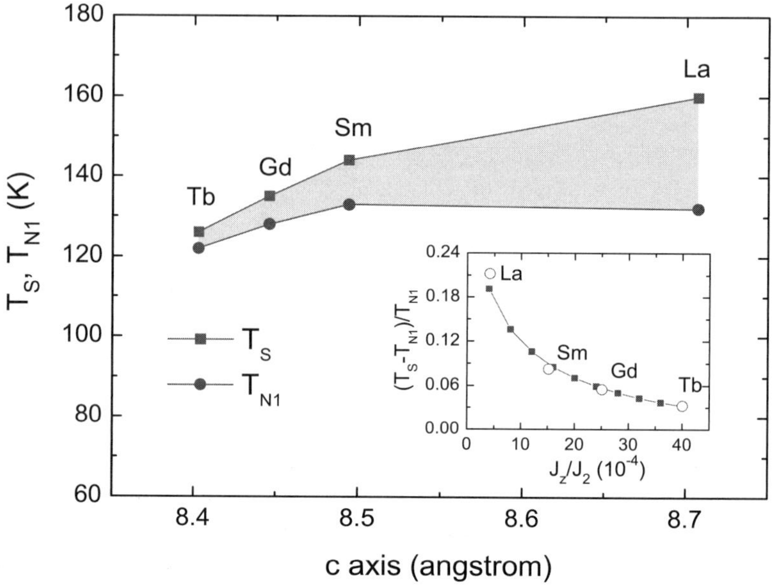

Figure 10.15 Plot of structural phase transition temperature T_s, and magnetic transition temperature T_N associated to the AF order of Fe ions, versus the c-axis for RFeAsO (R = La, Sm, Gd, Tb). Inset: plot of $(T_S - T_N)/T_N$ versus J_z/J_2. The open circles denote the experimental data of $(T_S - T_N)/T_N$, and the solid squares denote the theoretical values [47].

they become practically inseparable in experiments. This result is consistent with the fact there is only one phase transition in the parent compounds of the 122 families where J_z determined by spin wave excitations is rather large: in $CaFe_2Se_2$, J_z is almost one third of the in-plane coupling J_2 [28]. In $BaFe_2As_2$, J_z is about $0.015J_2$ [98]. The prediction is further confirmed in under-doped 122 materials. Upon electron doping in $BaFe_2As_2$, for example, $BaFe_{2-x}Ni_xAs_2$ [53], the spin excitations become much less three dimensional. The effective J_z is drastically reduced and two phase transitions were observed. Moreover, if the two transition is extremely close, it is most likely that the two transition becomes one first-order transition, i.e., the lattice distortion becomes finite immediately after lowering temperature to T_N and spins are ordered simultaneously. This can be easily generally argued using a phenomenological theory by considering the coupling betweenI the nematic order and the C-AF order [35].

In the case of B-CAF or B-AF states, the similar effective model for $J_1 - J_2 - J_3 - K - J_z$ model has also been derived in ref. [97, 102]. The complete field description of the $J_1 - J_2 - J_3 - K - J_z$ model near the B-CAF or B-AF phase is given by $H_T = H_B + H_K$ where

$$H_B = \int d^2\mathbf{r} \sum_{n,\alpha} \left[\frac{1}{2} J_3 |\nabla \vec{\phi}_{n,\alpha}(\mathbf{r})|^2 - J_z \vec{\phi}_{n,\alpha}(\mathbf{r}) \cdot \vec{\phi}_{n+1,\alpha}(\mathbf{r}) \right]$$
$$-g' \sum_n \left\{ \left[\vec{\phi}_{n,1}(\mathbf{r}) \cdot \vec{\phi}_{n,3}(\mathbf{r}) \right]^2 + \left[\vec{\phi}_{n,2}(\mathbf{r}) \cdot \vec{\phi}_{n,4}(\mathbf{r}) \right]^2 \right\}, \quad (10.31)$$

and

$$H_K = -K \int d^2\mathbf{r} \sum_n \left\{ \left[\vec{\phi}_{n,1}(\mathbf{r}) \cdot \vec{\phi}_{n,2}(\mathbf{r}) \right]^2 + \left[\vec{\phi}_{n,1}(\mathbf{r}) \cdot \vec{\phi}_{n,4}(\mathbf{r}) \right]^2 \right.$$
$$\left. + \left[\vec{\phi}_{n,2}(\mathbf{r}) \cdot \vec{\phi}_{n,3}(\mathbf{r}) \right]^2 + \left[\vec{\phi}_{n,3}(\mathbf{r}) \cdot \vec{\phi}_{n,4}(\mathbf{r}) \right]^2 \right\} \quad (10.32)$$

where $\vec{\phi}_{n,\alpha}$ with $\alpha = 1, 2, 3, 4$ are four Neel order parameters defined in the four sublattices of the tretragonal lattice specified by the J_3 coupling and $g' \sim 0.13 J_2^2/J_3$. In this model, there are four independent Ising orders $\sigma_1 = g' \vec{\phi}_{n,1}(\mathbf{r}) \cdot \vec{\phi}_{n,3}(\mathbf{r})$, $\sigma_2 = g' \vec{\phi}_{n,2}(\mathbf{r}) \cdot \vec{\phi}_{n,4}(\mathbf{r})$, $\sigma_+ = K(\vec{\phi}_{n,1}(\mathbf{r}) + \vec{\phi}_{n,3}(\mathbf{r})) \cdot (\vec{\phi}_{n,2}(\mathbf{r}) + \vec{\phi}_{n,4}(\mathbf{r}))$ and $\sigma_- = K(\vec{\phi}_{n,1}(\mathbf{r}) - \vec{\phi}_{n,3}(\mathbf{r})) \cdot (\vec{\phi}_{n,2}(\mathbf{r}) + \vec{\phi}_{n,4}(\mathbf{r}))$. σ_+ and σ_- describe the ising orders associated with the B-AF and B-CAF phases respectively. The analysis of this effective field theory has been carried out in [97, 102]. This effective theory also suggests that the lattice-spin coupling is extremely important in determining the true magnetic order in iron-chalcogenides due to intrinsic magnetic frustration.

10.5 Discussion

We have reviewed the magnetic orders and the effective models in the parent compounds of iron based superconductors and established a clear picture regarding the important local magnetic exchange couplings. The essential message of the picture is that the important AF exchange couplings in cuprates are J_1, in iron-pnictides are J_2 and in iron-chlcogenides are J_2 and J_3. In cuprates, the magnetic order is an AF order with an ordered wavevector (π, π). This state can be naturally derived from a Heisenberg model where only the NN AF interaction J_1 is important and longer range magnetic exchange interactions can be ignored. Iron-pnictides are in a C-AF state with an ordered wavevector $(\pi, 0)$. The C-AF order is mainly determined by the NNN AF J_2. Iron-chalcogenides are featured with several different magnetic orders. However, all these magnetic orders can be unifiedly understood by including a ferromagnetic (FM) J_1 and a NNNN AF J_3 as well as the AF J_2. The AF J_2 does not differ significantly from iron-pnictides.

Based on the picture of the local magnetic exchange couplings in all of these high T_c materials, a unified paradigm to understand high T_c has been proposed recently [66], which states that the key ingredients in the determination of high T_c and pairing symmetries are local AF exchange interactions in real space and Fermi surface topology in reciprocal space that can match to the pairing form factor provided by the AF interactions. This paradigm is essentially the summary of the theoretical studies for cuprates [103], iron-pncitides [104] and iron-chalcogenides [65]. It provides a clear theoretical roadmap to predict undiscovered high T_c materials by identifying local AF exchange and collaborative Fermi surfaces.

In our effective model, we ignore the orbital degree of freedom completely. This appoximation is only valid if the strength of Hunds coupling overcome the entire band width. In the parent compounds of iron-pnictides, the magnetic moment does not reach $4\mu_b$ which suggests that the contribution to magentism from different orbitals can be very different [105]. In order to understand how magnetism is suppressed and superconductivity is developed upon doping, a microscopic model including the orbital aspect is very important. Research in this direction will be a major theoretical effort in the future. Nevertheless, if we just consider low energy physics, we may simply combine a bare kinetic energy together with the effective exchange models to qualitatively address robust and generic properties, such as pairing symmetry which can be predicted in such an approach [65, 67, 104]. Determining pairing symmetries of iron-based superconductors thus can detrmine pairing mechanism and provide a test of the results in this paper.

Acknowledgments

I am grateful to Institute of Physics, Chinese Academy of Sciences for supporting my research in the past two years when I had difficulty in obtaining any other research fundings. I am also grateful to my collaborators, S. Kivelson, P. C. Dai, D. L. Feng, D. Hong, B. A. Bernevig, X. H. Chen, Q. K. Xue, X. Chen, R. Thomale, T. Xiang, H. Q. Yuan, Z. Y. Lu, Z. Xu, J. Zhao, H. Yao, H. H. Wen, N. L. Wang, G. Q. Zheng and C. Xu, my group current and former members, C. Fang, K. Seo, W. Li, C. Chandan, X. Zhou, B. Xu, N. N. Hao, and W. Tsai.

References

1. Y. Kamihara, T. Watanabe, M. Hirano, and H. Hosono, J. Am. Chem. Soc. **130**, 3296 (2008).
2. Z.-A. Ren *et al.*, Chin. Phys. Lett. **25**, 2215 (2008).
3. X. H. Chen *et al.*, Nature **453**, 761 (2008).
4. M. Rotter, M. Tegel, and D. Johrendt, Phys. Rev. Lett. **101**, 107006 (2008).
5. X. Zhu *et al.*, Phys. Rev. B **79**, 220512(R) (2009).
6. F.-C. Hsu *et al.*, PNAS **105**, 14262 (2008).
7. J. Guo *et al.*, Phys. Rev. B **82**, 180520 (2010).
8. M.-H. Fang *et al.*, Europhys. Lett. **94**, 27009 (2011).
9. R. H. Liu *et al.*, Europhys. Lett. **94**, 27008 (2011).
10. Y. J. Yan *et al.*, Sci. Rep. **2**, 212 (2012).
11. J. M. Caron *et al.*, Phys. Rev. B (Rapid) **84**, 180409 (2011).
12. A. Krzton-Maziopa *et al.*, J. Phys. Condens. Matter **23**, 2201 (2011).
13. D. G. F. Emma E. McCabe and J. S. O. Evans, Chem. Commun. **47**, 1261 (2010).
14. L. Sun *et al.*, Nature **483**, 67 (2012).
15. C. de la Cruz *et al.*, Nature **453**, 899 (2008).
16. M. A. McGuire *et al.*, Phys. Rev. B **78**, 094517 (2008).
17. J. Dong *et al.*, Europhys. Lett. **83**, 27006 (2008).
18. P. J. Hirschfeld, M. M. Korshunov, and I. I. Mazin, arxi:1106.3702 (2011).
19. W. Bao *et al.*, Phys. Rev. Lett. **102**, 247001 (2009).
20. S. L. Li *et al.*, Phys. Rev. B **79**, (2009).

21. F. J. Ma *et al.*, Phys. Rev. Lett. **102**, (2009).
22. C. Fang, B. A. Bernevig, and J. P. Hu, EPL **86**, (2009).
23. W. Bao *et al.*, Chin. Phys. Lett. **28**, 6104 (2011).
24. W. Bao *et al.*, arXiv:1102.3674 (2011).
25. D. Johnston, Adv. Phys. **59**, 803 (2010).
26. M. Wang *et al.*, Nat. Commun. **2**, 580 (2011).
27. O. J. Lipscombe *et al.*, Phys. Rev. Lett. **106**, 57004 (2011).
28. J. Zhao *et al.*, Nat. Phys. **5**, 555 (2009).
29. H. Q. Yuan *et al.*, arXiv:1102.1381 (2011).
30. R. J. McQueeney *et al.*, Phys. Rev. Lett. **101**, 227205 (2008).
31. K. Matan, R. Morinaga, K. Iida, and T. J. Sato, Phys. Rev. B **79**, 54526 (2009).
32. C. Zhang *et al.*, Sci. Rep. **1**, 115 (2011).
33. R. A. Ewings *et al.*, Phys. Rev. B **83**, 214519 (2011).
34. Q. Si and E. Abrahams, Phys. Rev. Lett. **101**, 76401 (2008).
35. C. Xu, M. Mller, and S. Sachdev, Phys. Rev. B **78**, 20501 (2008).
36. C. Fang *et al.*, Phys. Rev. B **77**, (2008).
37. M. Aichhorn *et al.*, Phys. Rev. B **82**, 64504 (2010).
38. A. L. Wysocki, K. D. Belashchenko, and V. P. Antropov, Nat. Phys. **7**, 485 (2011).
39. K. Kuroki *et al.*, Phys. Rev. Lett. **101**, 087004 (2008), pRL.
40. K. J. Seo, C. Fang, B. A. Bernevig, and J. P. Hu, Phys. Rev. B **79**, (2009).
41. I. Eremin and A. V. Chubukov, Phys. Rev. B **81**, 24511 (2010).
42. A. V. Chubukov, Physica C **469**, 640 (2009).
43. J. W. Lynn and P. Dai, Physica C **469**, 469 (2009).
44. J. Zhao *et al.*, Phys. Rev. B **78**, (2008).
45. J. Zhao *et al.*, Phys. Rev. Lett. **101**, (2008).
46. Y. Qiu *et al.*, Phys. Rev. Lett. **101**, 257002 (2008).
47. Y. K. Luo *et al.*, Phys. Rev. B **80**, (2009).
48. S. D. Wilson *et al.*, Phys. Rev. B **79**, 184519 (2009).
49. C. Rotundu and R. J. Birgeneau, arXiv: 1106.5761 (2011).
50. C. He *et al.*, Phys. Rev. Lett. **105**, (2010).
51. L. Ma *et al.*, Phys. Rev. B **83**, 132501 (2011).
52. M. Wang *et al.*, Phys. Rev. B(R) **83**, 220515(R) (2011).
53. L. W. Harriger *et al.*, Phys. Rev. Lett. **103**, (2009).

54. S.-H. Lee *et al.*, Phys. Rev. B **81**, 220502 (2010).
55. S. L. Li *et al.*, Phys. Rev. Lett. **105**, (2010).
56. H. A. Mook *et al.*, Phys. Rev. Lett. **104**, 187002 (2010).
57. E. E. Rodriguez *et al.*, Phys. Rev. B **83**, 134438 (2011).
58. E. E. Rodriguez *et al.*, arXiv:1105.1937(2011).
59. Y. Zhang *et al.*, Nat. Mater. **10**, 273 (2011).
60. X.-P. Wang *et al.*, Europhys. Lett. **93**, 57001 (2011).
61. D. Mou *et al.*, Phys. Rev. Lett. **106**, 107001 (2011).
62. S. Maiti *et al.*, Phys. Rev. Lett. **107**, 147002 (2011).
63. Y.-Z. You, H. Yao, and D.-H. Lee, Phys. Rev. B **84**, 20406 (2011).
64. T. A. Maier, S. Graser, P. J. Hirschfeld, and D. J. Scalapino, Phys. Rev. B **83**, 100515 (2011).
65. C. Fang *et al.*, Phys. Rev. X **1**, 011009 (2011).
66. J. Hu and H. Ding, Sci. Rep. **2**, 381 (2012).
67. R. Yu *et al.*, arXiv:1103.3259 (2011).
68. T. Saito, S. Onari, and H. Kontani, Phys. Rev. B **83**, 140512 (2011).
69. F. Chen *et al.*, Phys. Rev. B **81**, (2010).
70. M. D. Lumsden *et al.*, Nat. Phys. **6**, 182 (2010).
71. D. Wu *et al.*, Phys. Rev. B **79**, 155103 (2009).
72. Z. G. Chen, R. H. Yuan, T. Dong, and N. L. Wang, Phys. Rev. B **81**, 100502 (2010).
73. R. M. Fernandes and J. Schmalian, Phys. Rev. B **82**, 14520 (2010).
74. M. Nakajima *et al.*, Phys. Rev. B **81**, 104528 (2010).
75. L. X. Yang *et al.*, Phys. Rev. Lett. **102**, (2009).
76. C.-H. Li *et al.*, Phys. Rev. B **83**, 184521 (2011).
77. F. Chen *et al.*, Phys. Rev. X **1**, 021020 (2011).
78. P. Cai *et al.*, Phys. Rev. B **85**, 094512 (2012).
79. W. Li *et al.*, Nat. Phys. 10.1038/nphys2155 (2011).
80. Z. Wang *et al.*, Phys. Rev. B **83**, 140505 (2011).
81. R. H. Yuan *et al.*, Sci. Rep. **2**, 221 (2012).
82. L. Wei, S. Dong, C. Fang, and J. Hu, Phys. Rev. B **85**, 100407(R) (2012).
83. I. R. Fisher, L. Degiorgi, and Z. X. Shen, Rep. Prog. Phys. **74**, 124506 (2011).
84. J. J. Ying *et al.*, Phys. Rev. Lett. **107**, 067001 (2011).
85. M. Yi *et al.*, PNAS **108**, 6878 (2011).

86. F. Ma, Z.-Y. Lu, and T. Xiang, Phys. Rev. B **78**, 224517 (2008).
87. F. Ma and Z.-Y. LU, Phys. Rev. B **78**, 033111 (2008).
88. D. J. Singh and M. H. Du, Phys. Rev. Lett. **100**, 237003 (2008).
89. A. Subedi, L. Zhang, D. J. Singh, and M. H. Du, Phys. Rev. B **78**, 134514 (2008).
90. T. Yildirim, Phys. Rev. Lett. **101**, 57010 (2008).
91. M. J. Han, Q. Yin, W. E. Pickett, and S. Y. Savrasov, Phys. Rev. Lett. **102**, 107003 (2009).
92. X.-W. Yan, M. Gao, Z.-Y. Lu, and T. Xiang, Phys. Rev. B **84**, 54502 (2011).
93. X.-W. Yan, M. Gao, Z.-Y. Lu, and T. Xiang, Phys. Rev. B **83**, 233205 (2011).
94. W.-G. Yin, C.-H. Lin, and W. Ku, arXiv:1106.0881 (2011).
95. K. W. Lee, V. Pardo, and W. E. Pickett, Phys. Rev. B **78**, (2008).
96. P. Chandra, P. Coleman, and A. I. Larkin, Phys. Rev. Lett. **64**, 88 (1990).
97. J. Hu *et al.*, Phys. Rev. B **85**, 144403 (2012).
98. L. W. Harriger *et al.*, Phys. Rev. B **84**, 054544 (2010).
99. C. Fang *et al.*, Phys. Rev. B **85**, 134406 (2012).
100. W. Bao *et al.*, Phys. Rev. Lett. **102**, 247001 (2009).
101. P. Bonville, F. Rullier-Albenque, D. Colson, and A. Forget, arXiv: 1002.0931 (2010).
102. C. Xu and J. Hu, 0903.4477 (2009).
103. P. Anderson *et al.*, J. Phys. Condens. Matter **16**, R755 (2004).
104. K. J. Seo, B. A. Bernevig, and J. P. Hu, Phys. Rev. Lett. **101**, (2008).
105. W. G. Yin, C. C. Lee and W. Ku, Phys. Rev. Lett. **105**, 107004 (2008).

Index

Color Insert

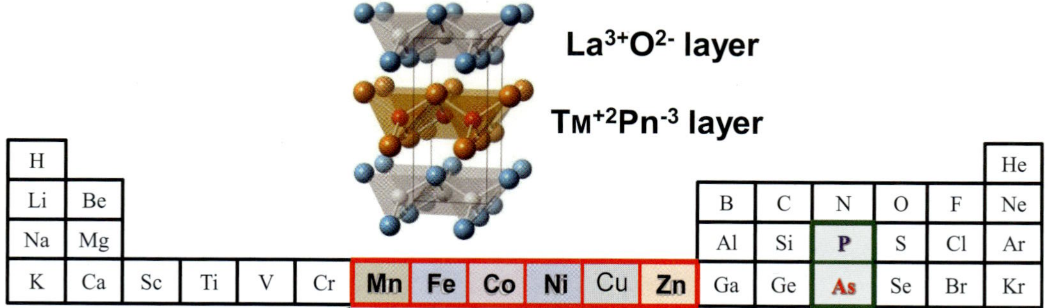

TM^{2+}(electron configuration)	Mn(3d^5)		Fe(3d^6)		Co(3d^7)		Ni(3d^8)		(Cu)	Zn(3d^{10})	
Pn	P	As	P	As	P	As	P	As		P	As
Elect. Prop.	Mott Insulator		Superconductor		Metal		Superconductor		–	Semiconductor	
Magnetism	AF				FM				–	Non-magnetic	
E_g	~1 eV		–		–		–		–	~1.5 eV	
T_C (SC) $T_{N,C}$(Mag.)	> 400 K		Undoped: 4 K	F-doped: 26 K	43 K	66 K	Undoped: 3 K	Undoped: 2.4 K		–	
Ref.	Yanagi *et al.* JAP (2009)		Kamihara *et al.* JACS (2006), Kamihara *et al.* JACS (2008)		Yanagi *et al.* PRB (2008)		Watanabe *et al.* IC (2007), Watanabe *et al.* JSSC (2007)		–	Kayanuma *et al.* PRB (2007), Kayanuma *et al.* TSF (2008)	

Figure 1.1

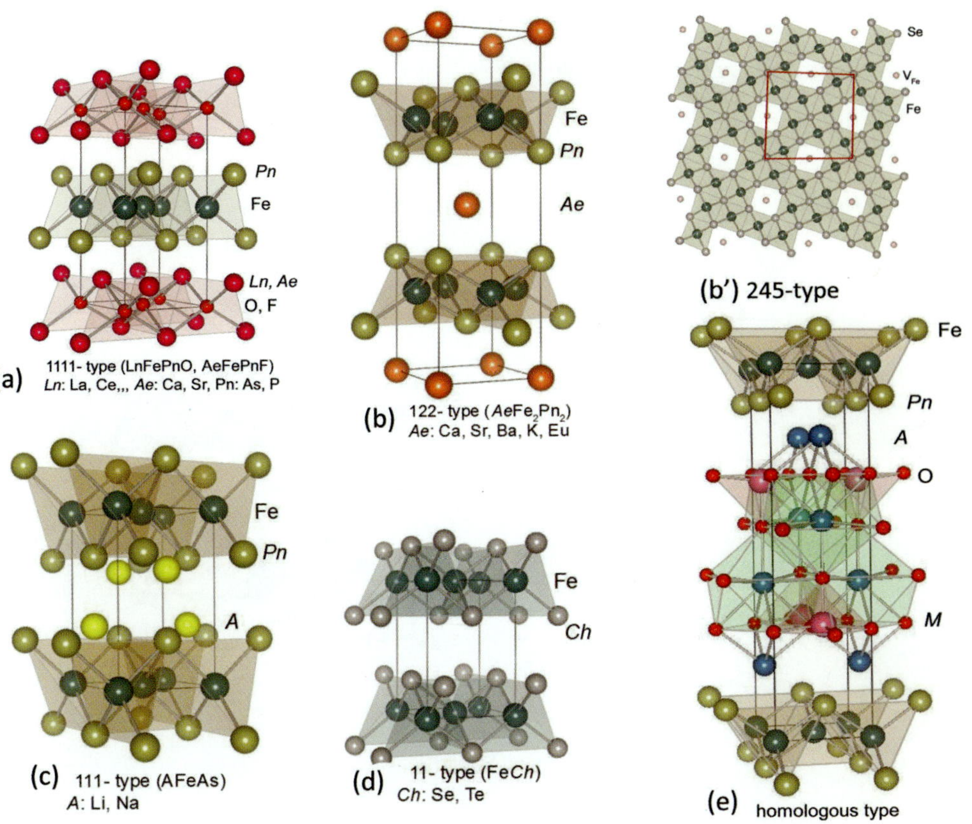
Pn
Fe
Ln, Ae
O, F
(a)
1111- type (LnFePnO, AeFePnF)
Ln: La, Ce,,, Ae: Ca, Sr, Pn: As, P
Fe
Pn
Ae
(b)
122- type ($AeFe_2Pn_2$)
Ae: Ca, Sr, Ba, K, Eu
Se
V_{Fe}
Fe
(b') 245-type
Fe
Pn
A
O
M
(e) homologous type
Fe
Pn
A
(c) 111- type (AFeAs)
A: Li, Na
Fe
Ch
(d) 11- type (FeCh)
Ch: Se, Te

Figure 1.4

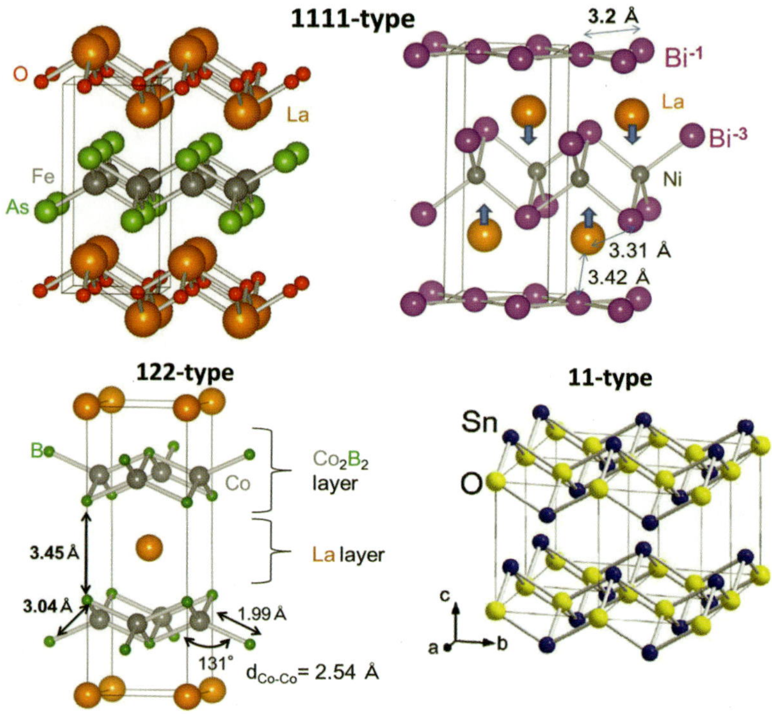
1111-type
O
La
Fe
As
3.2 Å
Bi-1
La
Bi-3
Ni
3.31 Å
3.42 Å
122-type
B
Co
Co2B2
layer
La layer
3.45 Å
3.04 Å
1.99 Å
131°
dCo-Co= 2.54 Å
11-type
Sn
O
c
a
b

Figure 1.24

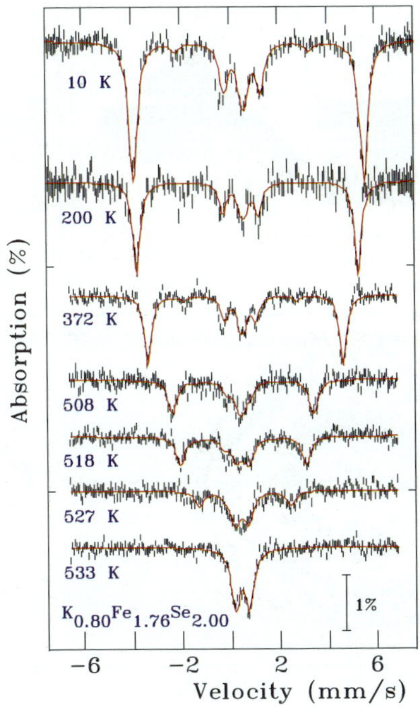

Figure 2.12

Figure 3.2

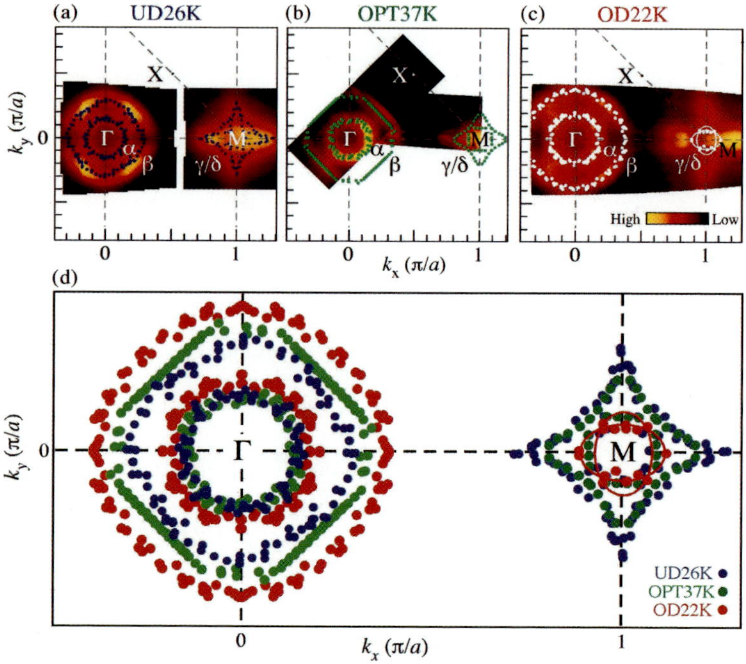

Figure 3.9

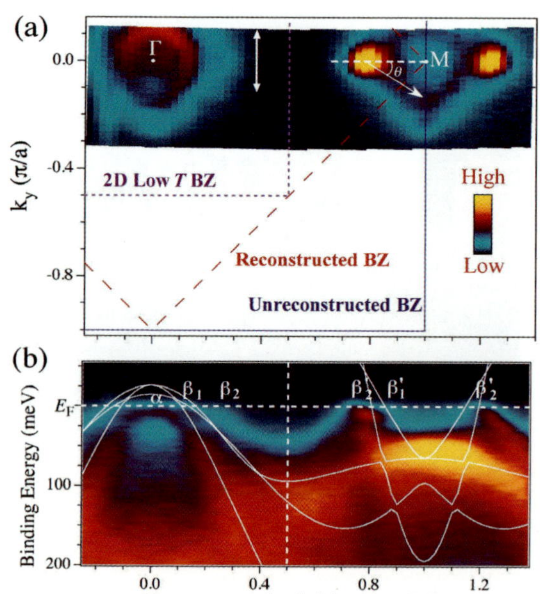

Figure 3.20

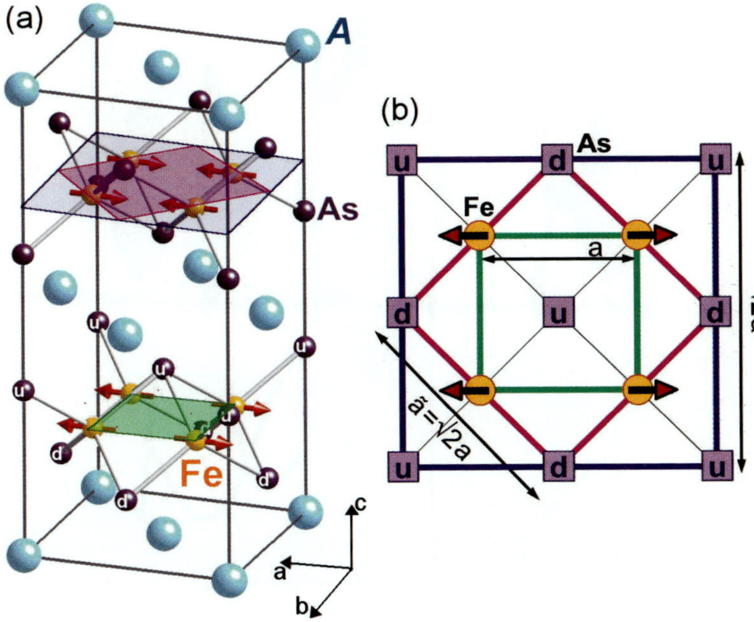

Figure 4.4

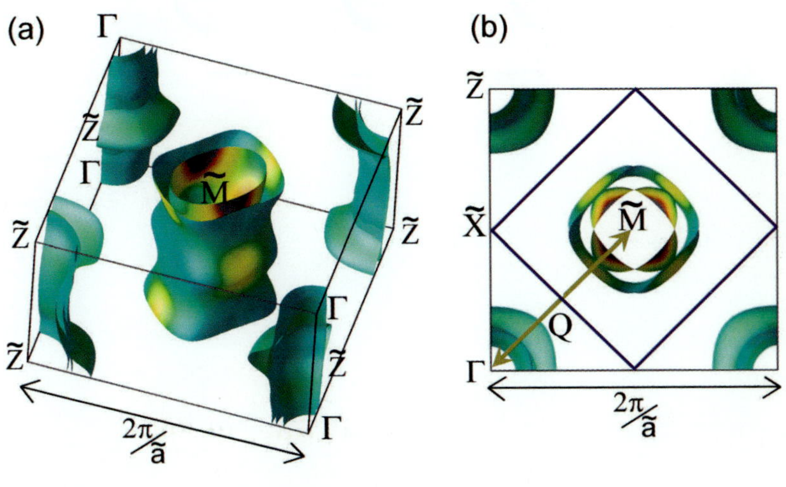

Figure 4.5

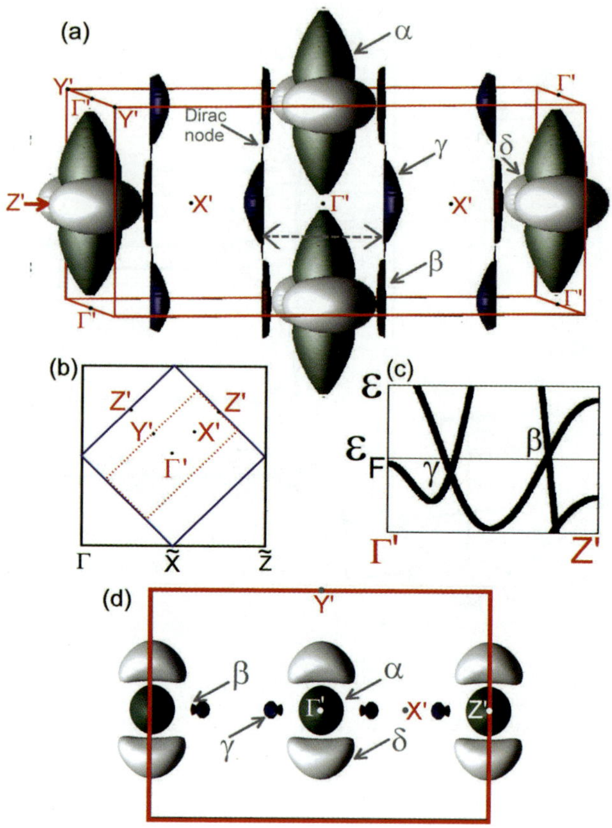

Figure 4.6

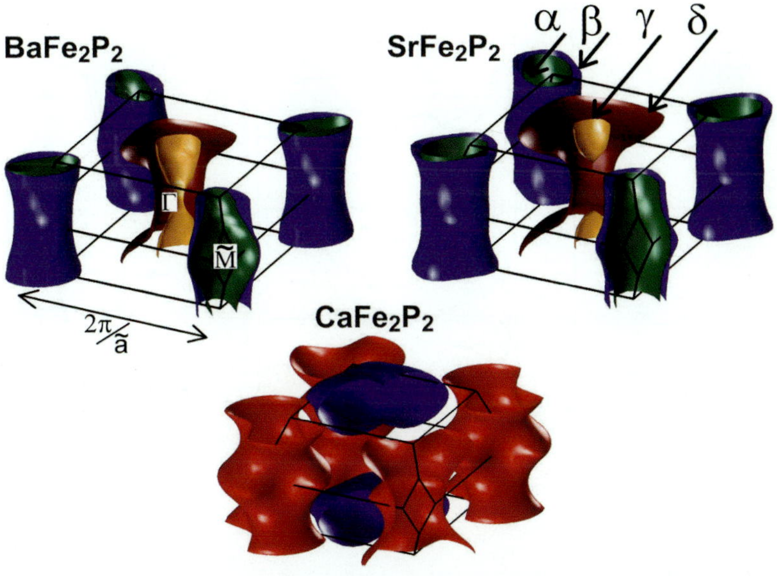

Figure 4.9

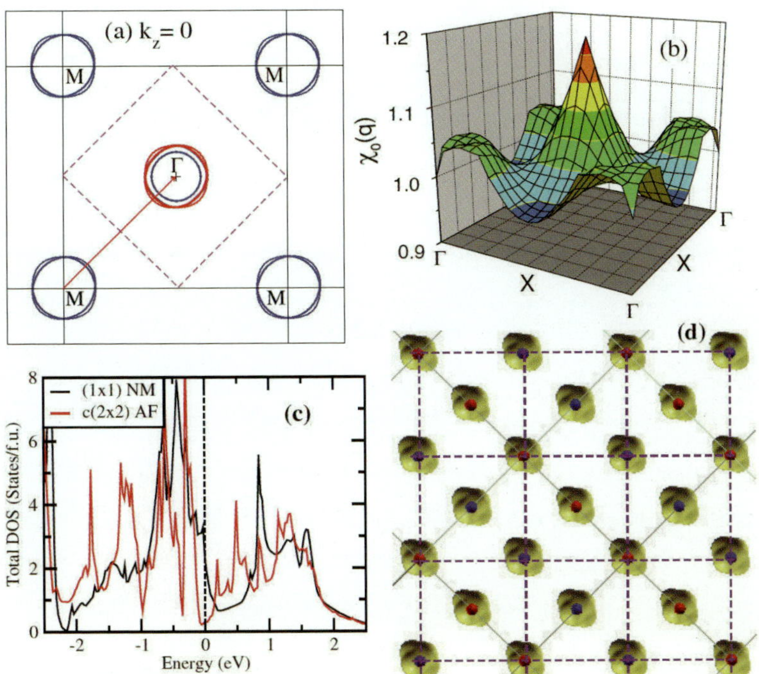

Figure 5.8

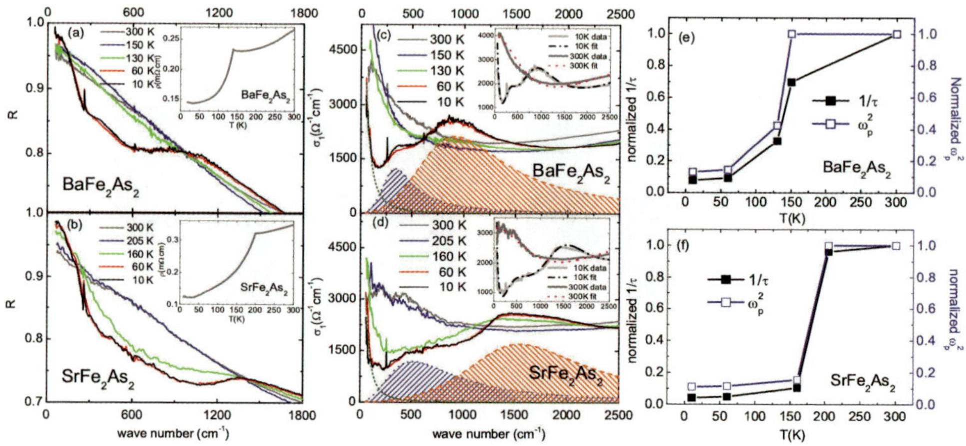

Figure 5.9

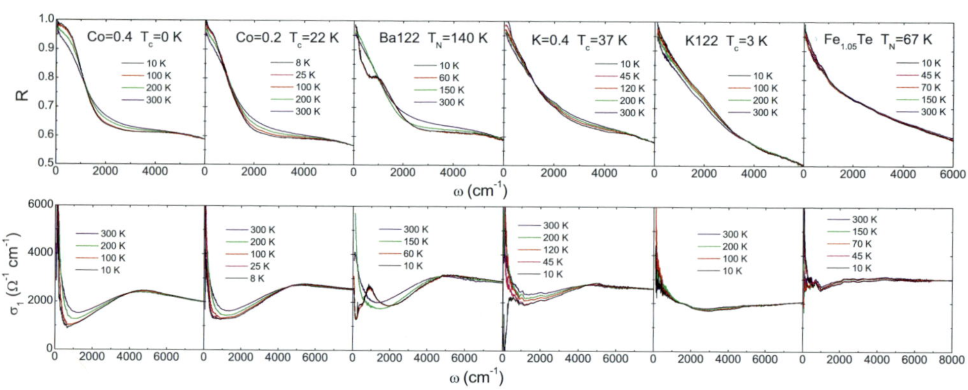

Figure 5.21

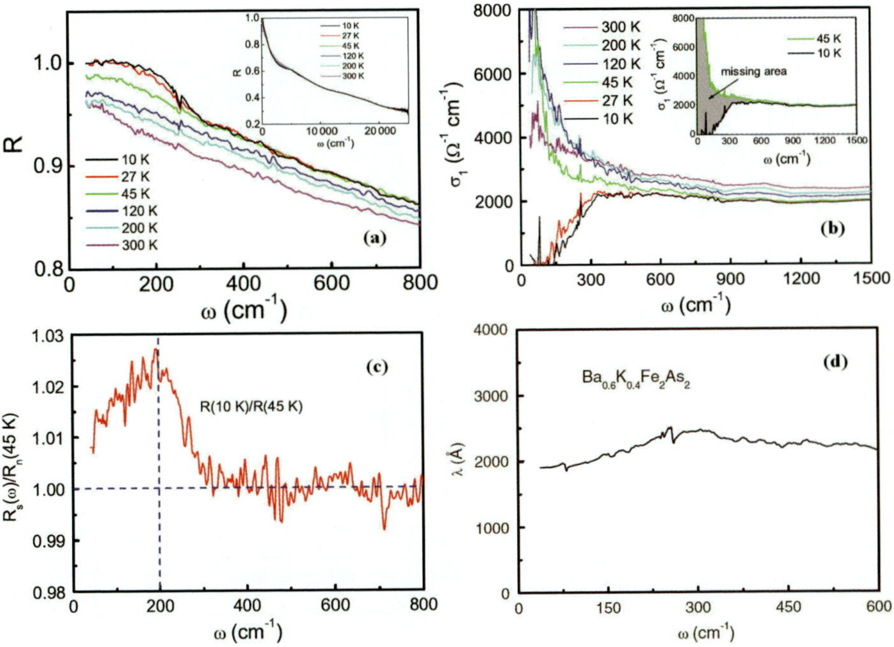

Figure 5.35

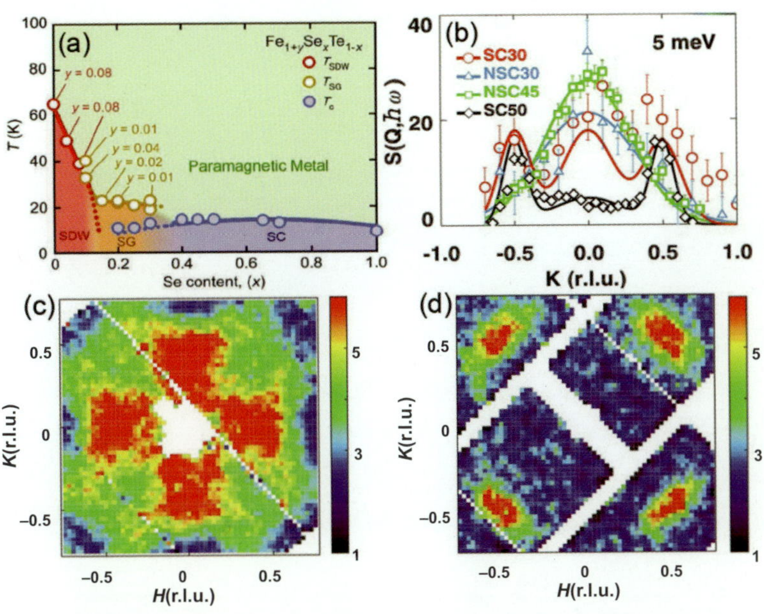

Figure 6.5

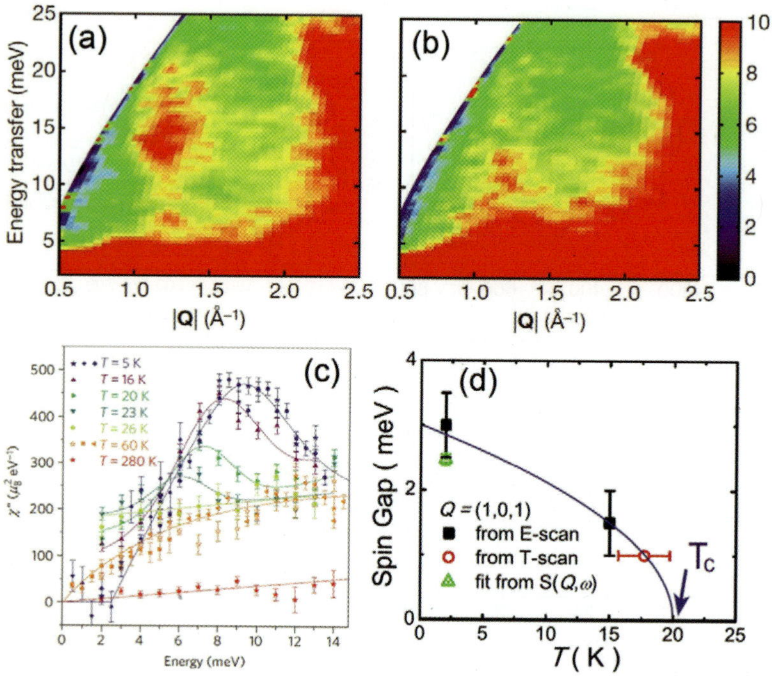
(a)
(b)
Energy transfer (meV)
|Q| (Å⁻¹)
(c)
T = 5 K
T = 16 K
T = 20 K
T = 23 K
T = 26 K
T = 60 K
T = 280 K
Energy (meV)
(d)
Spin Gap (meV)
Q = (1,0,1)
from E-scan
from T-scan
fit from S(Q,ω)
Tc
T (K)

Figure 6.6

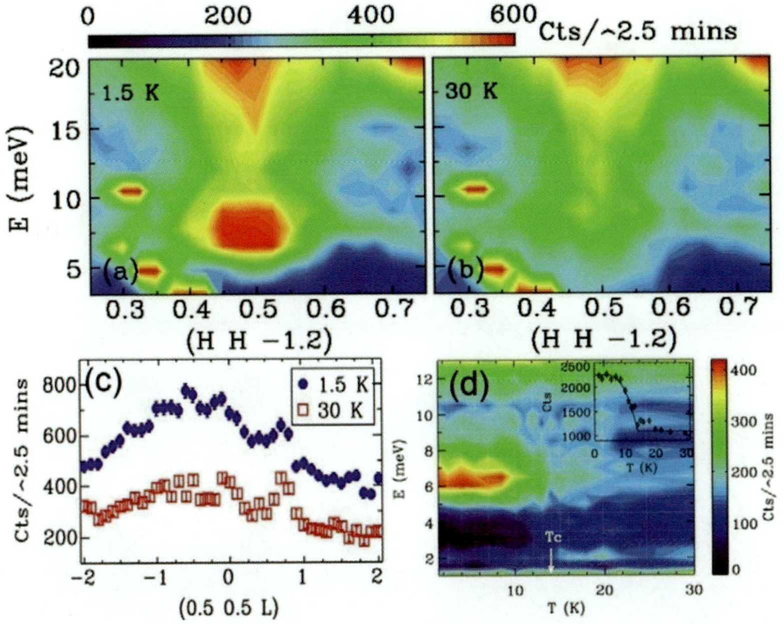

0
200
400
600
Cts/~2.5 mins
1.5 K
30 K
E (meV)
(a)
(b)
(H H −1.2)
(c)
1.5 K
30 K
Cts/~2.5 mins
(0.5 0.5 L)
(d)
E (meV)
T (K)
Tc
Cts
Cts/~2.5 mins

Figure 6.8

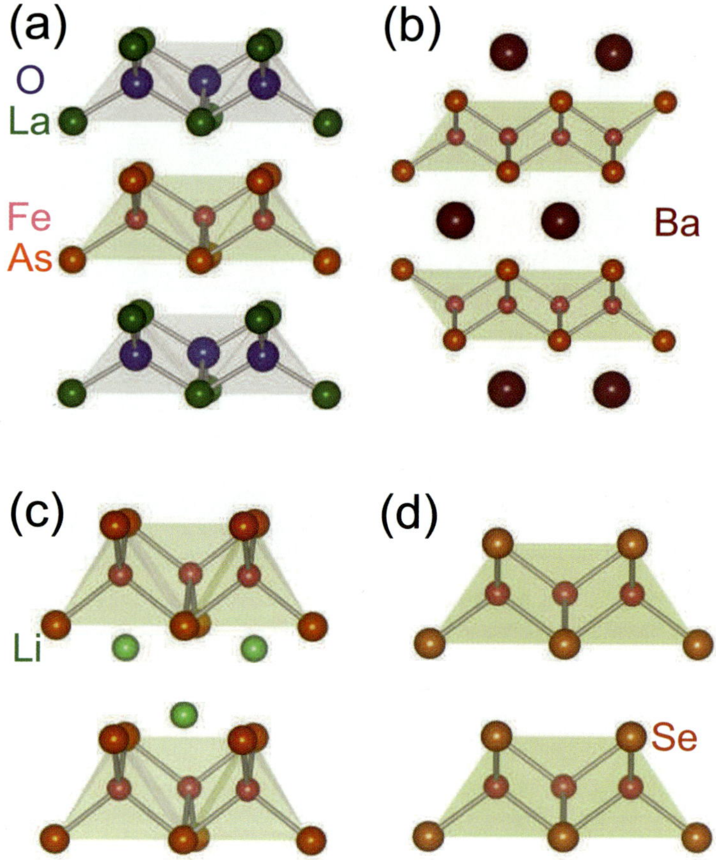

Figure 7.4

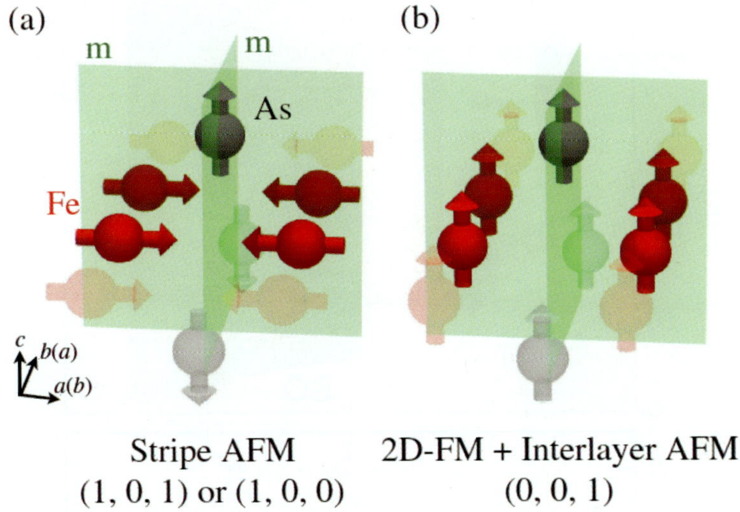
(a)
m
m
As
Fe
c
b(a)
a(b)
(b)
Stripe AFM
(1, 0, 1) or (1, 0, 0)
2D-FM + Interlayer AFM
(0, 0, 1)

Figure 7.21

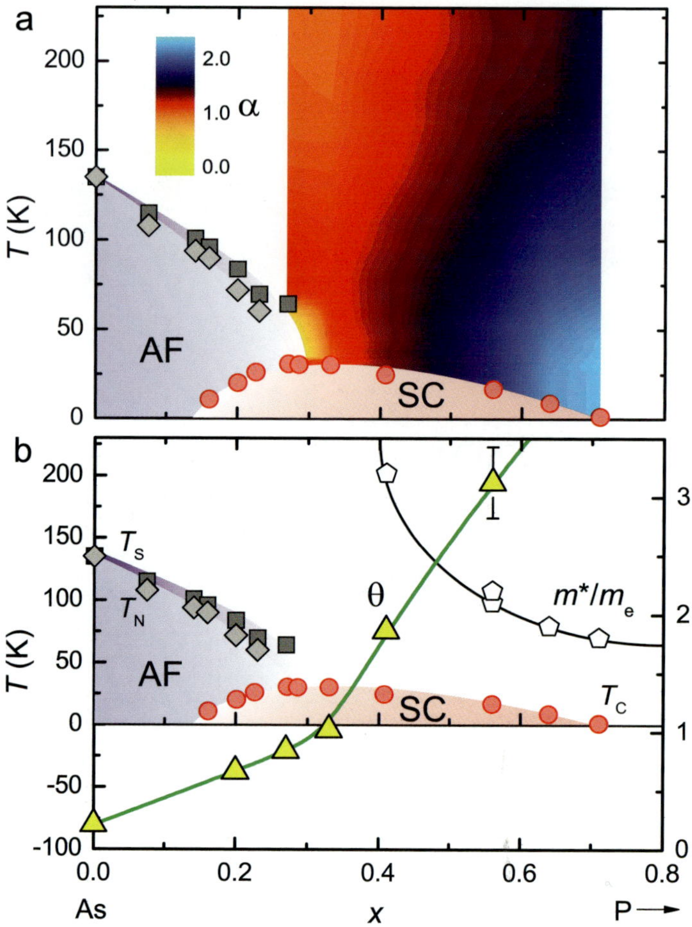

Figure 7.25

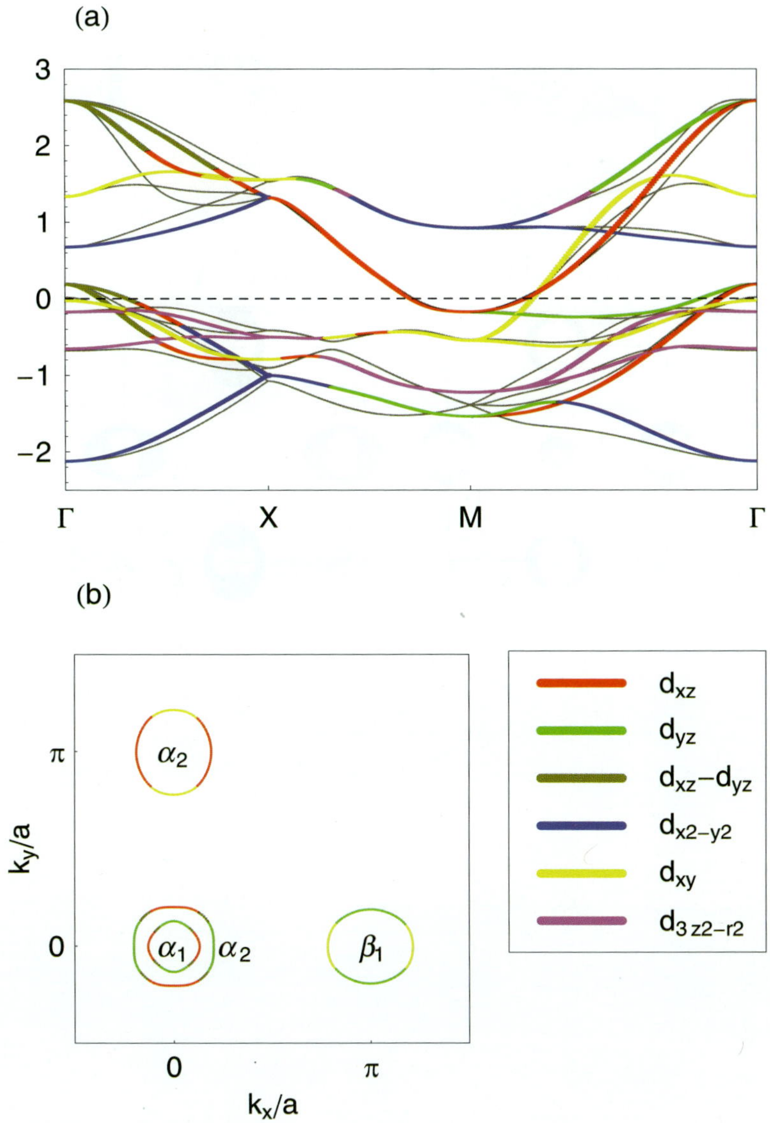
(a)
3
2
1
0
−1
−2
Γ
X
M
Γ
(b)
π
0
k_y/a
k_x/a
α_2
α_1
α_2
β_1
d_{xz}
d_{yz}
$d_{xz}-d_{yz}$
d_{x2-y2}
d_{xy}
d_{3z2-r2}

Figure 8.4

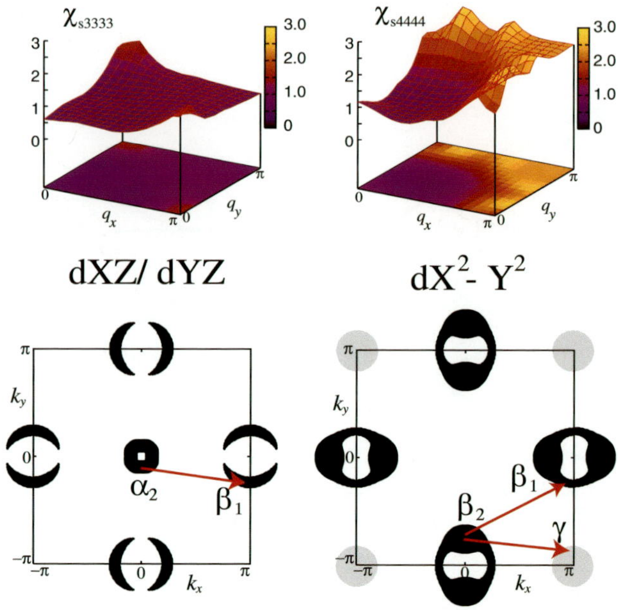

χ_{s3333}
3
2
1
0
3.0
2.0
1.0
0
q_x
q_y
χ_{s4444}
dXZ/ dYZ
dX^2- Y^2
k_y
k_x
α_2
β_1
β_2
γ
π
−π

Figure 8.7

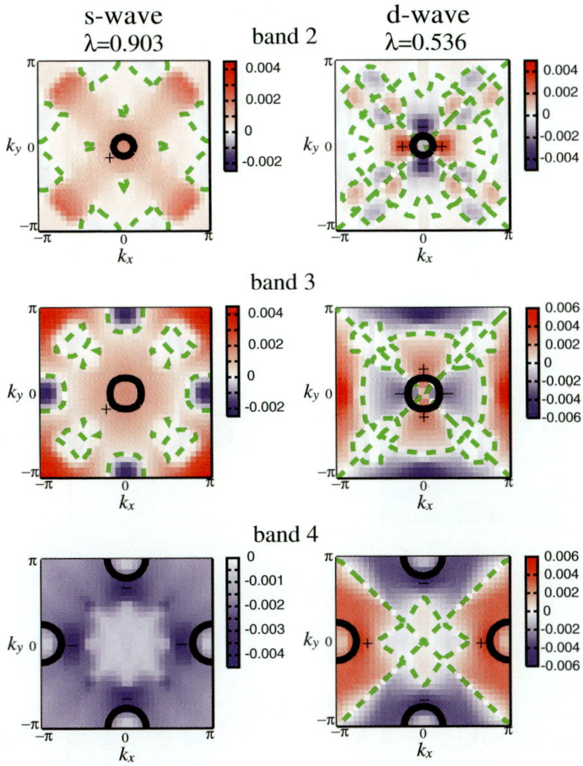
s-wave
λ=0.903
d-wave
λ=0.536
band 2
band 3
band 4
k_x
k_y
0.004
0.002
0
-0.002
-0.004
0.006
-0.006
-0.001
-0.003

Figure 8.14

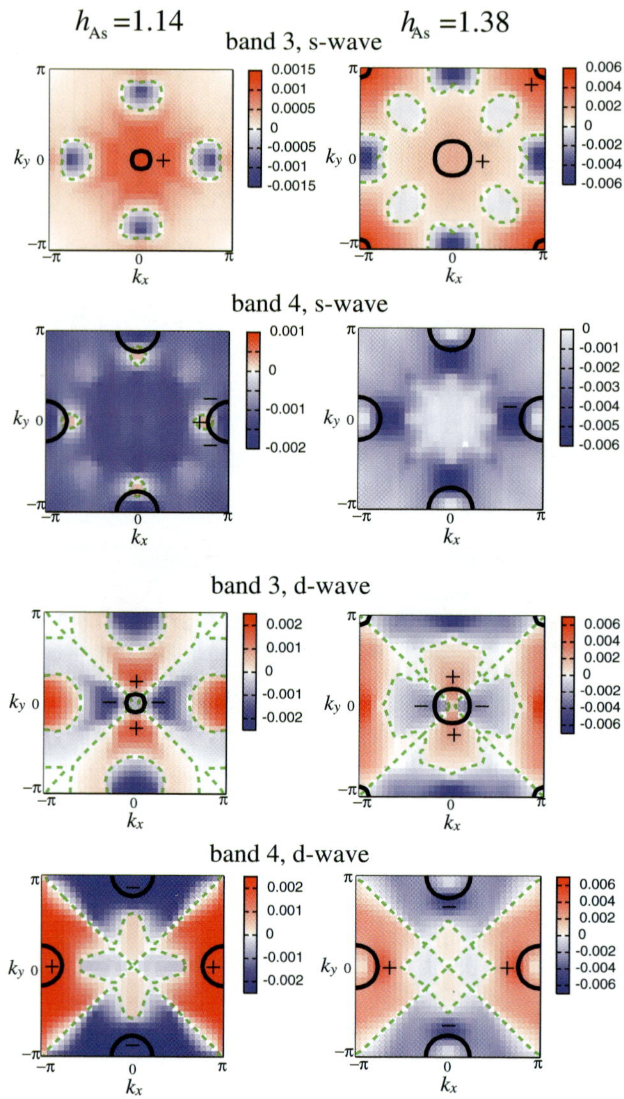

h_{As} =1.14
h_{As} =1.38
band 3, s-wave
band 4, s-wave
band 3, d-wave
band 4, d-wave
k_x
k_y
0.0015
0.001
0.0005
0
-0.0005
-0.001
-0.0015
0.006
0.004
0.002
0
-0.002
-0.004
-0.006
0.001
0
-0.001
-0.002
0
-0.001
-0.002
-0.003
-0.004
-0.005
-0.006

Figure 8.22

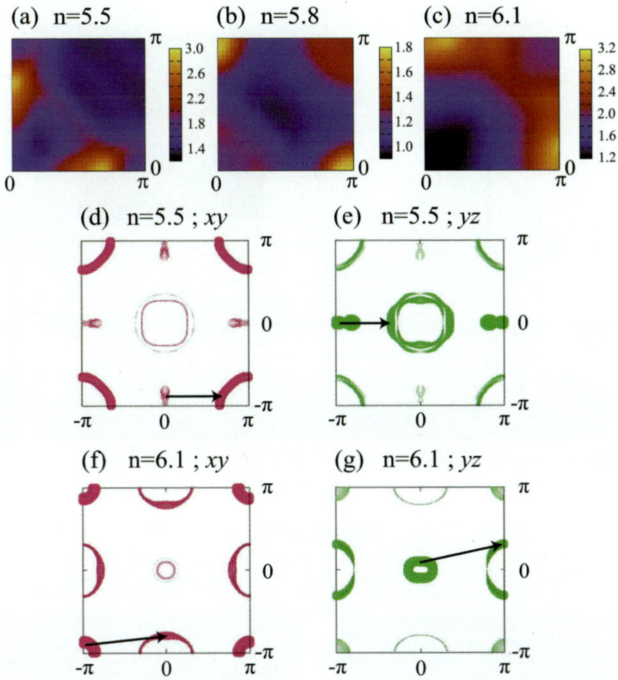
(a) n=5.5
(b) n=5.8
(c) n=6.1
3.0
2.6
2.2
1.8
1.4
1.8
1.6
1.4
1.2
1.0
3.2
2.8
2.4
2.0
1.6
1.2
π
0
(d) n=5.5 ; xy
(e) n=5.5 ; yz
(f) n=6.1 ; xy
(g) n=6.1 ; yz
-π

Figure 8.28

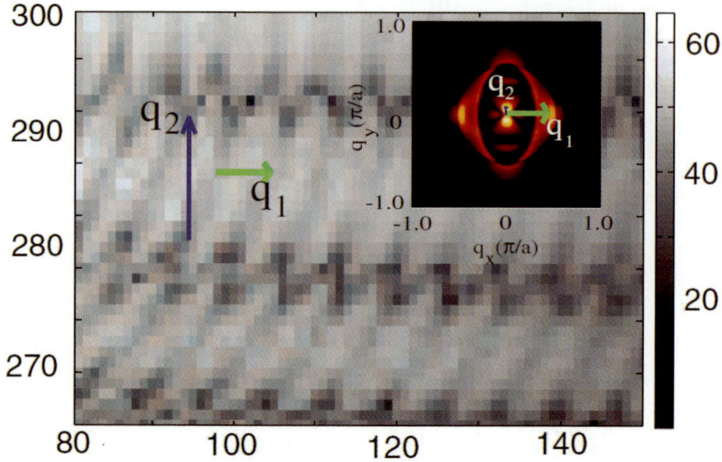

Figure 9.5

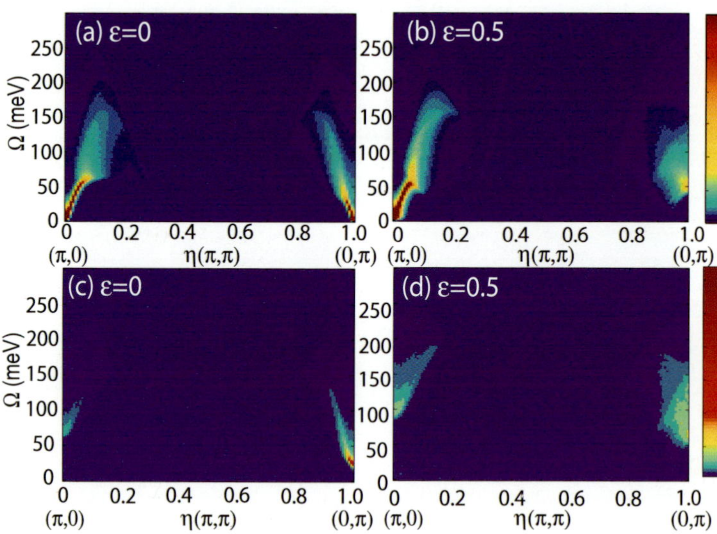

Figure 9.8

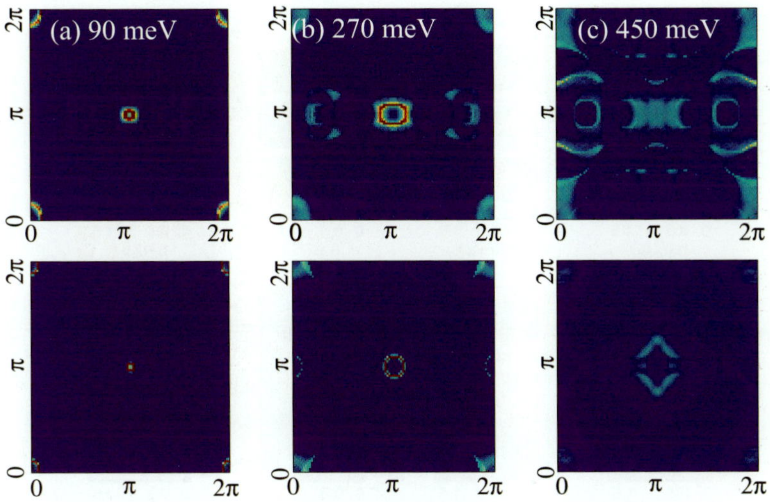

Figure 9.12

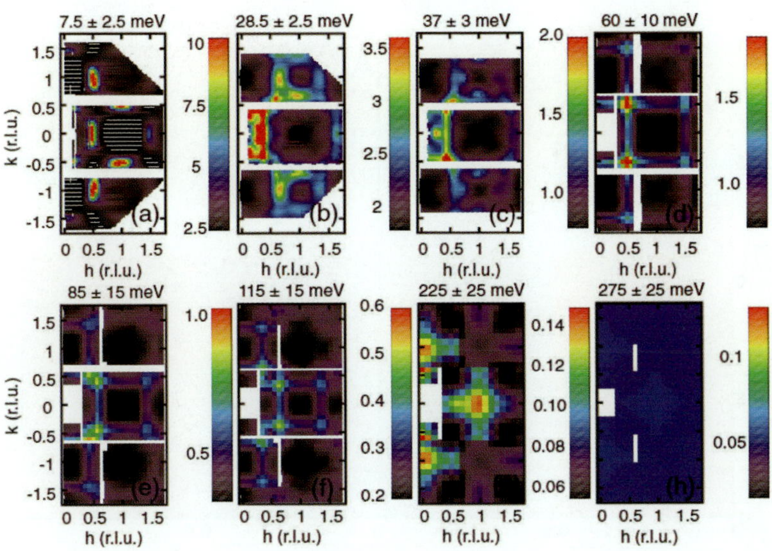

Figure 10.8

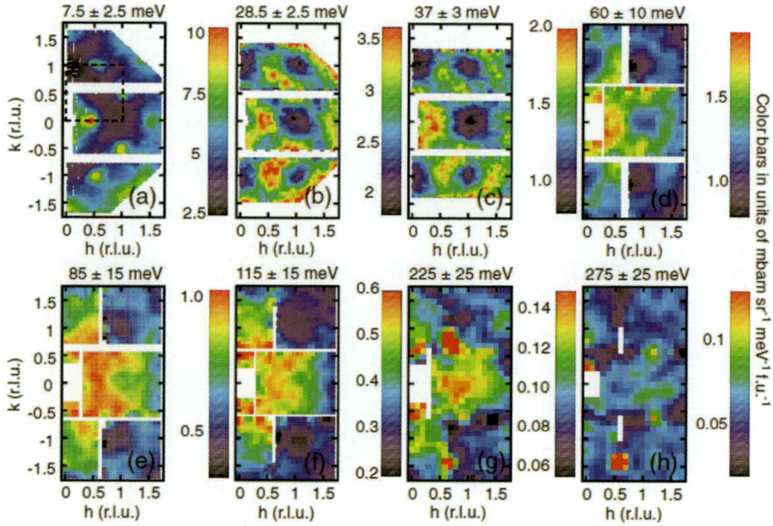

Figure 10.9

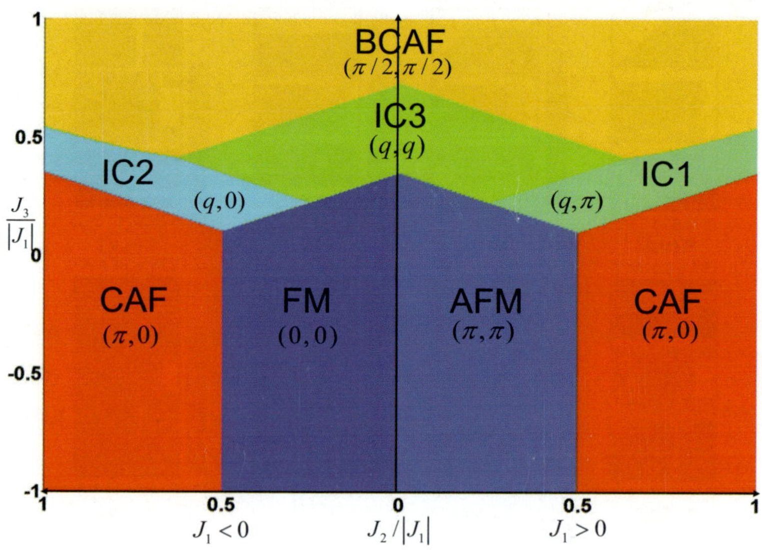

Figure 10.10